THE CRAB NEBULA

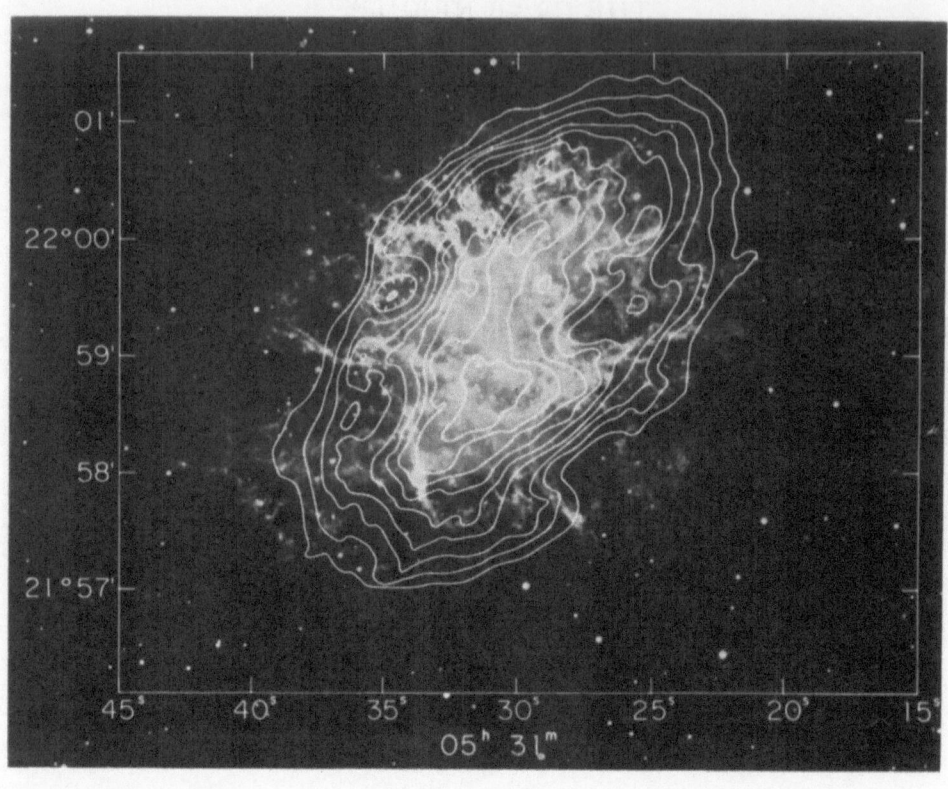

A 11 cm map of the Crab Nebula superposed on an optical photograph of the Nebula.

(By courtesy of Dr. R. G. Conway and the Astronomical Journal.)

INTERNATIONAL ASTRONOMICAL UNION
UNION ASTRONOMIQUE INTERNATIONALE

SYMPOSIUM No. 46

HELD AT JODRELL BANK, ENGLAND,
AUGUST 5–7, 1970

THE CRAB NEBULA

EDITED BY

R. D. DAVIES AND F. G. SMITH

Nuffield Radio Astronomy Laboratories, University of Manchester, U.K.

SPRINGER-SCIENCE+BUSINESS MEDIA, B.V.

1971

Library of Congress Catalog Card Number 73–154735

ISBN 978-94-010-3089-2   ISBN 978-94-010-3087-8 (eBook)

DOI 10.1007/978-94-010-3087-8

PREFACE

The Symposium on the Crab Nebula was held in the University of Manchester from 5 to 7 August, 1970. The meetings on the first day were held in the Physics Department on the University campus, and on the second and third days at the Nuffield Radio Astronomy Laboratories, Jodrell Bank. The 4th Symposium of the International Astronomical Union, convened in the University and at Jodrell Bank fifteen years earlier (25–27 August, 1955), dealt with the entire subject of radio and radar astronomy. Now the subject matter of this 46th Symposium of the International Astronomical Union was confined to one single object.

It is interesting to recall that even at the 1955 symposium the Crab Nebula figured prominently. In particular, J. H. Oort described the new measurements of the polarization of the light from the nebula and I. S. Shklovsky explained the light and radio emission in terms of the motion of relativistic electrons in the magnetic field of the nebula. No one could have foreseen the exciting discoveries of a decade later which stimulated the assembly of 172 participants to this 1970 Symposium. In addition to the lectures and discussions the visitors were able to tour the laboratories and telescopes at Jodrell Bank to see the various researches in progress. The demonstrations included a real-time display on a cathode ray tube of the pulses from pulsar CP 0328 received in the 250 ft steerable telescope.

With all my colleagues at Jodrell Bank I would like to record our deep indebtedness to the International Astronomical Union for the sponsorship of this Symposium, to the main Organizing Committee of Professors A. G. W. Cameron, F. D. Drake, V. L. Ginzburg, T. Gold, J. H. Oort, F. G. Smith, Dr. J. P. Wild and Professor L. Woltjer, and to the Local Organizing Committee of Drs. J. E. Baldwin, R. D. Davies, J. V. Jelley and Professors F. D. Kahn and F. G. Smith.

A. C. B. LOVELL

EDITORS' FOREWORD

The discovery of the pulsar in the Crab Nebula at once solved the mystery of the source of energy in the Nebula and stimulated interest in a whole new range of problems. The physical state of superdense stars, the electrodynamics of the medium round them, the pulsar radio emission, the transfer of energy into the surrounding nebula, all these represent new and exciting astrophysics. At the same time the Crab Nebula and its pulsar seem to provide a key to the mysteries of the other pulsars, whose origin and evolution were hard to understand before the clear association with remnants of supernovae. Both lines of thought led to the suggestion of an IAU Symposium on the Crab Nebula, in which the pulsar would clearly play the leading role.

A number of review papers were invited to begin each session and to cover the range of topics appropriate to the subject of study. Participants were also invited to submit papers of their own choice for inclusion in the programme. The papers submitted on the observational side ranged over the radio, infrared, optical, X-ray and γ-ray disciplines and theoretical papers covered the properties of the nebula and the pulsar. An attempt was made to classify the Crab Nebula amongst the variety of supernova remnants now known and to compare the Crab pulsar with the other pulsars.

The material in these Proceedings, like the Symposium, is divided into 8 sections each containing papers in a related field. Within each section the papers follow approximately the sequence in which they were presented at the Symposium; we have made some changes of order to bring together all those papers on a closely related topic. This, we hope will make the Proceedings more readable.

Several speakers did not submit manuscripts of their contributions. It was felt that the flavour of the Symposium, including the free discussion, could best be preserved if some note was made of the material of these talks. Short abstracts of the contributions have been prepared from a tape recording of the Symposium.

Contributors to the discussion after each paper or group of papers made a summary of their remarks which have been included in the Proceedings. Again in cases where no written summary was provided we made a version from the tape recording.

We wish to express our thanks to the session discussion secretaries, Dr. R. W. Clarke, Mr. D. A. Graham, Mr. G. C. Hunt, Drs. B. M. Lewis and P. Thomasson, for their help in collating the discussion contributions of the 8 sessions. Messrs. I. Morison, A. C. Pickwick and D. C. Wood made the successful tape recording of the symposium.

The organizers were sorry to learn on the first day of the Symposium that Professor Ginsberg, who was to introduce the final session, was unable to attend. One of us agreed at the last moment to give this introductory lecture.

R. D. DAVIES

F. G. SMITH

THE HISTORY OF THE CRAB NEBULA

1054 AD   Discovery by Chinese and Japanese. Visible for 650 days.

1731   Nebulosity discovered by John Bevis, English physicist and amateur astronomer.

1758   Charles Messier included it in his catalogue as M1.

~1850   Came to be known as 'Crab Nebula'.

1921   Lampland found expansion and noted variability in brightness of patches.

1942   Baade measured expansion rate – concluded it exploded 758 ± 36 years previously. The south-preceding star is a possible parent.

1948   Bolton and Stanley make the first identification of a galactic radio source with it.

1954   Discovery of optical polarization by Vashakidze and Dombrovsky, confirming the synchrotron mechanism of emission proposed by Shklovsky for the Crab Nebula.

1957   Radio polarization first measured by Mayer *et al.*

1963   Bowyer *et al.* find X-ray source near the Crab Nebula.

1964   Small diameter source found near centre of Crab Nebula by Hewish *et al.*

1968   Pulsar NP 0532 found in Crab Nebula by Staelin and Reifenstein.

1969   Optical pulsar NP 0532 found by Cocke *et al.* at position of south-preceding star in Crab Nebula. Infrared, X-ray and γ-ray emission subsequently detected from pulsar.

TABLE OF CONTENTS

SESSION 3 / OBSERVATIONS OF OTHER PULSARS

LIST OF PARTICIPANTS

D. J. Adams, Department of Physics, University of Leicester, U.K.

B. Agrinier, Commissariat à l'Énergie Atomique, Centre d'Études Nucleaires de Saclay, France.

L. H. Aller, Department of Astronomy, University of California, U.S.A.

B. Anderson, Nuffield Radio Astronomy Laboratories, University of Manchester, U.K.

J. R. P. Angel, Department of Astronomy, University of Columbia, New York, U.S.A.

W. D. Arnett, Institute of Theoretical Astronomy, Cambridge, U.K.

J. E. Baldwin, Cavendish Laboratory, University of Cambridge, U.K.

R. A. Batchelor, Research Laboratory of Electronics, MIT, Massachusetts, U.S.A.

J. Bergeron, Institut d'Astrophysique, Paris, France.

S. van den Bergh, David Dunlap Observatory, Ontario, Canada.

B. M. Bland, Department of Mathematics, University of Manchester, U.K.

N. Bonazzola, Observatoire de Paris, Section d'Astrophysique, France.

R. S. Booth, Nuffield Radio Astronomy Laboratories, University of Manchester, U.K.

G. Börner, Max Planck Institut für Physik und Astrophysik, Munich, Germany.

R. N. Bracewell, Radio Astronomy Institute, Stanford University, California, U.S.A.

N. J. B. A. Branson, Cavendish Laboratory, University of Cambridge, U.K.

J. W. Buckee, Department of Theoretical Physics, University of Oxford, U.K.

A. G. W. Cameron, Belfer Graduate School of Science, University of Yeshiva, New York, U.S.A.

V. Canuto, Goddard Institute for Space Studies, Maryland, U.S.A.

A. G. Cavaliere, Centre for Space Research, Cambridge, U.S.A.

G. Chanmugam, Institut d'Astrophysique, University de Liège, Belgium.

W. N. Charman, Nuclear Physics Division, Atomic Energy Research Establishment, Harwell, U.K.

H. Y. Chiu, Goddard Institute for Space Studies, Maryland, U.S.A.

W. N. Christiansen, School of Electrical Engineering, University of Sydney, Australia.

R. W. Clarke, Nuffield Radio Astronomy Laboratories, University of Manchester, U.K.

W. J. Cocke, Steward Observatory, University of Arizona, U.S.A.

J. M. Cohen, School of Natural Sciences, The Institute of Advanced Study, Princeton, U.S.A.

R. G. Conway, Nuffield Radio Astronomy Laboratories, University of Manchester, U.K.

B. A. Cooke, Department of Physics, University of Leicester, U.K.

A. M. Cruise, Mullard Space Science Laboratory, Department of Physics, University of London, U.K.

E. J. Daintree, Nuffield Astronomy Radio Laboratories, University of Manchester, U.K.

J. G. Davies, Nuffield Radio Astronomy Laboratories, University of Manchester, U.K.

R. D. Davies, Nuffield Radio Astronomy Laboratories, University of Manchester, U.K.

L. Davis, Jr., Department of Physics, California Institute of Technology, Pasadena, U.S.A.

F. D. Drake, Department of Astronomy, Cornell University, Ithaca, U.S.A.

R. W. P. Drever, Department of Natural Philosophy, University of Glasgow, U.K.

R. Ekers, California Institute of Technology, Pasadena, U.S.A.

A. Evans, Department of Physics, University of Wales, Aberystwyth, U.K.

A. C. Fabian, Mullard Space Science Laboratory, Department of Physics, University of London, U.K.

D. F. Falla, Department of Physics, University of Wales, Aberystwyth, U.K.

G. G. Fazio, Smithsonian Institution, Astrophysical Observatory, Cambridge, U.S.A.

P. A. Feldman, Queen's University, Ontario, Canada.

J. E. Felten, Institute of Theoretical Astronomy, Cambridge, U.K.

F. E. Ferrari, Consiglio Nazionale delle Richersche, Laboratorio di Cosmo-Geofisica, Torino, Italy.

B. Fitton, European Space Research Organization, Neuilly-sur-Seine, France.

R. A. E. Fosbury, Royal Greenwich Observatory, Herstmonceux, U.K.

P. R. Foster, Department of Natural Philosophy, University of Aberdeen, U.K.

W. A. Fowler, California Institute of Technology, Pasadena, U.S.A.

J. A. Galt, Dominion Radio Astrophysical Observatory, Penticton, Canada.

K. Gebler, Max-Planck-Institut für Radioastronomie, Bonn, Germany.

J. Gower, Department of Physics, University of British Columbia, Canada.

R. Griffiths, Department of Physics, University of Leicester, U.K.

J. E. Grindlay, Harvard College Observatory, Cambridge, U.S.A.

S. Grounds, Department of Theoretical Physics, University of Oxford, U.K.

D. ter Haar, Department of Theoretical Physics, University of Oxford, U.K.

O. Hachenberg, Radiosternwarte, Bonn, Germany.

J. P. Hagen, Pennsylvania State University, U.S.A.

C. G. Haslam, Nuffield Radio Astronomy Laboratories, University of Manchester, U.K.

D. J. Hegyi, NASA Institute for Space Studies, New York, U.S.A.

C. Heiles, Arecibo Ionospheric Observatory, Puerto Rico.

J. Heise, Space Research Laboratory, The Astronomical Institute at Utrecht, The Netherlands.

H. F. Helmken, Astrophysical Observatory, Cambridge, U.S.A.

H. Hesse, Department of Physics, Cambridge, U.K.

J. Heyvaerts, Institut d'Astrophysique, Paris, France.

A. M. Hillas, Department of Physics, University of Leeds, U.K.

R. R. Hillier, H. H. Wills Physics Laboratory, University of Bristol, U.K.

R. Hills, Astronomy Department and Radio Astronomy Laboratory, University of California, Berkeley, U.S.A.

P. Horowitz, Harvard University, Cambridge, U.S.A.

H. Hudson, University of California, San Diego, U.S.A.

V. A. Hughes, Queens University, Ontario, Canada.

R. A. James, Department of Astronomy, University of Manchester, U.K.

J. V. Jelley, Nuclear Physics Division, Atomic Energy Research, Harwell, U.K.

D. M. Jennings, Physics Department, University College, Dublin, Eire.

R. C. Jennison, Electronics Laboratory, University of Kent, U.K.

P. Kafka, Max Planck Institut für Physik und Astrophysik, Munich, Germany.

F. D. Kahn, Department of Astronomy, University of Manchester, U.K.

S. Karakula, Department of Physics, University of Durham, U.K.

E. M. Kellogg, American Science & Engineering, Massachusetts, U.S.A.

M. M. Komesaroff, CSIRO, Division of Radiophysics, N.S.W., Australia.

Z. Kopal, Department of Astronomy, University of Manchester, U.K.

J. Kristian, California Institute of Technology, Pasadena, U.S.A.

P. P. Kronberg, Department of Astronomy, University of Toronto, Canada.

F. K. Lamb, Department of Theoretical Physics, University of Oxford, U.K.

J. D. Landstreet, University of Columbia, New York, U.S.A.

M. I. Large, Nuffield Radio Astronomy Laboratories, University of Manchester, U.K.

J. P. Leray, Service d'Electronique Physique, C.E.A., France.

I. Lerche, Laboratory for Astrophysics & Space Research, University of Chicago, U.S.A.

B. M. Lewis, Nuffield Radio Astronomy Laboratories, University of Manchester, U.K.

J. L. Locke, Radio Astronomy Section, National Research Council, Ontario, Canada.

A. C. B. Lovell, Nuffield Radio Astronomy Laboratories, University of Manchester, U.K.

A. G. Lyne, Nuffield Radio Astronomy Laboratories, University of Manchester, U.K.

V. Malumyan, Biurakan Observatory, Armenia, U.S.S.R.

R. N. Manchester, National Radio Astronomy Observatory, Virginia, U.S.A.

P. G. Martin, Department of Theoretical Physics, Cambridge, U.K.

D. N. Matheson, Electronics Laboratory, University of Kent, U.K.

W. B. McAdam, National Radio Astronomy, Observatory Virginia, U.S.A.

M. R. McNaughton, Laboratory for Astrophysics & Space Research, University of Chicago, U.S.A.

J. Meaburn, Department of Astronomy, University of Manchester, U.K.

L. Mestel, Department of Mathematics, University of Manchester, U.K.

F. C. Michel, Institute of Theoretical Astronomy, Cambridge, U.K.

R. Minkowski, University of California, Berkeley, U.S.A.

A. T. Moffet, California Institute of Technology, Pasadena, U.S.A.

I. Morison, Nuffield Radio Astronomy Laboratories, University of Manchester, U.K.

J. E. Nelson, Lawrence Radiation Laboratory, University of California, Berkeley, U.S.A.

G. Neugebauer, Physics Department, California Institute of Technology, Pasadena, U.S.A.

P. S. Nicholson, Electronics Laboratory, University of Kent, U.K.

E. O'Mongain, Physics Department, University College, Dublin, Eire.

J. L. Osborne, Department of Physics, University of Durham, U.K.

J. P. Ostriker, Princeton University Observatory, New Jersey, U.S.A.

F. Pacini, Laboratorio Astrofisica, CP 67, Frascati (Rome), Italy.

C. G. Page, Cavendish Laboratory, University of Cambridge, U.K.

H. P. Palmer, Nuffield Radio Astronomy Laboratories, University of Manchester, U.K.

C. Papaliolios, Smithsonian Astrophysical Observatory and Harvard University, Cambridge, U.S.A.

J. C. B. Papaloizou, Astronomy Centre, University of Sussex, U.K.

R. B. Partridge, Princeton University, U.S.A.

A. Penny, Royal Greenwich Observatory, Herstmonceux, U.K.

G. H. Pettengill, Cornell University and Arecibo Observatory, Puerto Rico.

K. Pinkau, Max-Planck-Institut for extraterrestrische Physik, Munich, Germany.

L. Pointon, Nuffield Radio Astronomy Laboratories, University of Manchester, U.K.

J. E. B. Ponsonby, Nuffield Radio Astronomy Laboratories, University of Manchester, U.K.

N. A. Porter, Physics Department, University College, Dublin, Eire.

R. A. Porter, Nuffield Radio Astronomy Laboratories, University of Manchester, U.K.

K. A. Pounds, Department of Physics, University of Leicester, U.K.

V. Radhakrishnan, CSIRO Division of Radiophysics, N.S.W., Australia.

J. M. Rankin, Arecibo Observatory, Puerto Rico.

S. Rappaport, Department of Physics, Massachusetts Institute of Technology, U.S.A.

M. J. Rees, Institute for Theoretical Astronomy, Cambridge, U.K.

D. W. Richards, Arecibo Observatory, Puerto Rico.

F. E. Roach, National Bureau of Standards, Boulder, Colerado, U.S.A.

J. A. Roberts, Arecibo Observatory, Puerto Rico.

B. Rowson, Nuffield Radio Astronomy Laboratories, University of Manchester, U.K.

J. R. Ruffini, Institute for Advance Study, School of Natural Sciences, Princeton, U.S.A.

C. Ryter, Service d'Electronique Physique, CEA, France.

P. Sandford, Mullard Space Science Laboratory, University College, London, U.K.

L. Sartori, Department of Physics, Massachusetts Institute of Technology, U.S.A.

U. Schwartz, Kapteyn Laboratorium, The Netherlands.

D. W. Sciama, Department of Theoretical Physics, Cambridge, U.K.

E. R. Seaquist, Department of Astronomy, University of Toronto, Canada.

G. Share, Naval Research Laboratories, Washington, U.S.A.

S. F. Smerd, CSIRO, Division of Radiophysics, N.S.W., Australia.

D. F. Smith, Stanford University, California, U.S.A.

F. G. Smith, Nuffield Radio Astronomy Laboratories, University of Manchester, U.K.

R. E. Spencer, Nuffield Radio Astronomy Laboratories, University of Manchester, U.K.

D. H. Staelin, Department of Physics, Massachusetts Institute of Technology, U.S.A.

G. J. Stanley, California Institute of Technology, Pasadena, U.S.A.

P. Stewart, Department of Mathematics, University of Manchester, U.K.

J. M. Sutton, Institute of Theoretical Astronomy, Cambridge, U.K.

G. Swarup, Tata Institute of Fundamental Research, Bombay, India.

R. M. Tennant, Department of Physics, University of Leeds, U.K.

P. Thomasson, Nuffield Radio Astronomy Laboratories, University of Manchester, U.K.

A. Treves, Department of Applied Mathematics and Theoretical Physics, Cambridge, U.K.

V. Trimble, Institute of Theoretical Astronomy, Cambridge, U.K.

K. P. Tritton, Royal Greenwich Observatory, Herstmonceux, U.K.

J. Truemper, Max-Planck-Institut for extraterrestrische Physik, Munich, Germany.

S. Tsuruta, National Aeronautics & Space Administration, Goddard Space Centre, Maryland, U.S.A.

W. H. Tucker, American Science & Engineering, Massachusetts, U.S.A.

A. J. Turtle, School of Physics, University of Sydney, Australia.

N. Visvanathan, Harvard College Observatory, Massachusetts, U.S.A.

T. Walraven, University Observatory, Leiden, The Netherlands.

B. Warner, Radcliffe Observatory, S. Africa.

L. Webster, Royal Greenwich Observatory, Herstmonceux, U.K.

R. J. Weymann, Astronomy Department, University of Arizona, U.S.A.

C. H. Whitford, Department of Physics, University of Leicester, U.K.

J. P. Wild, CSIRO Division of Radiophysics, N.S.W., Australia.

R. Wielebinski, Max-Planck-Institut für Radioastronomie, Bonn, Germany.

D. T. Wilkinson, Physics Department, Princeton University, U.S.A.

R. V. Willstrop, University Observatories, Cambridge, U.K.

A. S. Wilson, Cavendish Laboratory, University of Cambridge, U.K.

J. G. Wilson, Department of Physics, University of Leeds, U.K.

M. M. Winn, Department of Physics, University of Leeds, U.K.

L. Woltjer, Department of Astronomy, Columbia University, New York, U.S.A.

W. W. Zuzak, Astronomy Department, University of Manchester, U.K.

OBSERVATIONS OF THE CRAB NEBULA

1.1 OPTICAL OBSERVATIONS OF THE CRAB NEBULA

VIRGINIA TRIMBLE

Institute of Theoretical Astronomy, University of Cambridge, Cambridge, U.K.

Abstract. The continuum radiation from the Crab Nebula has a great deal of structure, the majority of which is strongly polarized. Wisps in the vicinity of the pulsar at the centre of the nebula move noticeably in a few months. The appearance of the nebula changes markedly when photographed in different emission lines, as a result of the variations of physical conditions from place to place within the nebula.

1. Introduction

The optical emission from the Crab Nebula (Plate I) consists of two distinct components, line emission and continuum synchrotron emission. On a color photograph the two are readily separable. The line emission, coming from gas at densities near 10^3 cm^{-3} and temperatures near 10^4 K, is red due to the predominance of H α and [N II]. A more recent and probably more accurate color photograph by J. Miller at Lick Observatory shows many of the filaments (line emitting features) as green due to [O III] radiation at $\lambda 5007$. Other strong lines are produced by [S II], [O II], [O I], He I and II, [Ne III], and hydrogen. The white light, more strongly concentrated toward the center than the filaments, is the synchrotron continuum emission, which we also see in the radio, infrared, and (probably) X-ray regions of the spectrum.

2. The Continuum Emission

The outermost isophotes of the nebula as shown, e.g. by Woltjer (1957), are roughly elliptical. It has been suggested by Shklovskii (1966) that one would expect material to be ejected from an explosive event in the form of a prolate ellipsoid and by Münch (1958) that the interstellar magnetic field (given that the major axis of the ellipse is approximately parallel to the galactic equator) might have perturbed a symmetric explosion into the prolate shape. The most intense continuum emission regions trace out a crude S-shape. The arms of the S extend in the same general direction as continuum features are observed to move, and the shape is probably related in some way to the electrodynamics of the pulsar.

The continuum emission (Plate II) is frequently spoken of as amorphous, but this is far from the case. There are obvious large-scale features – a sort of hole near the center (this has perhaps been swept clean by agency of the pulsar), two indented bays on the east and west sides of the object, and regions of higher and lower intensity, especially in the northwest quadrant. Changes in these regions were first noted, with some surprise, by Lampland (1921). Near the center are the rapidly changing wisps, first mentioned by Baade (Oort and Walraven, 1956) and studied by Scargle (1969a). And, in very good seeing, the continuous emission appears to be largely

Davies and Smith (eds.), The Crab Nebula, 3–11. All Rights Reserved.

Plate I.   Photograph of the Crab Nebula (originally in colour) by W. Miller (courtesy Hale Observatories, Carnegie Institute of Washington).

concentrated in thin, thread-like features which are entangled in a sort of basket weave or cotton wool structure. Polarization data indicates that these fibers are aligned with the magnetic field lines. Their formation is not satisfactorily understood; neutral sheets in the magnetic field, thermal instability driven by the synchrotron radiation, and self-pinching relativistic current tubes have been suggested. At any rate, the traditional equipartition way of estimating magnetic field strength and relativistic electron energy from the synchrotron intensity should be modified to take this non-homogeneity into account. See M. Rees (this symposium, p. 407) concerning circular polarization and the possible absence of a magnetic field.

The work of Scargle (1969a) indicates that the nebula is rather bluer at the center than at the edges. This is what one would expect if relativistic particles are largely injected near the center, given that electrons responsible for optical radiation have synchrotron lifetimes of the order of the age of the remnant. The blueness of the center is also in accord with the difference between the radio and optical objects; although the outermost detectable contours in the two spectral regions are about the same size, the radio emission is less strongly peaked at the center than is the optical emission. It is possible that this correlation continues into the X-ray region and that the nebula is significantly smaller at high energies. This variation of color with position implies that the frequency dependence of the volume emissivity varies with location in the nebula. The apparent inflection of the integrated optical continuum after correction for reddening (see the spectrum given by Baldwin elsewhere in this volume, page 22) may, therefore, be the result of summing unlike spectra and not indicative of an inflection in the real volume emissivity. If, on the other hand, the spectrum really

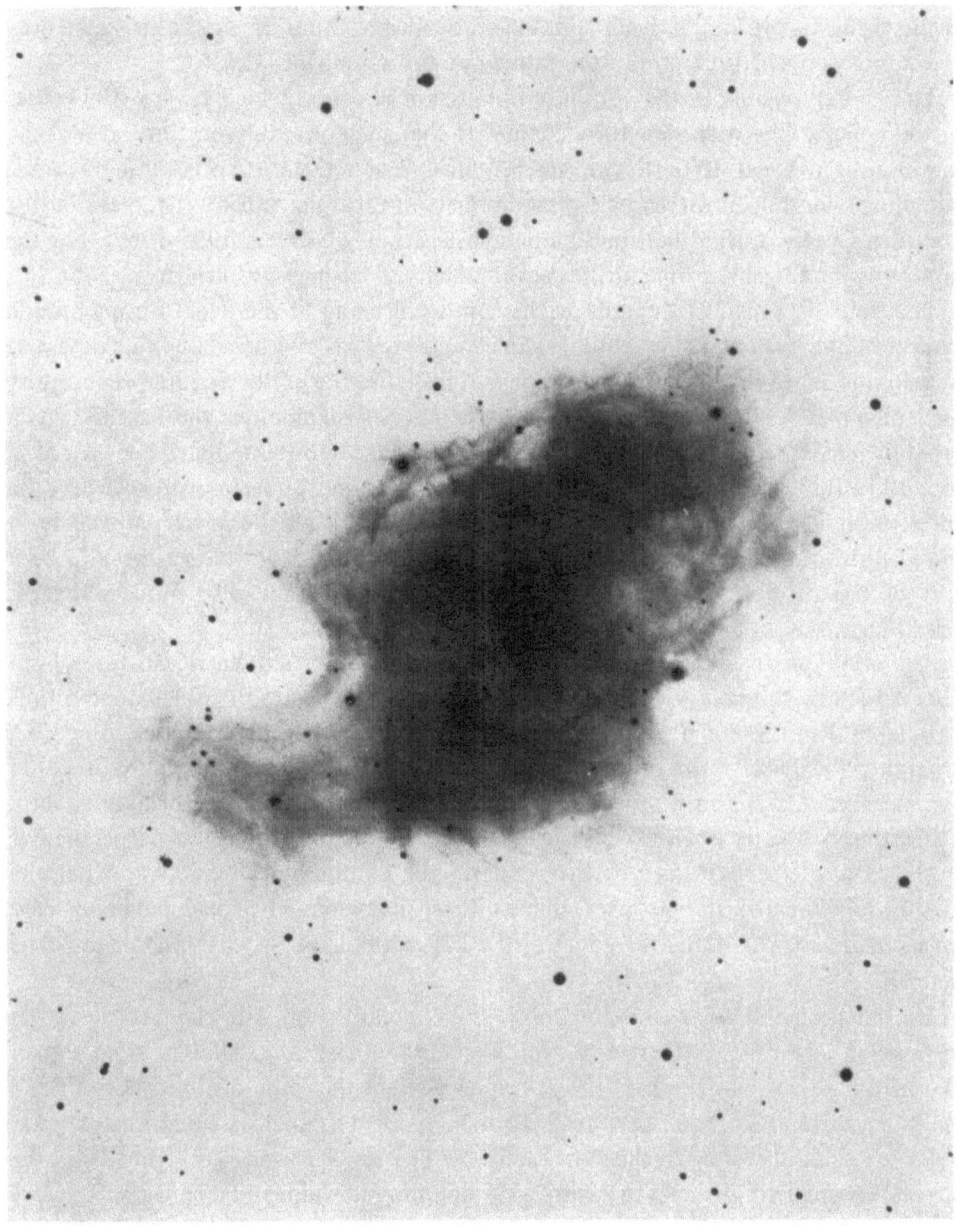

Plate II. Continuum emission photograph of the Crab Nebula taken by W. Baade (courtesy Hale Observatories).

has a relative maximum in the blue, this is most likely a feature of the volume emissivity in some part of the nebula (Minkowski, 1968).

The uncertainty in the shape of the optical spectrum results from our lack of knowledge of the reddening of the object. It ought to be possible to estimate this directly from the observed ratio of the [SII] lines in the red and the blue (whose emitted intensity ratio is known theoretically), but this has not been done. Guesses at the visual absorption derived from the reddening of O and B stars in the same area of the sky (O'Dell, 1962) are in the range one to two magnitudes.

The activity visible in the continuum emission is concentrated toward the center of the nebula. The wisp structure (Plate III) changes noticeably on time scales of a few months to a year. In particular, the brightest wisp in the north preceding quadrant has moved in and out in a quasi-periodic fashion (Scargle, 1969a). The period was about two years during the time the nebula was well observed (1955–1960), but the motion is no longer obvious in the plates taken less frequently since then.

The thin wisp, nearest the pulsar (the south preceding of the two 16th magnitude stars near the center of the nebula), comes and goes as well as changing shape and position. A new thin wisp may have appeared just after the pulsar period discontinuity in September 1969. The evidence for this is a series of pictures, the best of which are shown in Plate III (from Scargle and Harlan, 1970). In order to reach a length of 2″ (about 6×10^{16} cm) in a couple of months, the feature must have moved at nearly the speed of light. Less rapid changes can be seen further from the pulsar, especially in the northwest quadrant. Regions of strong and weak emission there appear to be a sort of extension of the series of wisps near the center and also move outward, but at speeds $\lesssim 0.1$ c.

A photograph taken through polaroid (e.g. Scargle, 1969b) emphasizes features, both fibers and larger regions, which are elongated perpendicular to the polaroid and thus parallel to the magnetic field in their vicinity. This must be trying to tell us something about how the features are produced. The detailed optical polarization maps prepared by Walraven (1957) and Woltjer (1957); see Figure 1 in Conway (1971) show several interesting correlations with intensities. The electric vectors tend to be perpendicular to the edges of the bays and to the wisps in the continuum emission and parallel to the directions of the stronger line-emitting filaments. The local magnetic field must, therefore, run along the edges of the bays and along the wisps and encircle the filaments.

The integrated nebular emission shows 9.2% polarization with the electric vector in position angle 159°.6 (Oort and Walraven, 1956). The interstellar polarization of O and B stars in this region of the sky and at about the same distance as the Crab amounts to about 2.4% in P.A. 148°.5 (Hiltner, 1956). The intrinsic integrated nebular polarization is, therefore, rather less than 7%. The smaller the region considered the larger the polarization can be; with a 5″ diaphragm, values higher than 60% are observed (Woltjer, 1957).

Unlike the radio continuum emission (for which a high resolution map discussed by Wilson elsewhere in this volume (page 68) shows intensity maxima near some of the

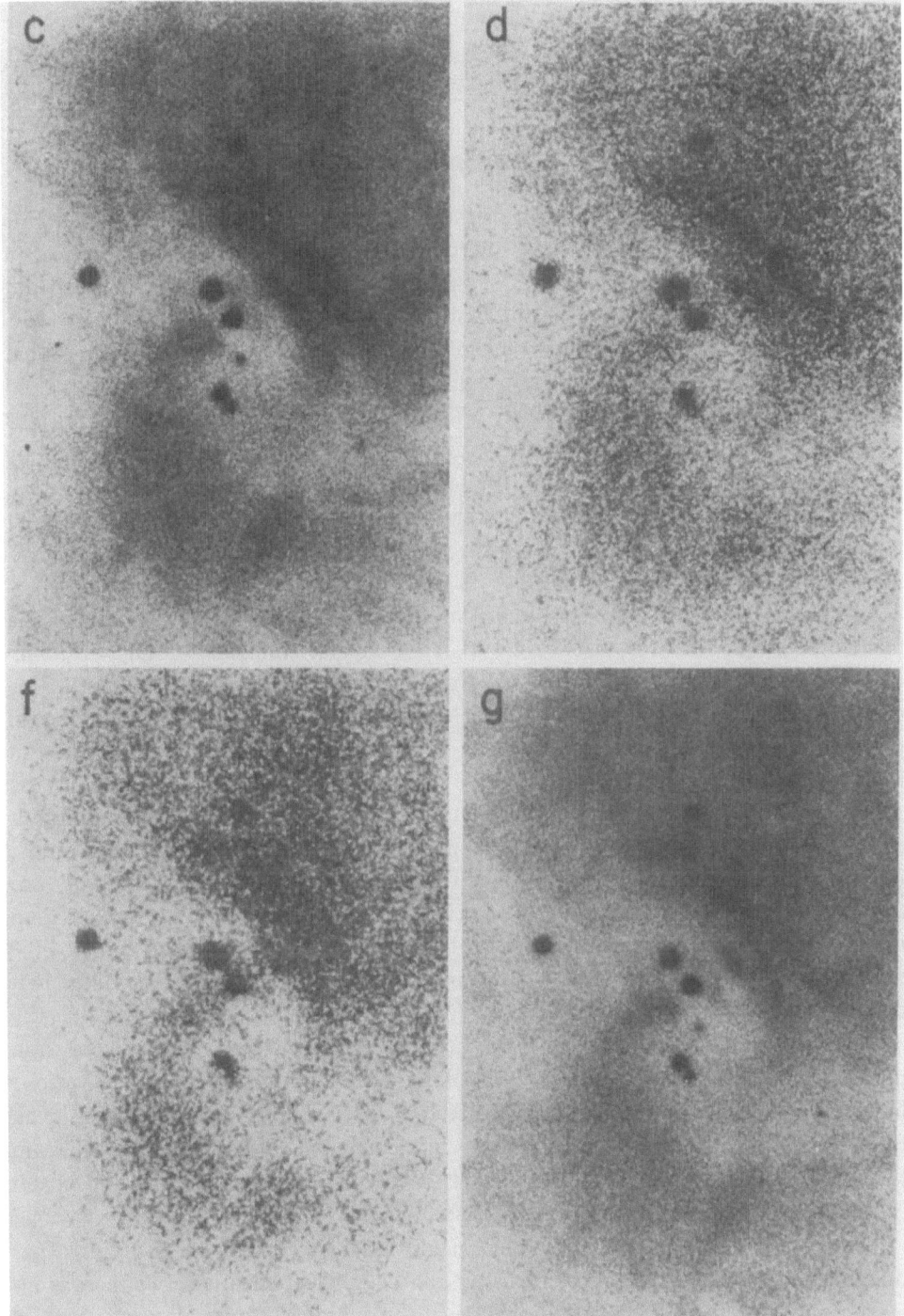

Plate III. Changes in the wisp structure at the center of the Crab Nebula
(from Scargle and Harlan, 1970).
(By courtesy of the Astrophysical Journal and University of Chicago Press)

stronger filaments) the optical continuum seems, if anything, to be somewhat anti-correlated with the line emission, though both are weak near the center. One of the bays, for instance, is almost bisected by an intense filament.

3. The Line Emission

A very heavily exposed photograph of the line emission (obtained using a suitable color or interference filter) shows a sharply defined edge, quite nearly elliptical in shape except for a bulge on the south preceding edge. An accidentally burned out [OII] $\lambda3727$ interference filter photograph, taken by the author in January 1967, also shows the faint jet on the north side of the Crab Nebula which was first reported by van den Bergh (1970) on the basis of a broader band photograph. This would appear, therefore, to be at least partly a line emission feature, but there seems also to be an indication of it in the outermost continuum emission isophote given by Woltjer (1957).

The line emission is concentrated in a filamentary structure (Plate IV), the strongest features being extremely irregularly distributed and the fainter ones scattered over most of the surface of the nebula.

It is possible to determine something of the three dimensional structure of the filamentary component by making use of radial velocity information and assuming that velocity is proportional to distance from the center of the nebula along the line of sight as it is in the plane of the sky. Mayall (1962) has in this way obtained a three dimensional reconstruction in which all the features appear to be connected in intertwined ribbons. The radial velocity data given by Trimble (1968) suggests, on the other hand, that the line-emitting material is at least partially in discontinuous clumps. It is clear in any case that the filaments are not confined to a thin shell on the outside of the object. Only a few faint filaments are found near the center, but about half of the features with radial velocities given by Trimble (1968) are less than two-thirds of the way from the center to the surface of the nebula, and some of the strongly emitting filaments extend more or less radially outward in the outer half of the object.

There are striking differences in the appearance of the nebula in various emission lines (Plate IVa, b, c), or alternatively, striking differences in the spectra of various filaments. The latter can be seen quantitatively from the line emission strengths given by Woltjer (1958) for a number of positions in the nebula and more qualitatively from the emission line ratios given by Trimble (1970). Woltjer also gives a mean spectrum of relative intensities averaged over the nebula.

The spectrum of a particular volume of gas must be explainable in terms of its unique composition, temperature, density, and degree of ionization, but there is no reason why the mean spectrum should be. Even within a single filament, there is probably a good deal of stratification (Miller, 1970). It is, therefore, not surprising that theoretical calculations (e.g. Davidson and Tucker, 1970) of the behaviour of a slab of gas of fixed density and composition exposed to a given flux of ionizing ultra-

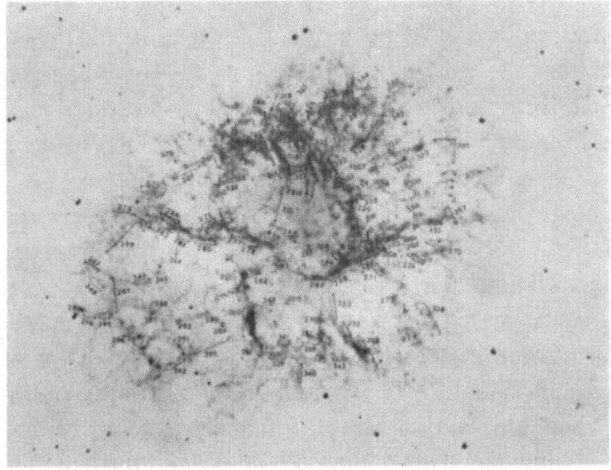

Plate IVa.

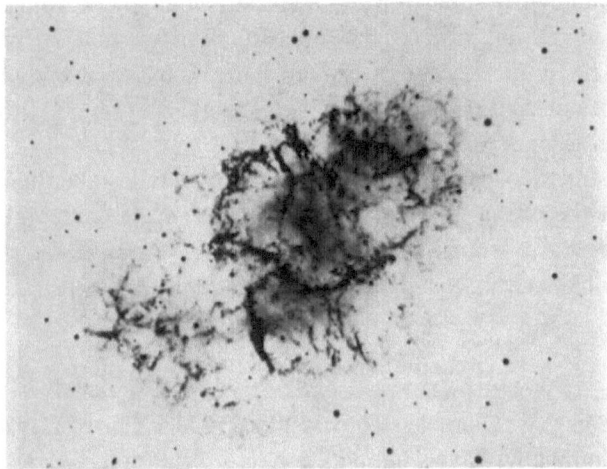

Plate IVb.

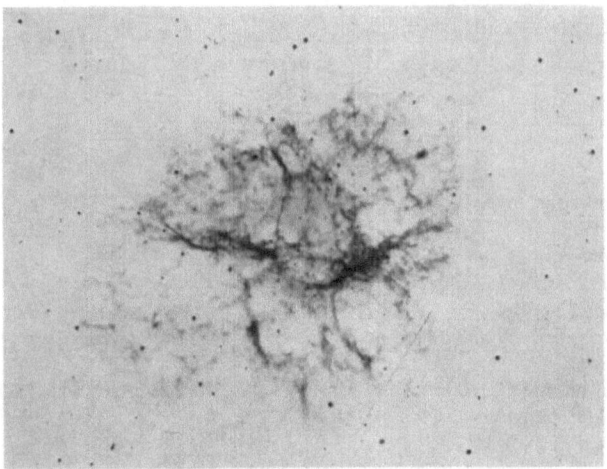

Plate IVc.

Plate IVa–c.   The Crab Nebula in the line emission of (a) Hα+(NII), (b) (SII), (c) (OII).
Taken by G. Münch (a and c) and W. Baade (b) on the 200″ telescope. (Courtesy Hale Observatories.)

violet radiation do not perfectly reproduce the average line intensities. The observation that the ratio of [OIII] to [OII] is on average larger near the center of the nebula than at the surface indicates that the general approach is right. There is weak evidence from the variation of Hα/[NII] with position in the object that there may be some additional heating due to interaction with the interstellar medium at the surface of the nebula. At any rate, there seems to be no difficulty, in principle, in accounting for the range of line intensities observed in terms of gas with (1) essentially solar composition, except that $N_{He}/N_H = 0.45$ (Woltjer, 1958) to 1.0 (Davidson and Tucker, 1970), but considerable variations may occur (and might be expected in gas which is a mixture of expelled stellar material and swept up interstellar material) and would not be easily detected; (2) densities within a factor of three or so of $10^3 \, \mathrm{cm}^{-3}$; (3) temperatures in the range 8000 to 27000 K; and (4) ionization due to the nebular ultraviolet continuum radiation. The surface of a filament exposed to this flux will contain H^+, He^{+2}, and doubly and triply ionized O, N, Ne and the like; intermediate regions will have predominantly singly ionized material, while at the center of dense filaments even hydrogen may be neutral. The total amount of material required is estimated from the intensity of Hβ, and is probably of the order of one solar mass (Minkowski, 1968), but considerable neutral material could be present without contributing much to the spectrum (except [OI] $\lambda 6300$).

Comparison of line emission photographs of the Crab Nebula taken ten or more years apart shows that the object is expanding with a time scale comparable to the known age of the supernova remnant and from a center within about 10″ of the pulsar. The dynamics of the Crab Nebula are discussed in the following article.

Acknowledgements

It was originally intended that this review be presented by J. Miller of Lick Observatory, who would undoubtedly have had additional information on the intrinsic spectrum of the Crab Nebula and its reddening. The author is grateful for a number of discussions of the Crab Nebula with Drs. G. Münch, R. Minkowski, J. Gunn, J. Scargle, M. Rees, T. Gold, L. Woltjer, W. Arnett, and D. Melrose.

References

Conway, R. G.: 1971, this symposium, Paper 5.2, p.292.
Davidson, K. and Tucker, W.: 1970, *Astrophys. J.* **161**, 437.
Hiltner, W. A.: 1956, *Astrophys. J. Suppl.* **2**, 389.
Lampland, C. O.: 1921, *Publ. Astron. Soc. Pacific* **33**, 79.
Mayall, N. U.: 1962, *Science* **137**, 91.
Miller, J.: 1970, *Publ. Astron. Soc. Pacific* (discussion at the Flagstaff Conference on the Crab Nebula). **82**, 386.
Minkowski, R.: 1968, 'Nonthermal Galactic Radio Sources' in B. Middlehurst and L. Aller (eds.), *Nebulae and Interstellar Matter*, Univ. of Chicago Press.
Münch, G: 1958, *Rev. Mod. Phys.* **30**, 1042.
Münch, G., Scargle, J., and Trimble, V.: 1971, *The Crab Nebula*, Gordon and Breach (in preparation).

O'Dell, C. R.: 1962, *Astrophys. J.* **136**, 809.
Oort, J. and Walraven, Th.: 1956, *Bull. Astron. Inst. Neth.* **12**, 285.
Scargle, J.: 1969a, *Astrophys. J.* **156**, 401.
Scargle, J.: 1969b, in P. Brancazio and A. Cameron (eds.), *Supernovae and Their Remnants*, Gordon and Breach.
Scargle, J. and Harlan, E.: 1970, *Astrophys. J.* **159**, L143.
Shklovskii, I. S.: 1966, *Soviet Astron. – AJ* **10**, 6.
Trimble, V.: 1968, *Astron. J.* **73**, 535.
Trimble, V.: 1970, *Astron. J.* **75**, 926.
Van den Bergh, S.: 1970, *Astrophys. J.* **160**, L27.
Walraven, Th.: 1957, *Bull. Astron. Inst. Neth.* **13**, 293.
Woltjer, L : 1957, *Bull. Astron. Inst. Neth.* **13**, 301.
Woltjer, L.: 1958, *Bull. Astron. Inst. Neth.* **14**, 39.

Discussion of this paper was deferred until after the following paper by the same author.

1.2   DYNAMICS OF THE CRAB NEBULA

VIRGINIA TRIMBLE*

Institute of Theoretical Astronomy, University of Cambridge, Cambridge, U.K.

Abstract. Studies of the proper motion of many filaments in the Crab Nebula show that the expansion centre does not coincide with the present position of the pulsar NP 0532. Possible reasons for this difference are discussed. Estimates of the turbulent velocity within the nebula indicate that it lies in the range 100 to 300 km sec^{-1}. An analysis of the present expansion rate of the nebula indicates a convergence at 1140 ± 10 AD. The acceleration implied in this result could arise from magnetic or relativistic electron pressure.

1. Introduction

Proper motions of features in the Crab Nebula have been reported by Duncan (1921, 1939), Deutsch and Lavdovsky (1940) and Trimble (1968). Radial velocities have been given by Sanford (1919), Mayall (1937, 1962), Woltjer (1958), Münch (1958), Trimble (1968), and Münch *et al.* (1971). Several kinds of information can be derived from this dynamical data. The distance to the nebula is almost certainly in the range 1030 pc (Woltjer, 1958) to 2170 pc (Trimble, 1970a), with the most probable value lying near the middle of the range (Woltjer, 1970). The kinetic energy of the expanding remnant is, to within a factor of two, 10^{49} $M/M_\odot$ ergs, but the value of the mass may be anywhere in the range one (Minkowski, 1968) to 10 (Gott *et al.*, 1970) solar masses. Limits to rotational and turbulent energy will be set below.

2. The Space Motion of the Nebula and NP 0531

The center of expansion of the Crab Nebula and the time scale of that expansion (in the sense of distance covered divided by present velocity) are two of the better determined properties of the object. This results from the large number of individual proper motions of filaments which have been measured. It is, therefore, significant that the expansion center does not coincide with the present position of the central star and that the time scale does not agree with the known age of the supernova remnant.

The position of the expansion center, determined by the intersection of the filamentary proper motion vectors (without regard to time) allows us to determine the proper motion of the central star and the nebula separately. It is evident that, in 1054, all the material must have been at the expansion center (provided only that supernovae occur in objects of more or less stellar dimensions), and that the star (NP 0531, Baade's Star, Star E of Trimble, 1968) is now where it is. Its proper motion is, therefore, just that change in position divided by the time elapsed since the Chinese saw their 'guest star'. The motion so found is

$$\mu_\alpha = -0.0116 \pm 0.0022''/\text{yr} \quad \text{and} \quad \mu_\delta = 0.0048 \pm 0.0022''/\text{yr} \qquad (1)$$

* NATO Postdoctoral Research Fellow.

Davies and Smith (eds.), The Crab Nebula, 12–21. All Rights Reserved.

where the uncertainties come largely from the 3″ uncertainty in the position of the expansion center (Trimble, 1968). This is in reasonable accord with directly measured values of the stellar proper motion (Minkowski, 1970).

The proper motion of the nebula can be found by analyzing asymmetries in the filamentary proper motions around the expansion center, where the center is determined from purely geometric considerations and not by requiring all the filaments to have been there at the same time. If, for example, the nebula were moving along its major axis, then $\mu - r$/(apparent age) ought to be systematically positive for filaments at one end of the axis and negative for those at the other (where r is the present distance of a filament from the expansion center). The proper motion data for individual filaments imply in this way:

$$\mu_\alpha = 0.0000 \pm 0.0007''/\text{yr} \quad \text{and} \quad \mu_\delta = -0.0016 \pm 0.0007''/\text{yr} \tag{2}$$

where the uncertainties are standard deviations found in a least squares solution for μ with all filaments weighted equally.

The motion of the system, nebula plus pulsar, is just the sum of (1) and (2), weighted by their respective masses, and is large only if most of the mass is in the star. This is shown in Table I.

TABLE I

Proper motions and space velocities of the NP 0531–Crab Nebula system for various values of $W = M_{0531}/M_{\text{CN}}$. The system is assumed to be 2000 pc from us.

W	μ_α	μ_δ	V_r	π	θ	z
0	−0.0000″/yr	−0.0016″/yr	0 km/sec	0 km/sec	−13 km/sec	−7 km/sec
0			+100	+99	−6	−17
0.2	−0.0018	−0.0005	0	−1	+4	−17
0.2			+100	+98	+11	−27
0.5	−0.0039	+0.0005	0	−4	+21	−31
0.5			+100	+95	+28	−41
1.0	−0.0058	+0.0016	0	−6	+39	−42
1.0			+100	+93	+46	−52
2.0	−0.0077	+0.0027	0	−9	+66	−54
2.0			+100	+90	+73	−64
4.0	−0.0093	+0.0035	0	−11	+71	−63
4.0			+100	+88	+78	−73
∞	−0.0116	+0.0048	0	−13	+92	−77
∞			+100	+86	+99	−87

It is not possible to determine the radial velocity of NP 0531 because its optical spectrum is featureless (Lynds et al., 1969). For the nebula itself, a radial velocity of −5.5 km/sec was suggested by Trimble (1968) predicated on some not-very-plausible assumptions. A direct measurement could be made, for example, using a diaphragm in the shape of an elliptical annulus, which, when placed in the focal plane of the telescope, would admit to the spectrograph only light from the edges of the nebula. This has not been done. The mean radial velocity of the 10% of the 418 features for which measurements are available nearest to the edge of the nebula is +66 km/sec. This is not necessarily indicative of what the result of the recommended operation might be.

Table I shows μ_α, μ_δ, and the space motion in galactic rotation coordinates of the nebula-pulsar system for various values of $W = M_{0531}/M_{CN}$ if their common radial velocity is 0 or +100 km/sec. For all reasonable values of W, the kinetic energy carried by star and nebula due to these motions will be at most 10% of the kinetic energy of the nebular expansion. Some unreasonable values of W are included in the table as being appropriate to other possible interpretations of what the two measured values may mean. All values of μ are uncertain by about 0.003″/yr, producing a 30 km/sec uncertainty in the space velocity, even if the radial velocity were well known.

Let us attempt for a moment that 'suspension of disbelief' so necessary for the appreciation of any work of fiction and inquire what, if anything, the numbers in the table are good for.

If both W and V_r are small, the object does not deviate greatly from circular motion. Galactic rotation 2 kpc further out is only about 12 km/sec slower than it is at our position (Schmidt, 1966). If W is large, on the other hand, the object is unusual in preceding the galactic rotation at its position by more than the 65 km/sec normally permitted (Mihalas and Routly, 1968).

A similar result, differently obtained, has prompted the suggestion by Woolf (cited by Gott *et al.*, 1970) that the star which became SN 1054 was a run-away star from the I Geminorum association. As this association has coordinates $\alpha = 6^\text{h}8^\text{m}$, $\delta = 23°31'$ and distance = 1400 pc (diameter $\sim 5° = 120$ pc), the suggestion is, at first sight, a very attractive one, given the range of possible proper motions and radial velocities for the pulsar-nebula system.

The catch is as follows: Known run-away stars are massive objects (Blaauw, 1961). If SN 1054 was a massive star, then, given an upper limit to stable neutron star masses of at most 2 $M_\odot$ or so, most of the mass must be in the nebula. That is, if the original star had a mass 12 $M_\odot$, $W \leqslant 0.2$. And for small values of W, the space motion is small enough that no particular explanation seems to be required.

The run-away hypothesis for pulsars in general and NP 0531 in particular is further discussed elsewhere in this volume.

3. Rotation and Turbulence in the Nebula

No rotation about any axis is detectable in the Crab Nebula to within the uncertainty of the determination. This is not surprising since, given conservation of angular momentum, even one km/sec of rotational velocity at the edge of the present nebula would correspond to a presupernova star which was rotationally unbound. The most stringent observational limit can be set to rotation about an axis parallel to our line of sight. An analysis of the deviations in position angle of the individual filamentary proper motion vectors from their corresponding radius vectors indicates that the nebula (as viewed from its own center) is rotating at a rate of 2.7±2.7″/yr. This corresponds to rotation energy of at most $8 \times 10^{45} (M/M_\odot)$ erg. For other possible rotation axes, the limits are about ten times larger.

There is, on the other hand, definite evidence for turbulence, that is, for *random* deviations from the general rule that the velocity vector of a filament is proportional to its radius vector. Some estimate of these deviations can be obtained from measured radial velocities. It is, of course, not possible to say in general whether the radial velocity of a given filament is precisely appropriate to its position along the line of sight or not, because that position can only be determined as the product of the velocity and the age of the nebula. On the other hand, since, in the plane of the sky, the filaments are confined so closely to an ellipse, it is reasonable to assume that, in three dimensions, they are confined within the ellipsoidal surface:

$$\frac{x^2}{a^2} + \frac{y^2}{b^2} + \frac{z^2}{b^2} = 1$$

where x and y are coordinates along the major and minor axes in the plane of the sky, z is the coordinate along the line of sight and a and b are the semi-major and semi-minor axes (5.4 and 3.6×10^{18} cm for a distance of 2 kpc). Thus the largest z a given filament should have can be found from its x and y positions. Any velocity excess, ΔV, given by

$$\Delta V = \left| V_r^{\text{observed}} - \frac{z}{\text{age}} \right|$$

must then be of the nature of a turbulent velocity, except that it will be only a lower limit since we have only an upper limit to z. Such a velocity excess is found for 37 of the 418 features for which radial velocities are available. (This excludes the features for which Mayall, 1962, gave only approximate velocities.) The mean value of the excesses is 102 km/sec.

An upper limit to the average turbulent velocity can be obtained from the proper motion data by comparing apparent deviations from purely radial motion with the uncertainties of the measurements. The quantities to be compared are the difference, D_1, of the proper motions as determined from two separate sets of measurements (of plates taken on the 100″ and 200″ telescopes),

$$D_1 = \mu(100'') - \mu(200'')$$

for each filament and the difference, D_2, between the average of the two measured μ's and the present radius vector divided by age,

$$D_2 = \bar{\mu} - r/t,$$

for each filament. Histograms of these two quantities show the same means and dispersions and are virtually indistinguishable. Their striking similarity may be taken to indicate that a major fraction of the apparent deviation from uniform radial expansion is due to measuring errors.

The deviations from radial motion in the plane of the sky, therefore, provide only an upper limit to the turbulent velocity of the filaments. The mean, median, and rms values of

$$V_{dev} = [(\mu_x - x/t)^2 + (\mu_y - y/t)^2]^{1/2}$$

all fall in the range 0.014 to 0.024″/yr. This corresponds to at most 225 km/sec, assuming the nebula to be at a distance of 2 kpc, or 275 km/sec, correcting for projection effects.

The lower and upper limits found from radial velocities and proper motions respectively thus indicate that turbulence must be in the range 100 to 300 km/sec. This corresponds to an energy of $1 - 9 \times 10^{47} \; M/M_\odot$ erg, that is, at most, about 10% of the expansion energy.

4. Acceleration of the Expansion

By extrapolating the measured proper motions backward in time, it is possible to find the time as well as the place at which they best converge. This was first done by Baade (1942), using Duncan's (1939) data. He found the rather surprising result that convergence occurred not in 1054 (as it would if the motions had been constant over the lifetime of the Crab Nebula) or earlier (as it would if the motions had been slowed by interaction with the interstellar medium), but later than 1054, implying that the present speeds are larger than the average ones over the nebular lifetime. Baade was not altogether convinced of the reality of this acceleration and, aside from pointing out that it could not be due to radiation pressure from any reasonable central star, deferred consideration of it until it should be confirmed by more accurate proper motion measurements.

His result was indeed confirmed. Convergence occurs in 1140 ± 10 AD, where the uncertainty is derived from the discrepancy of values obtained from proper motions measured independently on 100″ and 200″ direct photographs. If this is interpreted as meaning that there has been a constant acceleration over the history of the nebula, then the acceleration amounts to 0.0014 cm/sec^2 and the initial expansion velocity was 1700 km/sec along the major axis and 1100 km/sec^2 along the minor axis. The energy input required to maintain such an acceleration is $2.5 \times 10^{38} \; M/M_\odot$ erg/sec. This is of the same order as the electromagnetic radiation output of the object at all frequencies and the energy required to sweep up interstellar matter as the nebula expands, $8 \times 10^{38} \; N_H$ erg/sec, where N_H is the local density of the interstellar medium. This average acceleration will indeed be imparted to the nebula by outward pressure of a magnetic field of average strength $5 \times 10^{-4} (M/M_\odot)^{1/2}$ G or relativistic particles of total energy $3 \times 10^{48} \; M/M_\odot$ erg. These are very nearly equal to the minimum field strength and particle energy required to produce the observed synchrotron radiation (Woltjer, 1958), if $M \sim 1$.

There is, however, no particular reason to expect the acceleration to have been constant. On the one hand, the larger the nebula gets, the more interstellar matter it has to deal with per unit time, while ambient magnetic field and relativistic particles lose energy adiabatically in the course of the expansion, and so one might expect the acceleration to decrease with time. On the other hand, if the central neutron star is providing a continuing input of relativistic particles, their pressure will increase with time (because most of the energy is in electrons whose synchrotron lifetimes are long compared to the age of the nebula and, perhaps, protons), and so one might expect the acceleration to increase with time. It is, therefore, probable that the acceleration is some complicated function of time.

The observations do not help very much to define this function. It is possible, though, to set an upper limit to the acceleration going on now which is low enough to be of some interest. Comparison of proper motions determined over various stretches of the available baseline (1939–1966) indicates that the present acceleration is surely not more than three times the average value mentioned above. This corresponds to an upper limit on the magnetic field plus relativistic particle energy of $2 \times 10^{49} \, M/M_\odot$ erg. Now of the particles injected over the history of the nebula, all those with synchrotron lifetimes greater than 1000 years will still be there, aside from having lost energy adiabatically to the expansion. This means that, at most, about $3 \times 10^{49} \, M/M_\odot$ erg (present particle energy plus kinetic energy of the expansion) can have ever been injected into the nebula in the form of relativistic protons. The resulting constraints upon pulsar models are discussed by Trimble and Rees (1970). If the mass of the nebula is significantly greater than one solar mass, the limits are not, however, so stringent as they suggest.

5. The Mass of the Crab Nebula

It is clear that the size of a variety of quantities discussed here depends critically upon the mass in the Crab Nebula. The amount of material producing the optical emission lines has long been believed to be at most about one solar mass (Minkowski, 1968 and references cited therein). And the space between the filaments must contain much less material than this to prevent the dispersion measure of NP 0531 changing as the nebula expands (Drake, 1969). It has, however, recently been suggested that, in addition to the ionized material which produces most of the emission lines, the filaments might also contain large amounts of neutral material at their centers (Davidson and Tucker, 1970), increasing the total nebular mass to as much as 10 $M_\odot$ (Gott et al., 1970).

Some indication of the quantity of neutral material present can be obtained from the intensity of the $\lambda 6300$ radiation of [OI]. This radiation is necessarily produced in regions where hydrogen is neutral (and, therefore, does not contribute significantly to the intensity of $H\beta$, from which the mass of ionized material is obtained). This is a result of the large cross-section for the charge exchange reaction

$$H^+ + O^\circ \rightarrow H^\circ + O^+$$

which arises from the near identity of the ionization potentials of the two elements.

Unfortunately, neutral material has not yet had time to come into equilibrium with the synchrotron radiation field of the nebula in its 1000 year lifetime. The run of temperature through this part of the gas cannot, therefore, be calculated by the methods of Davidson and Tucker (1970). The temperature rather reflects cooling which occurred early in the nebular expansion and may thus be very low. One can easily calculate the amount of material at a given T_e required to produce the observed $\lambda 6300$ intensity:

$$I([O_I]) \simeq I(H\beta) = 1.24 \times 10^{-11} \text{ erg cm}^{-2} \text{ sec}^{-1}$$

(O'Dell, 1962). This has been done by Trimble (1970b). Table II shows the amount of matter required to produce $\lambda 6300$ as a function of T_e. The abundances assumed are $N_{He} = N_H$; $N_O = 6 \times 10^{-4} N_H$ (Davidson and Tucker, 1970). The distance to the nebula was taken to be 2 kpc. If it is really only 1.5 kpc, then the tabulated amount of

TABLE II

Amount of neutral gas in the filaments of the Crab Nebula required to produce the observed intensity of [O_I] $\lambda 6300$

$T_e(K)$	Mass (solar masses)
10^4	2.6
9×10^3	3.5
8×10^3	5.0
7×10^3	9.0
6×10^3	13
5×10^3	26
4×10^3	100

matter will produce $I(\lambda 6300) \simeq 1.8 \, I(H\beta)$. It is, therefore, by no means unlikely that the nebular mass is significantly larger than has usually been assumed, and it is, in any case, very uncertain. The amounts of energy in various forms which must be supplied to the nebula during and after the supernova event are thus uncertain by factors of about 10. The kinetic energy of the expansion, for instance, may be in excess of 10^{50} erg!

Trimble and Woltjer (1971) have recently presented a dynamical argument for the nebula mass not being much larger than $1 \, M_\odot$.

References

Baade, W.: 1942, *Astrophys. J.* **96**, 109.
Blaauw, A.: 1961, *Bull. Astron. Inst. Neth.* **15**, 265.
Davidson, K. and Tucker, W.: 1970, *Astrophys. J.* **161**, 437.
Deutsch, A. N. and Lavdovsky, V. V.: 1940, *Pulkovo Obs. Circ.* 30, 21.
Drake, F. D.: 1969, Talk at Rome Meeting on Pulsars and High Energy Phenomena.
Duncan, J. C.: 1921, *Proc. Nat. Acad. Sci.* **7**, 179.
Duncan, J. C.: 1939, *Astrophys. J.* **89**, 482.
Gott, J. R., Gunn, J. E., and Ostriker, J. P.: 1970, *Astrophys. J.* **160**, L91.
Lynds, R., Maran, S. P., and Trumbo, D. E.: 1969, *Astrophys. J.* **155**, L121.
Mayall, N. U.: 1937, *Publ. Astron. Soc. Pacific* **49**, 101.
Mayall, N. U.: 1962, *Science* **137**, 91.

Mihalas, D., and Routly, P.: 1968, *Galactic Astronomy*, W. H. Freeman & Co., p. 114–115.
Minkowski, R.: 1968, 'Nonthermal Galactic Radio Sources' in B. Middlehurst and L. Aller (eds.), *Nebulae and Interstellar Matter*, Univ. of Chicago Press, p. 637ff.
Minkowski, R.: 1970, *Publ. Astron. Soc. Pacific* **82**, 470.
Münch, G.: 1958, *Rev. Mod. Phys.* **30**, 1042.
Münch, G., Scargle, J., and Trimble, V.: 1971, *The Crab Nebula*, Gordon and Breach (in preparation).
O'Dell, C. R.: 1962, *Astrophys. J.* **136**, 809.
Sanford, R. F.: 1919, *Publ. Astron. Soc. Pacific* **31**, 108.
Schmidt, M.: 1966, in A. Blaauw and M. Schmidt (eds.), *Galactic Structure*, Univ. of Chicago Press, p. 528.
Trimble, V.: 1968, *Astron. J.* **73**, 535.
Trimble, V.: 1970a, *Observatory* **90**, 221.
Trimble, V.: 1970b, *Astron. J.* **75**, 926.
Trimble, V. and Rees, M.: 1970, *Astrophys. Letters* **5**, 93.
Trimble, V. and Woltjer, L.: 1971, *Astrophys. J.* **163**, L97.
Woltjer, L.: 1958, *Bull. Astron. Inst. Neth.* **14**, 39.
Woltjer, L.: 1970, *Publ. Astron. Soc. Pacific* **82**, 479.

Discussion (on Papers 1.1 and 1.2)

W. A. Fowler: How did you calculate mass of the nebula as 10 $M_\odot$? Did you use the solar abundance for oxygen?

V. Trimble: The amount of neutral material required to produce the observed intensity of [OI] $\lambda 6300$ radiation (1.24×10^{-11} erg cm^{-3} sec^{-1}; i.e. $I([OI]) \approx I(H\beta)$ as found by C. R. O'Dell, (*Astrophys. J.* **136** (1962) 809) was calculated as a function of electron temperature using formulae given by M. J. Seaton (*Monthly Notices Roy. Astron. Soc.* **114** (1954) 154). The electron density was estimated by assuming that atoms with I. P. < 13.6 eV are singly (radiatively) ionized and that hydrogen and helium are collisionally ionized. Useful formulae for the latter are given, e.g., by R. A. R. Parker (*Astrophys. J.* **139** (1964) 208). The abundances assumed are $N_{He} = N_H$; $N_O = 6 \times 10^{-4} N_H$, as suggested by Model 2 of K. Davidson and W. Tucker (*Astrophys. J.* **161**, 437, 1970). The distance to the nebula was taken to be 2 kpc. If it is really only 1.5 kpc, the tabulated amount of matter will produce $I(\lambda 6300) \approx 1.8 I(H\beta)$. The mass required cannot be more precisely estimated because the neutral material has not yet had time to come into equilibrium with the synchrotron radiation field in the 1000 year lifetime of the nebula (Davidson and Tucker, 1970). The temperature rather reflects cooling which occurred early in the nebular expansion (when the gas was perhaps much denser) and may thus be very low.

R. Minkowski: We do not know the proper motion of the nebula. The basic difficulty is that there seems to be no way to find the centre of mass of the nebula. Baade's ellipse is a rough fit to the outline of the nebula, but its centre is not and is not meant to be the centre of mass.

The proper motion of the pulsar has been measured, but there is a peculiar difficulty. Results obtained by different observers agree very poorly with each other, much poorer than the measuring errors admit. For the north-following star, which has about the same brightness and nebular background, the situation is quite different. All observations agree with each other quite as well as the measuring errors admit. The obvious interpretation of the poor internal agreement of the motions of the pulsar is that the measurements are affected by the presence of variable features of the nebulosity. The prime suspect is Scargle's 'thin wisp'. The observations with the largest telescopes are least strongly affected by this systematic error, but they cannot be expected to be free of it. The best that can be done at the moment is to take the mean of all observations with the 100-inch and 200-inch telescopes by van Maanen, Baade and Trimble. The position of the pulsar in $+1054$ computed with this mean value for the proper motion agrees reasonably well with the position of Trimble's convergence point of the filaments.

V. Trimble: In regard to the proper motion of NP 0531: I am pleasantly surprised to hear that the direct measurements (at least the large telescope ones) confirm the value found indirectly using the expansion centre and elapsed time.

In regard to the proper motion of the Crab Nebula: Dr. Minkowski is absolutely correct (as usual!) in saying that our lack of knowledge of the centre of mass of the present nebula prevents our finding its proper motion by the method used for 0531 – dividing change in position by time elapsed since 1054. There is, however, another possible approach to the problem, which was used to get the proper motion mentioned above. There are two ways of determining the expansion centre of the nebula: (1) geometrically – that is by finding the *intersection* (as nearly as possible, given measuring errors and perhaps turbulence) of the proper motion vectors of individual filaments.

(2) temporally – by moving backward along the proper motion vectors by the distance that each filament would cover in a fixed increment of time until, as nearly as possible, all features meet in the same place *at the same time*.

If the presupernova star had no proper motion of its own, the centre thus found is the same as in the first method.

But if there is some overall proper motion, then this method will give a different convergence centre.

Notice that in this latter case, the scatter of points at the time of best convergence might be expected to be somewhat larger than if there were no overall proper motion, but the effect is small compared to the scatter caused by measuring errors in the case of the Crab Nebula – that is, the scatter is the same for methods (1) and (2). In addition, the centres found by the two methods are virtually identical – no effect of this type is found to within the uncertainty of position of the two centres. It is evident that a variety of things, including asymmetrical acceleration of the nebular expansion, could invalidate the proper motion found in this way, but not knowing the centre of mass is not, a priori, one of them, provided that enough individual proper motions are available to represent the entire nebula.

R. Minkowski: Why does the neutral gas stay in filaments?

V. Trimble: The filaments as a whole appear to be kept together by a pressure balance which involves differences in density, temperature, magnetic field, and (perhaps) relativistic particle pressure across their boundaries. The 'neutral' gas will not be exempt even from the effects of the magnetic field, because at the temperatures and densities discussed the electron density will be several to ten percent of the total particle density.

L. Aller: It is extremely difficult to deduce the chemical composition of a gaseous nebula from an emission line spectrum if the gas contains numerous filaments and 'low densations'. Among less exotic nebulae, the effects are best exhibited in NGC 7027 where the available data clearly indicated a strongly inhomogeneous structure (Aller, *Astrophys. J.* **120**, 401, 1954). If we use density – sensitive line ratios of (SII), (OII), (ClIII), ArIV), with recent cross-section calculations (Czyzak, Seaton and their associates) we find that no single choice of density and temperature can represent the data. Either the atomic parameters or observed line intensities are grossly in error (which seems unlikely) or very substantial fluctuations in T and N_e must exist. These fluctuations must be taken into account in trying to estimate chemical compositions.

J. Kristian: Have you actually seen Scargle's original plates?

V. Trimble: I saw a better reproduction of the late 1969 Lick plate at the Rome pulsar meeting in December 1969 and was convinced at the time that there was a 'thin wisp' in the required position, but I agree that there is some room for doubt on the basis of the pictures presented here.

D. W. Richards: What is the origin of the figure 0″.009/yr for proper motion of the pulsar?

V. Trimble: My 'favourite' proper motion for NP 0531 is derived by taking the angular separation of the present pulsar position from the nebular expansion centre (after all, everything must have been in the same place when the supernova explosion occurred) and dividing by the time elapsed since 1054. Direct measurements of the proper motion made on plates taken with large telescopes, as discussed by R. Minkowski (*Publ. Astron. Soc. Pacific* **82**, (1970) 470 report from the Flagstaff Conferences, June 1969), confirm it to within the errors of the observations, despite the difficulties of the measurement as outlined by him elsewhere in this discussion.

L. Woltjer: These agree in magnitude but not in direction.

R. Minkowski: They do agree if you use the best measurements.

J. E. Baldwin: The proper motion of 0″.009/yr corresponds to a delay in the pulsar timing measurements of about 100 μsec. Is this accuracy easily achievable when all the corrections have been put into the observations?

J. A. Roberts: The behaviour of the Crab Nebula pulsar is so irregular that I doubt if such an effect could be disentangled from other effects.

P. Horowitz: I'd like to comment on the suggestion of measuring the proper motion of the Crab pulsar from timing measurements. The component of motion along the line of sight is manifested as a simple doppler shift, and is therefore unmeasurable since we don't know the unshifted pulsar period. The transverse motion (proper motion) would produce a yearly sinusoid in the observed arrival times, due to parallax, of about 25μsec amplitude (if the proper motion is 0.01 sec/yr). From our experience with optical timing measurements we can say that such variations are completely swallowed up by 'jumps' and other anomalies in the Crab pulsar period, and are therefore unmeasurable.

R. Hills: At Lick we are also making optical timing measurements and we hoping that when we have data covering about 2 years we will be able to estimate the component of the proper motion along the ecliptic. The problem is to separate out the term of one year period and find the rate of change of that term.

1.3 THE ELECTROMAGNETIC SPECTRUM OF
THE CRAB NEBULA

J. E. BALDWIN

Mullard Radio Astronomy Observatory, Cambridge, U.K.

Abstract. I shall discuss observations of the spectrum of the integrated emission from the Crab Nebula. The radio data with accurate calibrations lead to a flux density spectral index of -0.26. Discrepancies in the published fluxes at millimetre wavelengths can be resolved if appropriate angular dimensions are used. In the optical range the spectral index has increased to a value of -0.9 if $1^{\mathrm{m}}.0$ of absorption is used. At X-ray wavelengths the spectral index has increased further to -1.2.

1. Introduction

The spectrum of the integrated radiation from the Crab Nebula might be thought to be of very little importance. Evidently there are contributions from a wide variety of sources – the filaments, the amorphous nebula, the wisps, the compact low frequency source and the pulsar. The spectra of these components are likely to be, and in some cases known to be, very different from each other. But observers continue to attempt to measure the overall spectrum with greater accuracy and I believe this work to be important for two reasons

(1) The integrated radiation is the only quantity which can be measured with any ease.

(2) The combination of an integrated spectrum with detailed mapping at a number of fixed frequencies gives us the spectra of all points over the face of the nebula.

The separation of the two technical problems of absolute determinations of intensity and high resolution mapping of relative intensities seems at present to give us the best hope of obtaining the details of the spectrum as a function of position in the nebula – the primary data needed for any physical model.

I shall present data over the whole observed frequency range but I shall leave detailed discussion of the X- and γ-ray observations to those more competent to discuss them. There are still gaps in the observed spectrum and it is convenient to discuss the observations in separate groups.

2. The Radio Spectrum $10^7 < \nu < 2.5 \times 10^{11}$ Hz

In the radio range there is, relatively speaking, a very large power flux available. If radioastronomers ever thought of photons, they would find photon counting rates in their experiments which are typically $> 10^{10}$ sec^{-1}. They therefore use voltages and currents and I shall refer to flux densities in terms of Wm^{-2} Hz^{-1} and apply this in all parts of the spectrum.

The technical problems in the radio region are those of accurate calibration of equipment, especially over wide ranges of frequency. Measurements are usually made

Davies and Smith (eds.), The Crab Nebula, 22–31. All Rights Reserved.

in narrow frequency bands. The so-called absolute determinations rely on experimental or theoretical values of the scale of antenna temperature and the antenna gain. Relative determinations rest on comparisons with other sources such as planets, the Moon or an artificial Moon whose flux density is believed to be known. I include all observations in which absolute calibrations were made and most of those using relative calibrations where they have seemed to me to be in some sense independent. The difficulty here is that the various scales of flux density now in use are in some cases dependent on an assumption about the smoothness of the radio spectra of the sources which form the basis of the scale. The Crab Nebula is often one such basic source.

I take this opportunity to mention that many of the values obtained are not satisfactorily described by the authors. Some merely note that the Moon was used as a calibrating source without saying in what way. Values quoted by other authors change in successive papers without it being clear whether the later values supersede the former due to advances in technique or recalculation or whether they are just independent estimates.

In Table I and Figure 1 are assembled the radio flux densities I have found in the literature, nearly all of them post 1960. In Figure 1 I have omitted a few values whose error limits are too large to be useful now. It is probable that, to the accuracy of the present discussion, we can ignore any secular variations in the flux densities. It is clear that the range 0.1–10 GHz is the best for accurate measurements. At low frequencies it is difficult to detect the source with simple antennas of calculable gain and in any case ionospheric scintillation makes observations difficult. At high frequencies it is difficult to construct accurate antennas of large collecting area, there are no very good standards of noise power and atmospheric attenuation is very important.

The most serious discrepancies between observers are at mm wavelengths ($v > 3 \times 10^{10}$ Hz). One possible explanation would be in the use of Jupiter as a reference source. The disc temperature is frequently assumed to be 150 K. This may well be incorrect and perhaps by large amounts if there are absorption lines in the Jovian

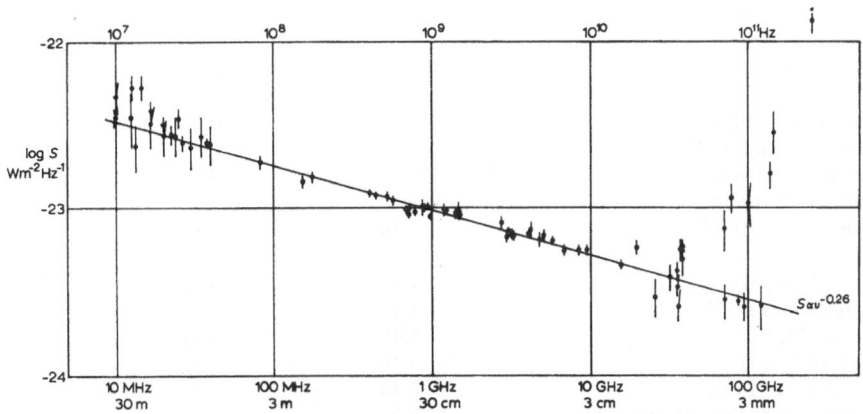

Fig. 1.   Radio spectrum of total radiation from the Crab nebula.

TABLE I
Measurements of the flux density of the integrated emission from the Crab Nebula

Author	Frequency	Flux density 10^{-26} W m^{-2} Hz^{-1}	Error %
Clark (1966)	10 MHz	3500	10
Bridle and Purton (1968)	10	4650	25
Baselyan et al. (1963)	12.5	3500	40
Braude et al. (1969)	12.6	5300	15
Andrew (1967)	13.1	2350	30
Braude et al. (1969)	14.7	5300	15
Braude et al. (1969)	16.7	3830	14
Baselyan et al. (1963)	16.7	3200	40
Baselyan et al. (1963)	20	2700	30
Braude et al. (1969)	20	3170	14
Roger et al. (1969)	22.25	2750	12
Baselyan et al. (1963)	24	2700	30
Braude et al. (1969)	25	3420	14
Erickson and Cronyn (1965)	26.3	2440	11
Baselyan et al. (1963)	30	2300	30
Baselyan et al. (1963)	35	2700	30
Williams et al. (1965)	38	2430	5
Baselyan et al. (1963)	40	2400	30
Parker (1968)	81.5	1880	10
Parker (1968)	152	1430	10
Kellermann et al. (1969)	178	1534	5
Baars et al. (1965)	400	1229	3.8
Baars et al. (1965)	440	1192	3.7
Alekseev et al. (1969)	518	1180	3
Lastochkin et al. (1963)	562	1110	6
Alekseev et al. (1969)	680	980	4.5
Lastochkin et al. (1963)	708	910	5
Alekseev et al. (1969)	714	980	4.5
Alekseev et al. (1969)	770	940	3
Lastochkin et al. (1963)	860	1016	10
Alekseev et al. (1969)	870	1030	3
Lastochkin et al. (1963)	876	1000	3
Razin and Fedorov (1963)	927	1025	4
Alekseev et al. (1969)	986	890	3
Lastochkin et al. (1963)	1.19 GHz	980	7
Baars et al. (1965)	1.20	971	3
Kellermann et al. (1969)	1.40	930	5
Mezger (1958)	1.419	968	12
Baars et al. (1965)	1.44	913	3
Altenhoff et al. (1961)	2.70	811	10
Sloanaker and Nichols (1960)	2.93	670	10

Tabel I (continued)

Author	Frequency	Flux density 10^{-26} W m^{-2} Hz^{-1}	Error %
Baars *et al.* (1965)	3.00 GHz	733	3.5
Medd and Ramana (1965b)	3.15	695	5
Broten and Medd (1960)	3.20	680	5
Wilson and Penzias (1966)	4.08	711	3
Yokoi (1966)	4.17	745	10
Golnev *et al.* (1965)	4.69	650	
Kellermann *et al.* (1969)	5.00	680	5
Dmitrenko and Strezhneva (1967)	5.68	646	5
Medd and Ramana (1965a)	6.66	565	5
Allen and Barrett (1966, 1967)	8.25	567	4.8
Lazarewski *et al.* (1963)	9.36	560	5.4
Allen and Barrett (1966)	15.5	461	5.9
Williams *et al.* (1965)	19.6	588	9
Staelin *et al.* (1964)	25.4	295	28
Hobbs *et al.* (1968)	32	387	18
Kalaghan and Wulfsburg (1967)	34.9	340	+19 −12
Tolbert and Straiton (1965)	35	420	14
Lynn *et al.* (1964)	35.3	260	25
Tolbert and Straiton (1965)	36.2	420	15
Matveenko and Pavlov (1967)	36.6	565	14
Kuzmin and Salomonovich (1962)	37.5	500	25
Barrett *et al.* (1965)	37.5	600	8
Hobbs *et al.* (1969)	69.75	281	26
Tolbert (1965)	70	750	28
Kisljakov and Lebsky (1967)	77.5	1130	21
Matveenko (1970)	85.6	280	7
Oliver *et al.* (1967)	93	260	19
Tolbert (1965)	100	1080	30
Efanov *et al.* (1969)	139	1600	13
Kisljakov and Naumov (1967)	142	2800	32
Zabolotny *et al.* (1970)	120	250	30
Beckman *et al.* (1969)	250	13400	15

atmosphere. A more likely explanation, as noted by Hobbs *et al.* (1969), lies in the corrections to the observed flux densities due to the finite angular size of the Crab Nebula when narrow beamwidths are used. In some cases the corrections applied have been certainly too large. If one takes the size as measured, for example, at 6 cm by Wilson (see paper 1.8, page 68, in this symposium) which can be represented approximately by Gaussians with half-widths of 2.2′ and 3.4′ respectively, then the short wavelength measurements when corrected are as shown in Figure 2. There is

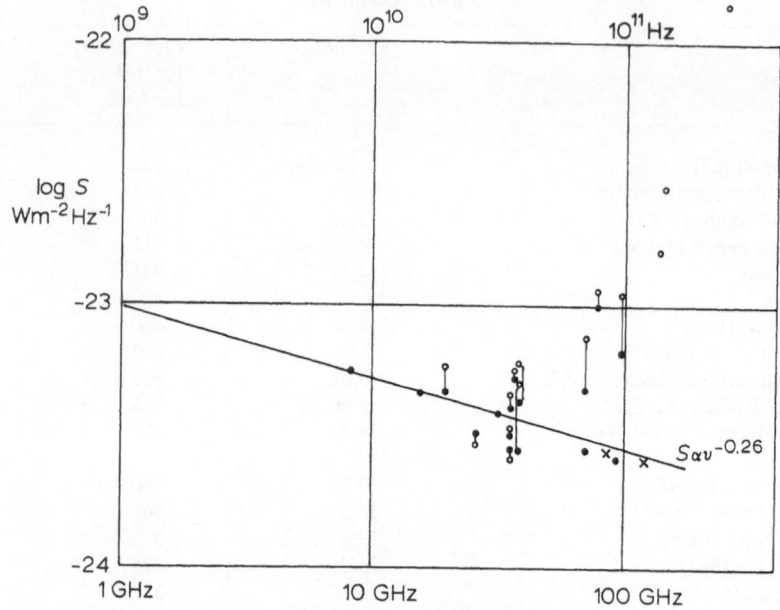

Fig. 2.   High frequency radio spectrum. ○ Published values. ● Values corrected for a
standard source size.

now much less evidence for an increase in flux density towards the high frequencies.
Of the three very discrepant points, that of Efanov *et al.* (1969) at 139 GHz (2.16 mm)
was made with a telescope operating at a much shorter wavelength than that for which
it was designed and that of Kisljakov and Naumov (1967) at 142 GHz (2.11 mm) was
made with a small telescope giving a very small signal to noise ratio. Only the measurement of Beckman *et al.* (1969) is unaccounted for. I take the now old fashioned view
that exciting results require further experimental confirmation.

The two points marked by × in Figure 2 are those of Matveenko and Zabolotny
which were announced at the IAU General Assembly subsequent to Symposium 46.
The present evidence seems to strongly favour a straight line spectrum over the whole
radio region having a spectral index α, defined by flux density $\propto$ (frequency)$^{\alpha}$, of -0.26.

3. The Infrared and Optical Spectrum

The evidence available has recently been reviewed by Scargle (1969, 1970). In the
optical range there is little disagreement between observers and the best line through
the points is straight with a spectral index α of -2.5. In the infrared the early values
by Moroz (1964) are probably superseded by those of Ney and Stein (1968) and
Becklin and Kleinmann (1968) slightly corrected for source size. The main uncertainty
lies in the correction for visual obscuration. Estimates of A_v of $1^{m}7$ have been most
popular. In Figure 3 are presented both the observations and curves corrected for
different values of the visual obscurations using the corrections as a function of

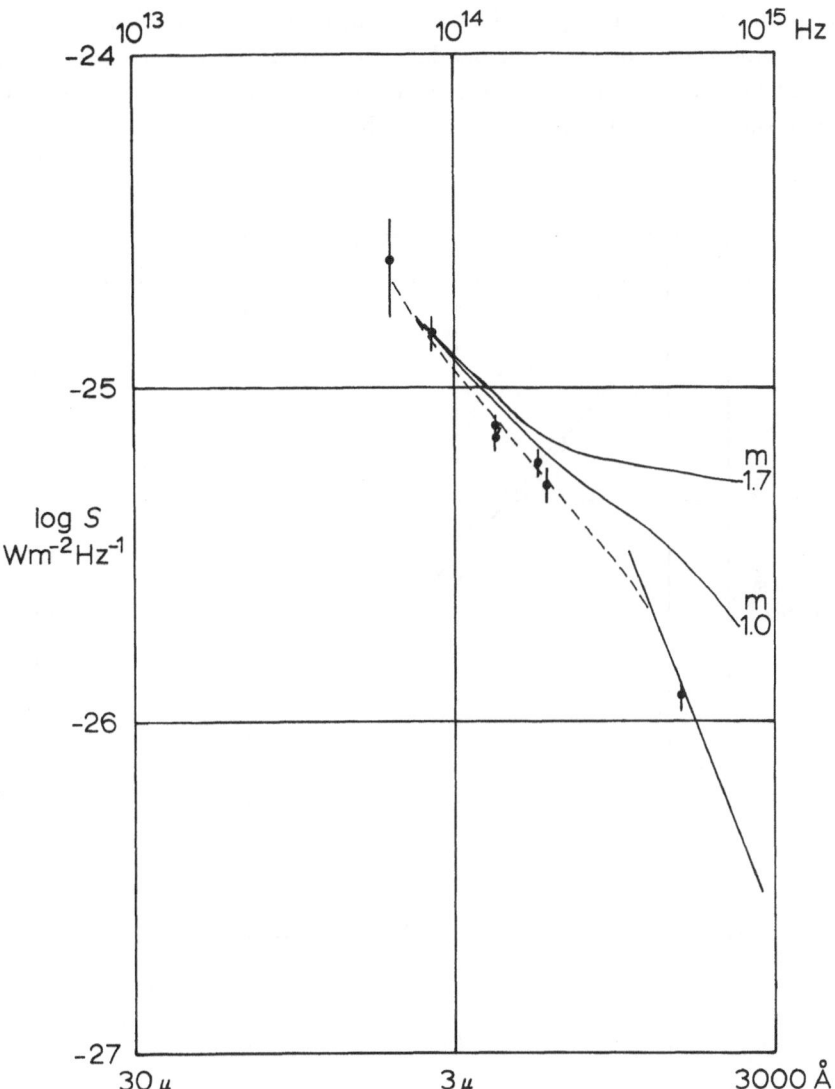

Fig. 3.   Infra red and optical continuum spectrum of the Crab nebula. The several curves correspond to different magnitudes of visual obscuration.

frequency given by Becklin and Neugebauer (1968). Note that $1^{m}.0$ corresponds to a fairly smooth unabsorbed spectrum with spectral index -0.9. There may be good reason to suppose that this is near the correct value since, line emission excepted, it is extremely difficult to account for sharp features in the synchrotron spectrum of the nebula.

4. The X-Ray and γ-Ray Spectrum

At X-ray wavelengths there is again a large amount of experimental evidence. Most of the observers quote results in the form of a photon flux at a particular energy together with a value for the spectral index. A recent review by Peterson and Jacobson (1970) assembles most of the available data and no details will be given here. The spectra are plotted in Figure 4 without error limits. The scatter of observations is somewhat outside the quoted error limits but it is clear that over the range

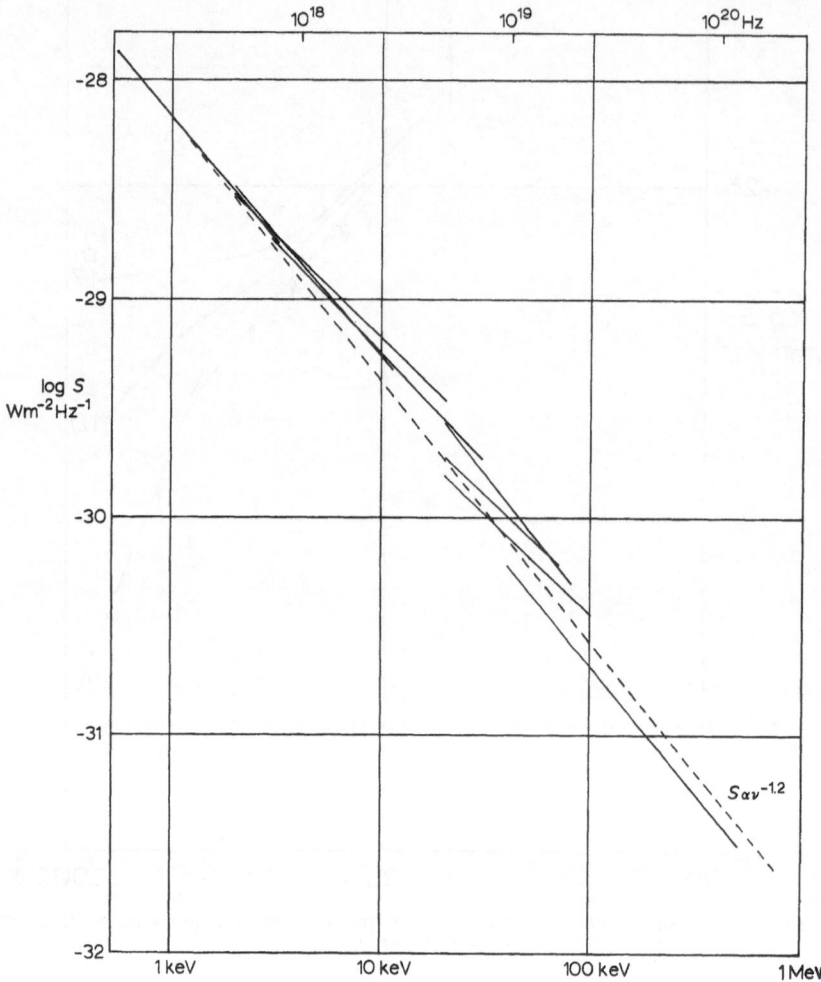

Fig. 4.   X-ray spectrum of the Crab nebula. The mean spectra from different authors are plotted without error bars.

$2 \times 10^{17} < \nu < 10^{20}$ Hz the observations can be represented by a spectrum with spectral index α of -1.2. At frequencies $> 10^{20}$ Hz there are, as yet, only upper limits to the flux density.

The separate parts of the spectrum are presented together in Figure 5. The widths of the lines cover the limits of uncertainty in different parts of the spectrum. I conclude from this figure that a simple smooth spectrum can be drawn through the observations over the whole observed range having a spectral index of -0.26 for $10^7 < v < 10^{11}$ Hz, -0.9 for $6 \times 10^{13} < v < 10^{15}$ Hz and -1.2 for $2 \times 10^{17} < v < 10^{20}$ Hz. It seems more important to attempt new observations in the gaps in the spectrum than to improve the accuracy in those parts already studied.

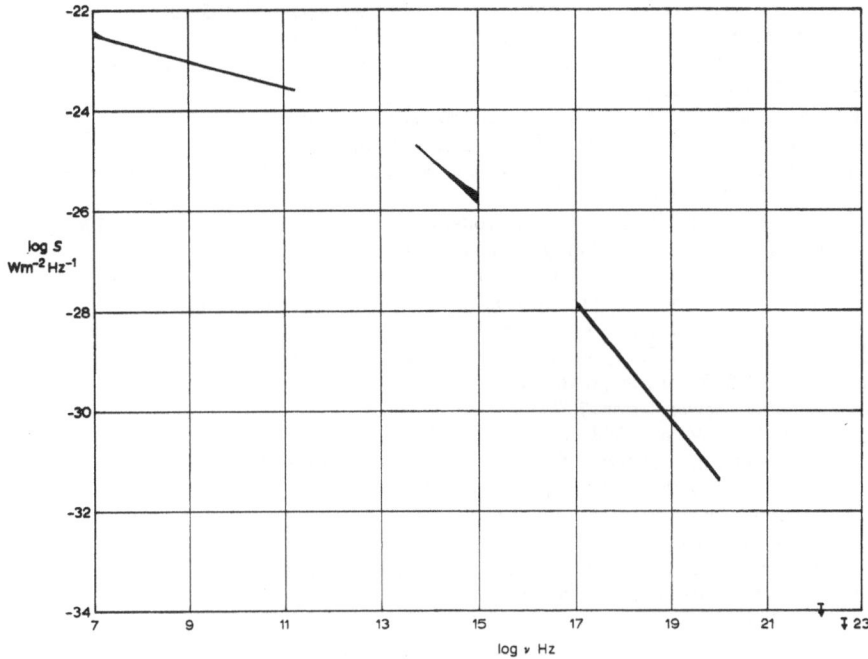

Fig. 5. The electromagnetic spectrum of the total radiation from the Crab nebula. The thickness of the lines covers the limits of uncertainty at any frequency. The upper limits at high frequencies are those of Frye and Wang (1969).

References

Alekseev, V. A., Gatelyuk, E. D., Dmitrenko, A. A., Romonychev, A. A., and Tseitlin, N. M.: 1969, *Radiofiz.* **12**, 168.
Allen, R. J. and Barrett, A. H.: 1966, *Astron. J.* **71**, 843.
Allen, R. J. and Barrett, A. H.: 1967, *Astron. J.* **72**, 288.
Altenhoff, W., Mezger, P. G., and Wendker, H.: 1961, *Veröff. Inst. Univ. Bonn* No. 59, Pt. IV.
Andrew, B. H.: 1967, *Astrophys. J.* **147**, 423.
Baars, J. W. M., Mezger, P. G., and Wendker, H.: 1965, *Astrophys. J.* **142**, 122.
Barrett, A. H., Kutuza, B. G., Matveenko, L. I., and Salomonovich, A. E.: 1965, *Soviet Astron.* **9**, 418.
Baselyan, L. L., Braude, S. Y., Krymkin, V. V., Men, A. V., and Sodin, L. G.: 1963, *Radiofiz.* **6**, 897.
Becklin, E. E. and Kleinmann, D. E.: 1968, *Astrophys. J.* **152**, L25.
Becklin, E. E. and Neugebauer, G.: 1968, *Astrophys. J.* **151**, 145.
Beckman, J. E., Bastin, J. A., and Clegg, P. E.: 1969, *Nature* **221**, 944.
Braude, S. Y., Lebedeva, O. M., Men, A. V., Ryabov, B. P., and Zhouck, I. N.: 1969, *Monthly Notices Roy. Astron. Soc.* **143**, 289.
Bridle, A. H. and Purton, C. R.: 1968, *Astron. J.* **73**, 717.
Broten, N. W. and Medd, W. J.: 1960, *Astrophys. J.* **132**, 279.
Clark, T. A.: 1966, *Astron. J.* **71**, 158.

Dmitrenko, D. A. and Strezhneva, K. M.: 1967, *Radiofiz.* **10**, 165.

Efanov, V. A., Kisljakov, A. G., Kostenko, V. I., Matveenko, L. I., Moiseev, I. G., and Naumov, A. I.: 1969, *Radiofiz.* **12**, 803.

Erickson, W. C. and Cronyn, W. M.: 1965, *Astrophys. J.* **142**, 1156.

Frye, G. M. and Wang, C. P.: 1969, *Astrophys. J.* **158**, 925.

Golnev, V. Y., Lipovka, N. M., and Parijsky, Y. N.: 1965, *Soviet Astron.* **9**, 690.

Hobbs, R. W., Corbett, H. H., and Santini, N. J.: 1968, *Astrophys. J.* **152**, 43.

Hobbs, R. W., Corbett, H. H., and Santini, N. J.: 1969, *Astrophys. J.* **155**, L87.

Kalaghan, P. M. and Wulfsburg, K. N.: 1967, *Astron. J.* **72**, 1051.

Kellermann, K. I., Pauliny-Toth, I. I. K., and Williams, P. J. S.: 1969, *Astrophys. J.* **157**, 1.

Kisljakov, A. G. and Lebsky, Y. V.: 1967, *Soviet Astron.* **11**, 561.

Kisljakov, A. G. and Naumov, A. I.: 1967, *Soviet Astron.* **11**, 1059.

Kuzmin, A. D. and Salomonovich, A. E.: 1962, *Soviet Phys. Dokl.* **6**, 745.

Lastochkin, V. P., Porfiriev, V. A., Stankevitch, K. S., Troitsky, C. S., Kholodilov, N. N., and Tseitlin, N. M.: 1963, *Radiofiz.* **6**, 629.

Lazarewski, V. S., Stankevitch, K. S., and Troitsky, V. S.: 1963, *Soviet Astron.* **7**, 8, 1963.

Lynn, V. L., Meeks, M. L. and Sohigian, M. D.: 1964, *Astron. J.* **69**, 65.

Matveenko, L. I. and Pavlov, A. V.: 1967, *Soviet Astron.* **11**, 300, 1967.

Matveenko, L. I.: 1970, preprint presented at IAU General Assembly, Brighton.

Medd, W. J. and Ramana, K. V. V.: 1965a, *Astron. J.* **70**, 327.

Medd, W. J. and Ramana, K. V. V.: 1965b, *Astrophys. J.* **142**, 383.

Mezger, P. G.: 1958, *Z. Astrophys.* **46**, 234.

Moroz, V. I.: 1964, *Soviet Astron.* **7**, 755.

Ney, E. P. and Stein, W. A.: 1968, *Astrophys. J.* **152**, L21.

Oliver, J. P., Epstein, E. E., Schorn, R. A., and Sotet, S. L.: 1967, *Astron. J.* **72**, 314.

Parker, E. A.: 1968, *Monthly Notices Roy. Astron. Soc.* **138**, 407.

Peterson, L. E. and Jacobson, A. S.: 1970, *Publ. Astron. Soc. Pacific.* **82**, 412.

Razin, V. A. and Fedorov, V. T.: 1963, *Radiofiz.* **6**, 1052.

Roger, R. S., Costain, C. H., and Lacey, J. D.: 1969, *Astron. J.* **74**, 366.

Scargle, J. D.: 1969, *Astrophys. J.* **156**, 401.

Scargle, J. D.: 1970, *Publ. Astron. Soc. Pacific* **82**, 388.

Sloanaker, R. M. and Nichols, J. H.: 1960, *Astron. J.* **65**, 109.

Staelin, D. H., Barrett, A. H. and Kusse, B. R.: 1964, *Astron. J.* **69**, 69.

Tolbert, C. W.: 1965, *Nature* **206**, 1304.

Tolbert, C. W. and Straiton, A. W.: 1965, *Astron. J.* **70**, 177.

Williams, D. R. W., Welch, W. J., and Thornton, D. D.: 1965, *Publ. Astron. Soc. Pacific* **77**, 178.

Wilson, R. W. and Penzias, A. A.: 1966, *Astrophys. J.* **146**, 286.

Yokoi, H.: 1966, *Publ. Astron. Soc. Japan* **18**, 271.

Zabolotny, V. F., Sholomitsky, G. B., and Slysh, V. I.: 1970, preprint presented at the IAU General Assembly, Brighton.

Discussion

V. Trimble: It now seems probable that the upper limit to the UV flux derived from the ratio of (O II) to (O III) is not in conflict with a smooth curve drawn from optical to X-ray observations. The rather low degree of ionization can be explained by a high helium abundance in the filaments (ionization of the helium 'soaking up' photons that would otherwise ionize O^+ to O^{++} further into the filaments).

R. C. Jennison: Dr. Baldwin offered some criticism of the high flux readings in the millimetre spectrum but did not discuss the reliability of the two readings just below the extrapolated mean flux line. The implication is that these readings are less subject to error and carry greater weight than all those above the line. Is this so or was Dr. Baldwin's argument conditioned by the beauty of the simple smooth spectrum?

J. E. Baldwin: There is little argument about the best curve at ~ 1 cm. At 8 mm I have shown that many of the observed points must be moved downwards towards the extrapolated spectrum from lower frequencies. The real argument concerns the points at yet shorter wavelengths. The values at 3.2 and 4.3 mm were both obtained with large telescopes designed for these short wavelengths and seem to be good values; that at 4.3 mm being particularly well documented by the authors.

M. M. Komesaroff: Is the optical flux density spectral index -0.25?

J. E. Baldwin: If we assume about $1^{m}.0$ of absorption, giving an essentially straight spectrum at optical wavelengths, then the index is -0.9.

W. J. Cocke: Most of the very low frequency radio spectrum comes from the compact radio source, presumably from an entirely different emission process. The synchrotron continuum emission flux density would then have to decrease sharply at low frequencies.

J. E. Baldwin: I agree that the compact source is very important at low frequencies. However, both its contribution to the total radiation and the optical depth of interstellar H II in the line of sight to the Crab nebula are very uncertain and it is therefore extremely difficult to make a good assessment of the spectrum of the nebula itself at low frequencies.

F. C. Michel: It would be very difficult for a synchrotron source to drop in flux rapidly with decreasing frequency, since the low frequency drop off goes as $\nu^{+0.33}$ regardless of the energy spectrum of the electrons with $\nu_{max} \rangle \nu$. In other words the source can decrease arbitrarily rapidly with increasing frequency, but can increase no more rapidly than $\nu^{+0.33}$, provided of course that self-absorption is unimportant. Thus the rapid rise beyond 1 cm would be difficult to reconcile with the synchrotron model.

R. Minkowski: Why should one expect a perfectly smooth spectrum? The size of the nebula depends on the frequency. Wherever the size decreases, the integrated flux must decrease. This decrease is superimposed on all changes of spectral index. I see no reason why the result should be a perfectly smooth spectrum. If, as seems probable, the visual interstellar absorption is between 1.5 and 2.0 magnitudes a perfectly smooth transition from the optical to the X-ray region may be impossible.

J. E. Baldwin: The size of the nebula depends on frequency, that is to say that the spectrum of the radiation varies from point to point in the nebula. The spectra of individual regions may contain sharp features but, unless these features are the same for all regions, the integrated spectrum of the whole nebula will be much smoother. If the visual obscuration is really 2.0 magnitudes and there is an interruption of the smooth spectrum it may be very difficult to explain.

J. E. Felten: The shape of the spectrum integrated over the entire nebula gives the first indication of how complicated one's theoretical model must be. Thus (to take a simple case) if the continuum emission from a source is entirely synchrotron radiation from electrons injected continuously in a power law at one point or several points in a uniform field, then the integrated spectrum, say in the optical and ultraviolet, may be a simple power law (resulting from equilibrium between injection and energy loss), even though the spectrum revealed by a smaller diaphragm is a function of position and presents more complications. Therefore it is good to look first for simplicity in the integrated spectrum, as Dr. Baldwin has done.

W. J. Cocke: But then you must take out the compact source component.

1.4 SEARCHES FOR γ-RAYS FROM THE CRAB NEBULA AND ITS PULSAR

J. V. JELLEY

Nuclear Physics Division, Atomic Energy Research Establishment, Harwell, Berkshire, U.K.

Abstract. The Crab Nebula has been regarded as the most promising celestial object to investigate for the detection of γ-rays. γ-ray emission might be expected from either the synchrotron or the inverse Compton mechanism. Periodic γ-ray emission could come from the pulsar, but no theory has yet been developed for such objects. Searches for γ-rays from both the Crab Nebula and the Crab pulsar made by a number of groups are described. Limits have been set to the γ-ray emission from both objects which are only a little above the extrapolated optical and X-ray fluxes.

1. Introduction and History

It is interesting to reflect that the idea of searching for γ-rays from celestial objects occurred as far back as the late fifties, some years prior to the discovery of X-rays, either from the Galaxy, or from point sources. Historically, of course, this arose because there were at that time no rockets or satellites to carry X-ray instruments above the atmosphere, and it was early appreciated that atmospheric absorption would preclude observations either from the ground or even from balloon altitudes. Over the years however, a very considerable effort has been expended in the development of γ-ray astronomy, using a wide variety of techniques, with equipment which has been flown on balloons, mounted in satellites, and also ground-based. In spite of these great efforts, γ-ray astronomy is only now just beginning to bear fruit, and interest has naturally been somewhat overshadowed by the impact of the impressive results obtained in the X-ray field.

While γ-rays are sufficiently penetrating to reach down to balloon altitudes, enabling at least relatively simple experiments to be conducted without recourse to rocket and satellite techniques, various factors emerge which combine to make γ-ray astronomy considerably more difficult than X-ray astronomy. First is the question of the low fluxes. Even on the basis of energy per unit bandwidth, most celestial sources of X-rays have spectra which fall away with increasing frequency, and since all X-ray and γ-ray detectors are essentially quantum counters, the effect is still more pronounced. Thus, to obtain even reasonable rates, short rocket flights are quite inadequate. It is essential to have large collecting areas and long integrating times. Rockets therefore cannot be used. The second problem is that of the cosmic-ray background of charged-particles; on average the γ-ray component of the cosmic-radiation is known to be $\leqslant 10^{-3}$ of the charged-particle component. It is therefore very important to have good discrimination against the charged-particle component.

γ-ray astronomy can be said to date from a paper by Morrison (1958) in which he outlined some of the processes which could be important for the production of γ-rays from the Galaxy and specific celestial objects. A large number of review articles on the subject are now available and we will mention just three. The first (Fazio, 1967) and the second (Ginsburg and Syrovatskii, 1965), cover the whole field, with emphasis on the physics of the production mechanisms and the astrophysical situations involved, while the third (Kraushaar, 1969) is primarily concerned with the instrumentation and detection aspects of the problems.

2. Techniques

The γ-ray spectrum extends approximately from ~ 500 keV to $\sim 5.10^{13}$ eV, a vast range, about eight decades. It is therefore not surprising that a wide variety of techniques have been developed to encompass this great span of energy (Kraushaar, 1969). There are however two broad subdivisions, the band 10 MeV–1 GeV being covered by instruments carried on balloons and in satellites, and a higher band, 10^{11} eV–10^{13} eV which is accessible by ground-based observations, using the Cherenkov night-sky technique (Jelley and Porter, 1963). As we shall see later, the most important work on the Crab Nebula carried out so far, has been in the region 30 MeV–1 GeV (balloons) and in the band 9.10^{10} eV–5.10^{12} eV (ground-based Cherenkov instruments). The most promising instruments under development at the moment are the vidicon spark chambers (and combinations of spark chambers and nuclear emulsions) and the enclosed gas Cherenkov detectors (Helmken and Hoffman, 1970). Space prohibits a more detailed discussion of techniques, but I have however attempted to summarise the situation by the diagram shown in Figure 1.

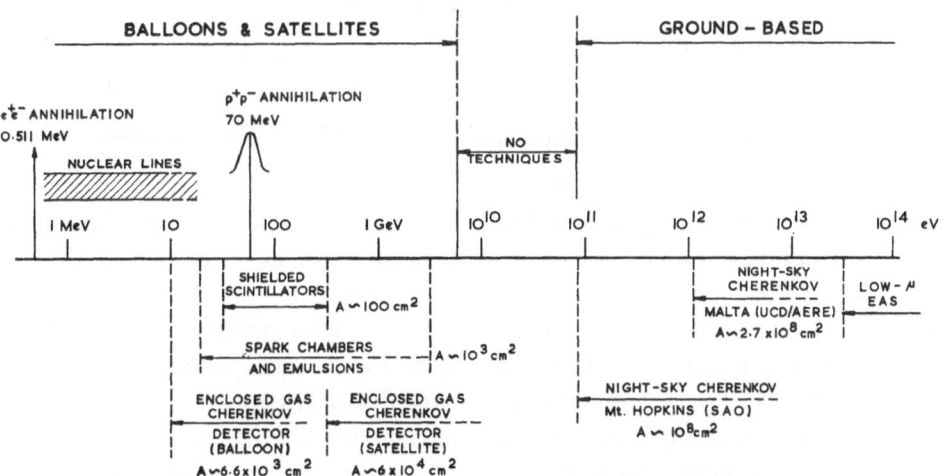

Fig. 1. The γ-ray spectrum and available techniques.

3. The Crab Nebula: Theoretical Models

It has long been realised (Ginsburg and Syrovatskii, 1964; Shklovskii, 1960) that the Crab Nebula is probably the most important celestial object in which high-energy cosmic-rays may be generated, and hence it has been accepted for a long time that it is likewise the most promising object from which to find γ-rays. All the optical and radio observations of the nebula alone, leaving out the additional considerations involved by the discovery of its pulsar, lead to these conclusions, and many sections of the work by Ginsburg and Syrovatskii (1964) are devoted to this theme. In most of these models both synchrotron radiation (Ginsburg and Syrovatskii, 1964) (magneto-bremsstrahlung) and the inverse Compton effect (Felten and Morrison, 1963) play important rôles, taken either separately or together. In most astrophysical situations and especially in the Crab, these two mechanisms are more important than nuclear processes and collisional bremsstrahlung, for their contribution to the generation of γ-rays.

Early predictions that the Crab might be expected to yield a measurable flux of high-energy γ-rays were due to Burbidge (1959) and to Cocconi (1960).

Assuming that all the continuum optical and radio emission from the nebula arises directly from synchrotron radiation from electrons, they suggested that if all these electrons were continually being generated by π–μ–e decays, then one could calculate a γ-ray flux from the decay of the accompanying π_0-mesons. It was proposed that both the $\pi^\pm$ and π_0-mesons were in turn generated by the collision of cosmic-ray protons with the nuclei of hydrogen gas known to be present in the nebula. The chain of reasoning on this model is illustrated as follows:

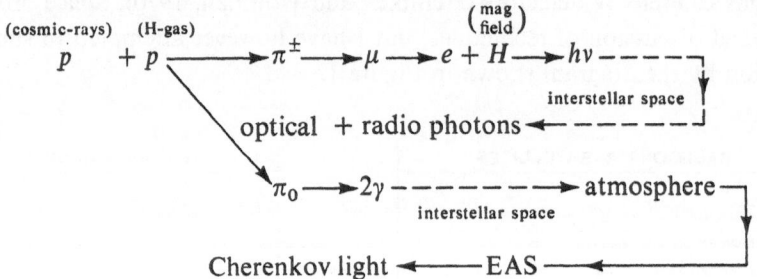

On this model a γ-ray flux of 1.6×10^{-7} photons cm^{-2} sec^{-1} at the Earth, for $E_\gamma \sim 10^{12}$ eV was predicted (Cocconi, 1960). Early observations (Chudakov *et al.*, 1962) however set an upper limit of 5×10^{-11} photons cm^{-2} sec, at 5×10^{12} eV, a factor at least 3000 below that predicted on Cocconi's model. Chudakov *et al.* (1962) were therefore able to show that the high-energy electrons in the nebula were not secondary products from π-meson decay, and since the lifetime of these electrons is so short compared to the age of the Crab, it was therefore deduced that the electrons must continuously derive their energy from some unknown source and undergo subsequent acceleration.

The Compton-synchrotron model of the Crab, by Gould (1965), was the next significant step, and served for some years as a basic model on which to hang the experimental results to be discussed later. Gould was able to show that the optical and radio photons generated by synchrotron radiation would be Compton-scattered in turn by the same electrons which themselves generated the synchrotron radiation, thus creating a photon spectrum extending right up into the γ-ray region. To simplify the calculations, Gould took a single value of $H = 10^{-4}$ G and two basic photon

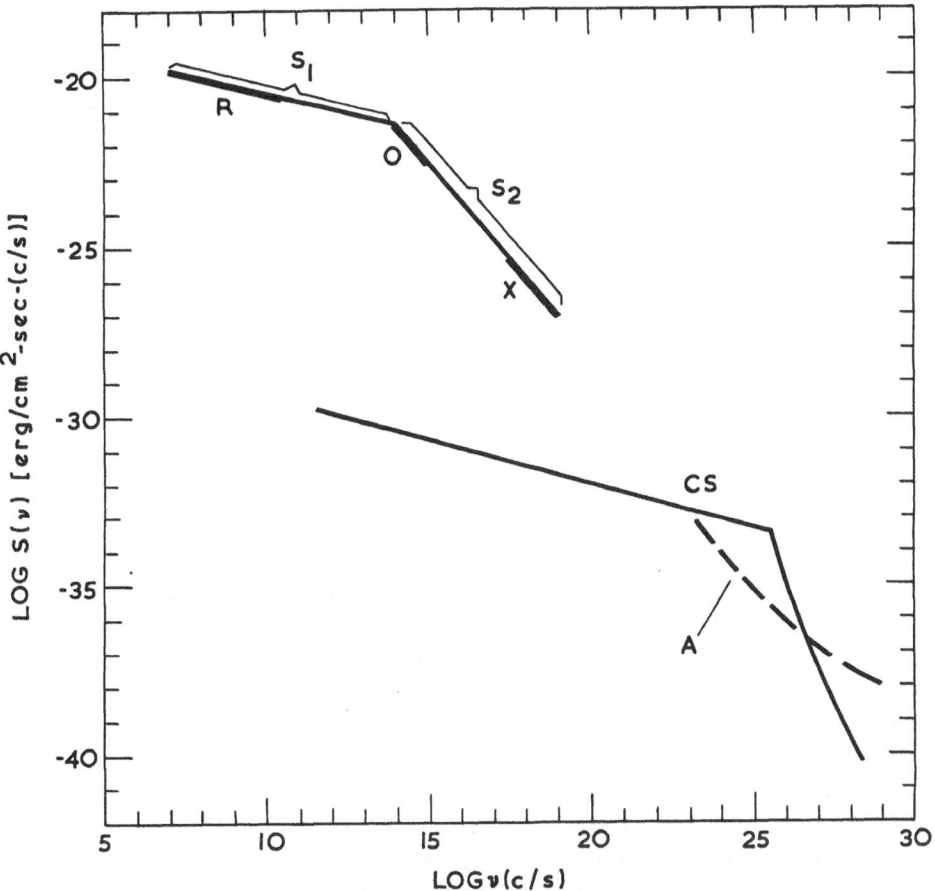

Fig. 2. The Compton-synchrotron spectrum *CS* calculated by Gould (1965) from the observed spectra S_1 and S_2, derived from measurements in the radio, optical and X-ray regions, *R*, *O*, and *X* respectively. The extension *A*, due to Apparao (1967) is a Compton spectrum deduced from interactions with the Universal Microwave Field.

spectra, Figure 2, S_1 corresponding to the radio-optical region, and S_2 the optical-X-ray region. The slopes of these power-law spectra are 0.27 and 1.1 respectively. We thus have two electron spectra interacting with two photon spectra. Assuming S_1 and S_2 are generated within spheres of angular diameter 4′ and 2′ respectively, Gould derived the high-energy photon spectrum *CS* shown in Figure 2.

A more refined calculation, with improved input data derived from a greater knowledge of the synchrotron spectra, has since been carried out by Rieke and Weekes (1969). In this development of the Gould model the authors calculated the γ-ray spectrum for two values of the magnetic field, 10^{-4} and 3×10^{-4} G respectively, and considered two specific models, one assuming the X-rays in the Crab followed a synchrotron spectrum, and the other, that the X-rays cut off sharply at 10^{16} Hz. They also considered the Compton scattered photons from the 3K background radiation, assumed to permeate the nebula throughout. The theoretical spectra

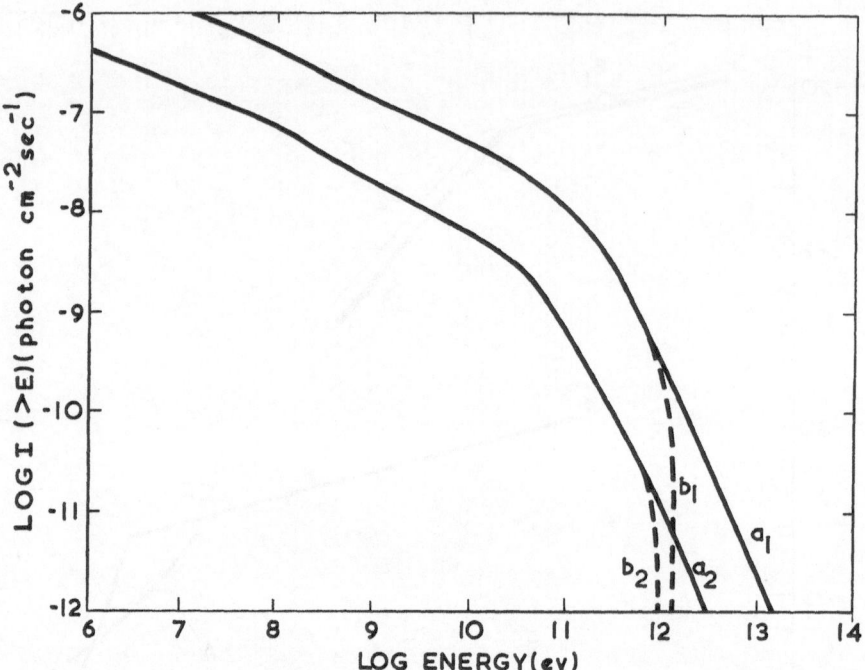

Fig. 3. Theoretical Compton-synchrotron spectra calculated by Rieke and Weekes (1969). Curves a_1 and b_1, $H = 10^{-4}$ G, a_2 and b_2; $H = 3 \times 10^{-4}$ G. The a-curves assume that the X-ray flux is synchrotron radiation. The b-curves are derived assuming the synchrotron spectrum cuts off at 10^{16} Hz.

derived by these authors is shown in Figure 3. We note that it differs in two respects from the earlier Gould model, Figure 2. The absolute fluxes are lower and the spectra are steeper, at the higher energies. While the Rieke-Weekes model represents the best available at the present time, it does not allow either (a) for a general variation of the magnetic field across the nebula, or (b), for spatial fluctuations of this field. A much greater knowledge of the distribution of the field within the nebula is clearly required before substantial improvements can be made.

In connection with the 3 K background radiation it should be mentioned that this was also considered by Apparao (1967). He showed that this would cause a substantial extension to the Gould spectrum, at the very highest energies, as shown in Figure 2.

4. The Crab Pulsar; Theoretical Models

All that has been said so far was based on our knowledge of the Crab nebula prior to the discovery of its pulsar, and its subsequent identification with the S_p-component of the central double star. However, the discovery of NP 0532 and the vast amount of data on its characteristics and spectrum, from radio frequencies right through to hard X-rays, has naturally caused one to ask what modifications may be necessary to the theories of the nebula as a whole. There are many facets to the problem and some of these are directly concerned with γ-rays, and at once raise the question of whether we might expect pulsed γ-rays from the Crab, either instead of or as well as steady γ-rays from the extended regions of the object.

The first point and a very satisfying conclusion, is that it has been found that the energy-loss of the pulsar, represented by its rate of slowing-down, $\sim 10^{38}$ ergs sec^{-1}, very closely matches the rate of energy-loss by radiation, from the nebula as a whole. This suggests first, that the pulsar is coupled tightly to the nebula, and secondly, that the pulsar is indeed the basic energy-source of the Crab, at least at the present epoch.

There are basically two models of the pulsars, both assuming that they are rapidly rotating neutron stars. In the first (Gold, 1968; 1969) it is assumed the radiation occurs out at a radius corresponding to the velocity-of-light circle, and in the second, that the radiations arise on the surface of the pulsar, and emerge from the magnetic poles, on the various forms of the oblique rotator model (Pacini, 1967; Gunn and Ostriker, 1969; Bertotti *et al.*, 1969).

On an energy basis it has become clear that NP 0532, and likewise other pulsars, may be adequate to accelerate all cosmic-rays, with perhaps the exception of those of the very highest energy. In particular Ostriker (1969) shows that the intensity of the 30 Hz low-frequency magnetic field is sufficient to produce direct acceleration of cosmic-rays by the electric fields induced by the varying magnetic fields, up to energies as high as 10^{18} eV. It has been suggested elsewhere at this Meeting (Rees, 1971) that the entire field throughout the Crab nebula may be modulated at 30 Hz and indeed that it may even be wholly an AC field. This suggestion immediately raises doubts as to whether the Gould (1965) or Rieke-Weekes models of the γ-ray emission are valid, for the synchrotron orbits will be so modified if the H-field is AC, that the calculations become extremely difficult. For example Fazio (private discussion) has suggested that we should perhaps take a dipole magnetic field which falls off as $(1/r^3)$ right from the surface of the star, where H is believed to be $\sim 10^{12}$ G, though again there are problems, for there is clearly some discontinuity at the velocity-of-light circle. Suffice it to say that at the time of this Meeting, no-one has so far presented a quantitative theory of the production of γ-rays from the Crab and its pulsar, which include these new considerations. In the absence of such models it is natural that the γ-ray astronomers have been searching for *periodic* γ-rays from the Crab, at all the energy bands accessible with the various techniques available.

In spite of my general remarks on these problems, some attempts have in fact already been made (Apparao, 1969), to calculate the γ-rays expected from the Crab, assuming now that its pulsar may be the essential and basic source.

A reliable theory cannot however be expected until it is established whether the periodic radio, optical and X-ray emissions arise from the velocity-of-light circle or from the magnetic poles of an oblique rotator; there are many other problems as well.

5. Searches for Steady γ-Rays from the Nebula

A very extensive list of groups have for many years been searching for a steady source of γ-rays from the Crab Nebula, and so far none have been found, at least not at a confidence level sufficiently high to be acceptable to all concerned. The principal investigators have been from MIT, AS and E, Rochester, Cornell, Rice, NASA, Bristol, Southampton, SAO, UCD and AERE. This is by no means a complete list. While no positive results have been obtained, there are numerous upper limits and these are listed in Table I. Even these upper limits however have already provided significant constraints to the theoretical models. We have previously mentioned that

TABLE I

Measured upper limits to the steady γ-ray flux from the Crab Nebula

Threshold γ-ray energy	Upper-limit steady flux photons cm^{-2} sec^{-1}	Experiment	References
30 MeV	1.5×10^{-4}	Frye and Smith (1966)	Duthie (1968)
50 MeV	6.6×10^{-4}	Kraushaar et al. (1965)	Duthie (1968)
100 MeV	5×10^{-5}	Cobb et al. (1965)	Duthie (1968)
1 GeV	2×10^{-5}	Rochester spark chamber	Duthie (1968)
5.10^{12} eV	5×10^{-11}	Chudakov et al. (1962)	Duthie (1968)
5.10^{12} eV	1×10^{-10}	Fruin et al. (1964)	Duthie (1968)
10 MeV	4.2×10^{-3}	Fichtel and Kniffen (1965)	Fichtel et al. (1969)
3×10^{11} eV	2.3×10^{-6}	Sekido et al. (1963)	Fazio (1967)
10^{15} eV	10^{-14}	Toyoda et al. (1965)	Fazio (1967)
2×10^{12} eV	1.5×10^{-10}	Long et al. (1966)	Long et al. (1966)
		Fegan et al. (1967)	Fegan et al. (1968)
4×10^{12} eV	8×10^{-11}	Fazio et al. (1968)	Fazio et al. (1968)
3×10^{12} eV	7×10^{-11}	Fazio et al. (1968)	Fazio et al. (1968)
2×10^{12} eV	1×10^{-10}	Fazio et al. (1968)	Fazio et al. (1968)
1.7×10^{11} eV	2.0×10^{-10}	Fazio et al. (1969)	Fazio et al. (1969 c)
1 GeV	1.2×10^{-5}	Delvaille et al. (1967)	Delvaille et al. (1968)
50 MeV	1.7×10^{-5}	Frye and Wang	Frye and Wang (1969)
150 MeV	9.0×10^{-6}	Frye and Wang	Frye and Wang (1969)
500 MeV	5.0×10^{-6}	Frye and Wang	Frye and Wang (1969)
30–100 MeV	2.7×10^{-4}	Fichtel et al. (1969)	Fichtel et al. (1969)
> 100 MeV	1.8×10^{-4}	Fichtel et al. (1969)	Fichtel et al. (1969)

Chudakov's limits directly implied that the fast electrons in the Crab cannot be secondary to cosmic-ray proton collisions with hydrogen gas. Another example is that limits can be set on the magnetic field, Fazio et al. (1969b) having shown that the average field H must be $\geqslant 1.2 \times 10^{-4}$ G; this in turn sets a constraint on the expected flux at 100 MeV, on the Compton-synchrotron model, at a level of $\phi_\gamma \sim 10^{-7}$ photons cm^{-2} sec^{-1}, a figure, we notice that is still nearly two orders of magnitude lower than the recent limits set by Frye and Wang (1969).

6. Searches for Periodic γ-Rays from NP 0532

It is of course only during the last two years that it has been feasible to consider experiments of this kind. The very low counting rates and γ-ray fluxes expected, require long integrating times, and this, combined with the high repetition frequency of

NP 0532, implies extremely high precision in the timing accuracy, and frequency-dividing circuits used in the periodicity analyses. In work of this type, particularly on this the fastest pulsar, it has been found to be exceedingly difficult to preserve phase in the analysing procedures over periods of more than a few hours. While ground-based observations can on occasion enjoy the luxury of on-line calibration from light-pulses (Fazio *et al.*, 1971) from the star, balloon-borne experiments necessitate continual corrections for the position and velocity of the balloon, at least for observations of several hours.

It is for just these reasons that the vast mass of earlier γ-ray data on the Crab taken prior to the discovery of the pulsar, cannot subsequently be analysed for periodicity, as the data were not in general recorded in real time, with the precision required, namely ~ 1 part in 10^8 over a few hours.

Several attempts have however now been made to detect γ-rays from the Crab and these, mostly upper limits, are listed in Table II.

TABLE II

Measured upper limits to the pulsed γ-ray flux from the Crab Nebula

Threshold γ-ray energy	(NP 0532) Pulsed γ-ray flux photons cm^{-2} sec^{-1}	Experiment
1.3×10^{13} eV	$\leqslant 3 \times 10^{-12}$	Charman *et al.* (1969)
$(1 \to 3) \times 10^{12}$ eV	$\leqslant 2.9 \times 10^{-11}$	Charman *et al.* (1969)
$(2 \to 4) \times 10^{12}$ eV	$\leqslant 2.7 \times 10^{-11}$	Charman *et al.* (1969)
$(2 \to 4) \times 10^{12}$ eV	$\leqslant 3.2 \times 10^{-11}$	Charman *et al.* (1969)
1.2×10^{11} eV	$\leqslant 2.6 \times 10^{-10}$	Fazio *et al.* (1969 a)
1.1×10^{12} eV	$\leqslant 3.0 \times 10^{-11}$	Fazio *et al.* (1969 a)
50 MeV	$\sim 10^{-5}$	Vasseur *et al.* (1970)

The reader will notice, once again, that nearly all the observations have only led to upper limits. Of the observations listed, one only, that made by a French-Italian group (Vasseur *et al.*, 1970) claimed a positive effect which appeared, at least superficially, to be significant. This claim was however subsequently discounted, by at least two groups (Charman and White, 1970; Delvaille and McBreen, 1970), on statistical grounds.

Subsequent however to this Symposium, but prior to the publication of its Proceedings, the author has heard that the Bristol group now have evidence, at a very reasonable level of significance, for periodic γ-rays from NP 0532 in the energy region 0.6 MeV–12 MeV; in this work they used a periodicity calculated by UCD, Dublin. If this observation can be confirmed, it represents the first detection of γ-rays from NP 0532.

7. Discussion

Considering the large efforts expended to find γ-rays from the Crab and/or its pulsar, the results have been disappointing indeed, especially as this object has always been cited as the most profitable one at which to peer, since it was expected to be the most likely source of cosmic-rays.

One naturally asks 'why no γ-rays'?, or, if they are produced, 'what becomes of them'? Photon-photon absorption has been considered but it seems rather unlikely. In the relatively short distances of ∼1.3 kpc, this absorption mechanism is believed to be negligible (Gould and Schréder, 1966; Jelley, 1966a) unless there are intense background radiations in the U.V. and shorter wavelengths, which have not yet been detected. It has however been suggested (McBreen, 1971) that if any γ-rays are generated very close to the surface of the pulsar, these may be absorbed within the source region itself (Jelley, 1966b). The problem here is that the γ-rays and the photons with which they collide will be travelling with respect to one another on almost parallel paths, thereby raising the threshold of the absorption process and lowering the effective cross-section, a point mentioned recently by Ginsburg (1970).

On the experimental side the efforts will clearly continue. Large-area spark-chamber detectors and large gas Cherenkov detectors will be flown on balloons and satellites, and it is therefore still hoped that high energy γ-rays from the Crab and/or its pulsar will eventually be found.

References

Apparao, M. V. K.: 1967, Tata Institute for Fundamental Research Report N. E. 66-8 (also *Proc. Indian Acad. Sci.* **65A**, 349).

Apparao, M. K. V.: 1969, *Nature* **221**, 645.

Bertotti, B., Cavaliere, A., and Pacini, F.: 1969, *Nature* **221**, 624.

Burbidge, G. R.: 1959, *Astrophys. J.* **127**, 48.

Charman, W. N., Fruin, J. H., Jelley, J. V., Fegan, D. J., Jennings, D. M., O'Mongain, E. P., Porter, N. A., and White, G. M.: 1969, *Proc. XIth Intern. Conf. on Cosmic Rays*, Budapest, Paper OG-10, p. 59.

Charman, W. N., Jelley, J. V., and Drever, R. W. P.: 1969, *Proc. XIth Intern. Conf. on Cosmic Rays*, Budapest, Paper OG-11, p. 63.

Charman, W. N. and White, G. M.: 1970, *Nature* **226**, 1233.

Chudakov, A. E., Dadykin, V. L., Zatsepin, V. I., and Nesterova, N. M.: 1962, *J. Phys. Soc. Japan* **17**, Suppl. A-III, 106.

Cocconi, G.: 1960, *Moscow Conf. on Cosmic Rays*, II, p. 309.

Delvaille, J. P., Albats, P., Greisen, K. I., and Ögelman, H. B.: 1968, *Can. J. Phys*, **46** [10], Part 3, S425.

Delvaille, J. P. and McBreen, B.: 1970, *Nature* **226**, 1234.

Duthie, J. G.: 1968, *Can. J. Phys.* **46** [10], Part 3, S401.

Fazio, G. G.: 1967, *Ann. Rev. Astron. Astrophys.* **5**, 481.

Fazio, G. G., Hearn, D. R., Helmken, H. F., Rieke, G. H., and Weekes, T. C.: 1969a, *Proc. XIth Intern. Conf. on Cosmic Rays*, Budapest.

Fazio, G. G., Helmken, H. F., Rieke, G. H., and Weekes, T. C.: 1968, *Astrophys. J. Letters* **154**, L83.

Fazio, G. G., Helmken, H. F., Rieke, G. H., and Weekes, T. C.: 1969b, in L. Gratton (ed.), 'Non-Solar X- and Gamma-Ray Astronomy', *IAU Symp.* **37**, 250.

Fazio, G. G., Helmken, H. F., Rieke, G. H., and Weekes, T. C.: 1969c, *XIth Intern. Conf. on Cosmic Rays*, Budapest. Also G. H. Rieke, Smithsonian Astrophys. Obs., Spec. Report No. 301. 25th June 1969.

Fazio, G. G., Helmken, H. F., Rieke, G. H., and Weekes, T. C.: 1971, this symposium, Paper 1.7, p. 65.

Fegan, D. J., McBreen, B., O'Mongain, E. P., Porter, N. A., and Slevin, P. J.: 1968, *Can. J. Phys.* **46** [10], Part 3, S433.

Felten, J. E. and Morrison, P.: 1963, *Phys. Rev. Letters* **10**, 453.

Fichtel, C. E., Kniffen, D. A., and Ögelman, H. B.: 1969, *Astrophys. J.* **158**, 193.

Frye, G. M. and Wang, C. P.: 1969, *Astrophys. J.* **158**, 925.

Ginsburg, V. L.: 1970, IAU General Assembly, Brighton, Sussex, Joint Discussion 'Pulsars and Cosmic Rays', 26 August.

Ginsburg, V. L. and Syrovatskii, S. I.: 1964, in D. ter Haar (ed.), *The Origin of Cosmic Rays*, Pergamon Press, Oxford.

Ginsburg, V. L. and Syrovatskii, S. I.: 1965, *Soviet Phys. (Uspekhi)* **7**, 696.

Gold, T.: 1968, *Nature* **218**, 731.

Gold, T.: 1969, *Nature* **221**, 25.

Gould, R. J.: 1965, *Phys. Rev. Letters* **15**, 577.

Gould, R. J. and Schréder, G.: 1966, *Phys. Rev. Letters* **16**, 252.

Gunn, J. E. and Ostriker, J. P.: 1969, *Nature* **221**, 454.

Helmken, H. F. and Hoffman, J.: 1970, *Nucl. Instr. Methods* **80**, No. 1.

Jelley, J. V.: 1966a, *Phys. Rev. Letters* **16**, 479.

Jelley, J. V.: 1966b, *Nature* **211**, 472.

Jelley, J. V. and Porter, N. A.: 1963, *Quart. J. Roy. Astron. Soc.* **4**, 275.

Kraushaar, W. L.: 1969, *Astronautics and Aeronautics*, July.

Long, C. D., McBreen, B., Porter, N. A., and Weekes, T. C.: 1966, *Proc. Intern. Conf. on Cosmic Rays*, London, Vol. 1, p. 318.

McBreen, B.: 1969, *Nature* **224**, p. 893.

Morrison, P.: 1958, *Nuovo Cimento* **7**, 858.

Ostriker, J. P.: 1969, *Proc. XIth Intern. Conf. on Cosmic Rays*, Budapest.

Pacini, F.: 1967, *Nature* **216**, 568.

Rees, M.: 1971, this symposium, Paper 7.4, p. 407.

Rieke, G. H. and Weekes, T. C.: 1969, *Astrophys. J.* **155**, 429.

Shklovskii, I. S.: 1960, *Cosmic-Radio Waves*, Harvard Univ. Press, Cambridge, Mass.

Vasseur, J., Paul, J., Parlier, B., Leray, J. P., Forichon, M., Agrinier, B., Boella, G., Maraschi, L., Treves, A., Buccheri, R., and Scarsi, L.: 1970, *Nature* **226**, 535.

Discussion

J. P. Ostriker: A model-independent question of fact: are the γ-ray upper limits above or below the extrapolated X-ray spectrum? Or is the extrapolation made over too great a range to be secure given the uncertainty of the X-ray spectral index?

J. V. Jelley: I think the γ-ray upper limits fall *below* the extrapolated X-ray line. I do not think this is meaningful however, as it is such a distant extrapolation.

J. E. Baldwin: The values I quoted lie above the extrapolated X-ray spectrum. They may not be the lowest limits presently available. There is also the problem of converting a total flux above some energy limit into a flux at a given energy. This may account for the discrepancy with Jelley's plot.

G. Fazio: Whether the 100 MeV upper limits lie above or below the extrapolated flux depends on the spectral index used (i.e. $\alpha = -2.0$ or -2.2). The extrapolation has to be done over 3 orders of magnitude at least. In either case the upper limits are within about one order of magnitude of the flux predicted by extrapolation.

G. Share: Our group in the Laboratory of Cosmic Ray Physics at NRL has flown a telescope sensitive to gamma rays with energies above 10 MeV. The telescope consists of a combination of emulsion, spark-chamber and counters, and is capable of attaining an angular resolution of 2° above 20 MeV. Successful balloon flights were performed on 25 September and 25 October 1969. In both flights the Crab Nebula was the object of our search. We had hoped to present results of the 25 October flight at this meeting but the analysis was not completed in time. Results from a search for pulsed radiation from the Crab will be available in the near future.

1.5 X-RAY OBSERVATIONS OF THE CRAB NEBULA

EDWIN M. KELLOGG

American Science and Engineering, 11 Carleton Street, Cambridge, Mass. U.S.A.

Abstract. This paper is a review of the X-ray observations made on the Crab Nebula They include angular size, location, energy spectrum as high as 560 keV, upper limits on time variability and line emission, interstellar absorption and polarization. Some data on the X-ray spectrum of the pulsar NP0532 are included, but for details on the X-ray pulsar, see the paper by Rappaport (1971). The X-ray luminosity versus radius of the Crab, compared with that of other known X-ray emitting supernova remnants is also discussed.

1. Introduction

The Crab Nebula has been observed in X-rays on more than thirty different occasions since 1964, using both sounding rocket and balloon experiments. The X-ray source in the Crab consists of a pulsed component, obviously associated with the radio and optical pulsar NP 0532, and a continuous emitter, which has a finite size of about 1–2 arc min. The continuous source emits about 90% of the X-ray power in the range 2–100 keV, which represents a large fraction of the total radiation from the Crab over all wavelengths.

Measurements of the angular structure have been very crude so far, allowing only approximate estimates of size, but no details on the X-ray brightness distribution. A good deal of data exist on the spectrum, which allow an extrapolation with reasonable confidence down through the frequency regions obscured by interstellar absorption to compare with the optical and radio data. This is perhaps the most useful result of the X-ray observations so far, for it allows comparison of the data with synchrotron models of the source over twelve decades in frequency.

Measurements of other parameters such as time variability, line emission, interstellar absorption, and polarization are still crude, not able to provide definitive answers about the nature of the Crab, but they are discussed here in order to indicate their potential for the future.

2. Angular Structure

The lunar occultation measurement on the Crab in 1964 (Bowyer *et al.*, 1964) showed that a large part of the X-ray source from 2–10 keV was extended, with a size of about 2′. The pulsating component, presumably a point source, would not have been statistically significant in the data. However, it is interesting that the centroid of the distribution along the direction of motion of the moon is located within 2–3 arc sec of the pulsar, after taking into account the parallax error due to the rocket's motion downrange during its flight* (Oda *et al.*, 1967). See Figure 1.

---

* The precision of that measurement is estimated to be ±5 arc sec.

Davies and Smith (eds.), The Crab Nebula, 42–53. All Rights Reserved.

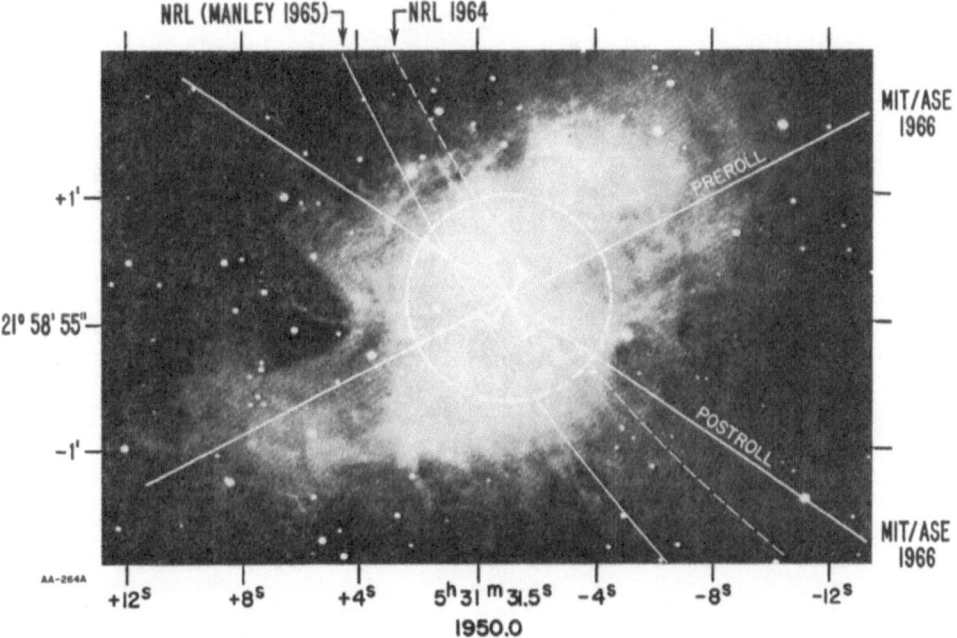

Fig. 1. Summary of correlation between visible light and X-rays from the Crab Nebula. The pulsar NP 0532 is at the origin of the coordinate system. Results from the lunar occultation of 1964 and the modulation collimator experiment of 1966 are shown, together with a visible light photo. The 100″ diameter circle is meant to show that the best estimate of the X-ray source distribution can only give a size parameter with no very significant further information on details of the source distribution.

The MIT/ASE observation of the Crab in 1966 used a modulation collimator to scan the Crab in two directions almost perpendicular. Counting data from 1–6 keV are shown in Figure 2, along with the best fit response to an extended source. A map of this source distribution is shown in Figure 3. The centroid is located about 20″ from the pulsar, one standard deviation in location. This result is consistent with the Crab source being extended, about 100″ in diameter. It could also be oblong, or it might be a thin line extending in a southeast-to-northwest direction. In fact, there are many source configurations consistent with this result (Oda *et al.*, 1967).

One phenomenon which could interfere with more detailed investigations of the Crab Nebula but perhaps interesting on its own is scattering of X-rays by interstellar grains. Slysh (1969) suggested that the extended source might be entirely the result of a point pulsar X-ray source whose emission is scattered into a 'halo' and loses time coherence. It appears that, among other arguments, the amount of dust required for that is too high (Naranan and Shah, 1970; Bowyer *et al.*, 1970; Ryter, 1970). Also, measurements of the pulsar fraction at high energies where scattering is not important argue against such a picture.

A measurement of size at higher energy for the unpulsed X-ray emission would be of great interest. It might allow us to trace the structure of the magnetic field and

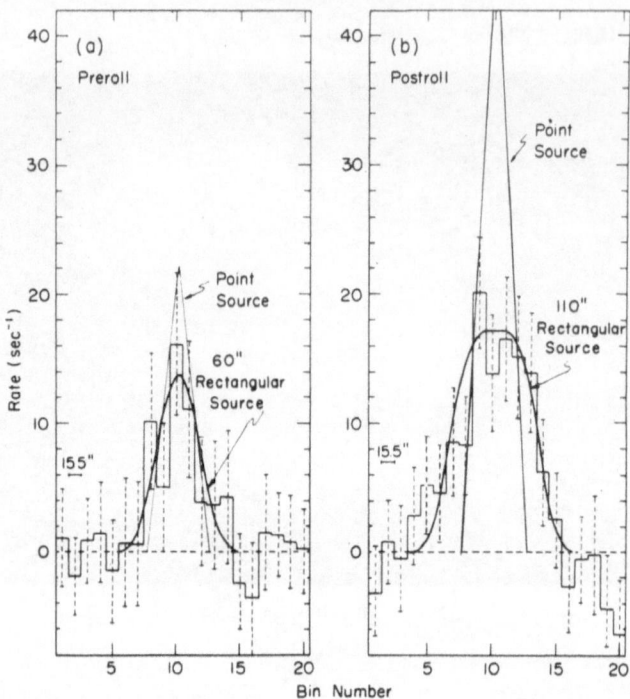

Fig. 2. Data obtained in the 1966 MIT/ASE modulation collimator experiment on the Crab. (a) and (b) are counting rate versus angle from two scans in nearly orthogonal directions across the Crab, each lasting 40 seconds. The solid lines show the count rate to be excepted from an assumed rectangular source intensity function 60″ and 110″ wide in the respective scan directions.

electrons in a region which may be closer and closer to the pulsar for higher energy X-ray emission, or it may show some other result, perhaps indicating that the X-ray emission is not so directly connected with the pulsar. In a recent modulation collimator experiment, Floyd (1970) attempted to measure the Crab's size at higher energies, 25–100 keV, from a balloon. While the detector observed X-rays from the Crab, no definite modulation of the counting rate was seen with a two-grid modulation collimator of 2′ period. The large background count rate compared to the signal from the Crab in that experiment leads Floyd to interpret the results as indicating a lower limit of 1′.1 on the size. This size result is a key point in our understanding of the Crab, and should be confirmed or studied in more detail.

Future experiments in preparation and in proposal stage at AS&E are expected to yield true images of the Crab in X-rays from ∼0.7 to 2 keV using a grazing incidence telescope. A 9.5 inch diameter telescope will be flown in September 1970 on an Aerobee 170 sounding rocket. This should yield a picture with about 20″ resolution. It will be extremely limited in contrast definition, however, due to the low efficiency of the telescope system and the short observing time available in a rocket flight. We expect to be able to distinguish major features of the source structure, however. Later versions of this instrument have been proposed, such as a 20 inch telescope for

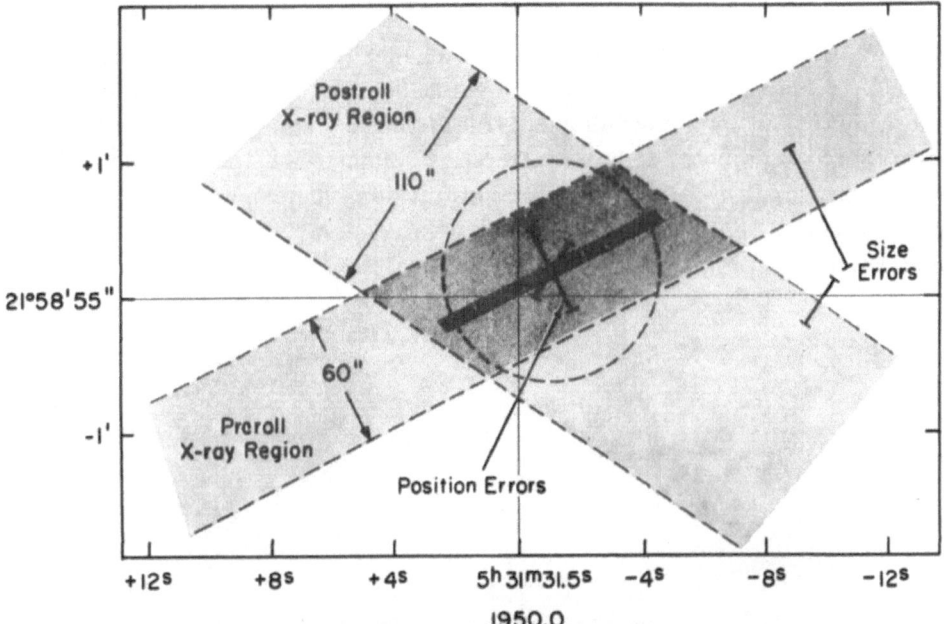

Fig. 3. X-ray source distribution indicated by the 1966 MIT/ASE experiment (Oda *et al.*, 1967). The coordinate system is again centred on the pulsar NP 0532. The combined results of the NRL occultation in 1964 and this experiment still cannot rule out a source which is a thin line extending in a southeast-to-northwest direction, as shown in this figure.

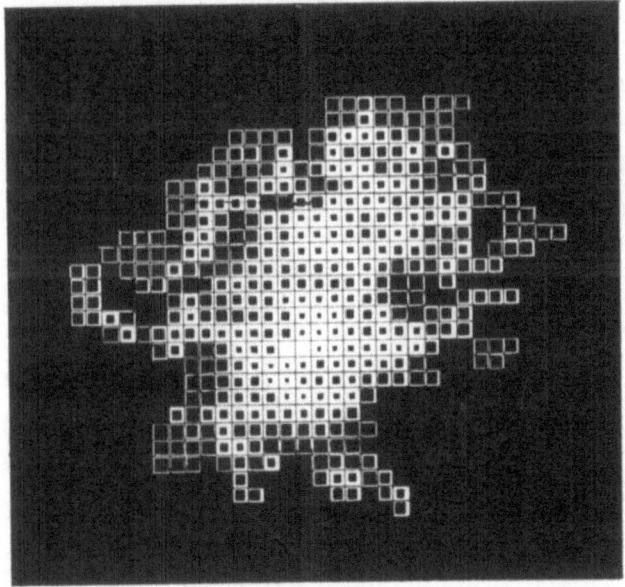

Fig. 4. Example of the type of X-ray image of the Crab Nebula to be obtained by a proposed X-ray telescope on a small satellite. The picture was derived from an optical photograph of the Crab broken into 5″ × 5″ resolution elements. The amount of unblackened area in each resolution element is representative of the relative brightness of that element. The pulsar is located midway between the two entirely unblackened squares.

use on a small satellite. This could yield a picture such as the one illustrated in Figure 4. This picture was derived from a visible-light picture of the Crab; the X-ray result will probably look much different. Further in the future is the Large Orbiting X-ray Telescope proposed jointly by AS&E, Columbia, MIT and Goddard Space Flight Center scientists. This instrument, illustrated in Figure 5, would be capable of imaging extended X-ray sources outside our Galaxy as well as making a variety of spectral and polarization measurements with high resolution.

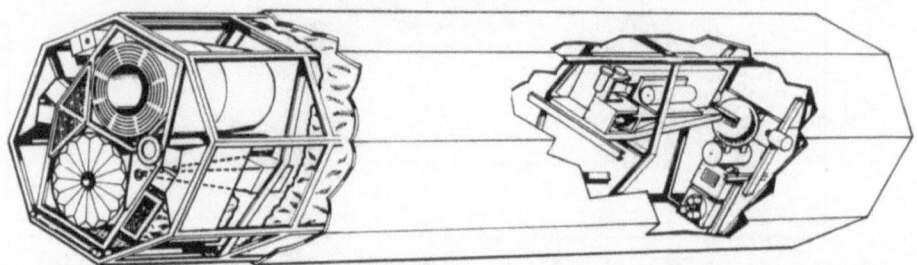

Fig. 5.   X-ray telescope observatory – Large Orbiting X-Ray Telescope – proposed for NASA's pointed High Energy Astronomy Observatory satellite by AS&E, Columbia, MIT and GSFC. Contains two telescopes, one with high resolution and the other with high efficiency, and several auxiliary instruments such as spectrometer and polarimeters. The experiment is about 8 ft diameter by 30 ft long.

3. Spectrum

A variety of observations of the Crab X-ray spectrum have been made. The sounding rocket measurements cover the energy range 0.15–20 keV and balloon measurements

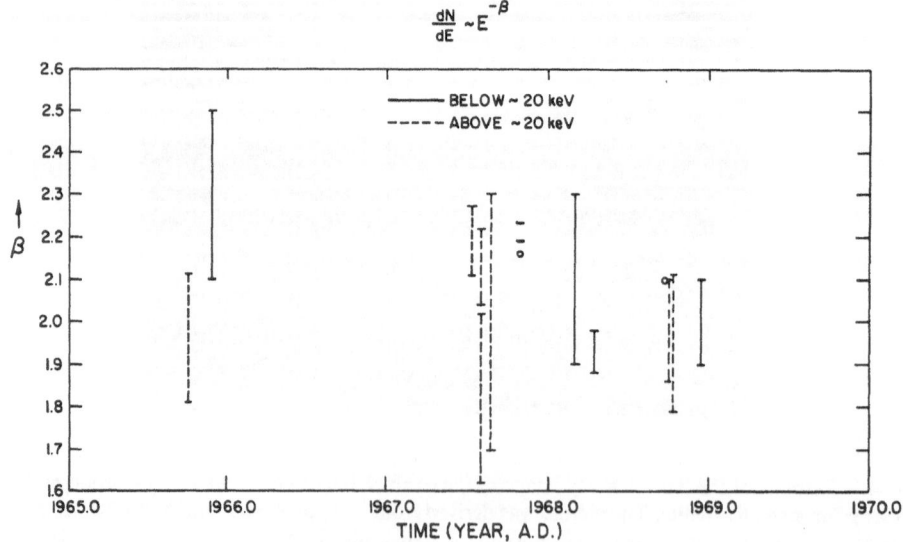

Fig. 6.   Selected measurements of the Crab's X-ray spectral index. See Appendix A for references to data used in plotting this figure.

cover the range 20–560 keV. The data above 1 keV fit an assumed power law spectrum
of the form

$$\frac{dN}{dE} = KE^{-\beta}$$

rather than an exponential spectrum. Selected measurements of the spectral index
are plotted against time in Figure 6. The scatter is fairly large; the data above 20 keV
seem to suggest close to 2.1, whereas the data below 20 keV may indicate $\beta \cong 2.0$.
The data do not indicate any secular change in β. However, the spectral results must
fit the intensity parameter, K as well. Measurements by Peterson *et al.* (1966, 1968)
and Gorenstein *et al.* (1969, 1970) at intervals of at least one year with similar
instruments show that the intensity of the X-ray flux from the nebula remains constant

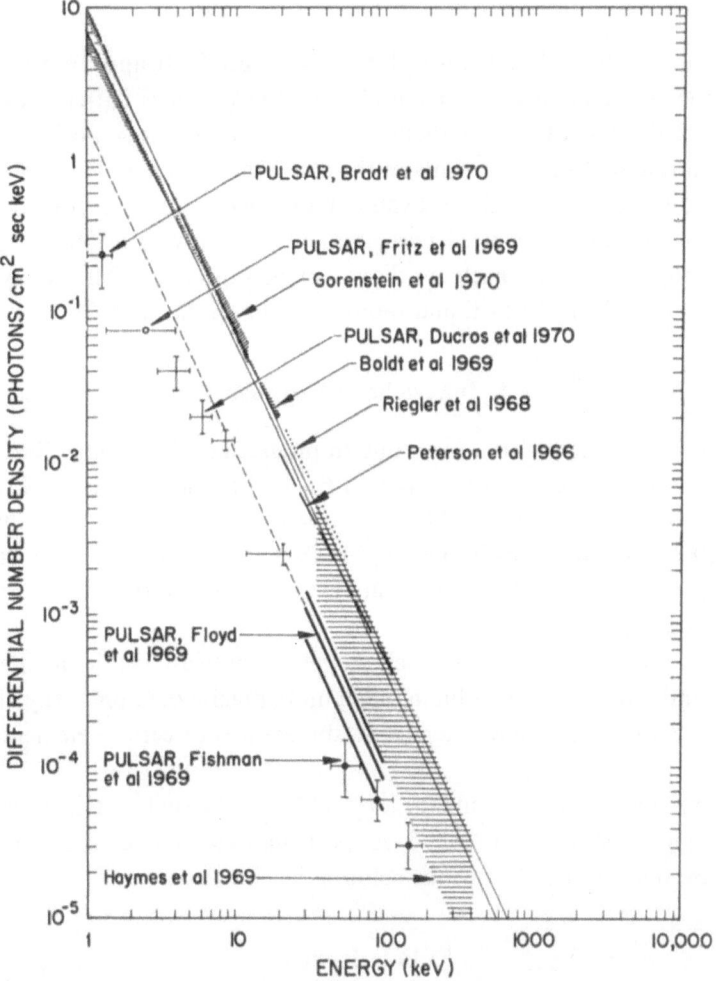

Fig. 7.   X-ray spectrum of the Crab Nebula and NP 0532.

to 5% both above and below 20 keV. If we construct a composite measured spectrum from the individual observations we obtain the results of Figure 7. A fairly close fit to the individual measurements can be obtained by a single power law

$$\frac{dN}{dE} = 9E^{-2.1} \text{ photons/cm}^2 \text{ sec keV}$$

or by a spectrum with one break at about 20 keV

$$\frac{dN}{dE} = \begin{cases} 5 & E^{-1.95} & E < 20 \text{ keV} \\ 9.1 & E^{-2.15} & E > 20 \text{ keV} \end{cases} \text{ photons/cm}^2 \text{ sec keV}.$$

Thus, if there is a break between 1 and 500 keV, the change of index is

$$\Delta\beta \lesssim 0.2.$$

The spectrum of the pulsar is also plotted in Figure 7. It appears to have about the same slope as the nebular spectrum above 10 keV, but is flatter between 1 and 10 keV. In both the case of the nebula and the pulsar, there are some discrepancies of detail between different measurements in the same energy range. However, these are not thought to be severe enough to invalidate the picture obtained in Figure 7. It is possible that future more accurate measurements may reveal a clearcut preference for a more complex nebular spectrum than a single power law. The X-ray spectrum will be compared with the optical and radio spectrum in Section 7.

4. Interstellar Absorption

The spectrum below 1 keV turns over, due to interstellar absorption. Radio data on 21 cm absorption give a column density of 1.6×10^{21} atoms/cm^2 to the Crab (Clark, 1965). X-ray absorption data at 1.2 keV (Rappaport *et al.*, 1969) set an upper limit of $N_H \lesssim 3 \times 10^{21}$. Measurements in the range 0.15–0.28 and 0.4–0.6 keV (Grader *et al.*, 1970) find $N_H = 1 - 2 \times 10^{21}$. The X-ray data are consistent with but less precise than the radio data at present.

Experiments now being designed, such as the X-ray telescope for a small satellite proposed recently by AS&E can obtain much more precise data on X-ray absorption. They should be able to measure interstellar abundances of certain elements, notably oxygen.

Experimental results that should be obtainable from such a telescope using an objective grating are shown in Figure 8. A spectrum like the Crab's is assumed, from a point source, and results for various column densities are shown.

5. Polarization

Novick and his group at Columbia have measured the Crab's polarization in X-rays

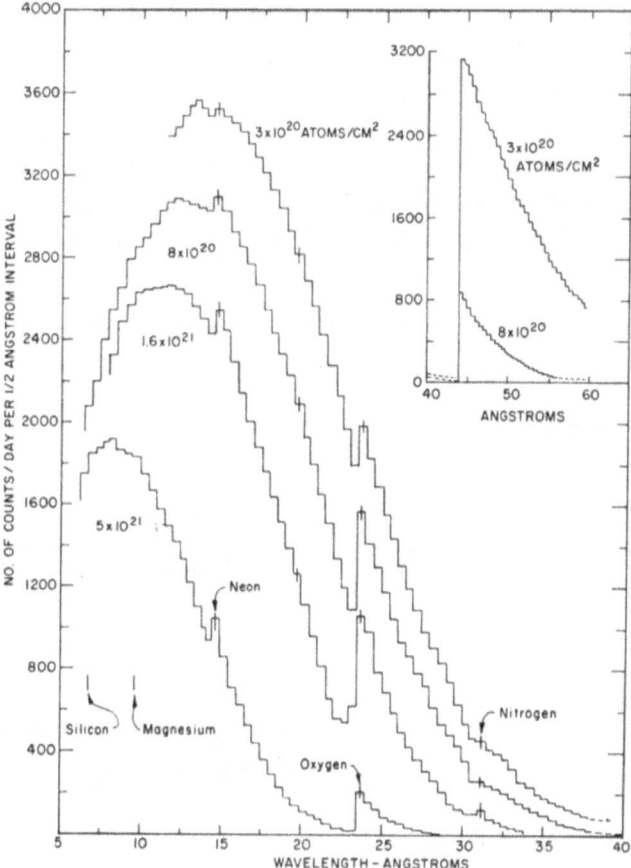

Fig. 8. Typical expected spectral data from a Small Satellite X-ray Telescope Experiment. A point source with a spectrum like the Crab's is assumed, and results are shown for various column densities of interstellar material.

to be $10\% \pm 10\%$. (Novick, 1969) One might expect the integrated polarization over the Crab to be of that order or less, so the present polarization data are not sufficient to make a crucial test on whether the X-rays are produced by the synchrotron process.

6. Line Emission

Line emission from the Crab might be possible in spite of the power law spectrum if the X-rays came from a hot gas with a complex of temperatures, densities and volumes (Sartori and Morrison, 1967). In that case, one might see emission lines. None have been found so far. Also, lines at several hundred keV might be present due to radioactive decay of heavy nuclides formed during the supernova explosion. The experimental upper limits set by Jacobson (1968) are a factor of four above the intensities predicted by Clayton and Craddock (1965).

7. Electromagnetic Spectrum of the Crab Nebula and NP 0532

If we assume the flux density spectrum indicated by the analysis of Section 3, we can extrapolate it back to optical and radio frequencies. The result is plotted in Figure 9. The nebular X-ray spectrum with a slope of $\nu^{-1.1}$ meets the optical result. The intersection with the extrapolated radio spectrum occurs at about 10^{14} Hz. The magnetic field can be derived from this, together with the known age for the nebula and is of order 10^{-4} G. The change of spectral index between the radio and X-ray spectra is 0.84.

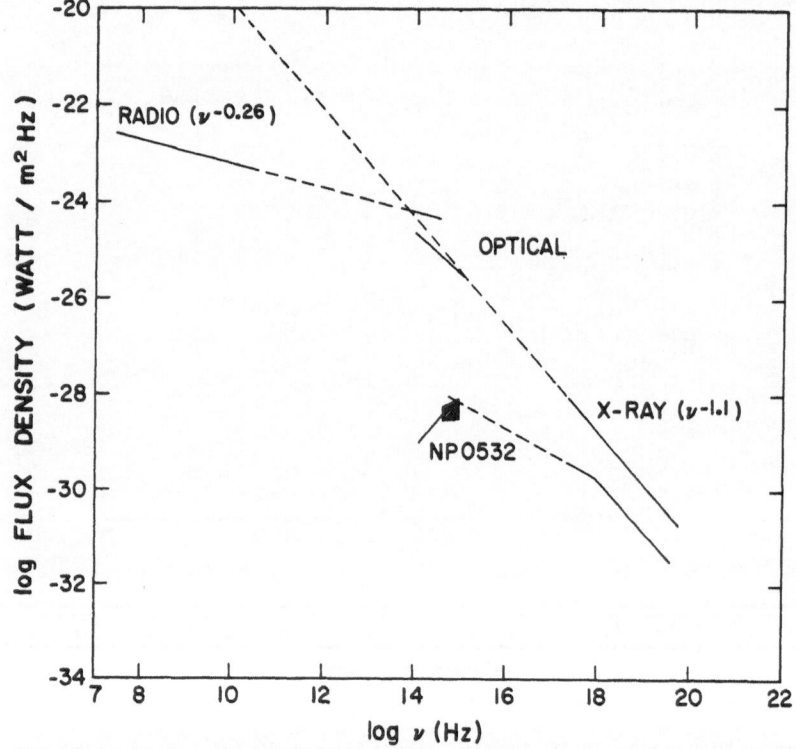

Fig. 9.   Electromagnetic spectrum of the Crab.

The X-ray pulsar spectrum extrapolation to optical frequencies meets the optical data fairly well. This suggests a break in the pulsar spectrum at about 10^{18} Hz, with a change of index of about 0.5.

8. X-Ray Luminosity with Size

Recently, the AS&E group have compared the X-ray luminosity of the Crab, Cas A and Tycho with their radio sizes (Gorenstein *et al.*, 1970). The plot is shown in Figure 10. Once again, the Crab shows up as a singular object, having the greatest intrinsic luminosity of any galactic supernova remnant. This plot also suggests that

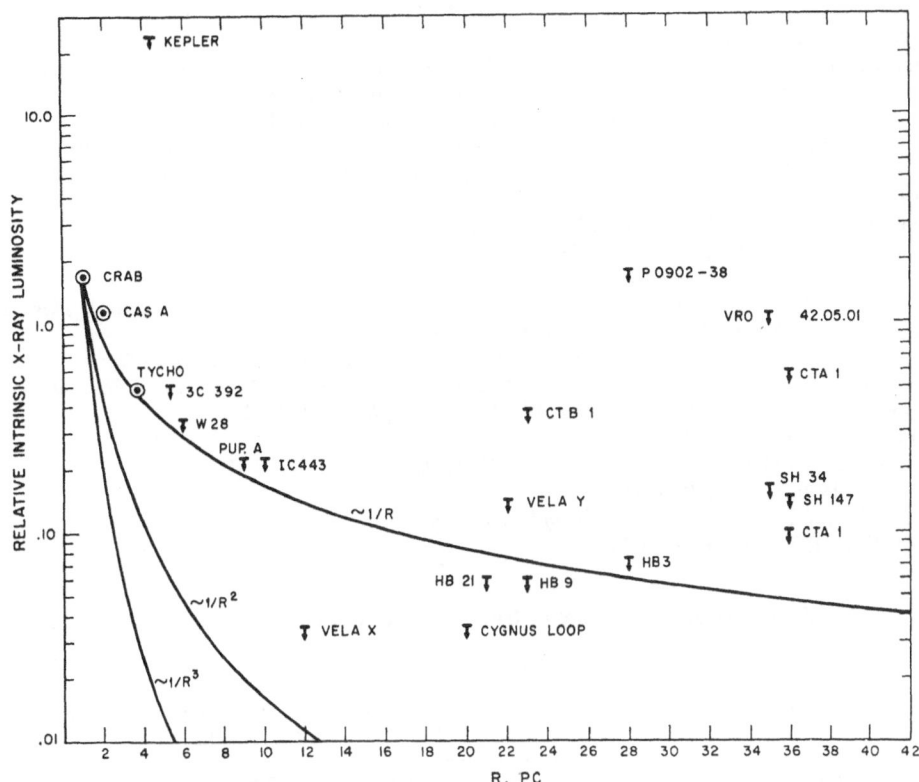

Fig. 10.   X-ray luminosity versus radio diameter for several supernova remnants.

the X-ray brightness is related to size in radio. The origin of this relation, if it is con-firmed by detection of weaker X-ray emission from the larger supernova remnants, is not clear.

The expansion of a magnetic field region containing high energy electrons would give decreasing luminosity with radius, if the magnetic field and electron energy decrease during expansion. A blast wave model would give increasing luminosity with radius until a peak is reached.

Acknowledgements

I wish to acknowledge helpful discussions with Wallace Tucker and Paul Gorenstein of AS&E.

References

Boldt, E., Desai, U., and Holt, S.: 1969, *Astrophys. J.* **156**, 427.
Bowyer, C. S., Bryam, E. T., Chubb, T. A., and Friedman, H.: 1964, *Science* **146**, 912.
Bowyer, C. S., Mack, J., and Lampton, M.: 1970, *Nature* **225**, 1125.
Bradt, H., *et al.*: 1970, *Bull. Am. Phys. Soc.* (in press).
Clark, B. G.: 1965, *Astrophys. J.* **142**, 1398.
Clayton, D. D. and Craddock, W. L.: 1965, *Astrophys. J.* **142**, 189.
Ducros, G., Ducros, R., Rocchia, R., and Tarrius, A.: 1970, *Nature* **227**, 152.
Fishman, G. J., Harnden, F. R., Jr., and Haymes, R. C.: 1969, *Astrophys. J.* **156**, L107.

Floyd, F. W., Glass, I. S., and Schnopper, H. W.: 1969, *Nature* **224**, 50.
Floyd, F. W.: 1970, *Nature* **226**, 733.
Fritz, G., Henry, R. C., Meekings, J. F., Chubb, T. A., and Friedman, H.: 1969, *Science* **164**, 709.
Gorenstein, P., Kellogg, E. M., and Gursky, H.: 1969, *Astrophys. J.* **156**, 315.
Gorenstein, P., Kellogg, E. M., and Gursky, H.: 1970, *Astrophys. J.* **160**, 199.
Grader, R. J., Hill, E. W., Seward, F. D., and Hiltner, W. A.: 1970, *Astrophys. J.* **159**, 201.
Haymes, R. C., Ellis, D. V., Fishman, G. J., Kurfess, J. D., and Tucker, W. H.: 1968, *Astrophys. J.* **151**, L9.
Jacobson, A. S.: 1968, Ph.D. Thesis, Univ. Calif., San Diego.
Naranan, S. and Shah, G. A.: 1970, *Nature* **225**, 834.
Novick, R.: 1969, *IAU Symp.* **37**, (ed. by L. Gratton).
Oda, M., Bradt, H., Garmire, G., Spada, G., Sreekantan, B. V., Gursky, H., Giacconi, R., Gorenstein, P., and Waters, J. R.: 1967, *Astrophys. J.* **148**, 45.
Peterson, L. E., Jacobson, A. S., and Pelling, R. M.: 1966, *Phys. Rev. Letters*, **16**, 142.
Peterson, L. E., Jacobson, A. S., Pelling, R. M., and Schwartz, D. A.: 1968, *Can. J. Phys.* **46**, 437.
Rappaport, S., Bradt, H. V., and Mayer, W.: 1969, *Astrophys. J.* **157**, L21.
Rappaport, S.: 1971, this symposium, Paper 2.2, p. 84.
Riegler, G. R., Boldt, E., and Serlemitsos, P.: 1968, *Astrophys. J.* **153**, L95.
Ryter, C.: 1970, *Nature* **226**, 1041.
Sartori, L. and Morrison, P.: 1967, *Astrophys. J.* **150**, 385.
Slysh, V. I.: 1969, *Nature* **224**, 159.

Appendix A. Summary of Spectral Index Data Used in Figure

$E \lesssim 20$ keV – Rocket Data

Date	Experiment	ΔE keV	Spectral Index $(dN/dE = KE^{-\beta})$*	K
10/65	Grader *et al.* *Science* **152** (1966) 1499	1–40	2.1–2.5	~ 15
2/68	Gorenstein *et al.* *Astrophys. J.* **156** (1969) 315	1–13	1.9 –2.3	7.5–12.7
3/68	Boldt *et al.* *Astrophys. J.* **156** (1969) 427	2–20	1.88–1.98	~ 8–9
12/68	Gorenstein *et al.* *Astrophys. J.* **160** (1970) 199	1–12	1.9 –2.1	8.0–9.7

* photons/cm² sec keV

$E \geq 20$ keV – Data Balloon

		ΔE keV	β	K
9/65	Peterson *et al.* *Phys. Rev. Letters* **16** (1966) 142	19–120	1.81–2.11	3.5
6/67	Haymes *et al.* *Astrophys. J.* **151** (1968) L9	35–560	2.11–2.27	4.3–9.9
6/67	Rocchia *et al.* *Astron. Astrophys.* **1** (1969) 48	15–100	1.62–2.02	7.7
7/67	Riegler, Boldt and Serlemitsos *Astrophys J.* **153** (1968) L95	22–60	1.7 –2.3	7.9
7/67	Jacobson UCSD Ph D Thesis (1968)	20–250	2.04–2.22	
10/67	Glass *Astrophys. J.* **157** (1969) 215	20–70	2.16 ± ? (2.19–2.23)	4.4
9/68	Rocchia *et al.* (Preprint, 1969)	15–100	1.86–2.10 1.79–2.11	4.8 5

Discussion

L. H. Aller: One of the most important programmes in space astronomy is to obtain observations from satellites with diffraction-limited optics. Are there any serious plans to use grazing incidence optics to obtain high resolution images of X-ray sources and spectroscopic observations?

E. M. Kellogg: The proposed large orbiting X-ray telescope will be able to take X-ray pictures with resolution of $\simeq 2$ arc sec, spectra in the X-ray range with $\lambda/\delta\lambda$ as good as 10^4, and polarization measurements with angular resolution of several arc seconds, enabling an X-ray polarisation map of the Crab to be obtained.

P. Sanford: Would you indicate the sensitivity of your forthcoming rocket flight?

E. M. Kellogg: We can obtain about 200 counts from the entire nebula, with a limiting resolution of 10–20 arc sec, from a 190 sec rocket observation. The grazing incidence optics, being designed originally for solar X-ray imaging, has a small effective area of $\sim 1\text{cm}^2$ at 8 Å. The image is detected by an X-ray image intensifier which has about 50 times the sensitivity of film, which has been used for obtaining solar X-ray pictures. The next generation of rocket X-ray telescopes, available in about a year or two from now, will have ten times as much sensitivity, allowing us to obtain perhaps 2000 counts from the Crab Nebula per flight.

1.6 UPPER LIMIT OF THE X-RAY POLARIZATION
OF THE CRAB NEBULA*

R. NOVICK, J. R. P. ANGEL and R. S. WOLFF

Columbia Astrophysics Laboratory, Columbia University, New York, N.Y. U.S.A.

Abstract. A rocket-borne X-ray polarimeter was flown to search for polarization in Taurus X-1. Although a result consistent with zero polarization was obtained, the statistics were such that X-ray polarization comparable in magnitude and direction to that of radio and optical continuum emission cannot be excluded.

1. Introduction

An X-ray polarimeter utilizing incoherent scattering was constructed and flown in an Aerobee-150 sounding rocket at 0327 UT on March 7, 1969, to search for polarization in the X-ray emission of Taurus X-1 (Wolff *et al.*, 1970). Four minutes of data obtained while the polarimeter was above 250000 ft and aimed at the source have been analyzed to obtain the Stokes parameters describing the magnitude and orientation of the polarization of the X-ray emission. The values obtained are $q = 7.26 \pm 9.5\%$ and $u = -5.0 \pm 9.3\%$, where u and q are the normalized Stokes parameters. The quoted uncertainties are 1σ standard deviations in these quantities. A number of systematic effects have been considered and found to be unimportant compared to these statistical uncertainties. Expressed in terms of polarization, magnitude P, and position angle θ, measured east from north, our results are $p = 8.8\%$ and $\theta = 163°$.

The existence of polarization in the optical emission of the Crab nebula was first discovered in 1953 (Vashakidze, 1954) and since has been extensively studied. The magnitude and orientation of the optical polarization varies considerably over the 6' extent of the nebula, but integrating over the entire object leads to a net result of about 9.3%. The degree of polarization increases as the field of view is narrowed, and has been measured as 19% with 1' aperture centered on the luminous central region of the nebula (Oort and Walraven, 1956). Polarization in the strong radio emission of the nebula at 9.55 mm was observed by Hobbs (1968) as 13.8%. Measurements with various antenna beam widths and at different wavelengths have been made, leading to a rather complete picture of the intensity and polarization of radiation emitted by the nebula over a broad energy range. As suggested by Shklovsky in 1953, the spectral behavior and polarization of the visible and radio emission can be explained as synchrotron radiation from a power-law distribution of electrons moving in a magnetic field. The observed spectrum and polarization of the optical

* This work was supported in part by the National Aeronautics and Space Administration under Grants NGR-33-008-102, NGR-33-008-012, NGR-33-008-125 and Contract NAS 8-24668, and in part by the Air Force Office of Scientific Research under Grant AFOSR-70-1945. It is Columbia Astrophysics Laboratory Contribution No. 33.

and radio radiation confirm this hypothesis, and the Crab nebula is generally accepted as a synchrotron emitter in this energy range.

The X-ray flux, first observed in 1963, originates in the central portion of the nebula (Bowyer *et al.*, 1964) and is coincident with the region of highest optical luminosity and polarization (Oda *et al.*, 1967). The X-ray data appear to be a natural extension of the synchrotron spectrum (Woltjer, 1964), in which case the X-ray and optical radiation would have comparable polarization. Serious questions have been raised regarding the source and lifetimes of relativistic electrons energetic enough to emit synchrotron radiation at X-ray wavelengths, and Sartori and Morrison (1967) have proposed a hot plasma model as an alternate X-ray production mechanism. The measurements of the spectral behavior of the X-ray flux cannot be used to distinguish between the two models, but a measurement of the X-ray polarization could serve as a definitive means of resolving these questions.

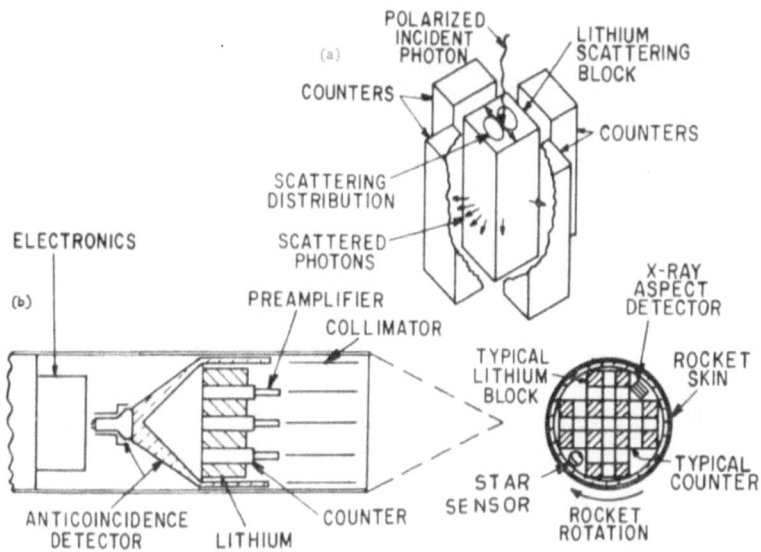

Fig. 1.   (a) Schematic representation of the polarimeter concept. (b) Mounting of the polarimeter and ancillary equipment in the rocket.

The instrument used in the present work exploits the polarization dependence of Thomson scattering. The probability of scattering at an angle α to the electric vector of the incident radiation is proportional to $\sin^2\alpha$. Metallic lithium scattering blocks are used, with 3-atm xenon-methane proportional counters arranged to detect the radiation scattered out through the sides of the blocks. This is shown schematically in Figure 1a; the mounting of the polarimeter in the rocket is shown in Figure 1b. In use, the polarimeter is pointed toward the source and rotated about the line of sight. If the incident radiation is polarized, the counting rate in each of the counters will be modulated at a frequency equal to twice the rotation frequency of the polarimeter; the depth and phase of the modulation provide a direct measure of the magnitude and position angle of the polarization vector (see Figure 2a). This mode of operation avoids false indications of polarization that would otherwise arise from differences in the counter sensitivities, amplifier gains, and pulse-height discriminator levels. Clearly, the modulation components of the orthogonal counters must be in antiphase; this fact allows us to discriminate against rapid changes in source strength which might otherwise appear as polarization. The size of the blocks is determined by the mean scattering length in lithium, about 10 cm. The blocks are 12.7 cm deep, sufficient to give a 70% probability of scattering, while the cross section of 25 cm² is small enough to avoid multiple scattering. The measured effective area of the polarimeter for unpolarized X-rays is shown in Figure 2b. The limits of sensitivity of the polarimeter are determined at low energies by the photoelectric absorption in the lithium and at high energies by the transparency of the counter gas to hard X-rays.

The response of a single polarimeter module to a beam of 100%-polarized brems-

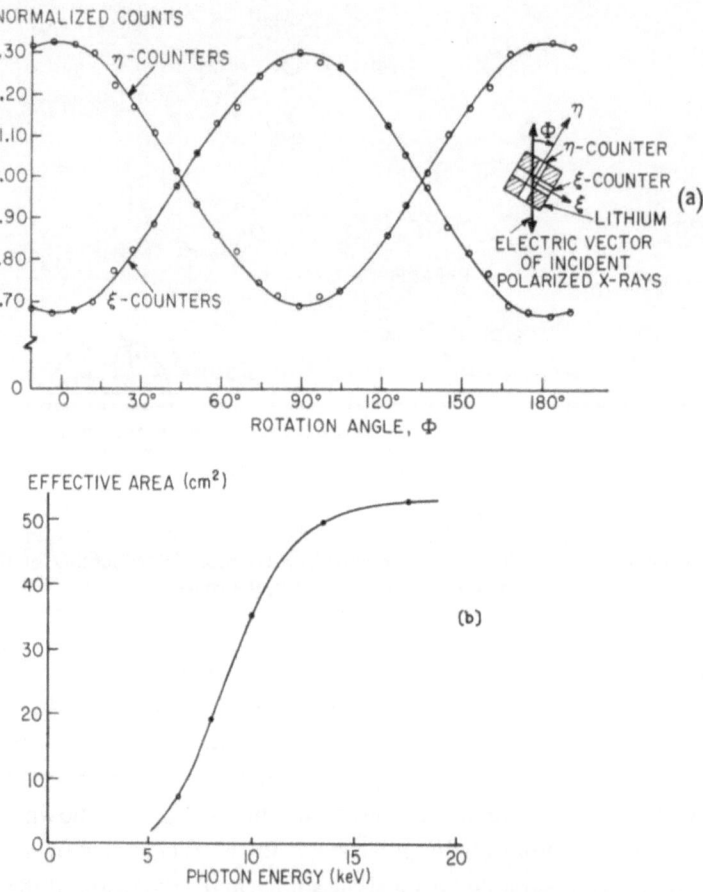

Fig. 2. (a) Variation in the counting rates for each of the orthogonal counters as the polarimeter is rotated with respect to a beam of 100%-polarized X-rays. The counting rates have been normalized to their average values. (b) Variation in the effective area of the polarimeter with photon energy. Note that the geometrical area of the polarimeter is about 900 cm².

strahlung X-rays with average energy of 15 keV is shown in Figure 2a. Here are shown the counting rates in each of the two sets of orthogonal counters as a function of the orientation. These rates have been normalized to their average values. As expected, the counts in the two sets of orthogonal counters vary harmonically with angle according to the equations

$$N_\xi = R_0 \left(1 - M_0 \cos 2\Phi\right),$$
$$N_\eta = R_0 \left(1 + M_0 \cos 2\Phi\right).$$

Here R_0 is the average counting rate, and M_0 is the depth of modulation for 100%-polarized X-rays; the axes ξ and η and the rotation angle Φ are defined in Figure 2a. Because each of the detectors subtends a large solid angle, the modulation depth M_0 has a value of only 31.6% for a 100%-polarized beam. Any attempt to increase the depth of modulation by decreasing the solid angle of the counters would reduce the efficiency and increase the minimum detectable polarization.

There are additional possible sources of systematic errors that must be considered. If the instrument were not pointed directly at the X-ray source, orthogonal detectors would not be equally illuminated, and a false indication of polarization would be obtained. Laboratory experiments showed that the instrument axis had to be pointed within 3° of the source to keep this effect small. To check the proper orientation of the rocket, an optical star sensor was used to determine the rocket orientation and a collimated forward-looking X-ray detector was built into the instrument. Systematic errors could be contributed by anisotropies in the background counts arising from cosmic rays and albedo gamma rays from the earth's atmosphere. Anisotropy of this radiation could result in orientation dependence of the background counting rate. A measurement of the background counting rate was carried out on a prototype instrument flown to a height of 95 000 ft in a balloon. The instrument was rotated about a vertical axis, and a search was made for apparent polarization effects that might arise from the known east-west anisotropy of the cosmic-ray protons. No effect large enough to affect the present results was observed.

The payload is illustrated in Figure 1b. After the nose cone is ejected, X-rays enter the scattering blocks through a collimator which gives a clear view up to 3° from the axis and is totally opaque for angles greater than about 12°. This collimator prevents illumination by other X-ray sources and severely limits the signal from the isotropic background. The counter-scatterer assembly is surrounded by an anticoincidence shield of scintillating plastic viewed by a 3-in. photomultiplier tube. The forward-looking X-ray detector is located in the corner of the main polarimeter. Charge-sensitive preamplifiers are mounted on each counter forming the lower part of the main collimator. In the main electronic processing unit, the heights of the pulses from each amplifier, which are not vetoed by the plastic scintillator, are analyzed separately into four bins, corresponding to detected photon energies of 5.5–11 keV, 11–16.5 keV, 16.5–22 keV, and above 22 keV. Four scalers for each channel accumulate the counts in each energy bin and are sampled every 12 msec by the telemetry system. A detailed description of the instrument will be given elsewhere.

Cosmic-ray background poses the greatest limitation in the detection of X-ray polarization in weak sources such as the Crab nebula. Although the polarimeter is surrounded by a plastic anticoincidence shield to veto energetic charged particles, leakage due to solid angle factors and the materialization of neutrals inside the shield lead to a background rate of 0.01 counts/cm² sec keV in each of the proportional counters measured above the atmosphere (Angel *et al.*, 1969). With recent data on the X-ray flux from the Crab and the measured spectral response of the instrument, a background-to-signal ratio of 4:1 was estimated. This would have imposed severe statistical limitations on the experiment. Substantial additional background suppression was accomplished by using rise-time discrimination which makes use of the different pulse shapes caused by charged particles and X-rays. Laboratory tests indicated that the background-to-signal ratio could be reduced to 1.2:1.

The aspect of the rocket during the flight was controlled with a system of gyros and was monitored by the forward-viewing X-ray detector and star sensor. During the 4-min data-acquisition period, the polarimeter remained pointed to within 1.5° of the source and rotated around the line of sight at 6.4°/sec. The energy levels of the pulse height discriminators were set using X-ray sources, and the calibration was checked 3 h prior to launch. The number of counts observed are listed in Table I. During the 4 min of flight while data from the source were being taken, the anticoin-

TABLE I

Source data: 240 sec

Bin	Energy range (keV)	X-ray counts	Counts/counter-sec
1	5.5 – 11	2499 ± 171	0.651 ± 0.045
2	11. – 16.5	2996 ± 186	0.782 ± 0.0485
3	16.5 – 22	1771 ± 176	0.461 ± 0.046

Background data: 53 sec

Bin	Counts	Counts/counter-sec
1	1095	1.284 ± 0.039
2	1303	1.5279 ± 0.042
3	1189	1.3942 ± 0.040

cidence rate was 1300/sec, leading to a negligible correction for dead time. The background counts were obtained by analyzing the data after the nose cone was ejected but before the instrument was pointed at the target and again after active control was lost until the Pfotzer cosmic-ray maximum was approached on re-entry; 53 sec of background data were obtained. The signal counting rate closely conformed to the predictions based on previous data.

The data were analyzed for polarization by looking for modulation in the counting rate in phase with the roll of the rocket. The data for each counter were fitted by least squares to the function

$$R(t) = S_0 (1 + M_1 \cos 2\omega t + M_2 \sin 2\omega t) \tag{1}$$

where $R(t)$ represents the observed counting rate, S_0 the background and the average X-ray intensity, and M_1 and M_2 are the modulation coefficients, related to the total modulation by $M = (M_1^2 + M_2^2)^{1/2}$. The phase of the modulation Ψ is defined by $\Psi = \frac{1}{2} \tan^{-1}(M_2/M_1)$. The modulation components for the signal-plus-background data were calculated for each energy bin of each counter, or a total of 64 pairs of numbers. The standard deviation of each component was also calculated, using the definition of variance for a single 1-sec sample of data and propagating this uncertainty through the equations of least squares.

The mean background rates were subtracted from the mean signal-plus-background rates S_0 to obtain a net mean signal rate S_0' for each counter and bin. The signal modulation components M_1' and M_2' were then calculated using the net rates, resulting again in 64 pairs of components. The standard deviations were also combined in the appropriate way. The 16 proportional counters are oriented in two orthogonal directions designated X and Y, forming two separate polarimeters. Modulation components for the eight counters in the X polarimeter were then weighted by their standard deviations and added, and the same procedure was followed for the Y counters. The weighted sums were computed for the lower three bins, and the fourth bin, corresponding to photons above 22 keV, was neglected as it contained virtually no signal. Since the X and Y counters are orthogonal, modulation in their counting rates due to signal polarization must display appropriate phase differences. The X and Y modulation components were then added by taking this phase difference into account. The resultant modulation components for the X-ray signal detected in the three bins, with their standard deviations, are listed in Table II, together with the weighted sum of the three bins. The components have been referenced to the celestial sphere using magnetometer aspect data, so that M_1' and M_2' now correspond to the same coordinate system in which the Stokes parameters are expressed. A striking feature of these results is the lack of a common direction among the modulations calculated for the three bins. Although the results for each bin when taken separately are suggestive of a nonzero modulation, their sum is consistent with a null result.

A check of the validity of the calculation procedure was accomplished by a Monte Carlo simulation of the experiment. Twenty-two thousand counts randomly distributed over 240 sec were generated and analyzed for modulation. The procedure was repeated 100 times. The standard deviations obtained by Monte Carlo simulation correspond closely to those achieved by calculation from the flight data, and its modulation is consistent with that of the sets of random data. There is a probability of obtaining a nonzero value of the modulation from purely random data since each of the components is statistically independent. The probability of obtaining modulation M, is given by

$$W(M)\,dM = \frac{MN}{2} \exp - \frac{M^2 N}{4}\,dM \tag{2}$$

TABLE II

Modulation components and normalized Stokes parameters

Bin	Energy range (keV)	M'_1	M'_2	q	u	Polarization P	Position angle θ (degrees)
1	5.5 – 11.0	0.082 ± 0.048	0.043 ± 0.047	0.262 ± 0.152	0.138 ± 0.150	0.30 ± 0.15	14
2	11.0 – 16.5	−0.014 ± 0.045	−0.032 ± 0.045	−0.046 ± 0.143	−0.262 ± 0.143	0.27 ± 0.14	130
3	16.5 – 22.0	−0.019 ± 0.069	0.010 ± 0.067	−0.061 ± 0.218	0.032 ± 0.216	0.07 ± 0.22	76
1 + 2 + 3	5.5 – 22.0	0.023 ± 0.030	−0.016 ± 0.029	0.072 ± 0.095	−0.050 ± 0.095	0.088 ± 0.095	163

with a mean value

$$\langle M \rangle = (\pi/N)^{1/2}.$$

Here N is the total number of counts obtained in the observation.

The data were analyzed for modulation at a range of frequencies close to the roll rate to examine the correlation between the signal modulation and frequency. If the apparent modulation is due to polarization, then the modulation should be a maximum at twice the roll rate; but if the modulation is due to a statistical fluctuation, the two quantities will be uncorrelated. The calculation was performed at rocket rotation periods ranging from 45 to 75 sec at intervals of 0.75 sec, and also at the true rotation period of 56 sec. Monte Carlo data were generated with varying amounts of modulation at the roll frequency, and subjected to the same analysis. The results, graphed as modulation versus period, are shown in Figure 3. Monte Carlo cases with less than

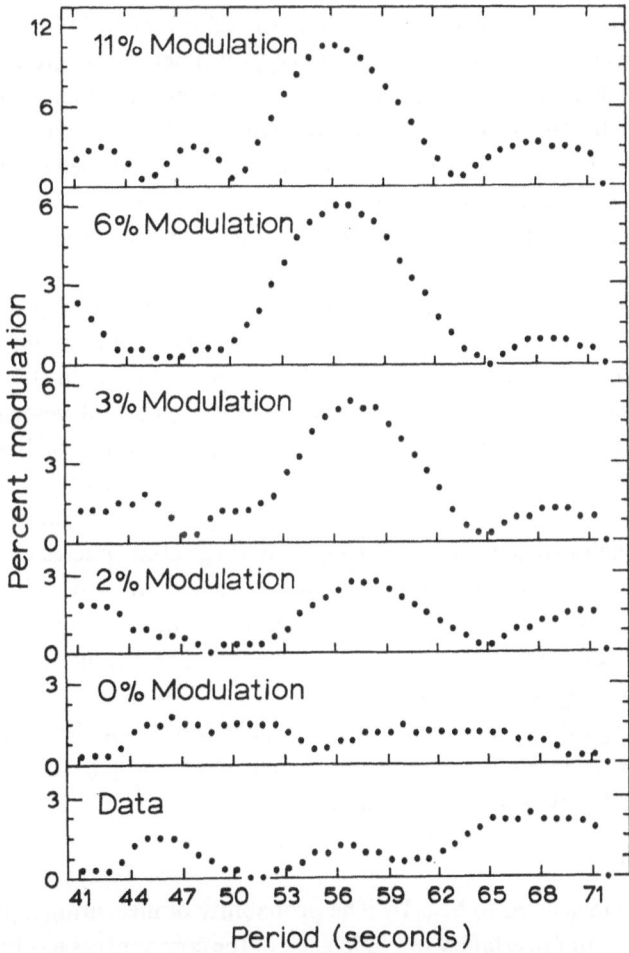

Fig. 3.   Percent modulation as a function of rotation frequency for flight data and Monte Carlo trials. Varying amounts of modulation with a period of 56 sec have been introduced in the Monte Carlo trials.

3% modulation are indistinguishable from purely random data. The maximum at the roll period begins to appear at 4% modulation, but is clearly missing in the real data.

Possible instrumental effects leading to cancellation of the source modulation were explored. Modulation in the background rate was assumed to be zero, but, if not, could add in antiphase to the signal modulation leaving a null result. Data from a balloon flight in 1967 and a rocket flight in 1968, using similar versions of the polarimeter, showed no evidence for background modulation. In the first flight the polarimeter was pointed toward the zenith, while in the second it was inclined at an angle of 45°. In the current flight, the instrument was aimed at 14° from vertical. Anisotropy in the albedo flux from the earth's atmosphere must be eliminated since neither of the earlier flights offer evidence for existence of this effect. An anisotropy in the primary cosmic-ray flux, such as the east-west effect, could manifest itself as spurious modulation. A search for this effect was made by analyzing the fourth bin ($E > 22$ keV) data for modulation. The events in the channel were entirely due to the cosmic-ray background. A very small modulation, $M = 0.007 \pm 0.0075$, consistent with zero, was discovered. The effectiveness of the polarimeter background suppression methods was tested using a monoenergetic and monodirectional charged-particle beam generated by the Princeton-Penn accelerator (PPA). Comprised of 1 BeV/c protons, the PPA beam was incident on the side of the polarimeter perpendicular to its longitudinal axis, and the counting rates at various azimuthal orientations monitored. No effects which would lead to spurious evidence for polarization were encountered.

If a cosmic-ray-induced modulation existed and were energy dependent, then the dispersion in direction of the modulations for the three bins might be explained. However, if cosmic-ray primary protons are interacting, they must be entering the polarimeter through the front, or they would be detected and vetoed by the anti-coincidence shield. This limits their trajectories to angles of 60° or less with respect to the line of sight of the instrument. The magnetic rigidity of the earth at 30°N latitude is substantial, and truncates the proton spectrum at about 2 BeV. Such minimum ionizing particles, passing through the three atmospheres of xenon, would deposit at least 40 keV in one orientation of a detector, and twice this in the other. Only very unusual paths, such as through the corners of the counters, would result in the deposition of a small enough energy for a primary cosmic ray to be recorded in one of the three signal bins.

Conclusions regarding the maximum likely polarization of the X-ray flux of the Crab nebula can be drawn from these results. The degree of polarization P is related to a detected modulation M by the equation

$$P = \mu M,$$

where μ has been measured to be 3.16. The probability of measuring a polarization of 10% or less, with a 1σ uncertainty of 9% in each of the components has been calculated for various true polarizations P. The probability of obtaining our result, or smaller, given a true source polarization of 27%, is less than 1%. This argument establishes an upper limit of 27% on the X-ray polarization of the nebula with a 99% confidence level. There is a 95% probability that the source polarization is not greater than 20%. The position angle of polarization vector obtained is 163°, which compares favorably with that of the radio and optical results, as can be seen in Figure 4. Although suggestive, this apparent correlation cannot be rigorously interpreted as evidence for a positive result, because the uncertainty in the polarization components is sufficiently large to include all angles in the 1σ circle of uncertainty.

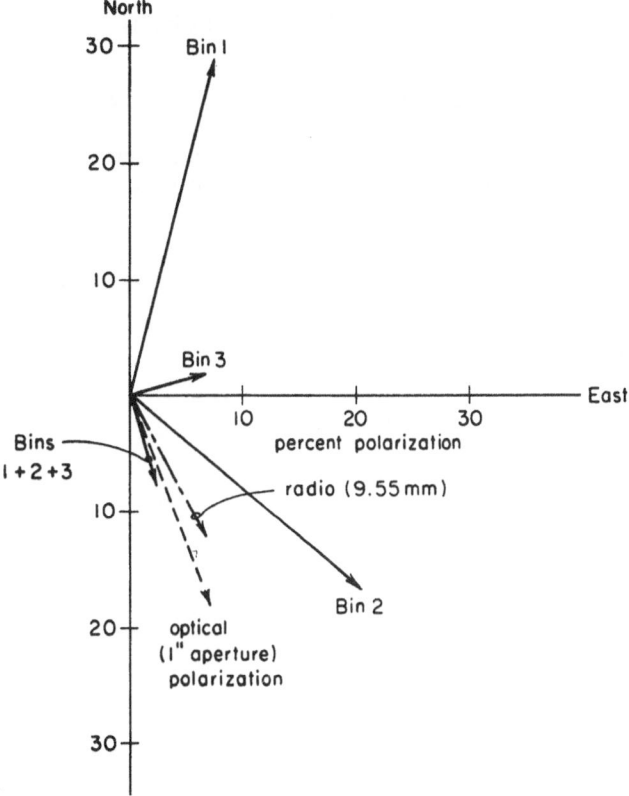

Fig. 4. The polarization vectors shown are in terms of celestial coordinates. The numbers beside the vectors correspond to the energy bin.

In summary, our results can best be interpreted as setting an upper limit of 27% on the polarization of the X-ray emission of Taurus X-1 with a 99% confidence level. The present result does not serve to distinguish between synchrotron and other processes suggested for X-ray production in the Crab nebula.

References

Angel, J. R. P., Novick, R., Vanden Bout, P., and Wolff, R.: 1969, *Phys. Rev. Letters* 22, 861.
Bowyer, S., Byram, E. T., Chubb, T. A., and Friedman, H.: 1964, *Science* 146, 912.
Hobbs, R. W.: 1968, *Astrophys. J.* 153, 1001.

Oda, M., Brandt, H., Garmire, G., Spada, G., Sreekantan, B. V., Gursky, H., Giacconi, R., Gorenstein, P., and Waters, J. R.: 1967, *Astrophys. J.* **148**, L5.
Oort, J. H. and Walraven, Th.: 1956, *Bull. Astron. Inst. Neth.* **12**, 285.
Sartori, L. and Morrison, P.: 1967, *Astrophys. J.* **150**, 385.
Vashakidze, M. A.: 1954, A. Ts., No. 147.
Wolff, R. S., Angel, J. R. P., Novick, R., and Vanden Bout, P.: 1970, *Astrophys. J.* **160**, L21.
Woltjer, L.: 1964, *Astrophys. J.* **140**, 1309.

Discussion

N. Visvanathan: Does the 19% polarization which you expect include an allowance for measuring efficiency?

R. Novick: Yes.

N. Visvanathan: The optical polarization varies across the Crab Nebula. Near the pulsar it is only 4%.

R. Novick: The signal which we are measuring is the integrated emission. Our limit suggests that if the X-radiation is synchrotron emission it is not coming from a region of very homogeneous field.

1.7 RECENT RESULTS ON THE SEARCH FOR
10^{11} eV GAMMA RAYS FROM THE CRAB NEBULA

G. G. FAZIO, H. F. HELMKEN, G. H. RIEKE,* and T. C. WEEKES

Smithsonian Astrophysical Observatory, Cambridge, Mass. U.S.A.

Abstract. The detection of Čerenkov light emitted by cosmic-ray air showers was used to search for cosmic gamma rays from the Crab Nebula. By use of the 10-m optical reflector at Mt. Hopkins, Arizona, the Crab Nebula was observed during the winter of 1969–1970 for approximately 112 hours, which was a significant increase in exposure time over previous experiments. Above a gamma-ray energy of 2.2×10^{11} eV, no significant flux was detected, resulting in an upper limit to the flux of 8.1×10^{-11} photon/cm^2 sec. In the synchrotron-Compton-scattering model of gamma-ray production in the Crab Nebula, this limit on the flux indicates the average magnetic field in the nebula must be greater than 3×10^{-4} G.

1. Introduction

At energies above 10^{11} eV, cosmic gamma rays interact with the atmosphere to generate a cascade of high-energy electrons and photons. This 'air shower' is highly directional and proceeds with relativistic velocity along the direction of the initial gamma-ray photon. Accompanying the shower is a cone of Čerenkov light about 2° wide and also directed along the shower axis. The duration of this light burst is about 10^{-8} sec, and its lateral spread at sea level is about 3×10^4 m^2. Cosmic-ray protons and nuclei of comparable energy also generate air showers and Čerenkov light, but primary gamma radiation from a discrete source can be distinguished by the presence of an anisotropy in the arrival directions of the Čerenkov light bursts.

2. Experimental Technique

The 10-m optical reflector at Mt. Hopkins, Arizona, was used to detect atmospheric Čerenkov light and to search for any anisotropy in the arrival direction of these bursts. The reflector was operated in a tracking mode, with two photomultiplier tubes (RCA 4522) in the focal plane. Each phototube had a full field of view of 1°, and the two phototubes were separated by 2.4° One field of view was centered on the suspected source, in this case the Crab Nebula, while the other viewed a nearby region of sky to monitor the background shower rate. Every 10 min, the fields of view were interchanged by slewing the reflector. Every 1 min, the count rate from each phototube was recorded.

Before this experiment, the 'drift-scan' technique had been used to search for an anisotropy. This technique consisted of positioning the reflector at a point in the sky ahead in right ascension and allowing the earth's rotation to bring the source through

* Present address: Lunar and Planetary Laboratory, University of Arizona, Tucson, Arizona.

Davies and Smith (eds.), The Crab Nebula, 65–67. All Rights Reserved.

the field of view. Such a scan required ∼40 min with only 8 min on the suspected source. The tracking mode, however, allowed the source and background regions to be observed simultaneously and greatly increased the observing time over a given period.

3. Results

From October 1969 to February 1970, the Crab Nebula was observed for 6698 min. A net positive effect was detected from the Nebula; further observations are planned, but at present this effect is not interpreted to be due to a flux of gamma rays. The number of showers observed from the direction of the source divided by the number from nearby sections of the sky was 1.0026 ± 0.0011. This ratio corresponds to an upper limit (at 3 standard deviations) of 8.1×10^{-11} photon/cm^2 sec for the gamma-ray flux from the Crab Nebula for energies above 2.2×10^{11} eV.

The objects M87, M82, Sag A, 3C273, and Cyg A were also observed and analyzed to determine if any systematic errors that could produce a positive effect were introduced into the data. The data from these other observations are presented in Table I. No net effect comparable to that for the Crab Nebula was observed. The results obtained on the Crab Nebula during the winter of 1968–1969 (Fazio *et al.*, 1970) are also given in Table I.

A theoretical spectrum of the gamma-ray flux from the Crab Nebula can be calculated on the basis of Compton scattering in the Nebula between synchrotron electrons and the observed synchrotron-emitted photons, assuming a uniform value of the magnetic field (Gould, 1965; Rieke and Weekes, 1969). The gamma-ray flux

TABLE I

Summary of gamma-ray observations

Object	Date	Observation time (min)	Source on/ Source off ratio	Threshold energy (10^{11} eV)	Flux limit (10^{-11} photon/cm^2 sec)
Crab Nebula	1969–1970	6698	1.0026 ± 0.0014*	2.2	8.0
Crab Nebula	1968–1969	776	1.0010 ± 0.0038	1.7	20
M87 (Virgo A)	1969–1970	2126	1.0013 ± 0.0021	2.1	10
M82	1969–1970	1549	$1 0010 \pm 0.0025$	3.2	7.1
Sag A	1969–1970	683	0.9886 ± 0.0073	22	0.46
3C273	1963–1970	1064	0.9947 ± 0.0033	2.8	3.6
Cyg A	1969–1970	518	0.9970 ± 0.0046	2.6	12

predicted is a sensitive function of the magnetic field. From these calculations and the upper limit to the flux set by this experiment, it is possible to place a lower limit of 3×10^{-4} G on the average strength of the magnetic field in the Crab Nebula.

* For the Crab Nebula (1969–1970), the standard deviation was based on experimental error; for all other sources, Poisson errors were used. The Poisson error for the Crab Nebula was ± 0.0011.

Acknowledgement

The authors wish to acknowledge the assistance of Ed Horine during this experiment.

References

Fazio, G. G., Helmken, H. F., Rieke, G. H., and Weekes, T. C.: 1970, in L. Gratton (ed.), 'Non-Solar X- and Gamma-Ray Astronomy', *IAU Symp.* **37**, 250.
Gould, R. J.: 1965, *Phys. Rev. Letters* **15**, 577.
Rieke, G. H. and Weekes, T. C.: 1969, *Astrophys. J.* **155**, 429.

1.8 HIGH RESOLUTION MAPS OF THE CRAB NEBULA
AT 2700 MHz AND 5000 MHz

A. S. WILSON

Mullard Radio Astronomy Observatory, Cambridge, U.K.

This paper describes some new maps of the Crab nebula made with the One Mile radio telescope. Branson (1965) used this instrument to map the nebula at 1400 MHz. It is now operating at 2700 MHz and 5000 MHz and, for the Crab nebula, the half power beamwidth is 6″ × 16″ at the higher frequency.

The map of total intensity at 2700 MHz shows considerable fine structure, particularly in the centre of the nebula. Features recognised at 1400 MHz reappear, together with further detail on a smaller scale. The total intensity at 5000 MHz shows still more fine structure.

In order to compare the distribution at radio wavelengths with that in the optical continuum, it is convenient to use the isophotes derived by Woltjer (1957), for which the contribution from the filaments is small. We notice the characteristic 'S' shape of the object and the bays to the west and east. The continuum emission of the central region of the nebula has been described recently by Scargle (1969a, 1969b). Apart from the thin wisp, there is relatively little continuum emission from the immediate surroundings of the pulsar. The other wisps appear to the NW of the pulsar, with less well defined activity to the SE.

Generally speaking, the radio distribution at 5000 MHz is remarkably similar to that of the optical continuum. Again there is an 'S' shaped ridge, but this is even more pronounced in the radio case. As well as this and the nearby 'bay', the valley to the east of the NW 'peak' of the 'S' and valleys to the south also appear on the radio map. Near the pulsar, we have a minimum in the radio emission. This 'valley' has a position angle of 22°. 'Ridges' of the radio emission appear both to the NW and SE of the pulsar, about 12″ from it, and in a direction approximately perpendicular to that of the valley. It may well be that these features are related to the wisp activity. The size of the radio nebula is larger than that of the optical by between $1\frac{1}{2}$ to 2, depending on the direction. The map at 2700 MHz is very similar to that at 5000 MHz but, of course, features are smoothed by the lower resolution.

In the region within a radius of about a minute of arc from the pulsar, there are a number of bright 'ridges' of radio emission which do not correlate with any optical continuum emission. Indeed, bright ridges protrude from the centre towards the 'bay' in the west and a valley in the east. If we now superpose the radio map and an optical photograph taken in a filamentary line, (kindly donated by Dr. V. Trimble) it is clear that some of these ridges are associated with bright filaments. An explanation of this enhancement may be that the magnetic field near the filaments is greater than that in the rest of the nebula. The absence of enhanced optical continuum emission could be attributed to the short half lives of the more energetic electrons. To account for

Davies and Smith (eds.), The Crab Nebula, 68–70. All Rights Reserved.
Copyright © 1971 by the IAU.

the observations, we require an increase in the magnetic field by a factor of 1.5 or more.

The distribution of linear polarization at 2700 MHz also shows much structure. The polarized power is greatest in the central regions of the nebula; at maximum it is about 12% of the total intensity. The maximum polarization at 5000 MHz is about 25%. Bright filaments with negative radial velocities are causing depolarization, as suggested by Burn (1966). Assuming that

$$B = 5 \times 10^{-4} \, G$$
$$n = 10^3 \, cm^{-3}$$

we find that the Faraday rotation in a filament $\simeq 29$ rad at $\lambda = 6$ cm, so that strong depolarization is to be expected.

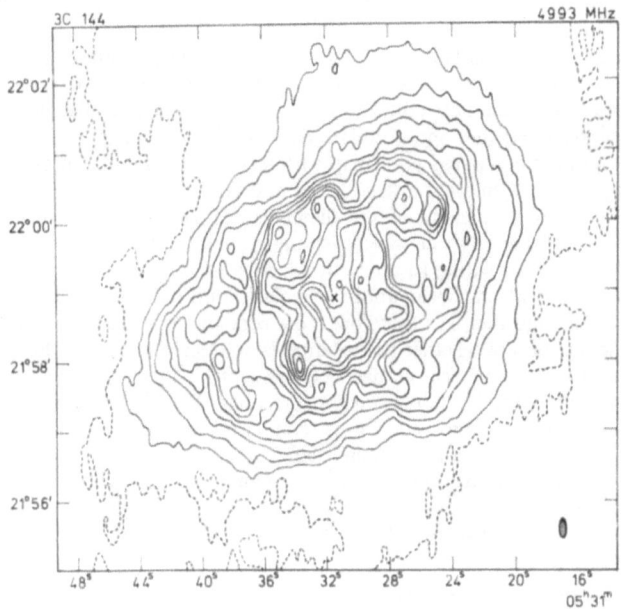

Fig. 1. The Crab Nebula at 6 cm wavelenght. The H.P.B.W. is shown in the bottom right hand corner and the cross marks the pulsar.

Apart from these features, the polarization at 5000 MHz is quite similar to the optical and there is little rotation in the centre of the map. Between 1400 MHz (Wright, 1970) and 5000 MHz the rotation in the centre is about 50°. It is hoped that a careful analysis of the distribution of rotation measure and depolarization across the nebula will define the amount of rotation in the nebula itself and in the interstellar medium and hence provide information on physical conditions in the nebula.

References

Branson, N. J. B. A.: 1965, *Observatory* **85**, 250.
Burn, B. J.: 1967, *Monthly Notices Roy. Astron. Soc.* **133**, 67.

Scargle, J. D.: 1969a, *Astrophys J.* **156**, 401.
Scargle, J. D.: 1969b, *Astrophys. Letters* **3**, 73.
Woltjer, L.: 1957, *Bull. Astron. Soc. Neth.* **13**, 302.
Wright, M.: 1970, *Monthly Notices Roy. Astron. Soc.* **150**, 271.

Discussion

J. E. Nelson: Are the correlations between 2.7 GHz intensities and the optical filament structure real? i.e. do the number of possible correlations decrease when the orientation of the 2 maps is deliberately changed by some arbitrary angle?

A. S. Wilson: I have not done this. Of course one would find correlations at some orientations. Nevertheless, I feel the correlation in the total intensity is strong enough, with sufficient bright filaments that it cannot be due to a chance coincidence. In this connection the strong correlation of the depolarised regions with negative radial velocity filaments is interesting.

C. Michel: How does your value for n_e compare with that from the pulsar observations?

A. S. Wilson: The pulsar value is < 0.5 cm^{-3}. I find 2.5×10^{-2} cm^{-3}, if the rotation is due to the nebula.

R. Minkowski: The four filaments that you consider as coincident with features of the continuum mass are neither very bright nor otherwise important. Should one not ask the inverse question: how many features of the continuum mass coincide with bright, conspicuous filaments?

A. S. Wilson: You may also find anti-correlations if you look for them.

R. G. Conway: You have compared your map with a photograph taken in one filamentary line – have you tried comparing it with photographs taken in other lines?

A. S. Wilson: The relative intensity of the filaments changes with the different lines they emit. Nevertheless the bright filaments recur whichever line you look in; it looks like the same object.

J. E. Baldwin: A comment relating to the remark concerning Drake's upper limit on the electron density in the nebula. The 21 cm observations of polarisation indicate a slow smooth variation of rotation over the nebula and that it is therefore most probably interstellar in origin.

A. S. Wilson: The question of whether the rotation is due to the nebula or the interstellar medium is not yet resolved. However Verschuur has detected a field of $3 \cdot 5$ μ G in between us and the Crab by Zeeman splitting of the 21 cm neutral hydrogen line. This is in such a sense as to cause rotation in an *opposite* sense to that observed. I must emphasise that my value for n_e in the amorphous mass assumes the rotation occurs in the nebula itself.

J. E. Baldwin: Verschuur's measurement was in one cloud only and is not necessarily characteristic of the interstellar medium.

A. S. Wilson: Yes. If the rotation is in the interstellar medium alone

$$n_e B \sim 1 \cdot 3 \times 10^{-2}$$

n_e in cm^{-3}, B in μG.

R. N. Manchester: For what area of the nebula have you measured the rotation measure?

A. S. Wilson: The centre only.

R. N. Manchester: What is the sense of the rotation?

R. G. Conway: The rotation measure of the Crab is -25 rad/m^2.

OBSERVATIONS OF THE CRAB PULSAR

2.1 RADIO OBSERVATIONS OF THE CRAB NEBULA PULSAR

F. D. DRAKE

Cornell-Sydney University Astronomy Center, Cornell University, Ithaca, N.Y., U.S.A.

Abstract. The radio properties of the Crab Nebula pulsar are reviewed. The pulsar lies at the centre of the Crab Nebula and has a period of 33 msec. Its increase in period with time releases an amount of energy which is equal in magnitude to the total radiated power. Instabilities in the period of the Crab pulsar have been discovered with timescales ranging from days to months. The length of the pulse increases at longer wavelengths due apparently to multipath propagation effects. A characteristic of the Crab pulsar is the great intensity of the occasional pulse.

The Crab Nebula pulsar is one of the most important objects in the history of astronomy. It alone has given us convincing evidence that the pulsars are rapidly rotating neutron stars in which the source of emissions we see is their rotational kinetic energy. These objects apparently convert this energy into electromagnetic radiation through the intermediary of a very intense magnetic field. With the Crab Nebula, its pulsar has proved the missing link so long sought to explain some of the most important features of the nebula, particularly its source of energy and the unique point source of low frequency radio emission observed in the centre of the nebula. This pulsar, of course, has led to the only optical and X-ray observations of any pulsar.

The existence of the pulsar became known in the latter part of 1968. Staelin and Reifenstein (1968) at the National Radio Astronomy Observatory observed brief pulses from the vicinity of the Crab Nebula which, although not apparently periodic, showed the famous dispersion effect characteristic of pulsar radiation. In fact, two dispersion measures were present which led Staelin and Reifenstein to hypothesize that there were either two pulsars or some new type of pulse emitting object in the vicinity of the Crab Nebula. Very shortly thereafter Comella *et al.* (1969) at the Arecibo Observatory showed that indeed the sources of the radiations were pulsars, one of which is now the famous Crab Nebula pulsar NP 0532 with its remarkably short period of about 33 msec. The other pulsar, NP 0527, is located nearly a degree and a half from the Crab Nebula and has the longest period of any known pulsar, 3.75 sec. The Arecibo data enabled Cocke *et al.* (1969) to detect optical pulsed radiation from the well known south-preceding member of the central pair of stars in the nebula. Thus the location and period of the pulsar were well established.

Shortly after the discovery of the radio pulsar, Richards and Comella (1969) at Arecibo detected in it the first known change in pulsar period. In this case, as in all subsequent cases, it was found that the period was increasing with time. In the case of the Crab Nebula pulsar, this increase in period is about 36 nsec per day. Though very small, this rate in one solar mass object of some ten kilometers radius rotating 30 times a second represents a rotational kinetic energy loss of some 10^{38} erg per sec. Thus, the pulsar is releasing energy at a rate equivalent to 10^5 solar luminosities. This

power output is remarkable not only for its magnitude but also because it is equal
to the total radiated power of the Crab Nebula at all wavelengths. The origin of this
power has long been a major puzzle; at last the pulsar seemed to offer a ready explanation of the source of energy. As was noted soon thereafter by Gold (1969), the fact
that so much of the rotational energy of the pulsar was released in the form of
relativistic particles suggests that pulsars may indeed be the dominant source of
cosmic rays.

   Radio studies of the pulsar pulse shape have shown it to be unique and remarkable.
In Figure 1 we see the mean pulse shape as observed at 430 MHz by Rankin *et al.*
(1970). Here the pulse shape is the mean of many thousands of pulses. As can be
seen, the pulse contains three strong components. There is an intense but brief 'main
pulse' component lasting some 250 μsec and a very similar although somewhat
less intense 'interpulse' component. Preceding both of these is a broad and weak
component known as the 'precursor'. The time interval between the well determined centroids of the main and interpulse components is 13.37±0.03 msec. This is
the same as the time interval between the two components of the optical pulse to
within 32 μsec as observed by Horowitz, Papaliolios, and Carleton. The pulse timing

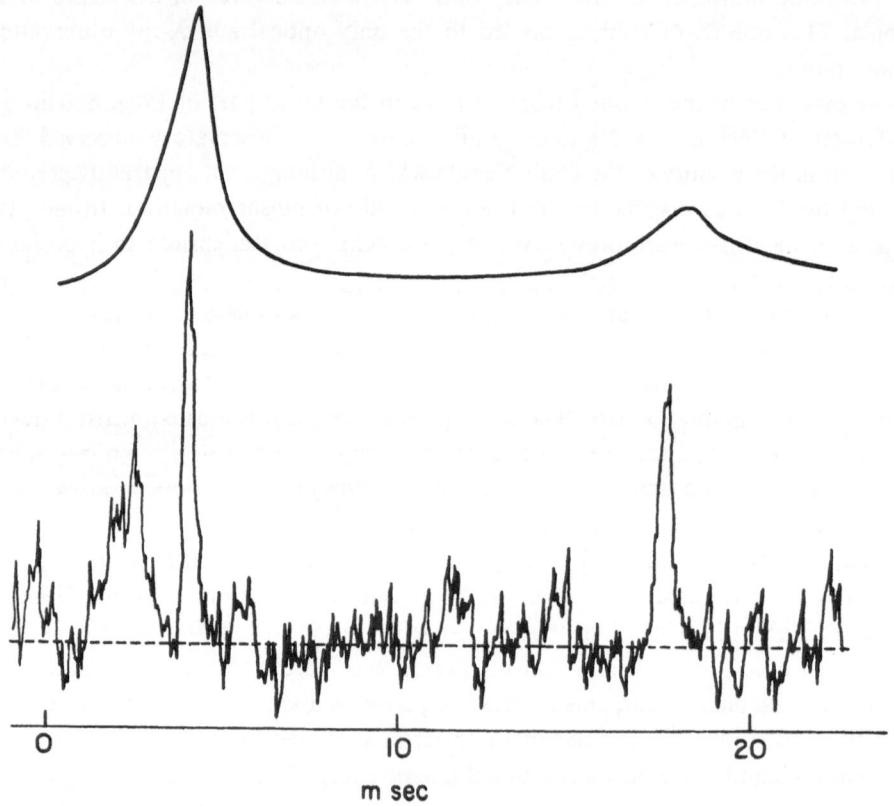

Fig. 1.   Below, observed average pulse shape of NP 0532 at 430 MHz.
Above, average optical pulse shape.

measurements have shown that the radio and optical pulses occur simultaneously after correction for dispersions to something better than 200 μsec. This implies that the radio and optical pulses are generated in regions that are separated by no more than 60 km. Since this distance is very much less than the dimensions of the magnetosphere of the object, some 1000 km, it implies strongly that the optical and radio pulses are indeed produced in the same material, although not necessarily by the same physical process.

Studies of the strengths of the various components of the pulse shape over time intervals of days have shown a small but significant relative variation of the pulse component intensities as shown in Figure 2, which is from the results of Campbell et al. (1970).

If the pulse shape is observed at many radio frequencies, it is found that major changes occur. In this respect, NP 0532 differs markedly from other pulsars. Figure 3a shows the pulse shapes measured at 74, 111, 196, 318 and 430 MHz by Rankin et al. (1970). The very well defined and brief components seen at the highest radio frequencies become broadened as one goes to lower frequencies. The relative strength of the

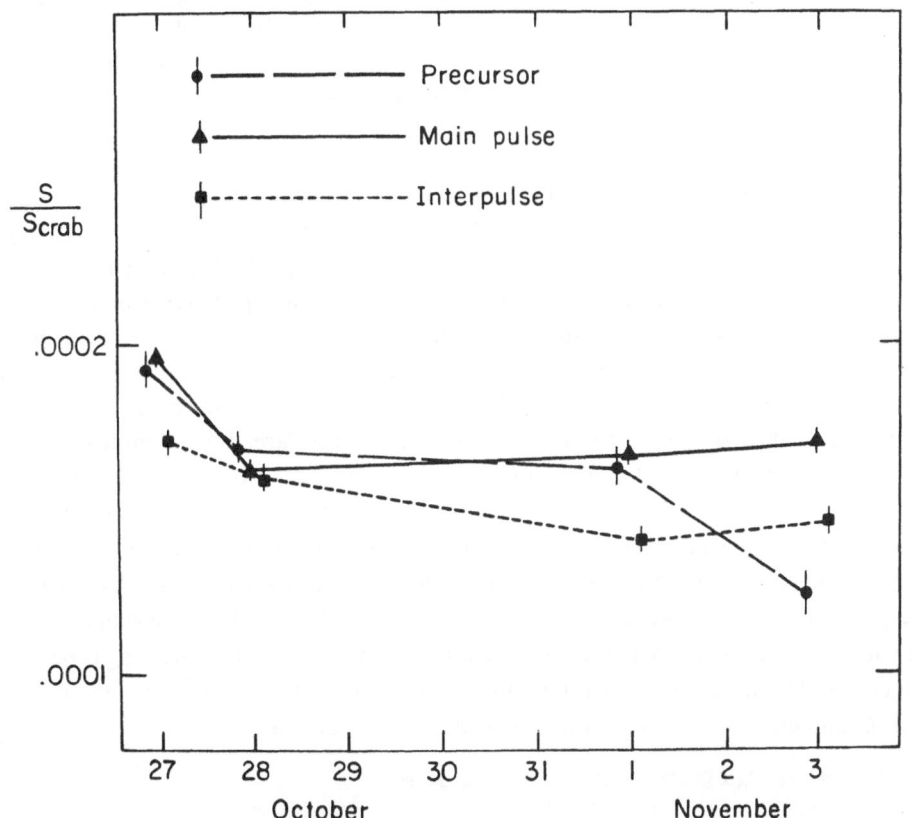

Fig. 2. Mean flux densities of three components of Crab pulsar pulse on several days in 1969. Observations were made at 430 MHz, and flux density is given in terms of the flux density of the Crab Nebula. Errors are probable errors.

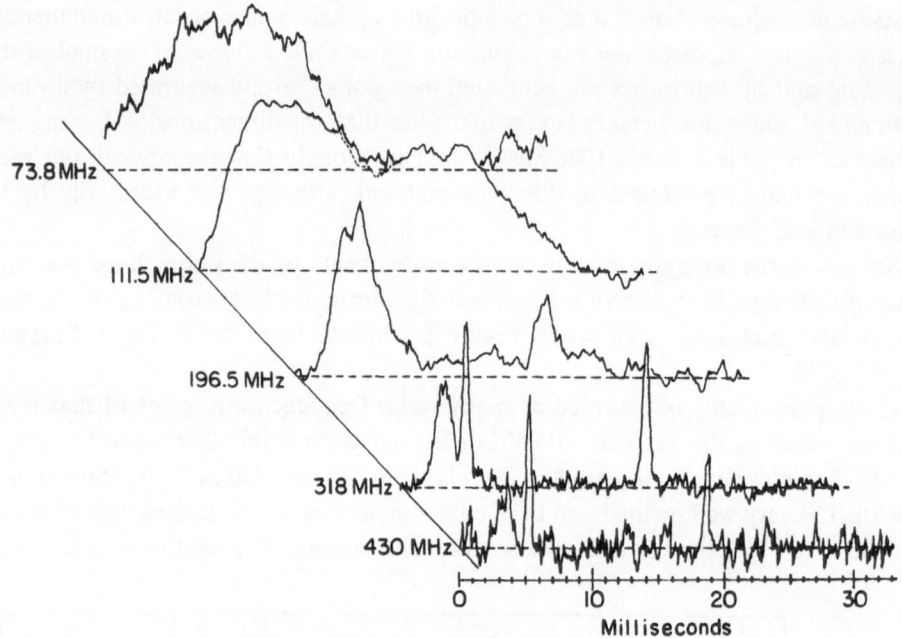

73.8 MHz

111.5 MHz

196.5 MHz

318 MHz

430 MHz

Milliseconds

Fig. 3.   Observed average pulse shape of Crab pulsar at frequencies shown. The phases of the various
pulse shapes have been shifted by arbitrary amounts.

precursor increases compared to the main pulse and interpulse at the lower frequencies.

A recent observation by Heiles and Rankin at Arecibo shows that the precursor is extremely weak or perhaps unobservable at 611 MHz. The fact that the main and interpulse components can be defined well at 430, 318 and 196 MHz permits a very accurate measurement of the dispersion measure. On 31 May 1969, the dispersion measure was 56.78 ± 0.02 parsec-electrons per cubic centimeter. As will be reported by Rankin, detectable and interesting changes in the dispersion measure occur.

The idea that the observed changes in pulse shape are due to multipath radio propagation in the interstellar medium has been proposed by many people, among them, Cronyn, Drake, Lang and Comella. In an attempt to study this quantitatively, Rankin *et al.* (1970) tried to fit various models of radio frequency dependent broadening functions to a hypothetical pulse shape to determine whether the observed pulse shapes could be reproduced. In doing this they assumed that the pulse shape observed at 430 MHz was very nearly the intrinsic pulse shape at the pulsar. Among broadening functions they have tried are the Gaussian function, a truncated exponential, and a half Gaussian function. The function which works best is

$$xe^{-x}(x \geqslant 0) \quad \text{where} \quad x = \sqrt{2}\,\Delta t/W. \tag{1}$$

This function gives the intensity that would be observed at time increment Δt after $t=0$, if the signal at the source is a delta function at $t=0$. $W = A_0\,(f_0/f)^\delta$ where A and δ are to be determined from the observations. $f=$ radio frequency. Figure 3b

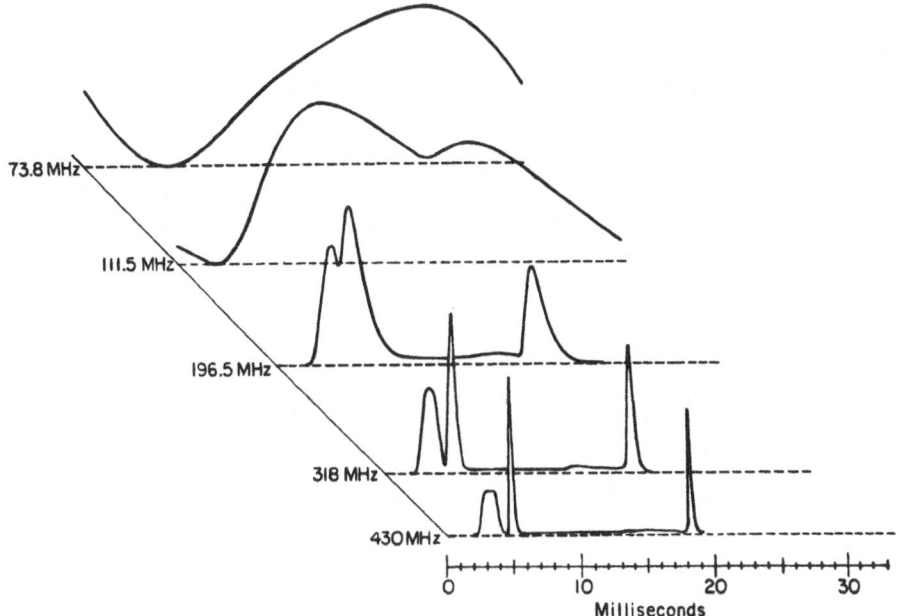

Fig. 3b.   Model pulse shapes predicted from model in which the intrinsic pulse shape is defined by the average pulse shape at 430 MHz, and this pulse shape is broadened by a function xe^{-x} as a result of multipath propagation effects in the interstellar medium. The broadening caused by dispersion and radiometer effects has also been included, so that these model pulse shapes are directly comparable to the pulse shapes of Figure 3.

gives the model pulse shapes found in this way when $\delta=4$ and $A_0=0.85$ at $f_0=196$ MHz. A value $\delta=4$ gives the best fit to the data, although $3.5<\delta<4.5$ is about as good.

It is interesting that the theory of Salpeter (1969) argues that the value of δ should equal 4 in a wide range of circumstances. Thus the value observed seems to support this contention, and in the process gives weight to the hypothesis that the observed pulse broadening is due to multipath propagation in the interstellar medium. Very recently, Lang has observed the same effect, with the same mathematical dependence, in the pulse shapes of JP 1933. Additional confirmation comes from the work of Staelin and Sutton (1970) who have observed single strong pulses of NP 0532 at frequencies near 100 MHz, and have found these broadened to a shape consistent with (1). There seems little doubt that gross pulse broadening at the lower radio frequencies is caused by multipath effects in the interstellar medium. The spectrum of the pulsar will be strongly affected at the lower radio frequencies as the pulse smearing leads to a conversion of pulsed radiation into continuous radiation. At a sufficiently low frequency, the pulsar will appear as a discrete continuous source and will no longer be recognized by the characteristic pulsed radiation of a pulsar.

The spectrum of the pulsar, as observed by Comella (reported in Rankin *et al.*) is shown in Figure 4. This is an unusual spectrum with a very steep slope on the high frequency side of -2.9 ± 0.4. Again on the low frequency side a very steep slope occurs such that the radiation falls off so rapidly that the spectral index can not be well defined. Peak intensity occurs at a frequency of about 100 MHz. Note that this spectrum is that of the pulsed radiation alone.

Figure 4 shows the pulsed flux density predicted by the models of pulse broadening constructed by Rankin *et al.* In particular, the models from relation (1) are shown for δ in the range 3.5 to 4.5, and it is seen that the spectrum follows the model predictions very precisely, with the assumption that the intrinsic spectrum at the pulsar

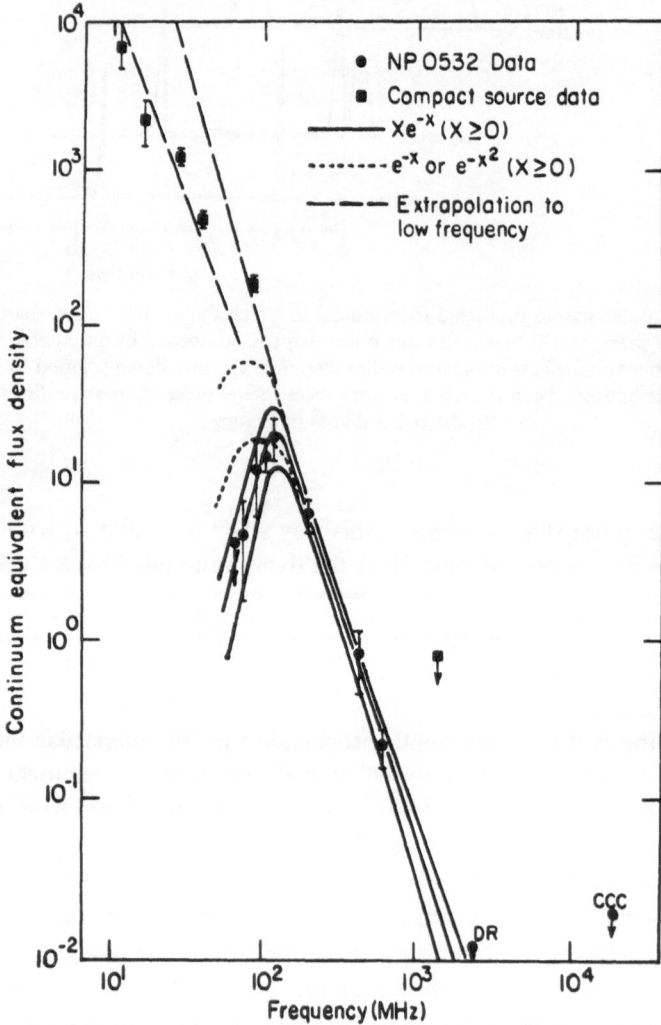

Fig. 4.   The radio spectra of NP 0532 and the compact radio source in the Crab Nebula. The points for the pulsar are from Rankin *et al.* (1970) except for the point DR which is by Downs and Reichley, and the point CCC by Counselman. Compact source values are by Bridle. The solid curves are the spectrum predicted by the model in which interstellar broadening causes a broadening proportional to xe^{-x}. The parallel solid curves give the error envelope for the fit.

at low frequencies is a continuation of the high frequency spectrum. It is very striking that the flux densities of the compact continuous source of radiation at the center of the Crab Nebula fall almost exactly on the extrapolation of the high frequency portion of the pulsar spectrum. This makes a very strong case that the compact source is the pulsar, specifically its radiation is the pulsed radiation converted to continuous radiation by the interstellar medium.

The picture of the situation leads to the additional prediction that the apparent angular size of the source will increase at the lowest frequencies where we receive radiation from ever more deviant rays at longer wavelengths. The theory gives the relation $\theta_{\text{scat}}^2 = 8c\,\Delta t/R$ between Δt and the apparent size of the source θ_{scat}, where R is the distance to the pulsar. At 81 MHz this relation predicts a value of $\theta_{\text{scat}} = = 1.1 \times 10^{-6}$ radian. Bell and Hewish (1967) report a measurement of the apparent size of the compact source of $1 \pm 0.5 \times 10^{-6}$ rad. Similarly at 26 MHz the relation predicts an apparent size of the source of about one second of arc and indeed Cronyn has observed an apparent size of about this value at that frequency.

There are now a number of arguments leading to the conclusion that the compact source radiation is nothing more than the pulsar radiation smeared in time so as to be continuous. These arguments are: (1) A very close proximity of the measured positions of the compact radio source and the pulsar. The difference in position is less than 15″. (2) The unusually high brightness temperature computed for the compact source. The compact source has presented a serious problem in this regard and has led to suggestions that unusual physics is involved. The hypothesis that the source is the pulsar solves the problem since the physics of the pulsar can easily produce the high brightness temperatures. (3) Compatibility of the spectra of the pulsar and the compact source as discussed. (4) The consistency between the measured appearant size of the compact source as a function of radio frequency as compared to the theoretical predictions of. the size to be expected from propagation effects.

A very striking characteristic of the Crab Nebula pulsar is its sporadic emission of individual very intense strong pulses at random intervals of several minutes. Approximately one in 10000 is one of these strong pulses. Individual pulses with a peak flux density of 20000 flux units have been observed by Heiles at 430 MHz; this is a flux density 20 times that of the Crab Nebula. For brief intervals, the Crab pulsar is the brightest cosmic radio source in the sky at this frequency. Figure 5 shows an example of one of these strong pulses as observed by Heiles, Rankin and Campbell. They observe strong pulses by recording digitally a 50 μsec average of the radiation with such a sample being taken every 50 μsec. The strong pulses are not simply the tail of the statistical distribution of numbers of pulses versus pulse intensity. They form an independent component in the statistical distribution.

In every such strong pulse observed it is the main pulse which is very intense, with there being no evidence of great intensity in either the precursor or the interpulse. This shows very clearly that the pulse enhancement is not caused by any effect in the interstellar medium, since this would affect all components of the pulse equally. The evidence is clear that these individual strong main pulses, typically a hundred times

F.D.DRAKE

stronger than the average pulse intensity, are created by a special physical mechanism in the pulsar itself.

The duration of the strong pulses is very short, being typically less than 100 μsec. In every strong pulse so far observed this has been true, except for a few instances

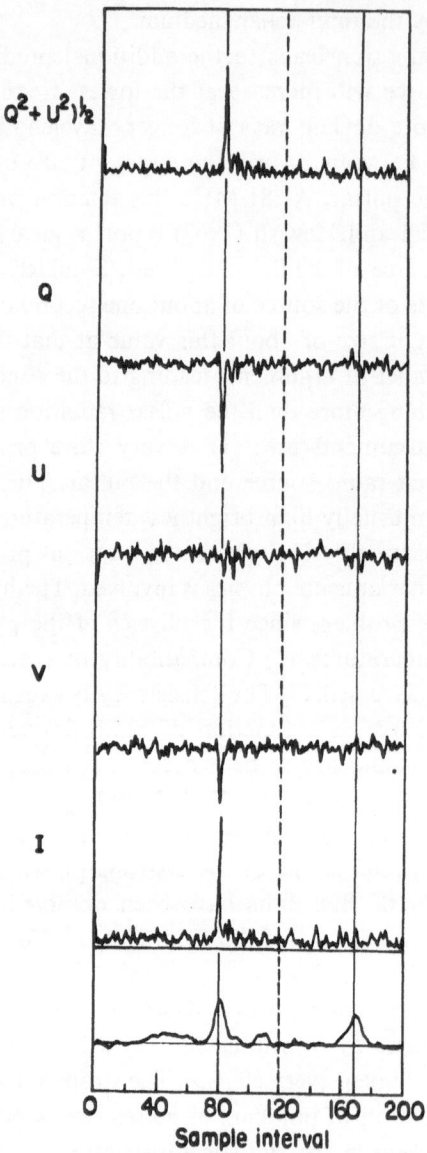

Fig. 5.   Observation at 430 MHz of four Stokes parameters of an individual strong pulse from NP 0532. The total intensity averaged over all pulses recorded on 3 November, 1969, is shown at the bottom. The sampling time is 50 μsec. Note that recording system ceased taking data for a time interval of 9 msec at sample interval 120. The solid vertical lines at sample interval 80 and 169 show the approximate centers of the main and interpulse, respectively. I is the total intensity, V is the circularly polarized Stokes parameter, and Q and U are the linearly polarized parameters. Note presence of strong linear and circular components.

where a double strong pulse consisting of two such short components has been seen. The individual strong pulses show a very large degree of polarization but with no apparent systematic correlation with the pulse phase at which the strong pulse occurs. The pulse of Figure 5 is strongly elliptically polarized. The pulse polarization may be strongly linear, but at times shows strong circular polarization. There seems to be no general rule describing the polarization of the strong pulses.

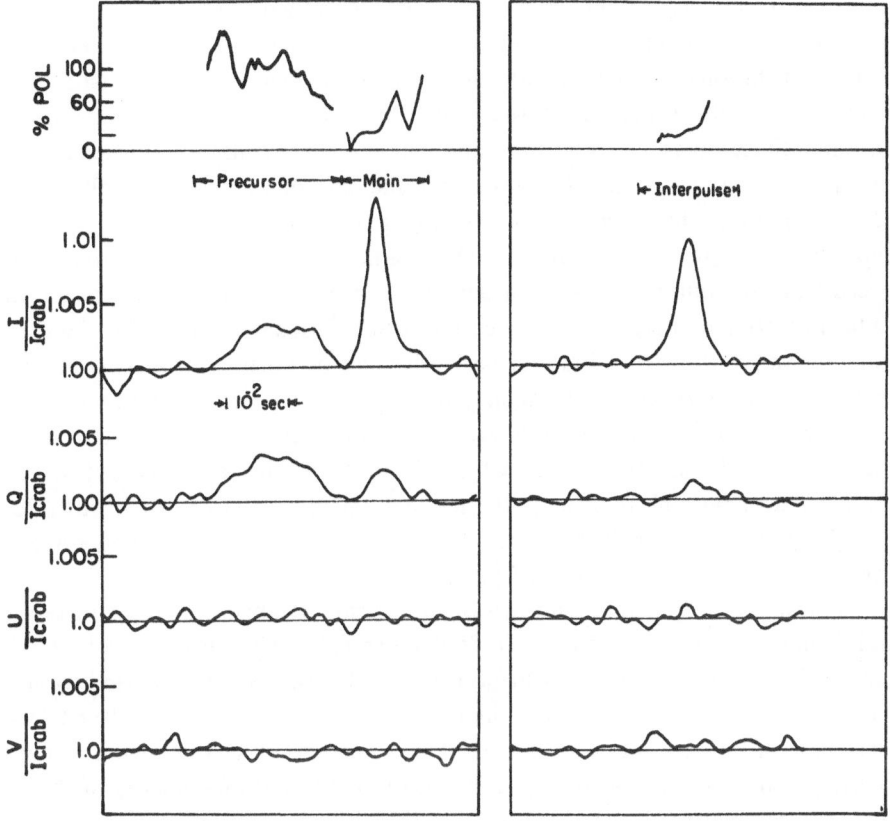

Fig. 6.   The mean polarization of the Crab pulsar. The Stokes parameters *I, Q, U* and *V* are again given in terms of the intensity of the Crab Nebula. At the top, the percentage linear polarization is shown.

On the other hand, measurements of the average polarization seen when the mean values of Stokes parameters for tens of thousands of pulses are combined give a large systematic polarization within a pulsar pulse. Figure 6 shows this result from Arecibo by Campbell *et al.* (1970). As can be seen in the plots of the four Stokes parameters, there is no evidence of circular polarization whatsoever. Only one of the linearly polarized Stokes parameters shows a finite intensity. It is to be noted that the plane of linear polarization of the receiving system has been intentionally rotated so that the polarization of the Stokes parameter is parallel to the evidently persistent plane of polarization of pulsar signals. The results seem to show that there is no rotation of the linear polarization in the course of the pulse, contrary to the effects seen in other pulsars. The polarization of the precursor is one hundred per cent linear throughout. The linear polarization of the main pulse is about 20% and of the interpulse 13%.

Accurate measurements of the time of arrival of pulses have been carried out for over a year at Arecibo primarily by Richards, Roberts and Rankin. Since these will be described in detail later, they will only be mentioned briefly here. As mentioned earlier, they have shown strong evidence that the pulsar is a rotating neutron star and that its rotational energy is being converted into the energy of relativistic particles and electromagnetic radiation in the radio, optical and X-ray regions. The short duration of the main and interpulse component at 430 MHz have permitted extremely good timing accuracies, typically 8 μsec from the data of a single day's observations. With this accuracy it has been possible to measure both the first and second derivatives in the rotation frequency of the pulsar. The first derivative leads directly to the energy released from the pulsar; the second derivative gives some evidence as to the mode by which rotational energy is converted into other forms of energy. The data on the second derivative show that the first derivative can be fairly accurately described as proportional to the rotational frequency raised to a power which is observationally in the range 2.2 to 2.4. This power is very close to the value of three predicted by the electromagnetic dipole braking theory for pulsars. As has been pointed out by Pacini, if the plasma of the pulsar distorts the magnetic field so as to draw it out into the equatorial plane of the pulsar, we would expect an exponent less than three. The observations are consistent with this picture.

In addition to the general spindown of the pulsar, an abrupt increase in the rotational frequency was observed on 28 September 1969. This was also observed, of course, by Boynton et al. (1969) at Princeton. The decrease in pulsar period was about 100 psec, much smaller than in the similar event observed in the Vela pulsar 0833–45. Nevertheless, if this discontinuity is interpreted to be a result of a change in a moment of inertia of the neutron star it implies a release of gravitational energy of the order of 10^{41} erg, equivalent to the luminosity of the sun over a year.

In addition to the abrupt change in period, instabilities in the pulsar period with time scales of the order of days to months have been observed. Prior to the 1969 discontinuity these deviations from a secular spindown were very nearly sinusoidal in form. After the 1969 event the quasisinusoidal variation continued for several months and then was transformed into an irregular variation of the pulsar with a time scale of the order of months. As will be seen in later papers in this symposium the interpretation of the irregularities in the pulsar timing is extremely complex and challenging.

References

Bell, S. J. and Hewish, A.: 1967, *Nature* **213**, 1214.
Boynton, P. E., Groth, E. J., III, Partridge, R. B., and Wilkinson, D. T.: 1969, *IAU Circular* 2179.
Campbell, D. B., Heiles, C. E., and Rankin, J. M.: 1970, *Nature* **225**, 527.

Cocke, W. J., Disney, M. J., and Taylor, D. J.: 1969, *Nature* **221**, 525.
Comella, J. M., Craft, H. D., Jr., Lovelace, R. V. E., Sutton, J. M., and Tyler, G. L.: 1969, *Nature* **221**, 453.
Gold, T.: 1969, *Nature* **221**, 25.
Heiles, C. E., Rankin, J. M., and Campbell, D. B.: 1970, *Nature* **226**, 529.
Rankin, J. M., Comella, J. M., Craft, H. D., Jr., Richards, D. W., Campbell, D. B., Counselman, C. C. III: 1970, *Astrophys. J.* **162**, 707.
Richards, D. W. and Comella, J. M.: 1969, *Nature* **222**, 551.
Salpeter, E. E.: 1969, *Nature* **221**, 31.
Staelin, D. H. and Reifenstein, E. C., III: 1968, *Science* **162**, 1481.
Staelin, D. H. and Sutton, J. M.: 1970, *Nature* **226**, 69.

Discussion

R. C. Jennison: Are any other compact low frequency sources known which nearly coincide in position with any other pulsars?

F. D. Drake: No. Craft observed the region of all the Arecibo observable pulsars, and found no associated continuous sources with an upper limit of about 2% of the mean pulsar flux density.

J. P. Ostriker: Is it difficult to detect distant pulsars with very short periods and with large dispersion measures?

F. D. Drake: Yes, the broadening prevents the detection of short period pulsars, say with periods less than about 100 milliseconds, when the dispersion measure is more than about 100.

J. Kristian: Do you know the absolute angle for the linear polarization of the Crab pulsar?

F. D. Drake: No, the measurement wasn't made at Arecibo.

J. E. Baldwin: Given a pulse shape at a high frequency where smoothing is small and also at a low frequency where smoothing is present, it is evidently possible to derive uniquely the smoothing function for that frequency Has this been done rather than attempting to fit algebraic functions to the data?

F. D. Drake: Counselman of MIT has attempted to fit a function at a single frequency and found that the xe^{-x} broadening function gives the best fit. Note however that this broadening is non-linear in frequency and so observations at two frequencies are not enough. The models I have discussed attempt to include this frequency dependence.

W. J. Cocke: Bridle has measured the diameter of the compact source to be 0.2 arc sec, at around 20 MHz. Have you any comment about that? (A. H. Bridle, *Nature* **225**, (1970) 1035.)

F. D. Drake: I did not know of the size limit of 0″.2 at low radio frequencies. This value disagrees with the observations of Cronyn.

2.2 X-RAY OBSERVATIONS OF NP 0532

SAUL RAPPAPORT

*Center for Space Research, Massachusetts Institute of Technology,
Cambridge, Mass., U.S.A.*

The Crab Nebula pulsar, NP 0532, is the only one of the 56 known pulsars which has been observed to pulse in the optical spectrum. Even more spectacular are the vast quantities of pulsed X-rays which it has been found to emit. Soft X-rays in the 1–10 keV region have been observed during four sounding rocket flights (Fritz *et al.*, 1969; Bradt *et al.*, 1969; Boldt *et al.*, 1969; Ducros *et al.* 1970) and harder X-rays, 25–200 keV, have been detected during two balloon flights (Fishman *et al.*, 1969a, b; Floyd *et al.*, 1969). From the results of these six experiments we summarize the properties of the X-ray pulsations from NP 0532.

The X-ray and optical pulse profiles for NP 0532 have striking similarities while a detailed comparison reveals several distinct differences. In X-rays the characteristic main pulse and interpulse are present, but unlike the optical case (Lynds *et al.*, 1969; Warner *et al.*, 1969; Papaliolios *et al.*, 1970) the energy in the broader and smaller amplitude interpulse is about equal to that in the main pulse. The separation of the two pulses is 13.2 ± 0.3 msec (Bradt *et al.*, 1969; Fritz *et al.*, 1970; Rappaport *et al.*, 1970). The full widths at half maximum of the main pulse and interpulse are 1.0 ± 0.2 msec (Fritz *et al.*, 1970; Rappaport *et al.*, 1970) and 2.8 ± 0.3 msec (Fritz *et al.*, 1970; Rappaport *et al.*, 1970) respectively. The interpulse has a slow rise and rapid decline while the opposite is true of the optical interpulse (Lynds *et al.*, 1969; Warner *et al.*, 1969; Papaliolios *et al.*, 1970). In the 13 msec interval between the two pulses the X-ray flux does not go to the dc background level but remains at about 10% of the peak rate in the main pulse (Bradt *et al.*, 1969; Rappaport *et al.*, 1970).

The MIT experiment of April 27, 1969 (Bradt *et al.*, 1969) featured nearly simultaneous optical and X-ray observations of the Crab pulsar. The time of the main optical pulse as determined at Mount Palomar and McDonald observatories differs from the time of the main X-ray pulse by only 0.4 ± 0.5 msec. Recently the Princeton group (Groth and Wilkinson, 1970) calculated from their data the arrival time of the main optical pulse at White Sands Missile Range nearest the time of the M.I.T. rocket launch. The time they obtain differs from the time of the main X-ray pulse by only 0.22 ± 0.3 msec. It is now clear that the radio (corrected to infinite frequency) (Rankin *et al.*, 1970), optical and X-ray pulses are all coincident to within ~ 0.3 msec.

The time averaged flux of pulsed soft X-rays (1–10 keV) from NP 0532 is about 8% of the soft X-ray flux received from the Crab Nebula (Bradt *et al.*, 1969). The X-ray flux from the nebula in this energy range is 3×10^{-8} erg cm^{-2} sec^{-1} (Kellogg, 1971) and the corresponding X-ray luminosity is 1.5×10^{37} erg sec^{-1}. The time averaged flux (1–10 keV) from the pulsar is 2.5×10^{-9} erg cm^{-2} sec^{-1} (Bradt *et al.*, 1969) and its X-ray luminosity depends on the geometry of the pulsar beam. When the

Davies and Smith (eds.), The Crab Nebula, 84–86. All Rights Reserved.
Copyright © 1971 by the IAU.

X-ray flux from the pulsar is maximum it is about half as bright as the non-pulsed X-radiation from the Crab (Bradt *et al.*, 1969; Rappaport *et al.*, 1970). The pulsar puts out ~ 100 times as much energy in X-rays as in the optical (Oke, 1969) and $\sim 10^5$ times as much as in the radio region ($> 10^8$ Hz) (Rankin *et al.*, 1970; Comella *et al.*, 1969).

The X-ray energy spectrum of the Crab Nebula in the region 1–560 keV can be well represented by a power law with spectral index $\alpha = 1.1$ (Kellogg, 1971). The pulsed component in the range 2–200 keV decreases with decreasing energy from about 15% of the flux density of the Crab to $\sim 5\%$ (Fritz *et al.*, 1969; Bradt *et al.*, 1969; Ducros *et al.*, 1970; Fishman *et al.*, 1969b; Floyd *et al.*, 1969). The MIT measurement in the interval 1–1.5 keV (Rappaport *et al.*, 1970) indicates that the pulsed component is only $3.5 \pm 1.2\%$ of the nebular radiation at these energies. This apparent decrease in the pulsed radiation is not due to interstellar photoelectric absorption since the same energy X-rays from the nebula are not significantly attenuated (Rappaport *et al.*, 1969). Scattering by dust grains in the interstellar medium has been suggested by Slysh (1969) but an earlier calculation (Overbeck, 1965) and more recent calculations (Naranan and Shah, 1969; Ryter, 1970; Bowyer *et al.*, 1970) show the optical depth for X-ray scattering at 1 keV for the Crab Nebula is small (~ 0.2). Absorption of the very soft X-rays could be taking place in the immediate vicinity of the pulsar or the intrinsic spectrum could be changing at 1 keV.

References

Boldt, E. A., Desai, U. D., Holt, S. S., Serlemitsos, P. J., and Silverberg, R. F.: 1969, *Nature* **223**, 280.

Bowyer, C. S., Mack, J., and Lampton, M.: 1970, *Nature* **225**, 1125.

Bradt, H., Rappaport, S., Mayer, W., Nather, R. E., Warner, B., MacFarlane, M., and Kristian, J.: 1969, *Nature* **222**, 728.

Comella, J. M., Craft, H. D., Lovelace, R. V. E., and Sutton, J. M.: 1969, *Nature* **221**, 453.

Ducros, G., Ducros, R., Rocchia, R., and Tarrius, A.: 1970, *Nature* **227**, 152.

Fishman, G. J., Harnden, F. R., Jr., and Haymes, R. C.: 1969a, *Astrophys. J. Letters* **156**, L107.

Fishman, G. J., Harnden, F. R., Jr., Johnson, W. N., III, and Haymes, R. C.: 1969b, *Astrophys. J.* **158**, L61.

Floyd, F. W., Glass, I. S., and Schnopper, H. W.: 1969, *Nature* **224**, 50.

Fritz, G., Henry, R. C., Meekins, J. F., Chubb, T. A., and Friedman, H.: 1969, *Science* **164**, 709.

Fritz, G., Henry, R. C., Meekins, J. F., Chubb, T. A., and Friedman, H.: 1971, preprint.

Groth, E. J. and Wilkinson, D. T.: 1970 (private communication).

Kellogg, E. M.: 1971, this symposium, Paper 1.5, p. 42.

Lynds, R., Maran, S. P., and Trumbo, D. E.: 1969, *Astrophys. J., Letters* **155**, L121.

Naranan, S. and Shah, G. A.: 1969, *Nature* **225**, 834.

Oke, J. B.: 1969, *Astrophys. J. Letters* **156**, L49.

Overbeck, J.: 1965, *Astrophys. J.* **141**, 864.

Papaliolios, C., Carleton, N. P., and Horowitz, P.: 1970, *Nature* **228**, 445.

Rankin, J. M., Comella, J. M., Craft, H. D., Jr., Richards, D. W., Campbell, D. B., and Counselman, C. C., III: 1970, *Astrophys. J.* **162**, 707.

Rappaport, S., Bradt, H. V., Mayer, W.: 1969, *Astrophys. J. Letters* **157**, L21.

Rappaport, S., Bradt, H., and Mayer, W.: 1971, *Nature* **229**, 40.

Ryter, Ch.: 1970, *Nature* **226**, 1040.

Slysh, V. I.: 1969, *Nature* **224**, 159.

Warner, B., Nather, R. E., and MacFarlane, M.: 1969, *Nature* **222**, 233.

Discussion

Sir Bernard Lovell: Is the ratio of pulsed to non-pulsed X-radiation a function of energy as indicated by the difference in the balloon and the rocket results or are there uncertainties in the two types of measurement?

S. Rappaport: It is a definite effect.

N. Visvanathan: In X-rays, the 8% pulsed intensity is compared to the total intensity from all over the Crab, while in optical the pulsed intensity is 18% compared to the 1 sec arc Nebula around the pulsar.

J. P. Ostriker: Can you quote limits for the periodic component of the X-rays observed from the two other supernova remnants emitting X-rays?

E. Kellogg: Upper limits on pulsed X-ray emission, 1–10 keV, from Cas A and Tycho, assuming a single pulse occupying one tenth of a period, with periods between 8 and 35 msec, with 99% confidence are

           Cas A   15%
           Tycho   19%

Results obtained by Lorenstein, Kellogg, and Gursky, are published in *Astrophys. J.* **160**, (1970) 199.

J. Kristian: By how much is your comparison with the optical FWHM affected by the fact that the optical peak is still unresolved with time resolution as short as 20 μsec?

P. Horowitz: Even though the turn-around near the optical peak hasn't been resolved, the FWHM estimate is not much affected since the rise to maximum is approximately linear from both sides and therefore the height of the peak is probably not in error by more than the rise during one timing channel, i.e. less than 5% of the total height.

D. M. Jennings: Ducros *et al.* have recently reported two further inter-pulses, one following both the main pulse and the well-known inter-pulse. Have you any information on these?

S. Rappaport: No.

F. G. Smith: There is another inter-pulse component observed at radio frequencies, which we will hear about later in a report by Mr. Schönhardt.

2.3 OPTICAL OBSERVATIONS OF THE CRAB PULSAR, AND SEARCHES FOR OTHER OPTICAL PULSARS

JEROME KRISTIAN

*Hale Observatories, Carnegie Institution of Washington and
California Institute of Technology*

Abstract. The optical properties of the Crab nebula pulsar are reviewed. The Crab nebula pulsar has a high degree of constancy at optical wavelengths. No time variations over short or long periods have been detected; the light curve is nearly the same in all colors. The intensity and color of the pulsar are $V = 16.5$, $B - V = +0.5$, $U - B = -0.45$ and $V - R = -0.75$. There is no precursor as found at radio wavelengths and the main pulse contains 65 per cent of the total energy. No lines have been detected.

Searches for other pulsars have been unsuccessful.

The Crab pulsar is still (July 1970) the only known optical pulsar, and its properties have been studied in some detail since it was first seen early in 1969 (Cocke, Disney and Taylor, 1969). I would like to briefly review some of the available data, as well as some results of searches for other optical pulsars. Details of the observations can be found in the papers listed at the end of the present one and in the references which they contain.

All of the visible emission from the Crab pulsar is in its pulsed radiation, as determined photographically (Miller and Wampler, 1969; Chiu *et al.*, 1970a) and photoelectrically (Kristian, 1970a) with an accuracy of a few percent. The light curve is very nearly the same in all colors (i.e., the color of the emitted radiation does not change through the pulses), although recent observations by Visvanathan and Kristian suggest that there may be a very small change through the main pulse. The shape of the pulses has remained remarkably constant. The groups doing optical timing measurements exploit this fact by fitting their observed pulses to a standard shape, which enables them to obtain arrival times with a precision several orders of magnitude better than the time scale of the pulse widths themselves. There is no indication of the splitting of the main pulse into two components which is seen at radio frequencies.

The intensity of visible radiation is secularly constant on all time scales investigated, with the following accuracy at present: successive pulses, a factor of two; 10 minutes to several hours, 5%; night-to-night over a year, 10%; year-to-year since 1920, 35%. There is no evidence for the substantial pulse-to-pulse variations or occasional extremely large pulses seen by the radio observers. As was pointed out very early by Gold, however, the very high brightness temperature of the radio emission requires that it be highly coherent, and changes in the degree of coherence could cause intensity fluctuations. If the radiation at optical and higher frequencies were incoherent, the difference might be qualitatively understood. Several X-ray measurements made at different times show a constant intensity, but the accuracy is quite low.

The broad-band intensity and colors of the optical emission, averaged over times of the order of minutes, are $V = 16.5$, $B - V = +0.5$, $U - B = -0.45$ and $V - R = -0.75$, with the main pulse containing 65% of the total energy in each pass band (Kristian et al., 1970). The detailed energy distribution for the main pulse and subpulse are the same. After correction for interstellar absorption, the flux density shows a broad maximum in the visible, with a decrease in both the ultraviolet and the red, the latter continuing into the infrared to at least 2μ (Oke, 1970; Neugebauer et al., 1970). The ultraviolet end and the observed X-ray intensity can be plausibly fit to the same curve, but this is not true for the red end and the radio observations. The radio spectrum has a slope steeper than 2 at 12 cm, while the slope in the infrared has the opposite sign. The general effect is of a two-humped energy distribution, with one hump at radio wavelengths and the other in the visible-to-X-ray region (see, for example, Kristian, 1970a). Good image-tube spectra of the pulsar show a smooth blue continuum, with no lines (Lynds, 1969; Van den Bergh, private communication).

The visible emission shows strong linear polarization, which changes through the pulses, but whose behavior is secularly constant (Kristian et al., 1970). The polarization decreases from about 20% in the leading edges of the pulses to near zero, then increases in a roughly symmetric way to about 10% in the trailing edges. This change in polarization is accompanied by a monotonic sweep of the plane of polarization through 150° during 60° of the pulsar's rotation. The polarization through the main pulse and subpulse is similar and is shifted in both cases with respect to the peaks of the pulses, by about 200 μsec for the main pulse and 1.5 msec for the subpulse. Circular polarization is absent, with an accuracy of a few percent.

Optical searches for other pulsars continue, both for visible counterparts of known radio pulsars and for radio quiet pulsars associated with a variety of galactic and extragalactic objects. The most extensive searches of the latter kind have been made at Harvard, and are reported by Horowitz at this symposium.

A radio pulsar of particular interest has been 0833–45, the Vela pulsar, because it is the fastest pulsar after the Crab, and the only other pulsar known to be associated with a supernova remnant, although searches have been made for pulsars near known supernova remnants and for supernova remnants near known pulsars. Negative results for 0833 have been reported by a number of groups, the faintest limit at present being $V \gtrsim 24$ (Chiu et al., 1970b; Kristian, 1970b). After corrections for distance and interstellar absorption, this implies an absolute visible luminosity for the Vela pulsar at least 4000 times fainter than that of the Crab.

Limits for visible counterparts of 15 radio pulsars searched for at Palomar (Kristian, 1970c) range from 3 to 9 magnitudes fainter than the Crab. Rough limits for the absolute luminosities of these objects, based upon their dispersion measures, range from 10^{-1} to 10^{-6} that of the Crab. The results appear to rule out the possibility that the visible pulsed emission of all pulsars scales as (frequency)[4] and is normalizable to the Crab. If all pulsars were normalizable to the Crab, the data indicate that the visible pulsed emission would have to scale by factors as large as (frequency)[8.4] for the best established case (Vela).

There is still no satisfactory theory for the pulsar emission mechanism. Even as basic a question as the location of the emitting region, whether at the surface of the star, the velocity of light cylinder, or some intermediate region, remains unanswered. A complete theory must account for a great deal of observational data, especially from the Crab pulsar, which has now been observed in some detail at accessible wavelengths from the radio to the X-ray regions. A few general remarks, however, can be made at present. There no longer seems to be a serious doubt that the pulsars are rotating objects, probably neutron stars. The optical measurements of the Crab imply that its pulses are due to a polarized emission pattern, locally fixed in the object, which is azimuthally scanned as it rotates. The data also provide fairly direct evidence for the existence in the pulsar of a long-lived non-axisymmetric magnetic field. Any emission mechanism may be expected to involve highly relativistic particles, and the secular constancy of the optical pulses requires that the flux of such particles be constant. It is attractive, on the grounds of economy of hypothesis, to suppose that the particle flux itself may be generated by the rotating magnetic field.

The polarization measurements can be interpreted in terms of a general geometrical model which requires that the pulsar's rotation axis must lie within 30° of the plane of the sky, and be either parallel or perpendicular to the magnetic field of the nebula in the immediate vicinity of the pulsar (Kristian, *et al.*, 1970). This suggests that the magnetic field of the nebula may have been generated by the pulsar, and provides evidence for a direct connection between the central star and the nebular field.

The coincidence in pulse arrival times of radio, optical and X-ray pulses from the Crab, and the similarity of their gross structure, indicate that they must be due to the same basic phenomenon, but the detailed differences in pulse shape, energy distribution, secular intensity and polarization are striking and complex, and will require considerable effort and ingenuity to unravel.

References

Chiu, H. Y., Lynds, R., and Maran, S. P.: 1970a, *Publ. Astron. Soc. Pacific* **82**, 660.
Chiu, H. Y., Lynds, R., and Maran, S. P.: 1970b, *Astrophys. J. Letters* **162**, 99.
Cocke, W. J., Disney, M. J., and Taylor, D. J.: 1969, *Nature* **221**, 525.
Kristian, J.: 1970a, *Publ. Astron. Soc. Pacific* **82**, 456.
Kristian, J.: 1970b, *Astrophys. J. Letters* **162**, 103.
Kristian, J.: 1970c, *Astrophys. J.* **162**, L173.
Kristian, J., Visvanathan, N., Westphal, J. A., and Snellen, G. H.: 1970, *Astrophys. J.* **162**, 475.
Lynds, C. R.: 1969, *Astrophys. J. Letters* **157**, L11.
Miller, J. S. and Wampler, E. J.: 1969, *Nature*, **221**, 1037.
Neugebauer, G., Becklin, E. E., Kristian, J., Leighton, R. B., Snellen, G., and Westphal, J. A.: 1969, *Astrophys. J. Letters* **156**, L115.
Oke, J. B.: 1969, *Astrophys. J. Letters* **156**, L49.

Discussion

W. J. Cocke: I would like to report briefly on some optical polarisation observations of NP 0532 done by myself, M. J. Disney, T. Gehrels, and G. Muncaster. Last year we published results which we believed showed that the polarisation was variable, on a time scale of an hour or so. We have since discovered that the variability was most likely due to a time-dependent malfunction in our photomultiplier electronics.

We repeated our observations in October, 1969, at the Steward Observatory 90-inch telescope, and have obtained polarisation curves that agree very nicely with those of Kristian and Visvanathan. However, we have measured a somewhat different value for the interstellar polarisation, and therefore our final corrected curves do look somewhat different.

2.4 THE RELATION OF THE LOW FREQUENCY SOURCE TO THE CRAB PULSAR

KENNETH R. LANG

Cornell-Sydney University Astronomy Center, Arecibo Observatory,
*Cornell University, Ithaca N.Y., U.S.A.**

Abstract. The view that the compact low frequency source and the pulsar NP 0532 are the same object is substantiated by an examination of the general properties of interstellar scattering. This scattering accounts for the observed angular size of the compact source, the observed pulse broadening of NP 0532, the continuum nature of the compact source, and the observed spectrum of both the pulsar and the compact source.

1. Introduction

Observations using an interferometer at 38 MHz (Hewish and Okoye, 1964) and the lunar occultation technique at 26 MHz (Andrew *et al.*, 1964) first showed that the Crab Nebula contains a compact continuum source. Interplanetary scintillations of the source (Hewish and Okoye, 1965 and Bell and Hewish, 1967) indicated that it was unusually small ($\approx 0.1''$ or 10^{-3} pc) and had a very high brightness temperature ($\approx 10^{14}$ K). Gower (1967) found that the position of the compact source coincides with the south-preceding central star of the nebula, within the positional accuracy of $\pm 12''$ in α and $\pm 1'$ in δ. This central star was thought to be the neutron star remnant of the supernova explosion. The theoretical models which were thought to explain pulsars incorporated an object whose properties – small size, high brightness temperature, and association with a neutron star – are very similar to those of the compact source. Noting this similarity, Woltjer (1968) suggested that the compact source might be found to be a pulsar. When the pulsar NP 0532 was subsequently discovered and found to coincide with the central star of the supernova, there was naturally considerable speculation that the pulsar and the compact source were the same object.

2. Angular Scattering by the Interstellar Medium

One of the major obstacles to the view that the compact source is the pulsar NP 0532 is that the compact source is considerably larger than the pulsar. Angular scattering in the interstellar medium might, however, cause the pulsar to appear to be larger at lower frequencies. As long as the pulsar was much smaller than the rms scattering angle, θ_{scat}, it would appear to have the size θ_{scat}. The presence of such scattering has been inferred from the observed fluctuations in pulsar radiation intensity (Rickett, 1969; Lang, 1969). For example, if the scattering takes place midway between the pulsar and the Earth, the path difference between the direct and scattered rays from a pulsar at a distance, D, will be $D\theta_{scat}^2/4$. Consequently, when pulsar signals are

---

* The Arecibo Observatory is operated by Cornell University under contract to the National Science Foundation and with partial support from the Advanced Research Projects Agency.

observed over bandwidths larger than the decorrelation frequency

$$f_v = 4c/D\theta^2_{\text{scat}}, \tag{1}$$

the scintillation patterns at different observing frequencies will interfere, and the observed intensity fluctuations will be considerably reduced.

Rickett (1969) and Lang (1971) have measured decorrelation frequencies for many nearby pulsars (Figure 1). These data indicate that

$$f_v \approx 5 \times 10^{-9} v^4 / \left(\int n_e \, dl \right)^2, \tag{2}$$

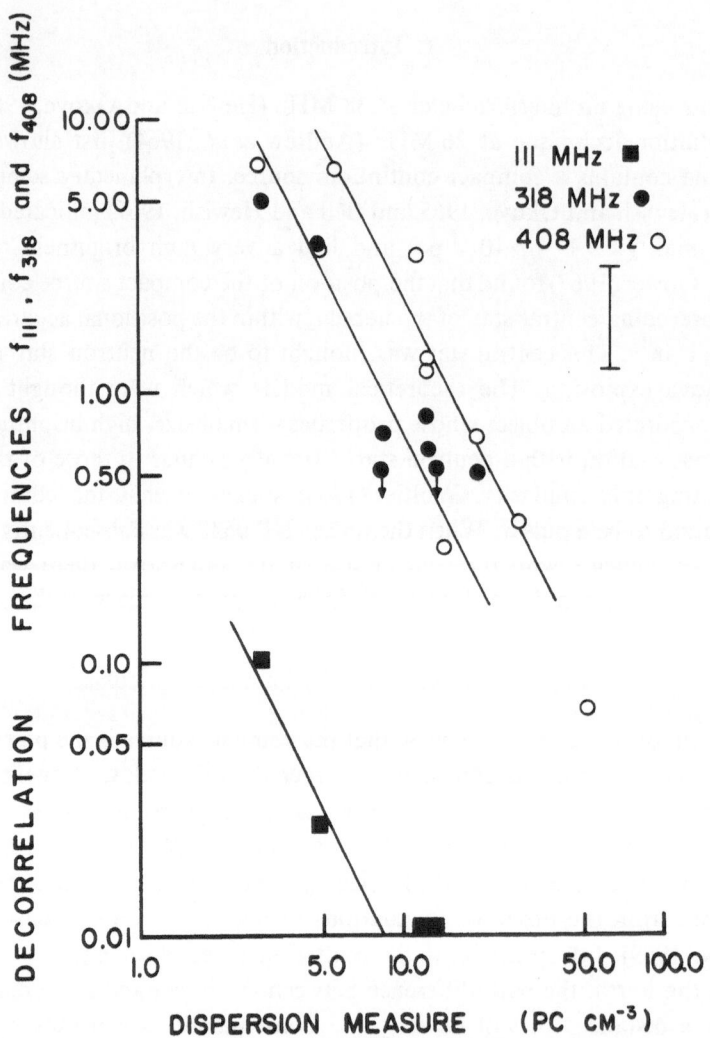

Fig. 1.   Decorrelation frequencies, f_v, at $v = 111$, 318, and 408 MHz. The solid lines denote decorrelation frequencies which are proportional to the fourth power of observing frequency and inversely proportional to the square of the dispersion measure. The 408 MHz data is taken from Rickett (1969) with $f_{408} \approx 0.3\ B_{408}$.

where f_v is in MHz, v is the observing frequency in MHz, and $\int n_e \, dl$ is the dispersion measure in pc cm^{-3}. Using Equations (1) and (2), the data shown in Figure 1 indicate that

$$\theta_{\text{scat}} \approx 10^3 \frac{\int n_e \, dl}{D^{1/2} v^2}, \tag{3}$$

where θ_{scat} is in seconds of arc and D is in pc. Using $\int n_e \, dl \approx 56.81$ pc cm^{-3} and $D \approx 2200$ pc for NP 0532, Equation (3) indicates $\theta_{\text{scat}} \approx 0.2''$ at 80 MHz and $1''$ at 26 MHz. These angular sizes agree well with recent measurements of the angular size of the compact source in the Crab Nebula (Bell and Hewish, 1967; Antova and Vitkevich, 1969; Cronyn, 1970a).

3. Pulse Broadening by the Interstellar Medium

The other major obstacle to the view that the compact source and the pulsar NP 0532 are the same object is that the compact source has not been seen to pulsate. Pulse broadening caused by interstellar scattering would, however, result in a reduction of pulsed emission and an increase in the continuum emission observed at low frequencies. Because the distribution function for the angular scattering is probably a Gaussian function, and because the scattered radiation is delayed in time relative to the unscattered radiation by $D\theta_{\text{scat}}^2/4c$, the time profile of the scattered power will be the convolution of the emitted pulse profile with an exponential function whose $1/e$ decay time is f_v^{-1}.

It follows from Equation (2) that the pulse broadening will be proportional to v^{-4}, and that $f_{111}^{-1} \approx 10$ msec for the Crab pulsar. Both Staelin and Sutton (1970) and Rankin et al. (1970) have shown that $f_{111}^{-1} \approx 10$ msec for NP 0532. They also show that f_v^{-1} is proportional to $v^{-4 \pm 1}$ for this pulsar (cf. Drake, Figure 3). When this scattering data is combined with flux density measurements of both the pulsar and the compact source, the data indicate that the emitted pulse spectrum is straight over the region 10 MHz to 600 MHz, with a slope of -2.9 ± 0.4 (Cronyn, 1970b and Rankin et al., 1970). Pulse broadening due to interstellar scattering accounts for the cutoff in the apparent spectrum of the pulsed radiation and for the apparent increase in the continuum flux density at low frequencies.

Although measurements of the broadening of the Crab pulses agree with extrapolations from the measurements of the f_v of other pulsars (Figure 1), the absence of similar broadening of other pulsars caused some scientists to view the scintillation argument with skepticism. Most pulsars, however, emit wide pulses and are sufficiently close that pulse broadening due to interstellar scattering would not be detectable. In order to examine pulse broadening further, the average pulse profiles of the distant pulsar JP 1933 have been obtained (Figures 2 and 3).

The observed profiles indicate a progressive broadening at the lower frequencies which is proportional to v^{-4}. The $f_{111}^{-1} \approx 78$ msec for this pulsar, which has an $\int n_e \, dl$

of 158.6 pc cm^{-3}. It is especially interesting that the f_{111}^{-1} for JP 1933 is wider than that of NP 0532 by a factor which goes as the square of dispersion measure, in agreement with Equation (2). These observations show that the broadening of pulses from NP 0532 is not an isolated phenomenon, and lend additional support to the view that the compact source is the pulsar NP 0532.

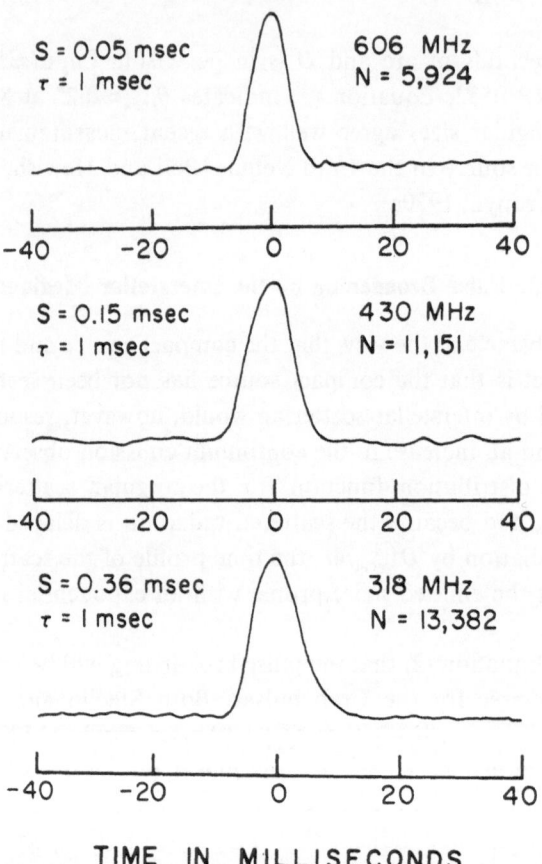

TIME IN MILLISECONDS

Fig. 2.   Average pulse profiles of JP 1933 at high radio frequencies. The IF bandwidth of 10 kHz caused a dispersion smearing of S, the post detection RC time constant was τ, and the number of pulses averaged was N. The zero corresponds to the zero time phase when the delay due to dispersion, $\int n_e \, dl = 158.60 \pm 0.05$ pc cm^{-3}, is corrected for.

4. Conclusions

The view that the compact source is the pulsar NP 0532 is consistent with measurements of the angular size of the compact source, with measurements of the pulse broadening of NP 0532, with measurements of the spectrum of both the compact source and the pulsar, and with measurements of the interstellar scintillation parameters of many other pulsars.

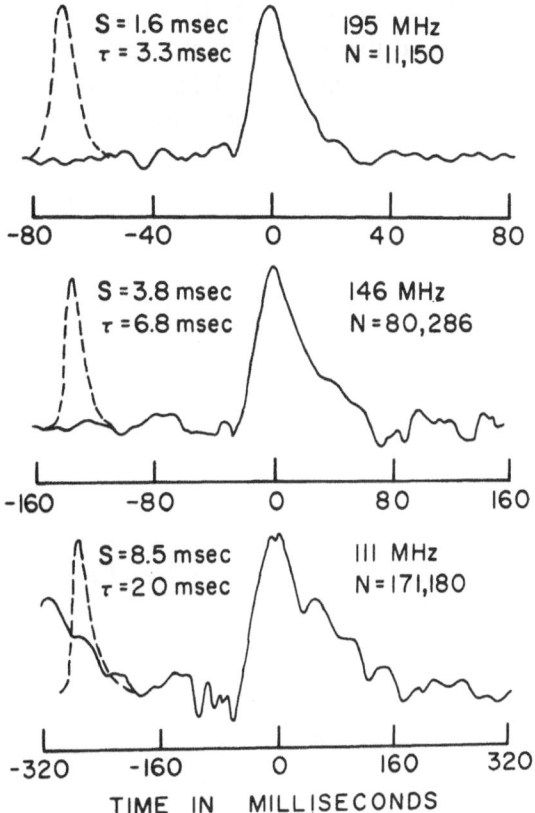

Fig. 3. Average pulse profiles of JP 1933 at low radio frequencies (solid lines). The dotted profile is that profile which would be observed if the pulse profile were independent of frequency. The IF bandwidth of 10 kHz caused a dispersion smearing of S, the post detection RC time constant was τ, and the number of pulses averaged was N. The zero corresponds to the zero time phase when the delay due to dispersion, $\int n_e \, dl = 158.60 \pm 0.05$ pc cm^{-3}, is corrected for.

References

Andrew, B. H., Branson, N. J. B. A., and Wills, D.: 1964, *Nature* **203**, 171.
Antova, T. D. and Vitkevich, V. V.: 1969, *Soviet Astron.* **12**, 788.
Bell, S. J. and Hewish, A.: 1967, *Nature* **213**, 1214.
Cronyn, W. M.: 1970a, Ph.D. Thesis, Astronomy Program, University of Maryland.
Cronyn, W. M.: 1970b, *Science* **168**, 1453.
Gower, J. F. R.: 1967, *Nature* **213**, 1213.
Hewish, A. and Okoye, S. E.: 1964, *Nature* **203**, 171.
Hewish, A. and Okoye, S. E.: 1965, *Nature* **207**, 59.
Lang, K. R.: 1969, *Science* **166**, 1401.
Lang, K. R.: 1971, *Astrophys. J.* **164**, 249.
Rankin, J. M., Comella, J. M., Craft, H. D., Richards, D. W., Campbell, D. B., and Counselman, C. C.: 1970, *Astrophys. J.* **162**, 707.
Rickett, B. J.: 1969, *Nature* **221**, 158.
Staelin, D. H. and Sutton, J. M.: 1970, *Nature* **226**, 69.
Woltjer, L.: 1968, *Astrophys. J. Letters* **152**, L179.

Note added in proof. W. C. Erickson, T. B. H. Kuiper, S. H. Knowles, and J. J. Broderick have recently found that the position of the pulsar NP 0532 and that of the compact source agree to 0.17″ at 121.6 MHz. They also find the pulsating flux of the pulsar is 14 fu whereas the unpulsating flux is 30 fu (reported at IAU XIV General Assembly).

2.5 INDIVIDUAL RADIO PULSES FROM NP 0531

J. M. SUTTON

NRAO, Green Bank, W. Va., U.S.A.* and IOTA, University of Cambridge, U.K.

and

D. H. STAELIN and R. M. PRICE

Massachusetts Institute of Technology, Cambridge, Mass., U.S.A.

Abstract. Observations with a swept frequency machine show that the shapes of individual pulses from NP 0531 are truncated exponentials with widths proportional to $\lambda^{\sim 4}$, consistent with the broadening arising from scattering in the interstellar medium. A histogram of pulse intensity reveals two different populations. The very strong pulses are associated only with the main pulse and interpulse.

We present further observations of individual pulses from the Crab Nebula pulsar (NP 0531) obtained using a swept frequency machine on the 300 ft radiotelescope at Green Bank. The technique has been described previously by Staelin and Sutton (1970) and Sutton *et al.* (1970). Briefly, to achieve a given time resolution it is necessary to use a bandwidth narrower than a critical value determined by the dispersion measure and frequency of observation. The combination of bandwidth and time resolution restricts the sensitivity such that observations of individual pulses are limited to strong pulses from particular pulsars. The sensitivity can be considerably improved by tracking the pulses as they sweep in frequency using individual channels of a filter bank to track different parts of the pulse. For all our observations the polarization is circular.

Radiation from NP 0531 at metre wavelengths is characterized by the occasional very strong pulses which first led to its discovery (Staelin and Reifenstein, 1968). We have already reported observations of individual strong pulses at 115 MHz and 157 MHz (Staelin and Sutton, 1970) obtained by triggering the swept frequency machine on strong pulses previously detected at a slightly higher frequency. Figure 1 shows some further pulse shapes of higher resolution measured at 115, 157 and 230 MHz using contiguous filters of width 10 kHz, and sweeping in frequency for 3, 5 and 9 MHz respectively. The shapes are adequately described as truncated exponentials of width Δt to $1/e$. The rise time is essentially instantaneous – the apparent rise times of about one channel can be attributed to arbitrary positioning of the channels relative to the leading edge of the pulse. Δt was measured by comparing relative intensities of adjacent channels on the trailing side of the pulse.

In Figure 2 Δt is plotted as a function of frequency. No correction has been made for the intrinsic pulse width which is ~ 0.1 msec at higher frequencies (Rankin *et al.*, 1970). Between 115 and 230 MHz Δt varies as λ^β where $\beta = 3.6 \pm 0.2$ (rms). The pulse shapes and wavelength dependence of the width are consistent with the truncated

* Operated by Associated Universities, Inc., under contract with the U.S. National Science Foundation.

exponential shape and $\beta = 4$ predicted by the theory of electron scattering in the interstellar medium (Cronyn, 1970). As to whether the scattering occurs uniformly along the line of sight, or is dominated by one or two regions of inhomogeneity or by irregularities within the Crab Nebula itself can be determined by comparing Δt with the observed angular diameter of NP 0531. Assuming that the small diameter source in the Crab Nebula is identical with NP 0531, present results indicate that the scattering is interstellar.

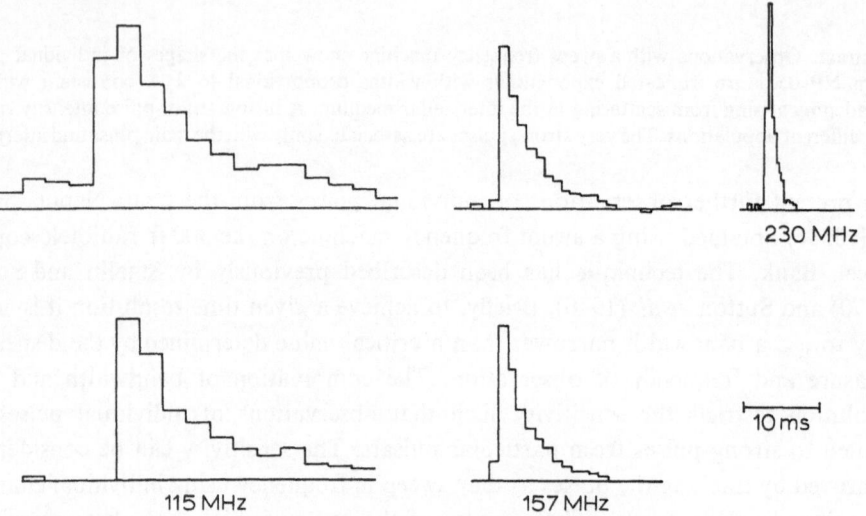

Fig. 1.   Shapes of different strong pulses from NP 0531 measured at different frequencies. All pulses are plotted with the same time scale.

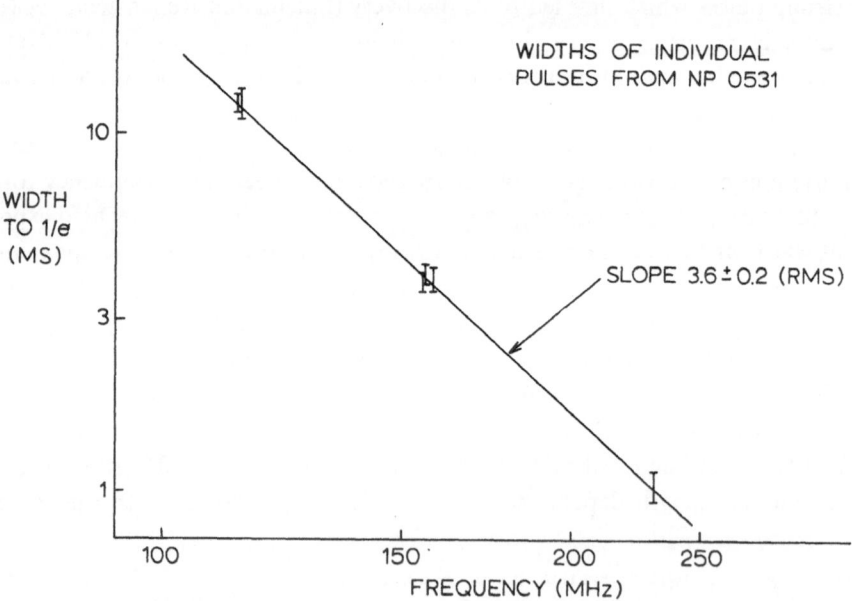

Fig. 2.   Variation of $1/e$ decay time with frequency for strong pulses from NP 0531.

In a second experiment we used the swept frequency machine to record 59000 consecutive pulses from NP 0531. Observing at 160 MHz with filters of resolution 30 kHz, the time resolution was 3.46 msec per channel, so that 5 pulses were observed simultaneously across the 50 channels of the filter bank. These 5 pulses were tracked for 0.56 MHz and the sweep was recommenced every 3.500 pulsar periods. In this way 30% of the pulses were observed on two sweeps, while the remainder were observed on only one. The choice of 'pseudo-period' for the observations (2×3.500 pulsar periods) ensured that in the subtraction of the comparison sweep from the pulsar sweep (see Sutton *et al.*, 1970), corresponding parts of successive pulses were not subtracted from one another. However there is a partial subtraction of interpulses from main pulses, and vice versa.

Rankin *et al.* (1970) have shown that the average pulse shape of NP 0531 at 430 MHz contains three major components – the main pulse of width 300 μsec which is followed 13.4 msec later by the interpulse of width 400 μsec, and is preceded 1.6 msec earlier by the 100% linearly polarized precursor of width $\sim$1 msec. At lower frequencies the average pulse profile is greatly smoothed by the interstellar scattering described above.

Figure 3 shows a histogram of observed intensity for the main pulse at 160 MHz based on 27000 pulses of interference-free data. By restricting attention to 33 of the 50 filter bank channels, each pulse was counted only once. The intensities are integrations over a 3.46 msec window whose centre varies with equal probability over a range of ± 1.73 msec about the main pulse. Thus they are not true measures of intensity for individual pulses but are quite adequate for statistical purposes. Due to the pulse broadening there is some contamination from the precursor, and also from interpulses in the comparison sweep which will tend to produce negative intensities.

The combined contributions of the sky, the receiver and the Crab Nebula to the system temperature are about 90 units of intensity, of which the Crab Nebula contributes about 50 units.

Observations of strong radio sources showed that the recorded intensities were proportional to the power received. As described by Lovelace and Craft (1968), the observed histogram is then the convolution of the true distribution with some noise distribution. Also, because of the comparison sweep technique used, the noise distribution must be symmetric.

The histogram consists of two populations:

(1) A low intensity distribution which is symmetric about 1.2 units of intensity. The broadening can be attributed entirely to a Gaussian noise distribution of standard deviation 2.8 units, which is consistent with the system temperature. The pulse distribution is clearly narrow and includes many pulses with non-zero energy. Beyond this its shape is uncertain.

(2) A high intensity tail, which can conveniently be described as jumbo pulses. Assuming an exponential distribution extrapolated back to zero intensity, this contains $\sim 0.4\%$ of the pulses and $\sim 5\%$ of the pulsed energy.

This histogram is quite unlike those published for other pulsars which typically show a single population of exponential shape (Lovelace and Craft, 1968; Lang, 1969). However it should be pointed out that this histogram is produced by averaging pulse intensities over a frequency range considerably greater than the expected decorrelation frequency, thus smoothing out the effects of interstellar scattering. Most published histograms have been derived from observations with a narrow bandwidth in experiments purposely designed to measure the effects of interstellar scattering. For all pulsars there is clearly a need for measurements of intrinsic intensity histograms as

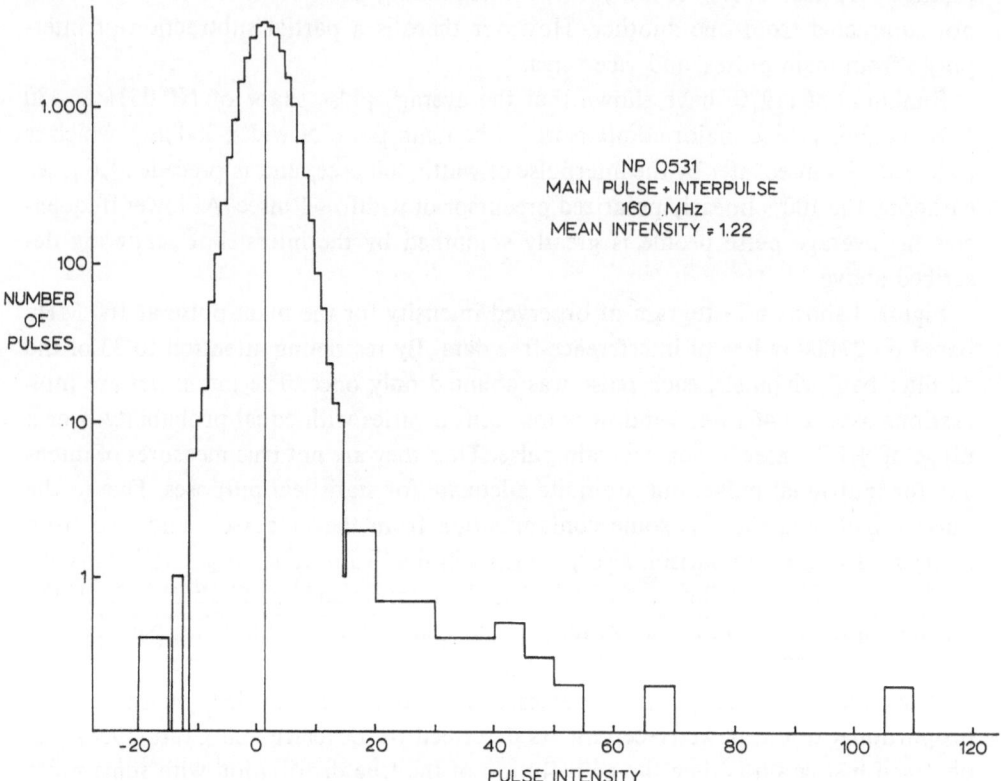

Fig. 3.   Histogram of pulse intensity for 27000 main pulses of NP 0531 observed at 160 MHz. Intensities are integrals over a window of 3.46 msec centred approximately on the main pulse, and are measured in arbitrary units. The mean intensity is 1.22 units. The cell size is 1 unit, but beyond 15 units the value plotted is the average over 5 cells. The vertical line indicates the axis of symmetry for the low intensity part of the distribution.

opposed to those produced by the interstellar medium. For example, Jones and Wielebinski (1969) overcame the difficulty of observing with a relatively narrow bandwidth by normalizing pulse intensities relative to the average of the adjoining pulses. Ten thousand pulses should be sufficient to establish the existence of a jumbo population similar to NP 0531.

Figure 4 shows arrival times of all pulses stronger than 16 units of intensity on the scale of Figure 3. All 59 000 pulses were used. Although the time resolution was 3.46 msec, arrival times for the sharp leading edge could be estimated to the nearest 0.4 msec using the relative intensities of the first two samples of the pulse, and assuming a truncated exponential pulse shape. There are clearly two groups of pulses, centred on the main pulse and the interpulse. No strong precursor pulses were observed although, within the limits of measurement, the possibility of one or two cannot be excluded. These results are a further example of the similarity of the main pulse and interpulse, and their differences from the precursor as regards pulse widths, optical radiation, X-radiation and average linear polarization.

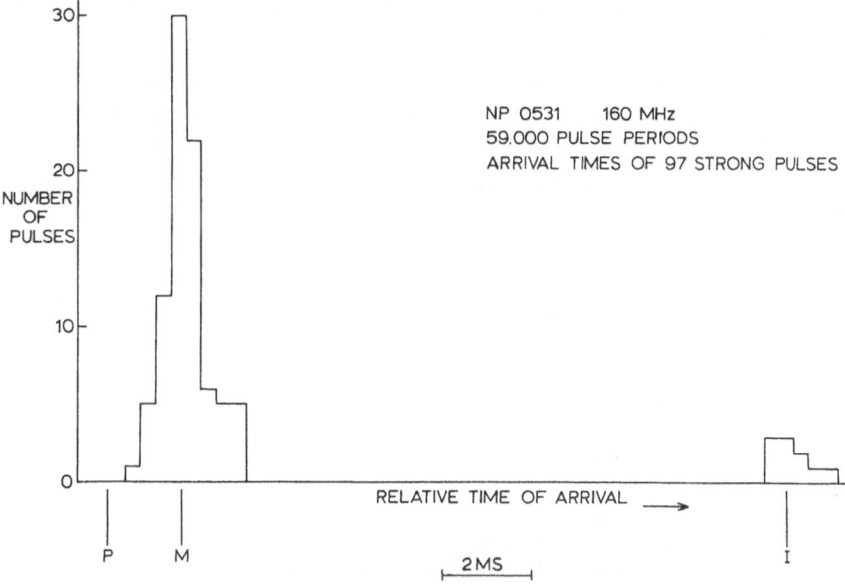

Fig. 4.   Arrival times of 97 strong pulses, corresponding to all those stronger than 16 units of intensity (see Figure 3) but using all 59 000 pulses. The relative arrival times of the precursor (P), main pulse (M) and interpulse (I) are also shown.

The arrival times of these jumbo pulses were examined for periodicities, for bunching in time, and for consecutive main pulses and interpulses. For this purpose all intensities were corrected for the positioning of the window relative to the main pulse, and for small variations of antenna gain with hour angle. The resultant sample contained 50 pulses and was complete above a specified threshold. There is no evidence of any correlation in intensity between adjacent or nearby pulses or interpulses, or over longer intervals of time. It appears that the arrival times of jumbo pulses are independent of one another and are random in a Poissonian sense.

The pulse broadening due to scattering precludes measurements of intrinsic pulse widths or peak temperatures. However, the integrated energies of the strongest pulses can be compared with those at 430 MHz (Heiles *et al.*, 1970). In a sample of 174 500 pulses Heiles *et al.* detected two pulses with energies $>2 \times 10^{-26}$ J m^{-2} Hz^{-1}, and several with energies slightly less. In 59 000 pulses at 160 MHz the three largest have energies of 39, 32 and 28×10^{-26} J m^{-2} Hz^{-1}. Rankin *et all* (1970) measured the spectral index of NP 0531 as -2.6 ± 0.3. It follows that, within the errors of measurement, the ratios of peak pulse energies at the two frequencies are the same as the ratios of average pulse energies. Although there is evidence of wideband enhancement of NP 0531 pulses from 111 to 74 MHz (Comella *et al.* 1968) and from 172 to 112 MHz (Goldstein and Meisel, 1969), there is no evidence that the largest pulses at 430 MHz are also the largest at 160 MHz. Finally, assuming that the intrinsic pulse width at 160 MHz is ~ 0.1 msec, the same as that observed by Heiles *et al.* at 430 MHz, the peak brightness temperature at 160 MHz is $\sim 10^{33}$ K.

The two populations of pulse intensity can be explained qualitatively by a beaming process in which the beam contains a strong narrow peak superimposed on a broad base and the orientation of this feature changes randomly from pulse to pulse in such a way that only occasionally is it directed at the Earth. This could arise from wideband maser amplification in which the direction of maximum amplification varies due to refraction in the amplifying plasma. In such a case it is unlikely that the largest pulses at 160 and 430 MHz will be correlated.

The work of D. H. Staelin and R. M. Price is supported principally by grants from the U.S. National Science Foundation.

References

Comella, J. M., Craft, H. D., Lovelace, R. V. E., Sutton, J. M., and Tyler, G. L.: 1969, *Nature* **221**, 453.
Cronyn, W. M.: 1970, *Science* **168**, 1453.
Goldstein, S. J. and James, J. T.: 1969, *Astrophys. J.* **158**, L179.
Heiles, C., Campbell, D. B., and Rankin, J. M.: 1970, *Nature* **226**, 529.
Jones, B. and Wielebinski, R.: 1970, *Astrophys. Letters* **5**, 17.
Lang, K. R.: 1969, *Science* **166**, 1401.
Lovelace, R. V. E. and Craft, H. D.: 1968, *Nature* **220**, 875.
Rankin, J. M., Comella, J. M., Craft, H. D., Richards, D. W., Campbell, D. B., and Counselman, C. C.: 1970, *Astrophys. J.* **162**, 707.
Staelin, D. H., Reifenstein, E. C.: 1968, *Science* **162**, 1481.
Staelin, D. H., Sutton, J. M.: 1970, *Nature* **226**, 69.
Sutton, J. M., Staelin, D. H., Price, R. M. and Weimer, R.: 1970, *Astrophys. J.* **159**, L89.

2.6 PULSAR NP 0532: RECENT RESULTS
ON STRONG PULSES OBTAINED AT ARECIBO

CARL HEILES*

Arecibo Observatory, Arecibo, Puerto Rico

and

JOHN M. RANKIN

Department of Physics and Astronomy, University of Iowa, Iowa City, Iowa, U.S.A.

Abstract. New observations have been made of strong pulses from NP 0532. Linear polarization is typically 25 per cent and circular polarization is less than 10 per cent but definitely not zero. A comparison of pulse shapes at 429.9 and 430.1 MHz showed significant differences in detailed structure and polarization. In another study strong pulses at 318 and 111 MHz did not occur together although a strong pulse at one frequency was accompanied by an above average intensity at the other frequency. Similar behaviour was found at 111 and 74 MHz. We conclude that strong pulses do not have a smooth continuous frequency spectrum.

1. Clarification of Previously Published Results

The properties of strong pulses from NP 0532 have been investigated by Heiles *et al.* (1970) at 430 MHz using a bandwidth of 8 kHz. They found that the pulses were sometimes $\leqslant 100$ μsec in width and showed circular polarization whose sense varied systematically with the arrival time of the strong pulse within the average pulse envelope, whose half-width is about 250 μsec. However, observations with a 300 kHz bandwidth (which yields a time resolution of about 2 msec due to dispersion) by Graham *et al.* (1970) showed little circular polarization but did show 70 per cent linear polarization in a typical pulse. We have performed two new sets of observations to clarify the situation.

One observation used 300 kHz bandwidth and a time constant of 1 msec, and was basically a repetition of the work done at Jodrell Bank (Graham *et al.,* 1970). These observations have been published elsewhere (Rankin and Heiles, 1970) and will not be described in detail here. Only strong main pulses were observed – no precursors or interpulses. Most of the strong pulses did not show the 70 percent linear polarization implied to be typical by Graham *et al.* (1970). Linear polarizations of order 25 per cent were typical but many pulses showed much smaller polarization. There was no systematic trend in position angle from one pulse to another. Circular polarization was nearly always smaller than linear polarization, usually less than 10 per cent, but definitely not always zero.

The other observation was done on 11 July 1970, and used both senses of circular polarization, recording each simultaneously in two bandwidths centered at precisely 429.9 and 430.1 MHz. Each of the four channels was mixed to baseband and the

---

* Usual address: Department of Astronomy, University of California, Berkeley.

Davies and Smith (eds.), The Crab Nebula, 103–109. All Rights Reserved.

bandwidths were defined by RC time constants which provided 3 dB total widths of about 7.5 kHz, or integrated bandwidths of 12 kHz. The signals were then detected, passed through approximately rectangular 100 μsec time filters, and recorded at 50 μsec intervals in synchronism with the pulsar period.

As is our usual practice, we computed the energy distribution on and off the main pulse; the strong pulses were selected on the basis of these distributions and were positively substantiated by detection with approximately the correct time delay at both frequencies. Previous 430 MHz distributions have been given previously (Heiles *et al.*, 1970; Rankin and Heiles, 1970), and we will not repeat them here.

The average of all the strong pulses was obtained and is shown in Figure 1. It does

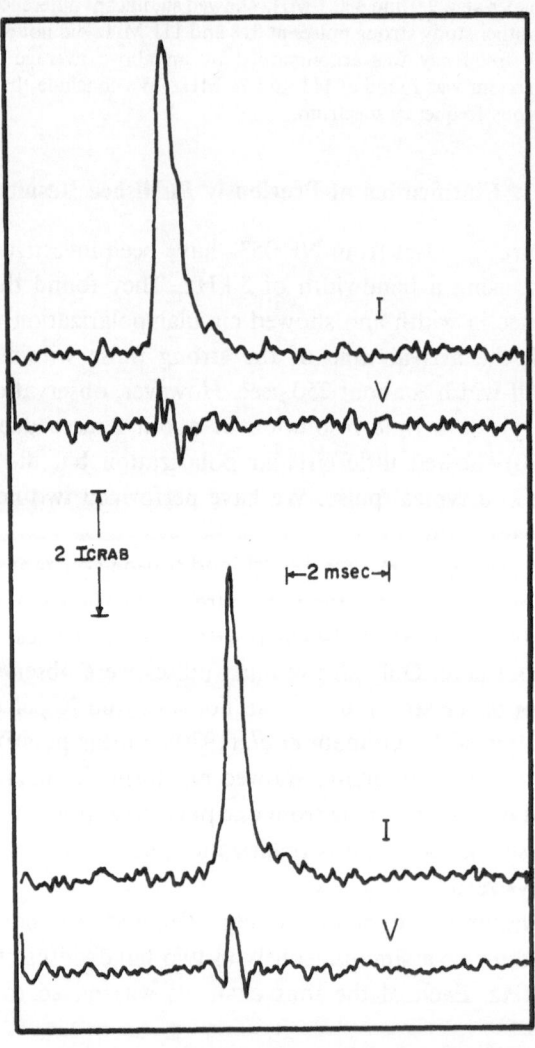

Fig. 1.   Average of the strong pulses at 430.1 MHz (top) and 429.9 MHz (bottom) in total intensity, *I* and circularly-polarized Stokes parameter, *V*. The relative time delay due to dispersion is evident. Note the absence of circular polarization reported earlier!

not show the circular polarization reversal that was found earlier (Heiles *et al.*, 1970). We know of no reason why there should be any systematic effects in either experiment. We can only conclude that our previous result was a statistical fluke due to the small number of pulses involved.

Individual strong pulses appeared different in the two channels. When the pulse was narrow, it would often arrive at different phases of the pulsar period at the two frequencies. Also, differences in pulse width were common and the sense of circular polarization was often different. Examples can be seen in Figure 2.

Some differences in pulse shapes observed at the two frequencies are clearly expected because the time-bandwidth product for each plotted point in Figure 2 is only about 1.2. However, we question whether this can account for the differences

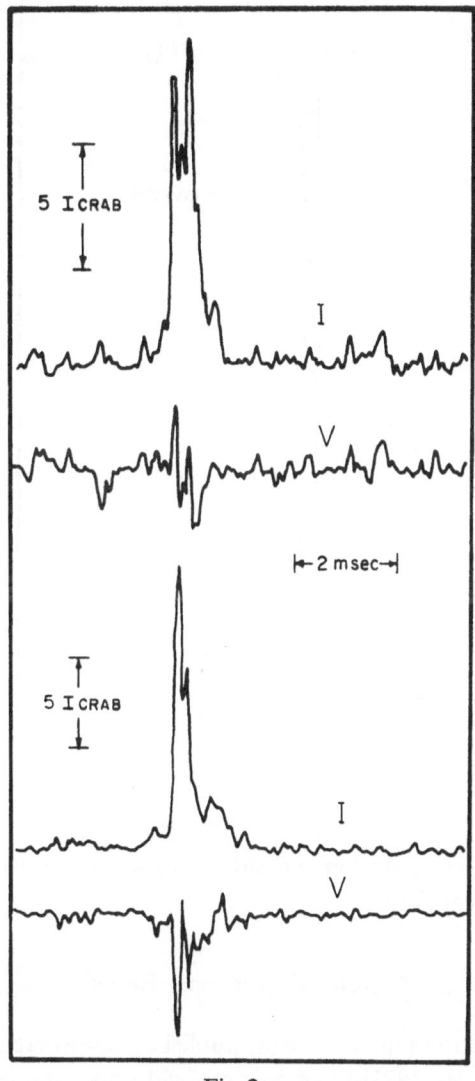

Fig. 2a.

in pulse width we often observe, or for the differences in circular polarization, especially in cases where the pulse is wide enough to substantially increase the effective time-bandwidth product. Furthermore, in the cases where the pulse or its components are narrow, they should at least arrive at the same pulsar phase at the two frequencies.

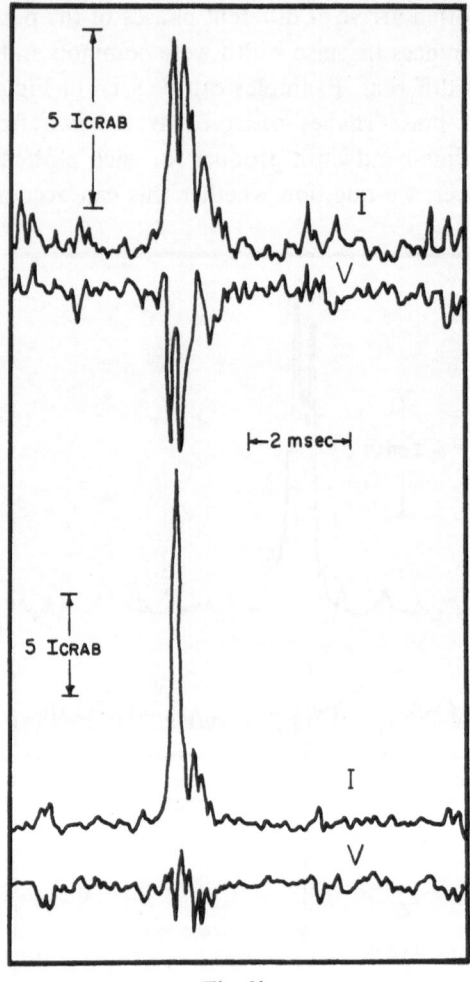

Fig. 2b.

This is often not the case. Therefore the pulse shape differences are thought to be significantly real. However, further measurements are obviously called for and will be performed in the near future.

2. New Observational Results

Most of our new observations are being published elsewhere (Heiles and Rankin, 1971; Rankin and Heiles, 1971b) and here we will only summarize the results. The new results involve simultaneous observations at various frequencies and an approximate determination of the spectral behavior of strong pulses.

Simultaneous observations of individual strong pulses at 318 and 111 MHz reveal that a detectable strong pulse at one frequency does not usually give rise to an independently detectable strong pulse at the other frequency. However, a strong pulse detected at one frequency does tend to produce a pulse which is stronger than the average pulse intensity at the other frequency. A strong pulse which is detected at

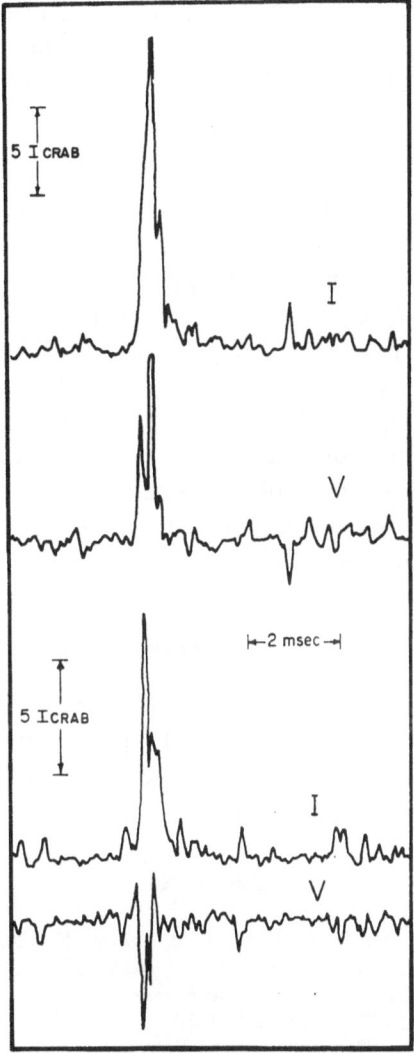

Fig. 2c.

Fig. 2a–c.   Individual 430 MHz strong pulses at 430.1 MHz (top) and 429.9 MHz (bottom) in total intensity, I, and circularly-polarized Stokes parameter, V. The relative time delay due to dispersion has been removed by shifting the graphs by the appropriate amount. The intensity scales for V and I at each frequency are the same, but the scales at the two frequencies are different and are shown.
Note the different pulse shapes and widths, and the differences in V at the two frequencies.

318 MHz tends to produce a pulse at 111 MHz which is about eight times stronger than the average 111 MHz pulse, and *vice versa*. However, detectable strong pulses at any frequency are typically 1500 times stronger than the average pulse. Thus there is a discrepancy of more than two orders of magnitude which can only be the result of real frequency structure in the spectrum of an individual strong pulse.

Similar behavior is exhibited between 111 and 74 MHz as revealed by further simultaneous measurements. About half of the strong pulses are independently detectable at both frequencies, but the other half behave in a manner akin to the usual one seen in the 318–111 MHz data described above. These data lead to the conclusion that the strong pulses are not received with a smooth, continuous frequency spectrum. Our 111–74 MHz observations imply a frequency width of about 30 MHz, perhaps implying that

$$\Delta f/f \approx \tfrac{1}{3}.$$

On the other hand, the spectral behavior may be more complicated. Further observations are required to determine the spectral behavior more accurately. Additional spectral information is given in our other papers (Heiles and Rankin, 1971).

These observations were obtained with large bandwidths and do not suffer from a small time-bandwidth product. Current knowledge about interstellar scintillation (Heiles and Rankin, 1971) allows us to rule out interstellar scintillation as a significant contributing cause to the spectral behavior described here. The pulse emission mechanism itself, or scattering near the pulsar itself, could be responsible for the frequency behavior. In any case, the existence of occasional huge pulses which are characterized by complex spectral behavior and random polarization behavior contrasts with other pulsars, which are not thought to show these properties.

Acknowledgement

The Arecibo Observatory is operated by Cornell University under contract to the National Science Foundation and with partial support from the Advanced Research Projects Agency. This work was supported in part by the Air Force Office of Scientific Research, Office of Aerospace Research, U.S. Air Force, under contract F44620-69-C-0092.

References

Graham, D. A., Lyne, A. G., and Smith, F. G.: 1970, *Nature* **225**, 526.
Heiles, C., Campbell, D. B., and Rankin, J. M.: 1970, *Nature* **226**, 529.
Heiles, C., and Rankin, J. M.: 1971, submitted to *Nature*.
Rankin, J. M. and Heiles, C.: 1970, *Nature* **227**, 1330.
Rankin, J. M. and Heiles, C.: 1971b, submitted to *Nature*.

Discussion

F. G. Smith: It is important to distinguish between the averages of Stokes parameters over many pulses and the measurements on single pulses. If a single pulse is measured, the radiation is necessarily 100% elliptically polarised during a time which is the inverse of the receiver bandwidth. A time average must be made, which must be over a much longer time. Some of Dr. Heiles' results seem to show an inadequate time smoothing, and without more detail of receiver parameters we should not accept his measurements of pulse structures or circular polarisation.

C. Heiles: The time-bandwidth product for each plotted point $= 1.2$. However, there are many cases in which the pulse is, for example, 100% circularly polarised throughout the whole strong pulse in one bandwidth, while no circular polarisation, or the opposite sense at the 100% level, occurs in the other band-width which is 200 kHz away. This occurs even for broad strong pulses, for which the effective time-bandwidth product is 4 or 5. Furthermore, in such cases the Stokes parameter V looks like the total intensity over the whole pulse width, which is clearly improbable if V results simply from statistically random variations in the radiation.

2.7 RADIO OBSERVATIONS OF THE CRAB PULSAR AT 408, 240 AND 151 MHz

RUDOLF E. SCHÖNHARDT

University of Manchester, Nuffield Radio Astronomy Laboratories,
Jodrell Bank, U.K.

The radio pulses from the Crab pulsar have been observed at Jodrell Bank on several days in June, July and August 1970 at frequencies of 408 MHz, 240 MHz and 151 MHz. The integrated pulse shapes and the linear polarization of the average pulses have been investigated.

Most of the measurements at 408 MHz were in general agreement with earlier observations performed at Arecibo (Campbell *et al.*, 1970). The receiver bandwidth used at Jodrell Bank was 13 kHz. The linear polarization in the main pulse and inter-pulse is roughly 18% and 10% respectively. The degree of linear polarization in the precursor is near 100%, but may be somewhat variable. The position angle of the linearly polarized radiation is fairly constant throughout the precursor; any possible swing is less than about 15° across the pulse width.

The interesting question of a possible swing of the polarization angle in the main pulse was investigated. Four integrations of about 1.5 hours integration time each were obtained on different days in June, July and August. In three of these integrations a swing of roughly the same rate and of the same direction seemed to be visible. As the signal to noise ratios in these integrations were not very satisfactory, all the four integrations were added together, yielding a total integration time of 6.7 hours (see Figure 1). Within the errors calculated from the rms noise one obtains a swing rate of 100° ($\pm 40°$) per millisecond. This rate is in agreement with the optical swing rate near the peak of the optical pulse of about 90° per millisecond (Wampler *et al.*, 1969). Unfortunately the question of the direction of the rotation in the radio pulse cannot yet be settled.

In some of the integrations a small pulse was detected, occurring 5.2 (± 0.3) msec after the main pulse. This 'postcursor' was detected at 408 MHz and 240 MHz (Figure 2). At both these frequencies its energy was roughly 7.5% ($\pm 2.5\%$) of the sum of the energies of the main pulse and precursor.

At 240 MHz the degree of linear polarization in the precursor is about 50% (Figure 3a), but it seems to vary as seen by comparing Figure 3b and Figure 3c.

At 151 MHz the stronger pulse could no longer be resolved into a precursor and main pulse, but its linear polarization of the order of 15% to 20% could still be clearly detected.

References

Campbell, D. B., Heiles, C., and Rankin, J. M.: 1970, *Nature* **225**, 527.
Wampler, E. J., Scargle, J. D., and Miller, J. S.: 1969, *Astrophys. J. Letters* **157**, L1.

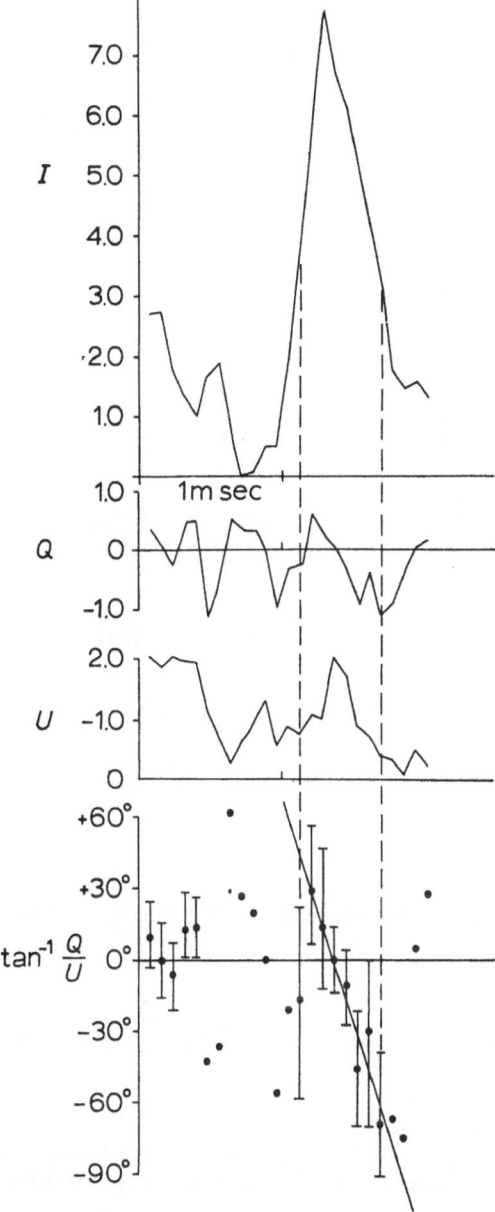

Fig. 1. Main Pulse at 408 MHz; integration time = 6h 45 min.

Discussion

C. Heiles: I would like to report a new result obtained by Rankin and me at Arecibo concerning the average pulse shape at 606 MHz, namely that in two days (6 h) of integration we were unable to see the precursor. If it was as strong in total intensity relative to the main pulse as it is at 430 MHz its amplitude would have been more than 10 times the noise on our average. Our null detection is possibly the result of our feed, which is linearly polarised, being unfavourably oriented with respect to the position angle of the linearly polarised precursor. Assuming that this is the case, we are able

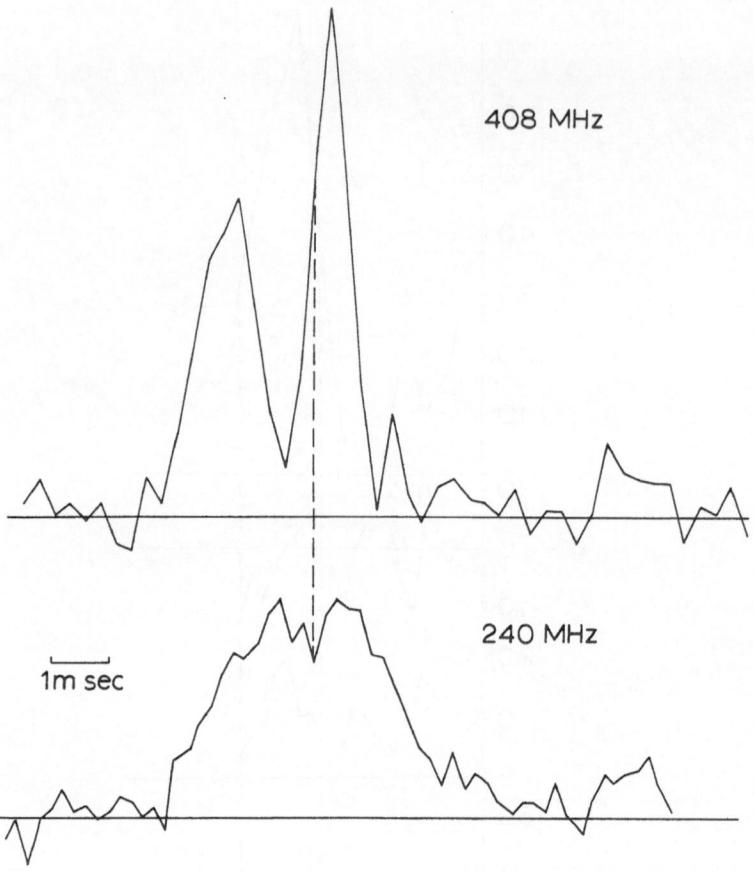

Fig. 2.   The postcursor at 408 MHz and 240 MHz.

to say that the relative intensity of the precursor at 606 MHz is less than 60% of its value at 430 MHz.

J. Rankin: Although we observe a good deal of emission between the main pulse and interpulse at frequencies between 430 and 196.5 MHz, we have never noted a specific short component in the average of several hundred thousand pulses.

M. M. Komesaroff: Was the swing in position angle of the polarisation of the radio emission of the main Crab pulsar pulse as seen by Mr. Schönhardt similar to that previously reported as seen optically?

R. Schönhardt: I may remark that in our integrations of high resolution the broadening due to dispersion within the bandwidth was 90 μsec. For our bandwidth of 13 kHz, the inverse of the bandwidth is 77 μsec.

A. T. Moffet: Arecibo results show no swing of position angle within the main pulse of the Crab pulsar. Can this be reconciled with the Jodrell results?

F. D. Drake: At Arecibo it has not been possible to measure the rotation of the position angle of the linear polarisation of the main pulse due to inadequate time resolution. Time resolution is limited by dispersion effects, which require one to reduce bandwidth as much as possible. However, it is only worthwhile to reduce the bandwidth to the value whose reciprocal is equal to the time smearing introduced by dispersion over that bandwidth. At 430 MHz this limiting time resolution is 130 μsec in NP 0532. With this resolution one can only resolve the main pulse into two intervals.

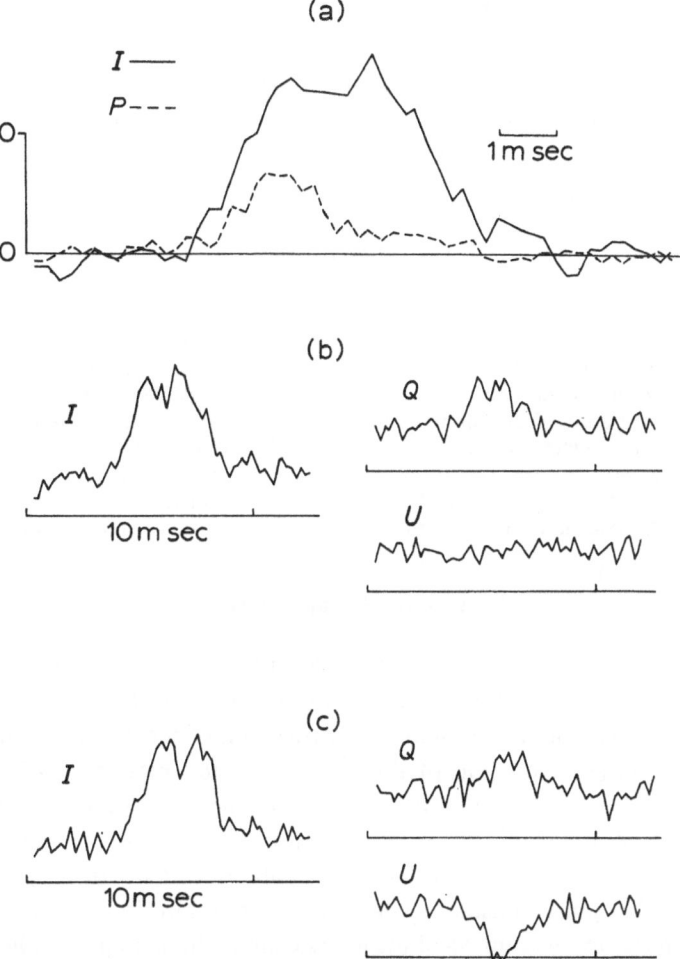

Fig. 3.   (a) 240 MHz, main pulse and precursor (*I* and *P*). (b) 240 MHz, main pulse and precursor (*I*, *Q* and *U*). (c) 240 MHz, main pulse and precursor (*I*, *Q* and *U*), at a later date.

With these few intervals, the small intrinsic polarisation, and the small signal/noise ratio, no significant measurement of polarisation rotation is possible.

In view of the fact that the intrinsic dispersion limit on time resolution also applies to Jodrell Bank data, it is hard to understand how a definitive polarisation measurement could be made. Multichannel radiometers are required to make this measurement properly.

2.8 TIME VARIABILITY OF THE DISPERSION OF THE
CRAB NEBULA PULSAR

J. M. RANKIN

Department of Physics and Astronomy, University of Iowa,
Iowa City, Iowa, U.S.A.

and

J. A. ROBERTS*

Arecibo Observatory, Arecibo, Puerto Rico.

Abstract. The dispersion of the Crab nebula pulsar was measured as a function of time from 10 May 1969 to 24 July 1970. Transient events occurred in the middle of June each year which coincided with the occultation of the pulsar by the solar corona. In addition there were 2 or 3 distinct events which produced enhancements of several times 10^{16} electrons cm^{-2}; these were characterized by rise times of about 50 days and decay times several times longer. One event correlated with the frequency jump of the pulsar at the end of September 1969 and with the observation of optical activity in the nebula. A discussion is given of the interpretation of the variations in dispersion measure.

1. Experimental Results

The routine multi-frequency timing observations of the Crab Nebula pulsar which have been made at the Arecibo Observatory (Richards and Roberts, 1970) since May, 1969, have also been used to monitor the dispersion of the pulsar. Figure 1 is a plot of the dispersion as a function of time for the period between 10 May 1969 and 24 July 1970. The data are presented in terms of the dispersion constant in units of sec MHz^2 as it is this quantity which is most accessible experimentally. Conversion to dispersion measure in the more common units of e^-/cm^2 or e^- pc/cm^3 involves both the use of physical constants whose values are imprecisely known and assumptions about the ionic constitution of the interstellar medium. In providing the auxilliary scale in Figure 1 we have used the conversion factor 7.43366×10^{14} $e^-/cm^2/sec$ MHz^2 which assumes an interstellar medium of pure hydrogen and the physical constants in Allen (1964).

The most striking features in Figure 1 are the transient events occurring in the middle of June each year. These are the result of the occultation of the pulsar by the solar corona. The interpretation of these events, which is complicated by scattering from irregularities in the corona, has been considered elsewhere (Rankin, 1970; Counselman and Rankin, 1971) and will not be discussed here.

Apart from the coronal events, Figure 1 shows two or three distinct events which produce enhancements of several times 10^{16} e^-/cm^2. Each event is characterized by a rise time of the order of 50 days and a decay which is several times longer. The total dispersion increases with time. This might result from the summation of the separate

---

* On leave from C.S.I.R.O. Radiophysics Laboratory, Epping, N.S.W., 2121, Australia.

Davies and Smith (eds.), The Crab Nebula, 114–117. All Rights Reserved.

events, or the events could be superposed on a quasi-linear increase resulting from some separate mechanism. These events are very much larger than the random errors which are indicated by the rms errors of the daily means shown in the figure. It should be noted, however, that the zero point of the dispersion scale may have a systematic error as large as ± 15 sec MHz^2 due to the failure to correctly separate the f^{-2} plasma group delay (plotted in Figure 1) from the f^{-4} delay associated with the interstellar multi-path broadening (Rankin *et al.*, 1970; Rankin and Roberts, 1970).

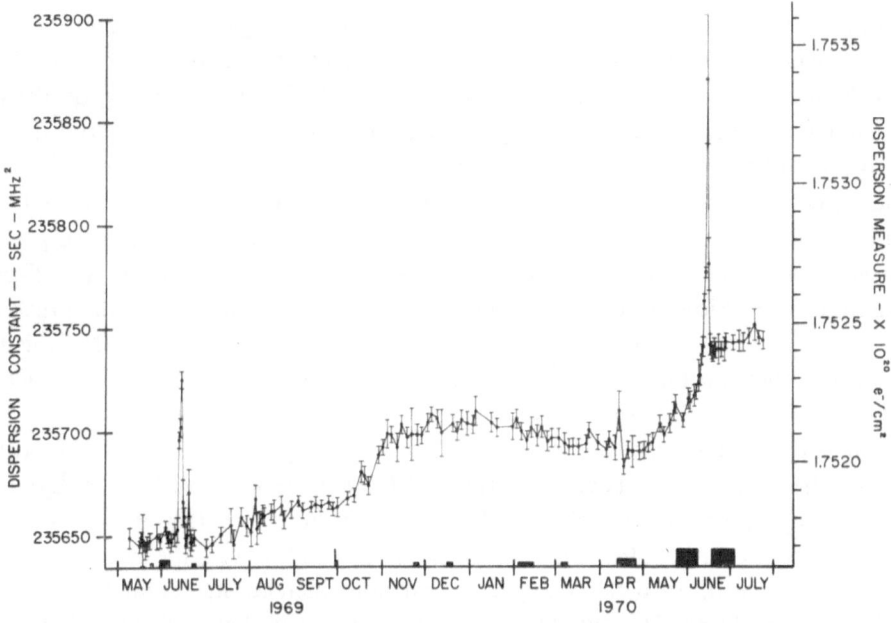

Fig. 1.   Dispersion constant as a function of time for the period between May 10, 1969 and July 24, 1970. An auxiliary scale in dispersion measure is also provided (see text). The filled blocks along the abscissa denote times when the pulsar repetition frequency was irregular; the height and width of the blocks indicate the severity and duration of the irregularity, respectively.

The filled blocks along the abscissa of Figure 1 denote times when the pulsar repetition frequency was irregular (Richards and Roberts, 1970). The height and width of the blocks indicate the severity and duration of the irregularity, respectively. The frequency jump at the end of September 1969 is well correlated with the onset of one of the events in Figure 1, and Scargle and Harlan's (1970) observation of activity in the nebula following this frequency jump makes this coincidence even more tantalizing. However, the other main increase in Figure 1 does not appear to be associated with any definite frequency jump in the pulsar.

2. Discussion

If the underlying increase in Figure 1 is produced in some other way, then the events could be caused by clouds of ionization drifting across the line of sight. An interstellar cloud with a transverse velocity of 10 km/sec, a scale size of 0.5 AU, and a peak density of 4000 e^-/cm^3 could produce the central event in Figure 1. A more likely explanation is a cloud within the nebula with a velocity $\sim$500 km/sec, a scale size $\sim$25 AU and a central density of 75 e^-/cm^3.

If the dispersion changes are associated with irregularities in the rotation of the pulsar, one might suppose that additional ionization is ejected from the star or its magnetosphere, or that an ionizing disturbance travels outward from the pulsar ionizing previously neutral material in the nebula. Two difficulties arise with the simple ejection model. Firstly, the timing observations suggest a sudden occurrence at the neutron star occupying a day or two, whereas the dispersion change begins slowly and continues to increase for of the order of 50 days. Secondly, on the basis of simple models, the observed dispersion changes would require electron densities in the vicinity of the neutron star which are so great as to be opaque to 111 MHz radiation (the lowest frequency at which the present observations were made). Both of the difficulties could be circumvented if, for example, a cloud of ionized gas were ejected at the time of the frequency jump, but in such a way that it did not affect the measured dispersion until it was a considerable distance from the star, and then only slowly attained its maximum effect. Pacini (1970) has discussed a model of this type.

Both of these problems are also avoided if one invokes an expanding ionizing front. This would produce a slow increase in dispersion beginning near the time of the frequency jump and since the ionization is nowhere highly concentrated the problem of high densities would not occur. Scargle and Harlan's report (1970) of changes in the optical appearance of the nebula, which they attribute to a travelling disturbance, provides some support for this type of mechanism. However, the occurrence of repeated increases in dispersion before the previous increase has appreciably decayed presents a problem. This would seem to require either that each disturbance only partially ionizes the neutral material, or else that the successive events produce their ionization at increasingly greater distances from the star.

If extra ionized gas is produced, as in the last two models, it is necessary to consider the decay of this ionization. At the densities one is likely to encounter, it would seem that recombination is completely negligible. Spreading of the ionization in directions perpendicular to the line of sight is a much more likely cause of a decrease in the dispersion measure.

Finally, the nature of the dispersion events may be illuminated by a further experiment. The dispersion provides a measure of $\int N \, ds$, while Faraday rotation measures $\int N\mathbf{B} \, ds$, where $\mathbf{B}$ is the magnetic flux density along the path. The magnetic field in the Crab Nebula is very much stronger than that in the intervening interstellar medium, about 10^{-3} G as compared to about 10^{-6} G. Close to the neutron star the field will be even greater. If the dispersion increases are caused by electrons within the Crab Nebula they should be accompanied by measurable increases in the Faraday rotation. For a mean longitudinal field of 10^{-3} G, the total measured change in dispersion would cause a change of the rotation measure of 16 rad/m^2.

This report is a brief summary of a paper in preparation (Rankin and Roberts, 1970),

We would like to thank Counselman and Richards for their invaluable assistance in planning and carrying out the experiment on which this report is based. The Arecibo Observatory is operated by Cornell University under contract to the National Science Foundation with partial support from the Advanced Research Projects Agency. This work was also supported in part by the Air Force Office of Scientific Research under contract F44620-69-C-0092 and also in part by the National Aeronautics and Space Administration under Grant NGL 16-001-002.

References

Allen, C. W.: 1964, *Astrophysical Quantities*, University of London, Athlone Press, London.

Counselman, C. C., III and Rankin, J. M.: 1970, in preparation.

Pacini, F.: 1970, 'Physical Processes and Parameters in the Magnetosphere of NP 0532', this symposium, Paper 7.3, p. 394.

Rankin, J. M.: 1970, 'The 1969 Solar Occultation of the Crab Nebula Pulsar, NP 0532', Ph.D. Thesis, University of Iowa.

Rankin, J. M., Comella, J. M., Craft, H. D., Jr., Richards, D. W., Campbell, D. B., and Counselman, C. C., III: 1970, *Astrophys. J.* **162**, 707.

Rankin, J. M. and Roberts, J. A.: 1970, 'Time Variability of the Dispersion of the Crab Nebula Pulsar', to be submitted to *Astrophys. J. Letters*.

Richards, D. W. and Roberts, J. A.: 1970, 'Timing Observations of the Crab Nebula Pulsar at the Arecibo Observatory', this symposium, Paper 2.11, p. 125.

Scargle, J. D. and Harlan, E. A.: 1970, *Astrophys. J.* **159**, L143.

Discussion

F. C. Michel: Is this the same data on which the limit: $n_e < 0.25/cm^3$ was deduced. If so, the *rise* in DM, which is not well understood, could mask any secular decrease due to nebular expansion.

J. Rankin: The earlier limit on electron density from secular changes in the dispersion measure was deduced by H. D. Craft, Jr., from other less accurate data. Since we now see that there are large unexplained phenomena in the time variation of the dispersion measure, it is apparent that no information can be deduced on the electron density in the Crab Nebula.

J. Kristian: Would you expect to see the geometrical dilution factor on the same scale as the data that you have shown?

J. Rankin: My recollection is that the rate of decrease of dispersion due to geometrical effects (i.e. expansion) is comparable to the linear rate of increase of dispersion measure that might be deduced from the data presented. Thus such an effect should easily be observable.

2.9 FARADAY ROTATION OF THE CRAB PULSAR RADIATION

R. N. MANCHESTER

National Radio Astronomy Observatory*, Green Bank, W. Va., U.S.A.

During April, 1970, the 300-ft telescope of the National Radio Astronomy Observatory was used to determine the mean polarisation of the Crab Nebula pulsar radiation at several frequencies around 400 MHz. The position angle of the highly polarised precursor measured at each frequency, corrected for ionospheric Faraday rotation and plotted against inverse frequency squared is shown in Figure 1. The observed variation of the position angle with frequency is consistent with Faraday rotation of the plane of polarisation with a rotation measure of -40.5 ± 4.5 rad/m^2. This value is of the same sign but larger than the rotation measure for the nebular radiation in the vicinity of the pulsar.

The intrinsic angle (position angle at infinite frequency) of the radio radiation is an important quantity, particularly since polarised optical pulses are observed from this pulsar. Unfortunately, the present accuracy of the rotation measure determination is insufficient to fix this angle, the formal value being $90° \pm 140°$. However even when an improved value for the intrinsic angle is obtained, comparison of the radio and optical results will not be straightforward as pulse shapes and polarisation characteristics are different for the radio and optical pulses.

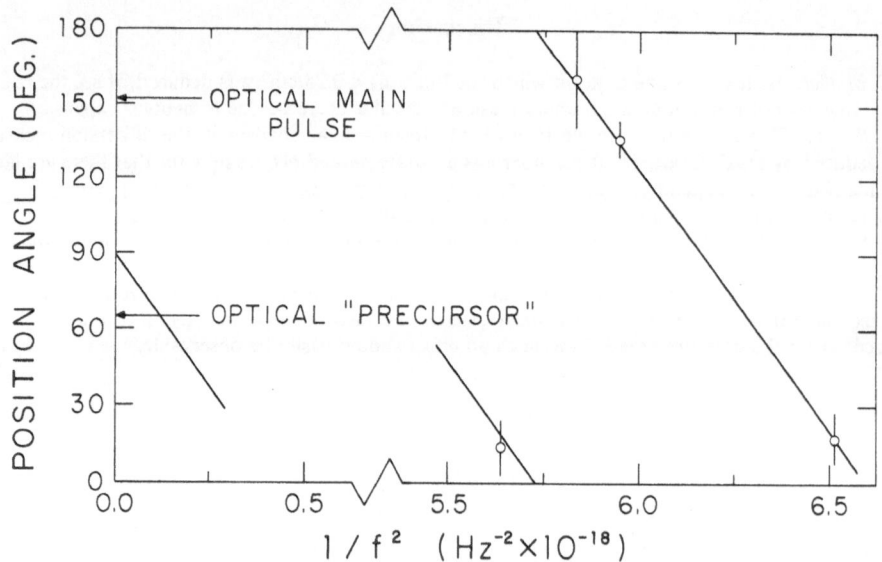

Fig. 1. Position angle of the Crab pulsar precursor measured at four frequencies between 392 MHz and 421 MHz plotted against inverse frequency squared. A line corresponding to a rotation measure of -40.5 is drawn through the points. The position angle of the pulsed optical radiation at the peak of the main pulse and at pulse phase corresponding to the radio precursor are indicated.

* Operated by Associated Universities, Inc., under contract with the National Science Foundation.

2.10 TIMING OF OPTICAL PULSES FROM THE CRAB NEBULA PULSAR*

P. E. BOYNTON**, E. J. GROTH, R. B. PARTRIDGE[†] and
DAVID T. WILKINSON[††]

*Joseph Henry Laboratories, Physics Department, Princeton University,
Princeton, N.J., U.S.A.*

Abstract. Timing the arrival of optical pulses from NP 0532 is a potentially important tool for studying the physics of this fascinating object. However, there are some difficulties in interpreting the data in terms of physical models. Some progress has been made on understanding the largest effect – the pulsar braking mechanism. The glitch of late September, 1969 can be interpreted as the speed-up, and subsequent relaxation, of the rotation of a neutron star crust. An alternate explanation is that of a planet in an eccentric orbit. Both models fit the rather meager data near the event. A small sinusoidal effect is indicated in a relatively quiet period of the data.

1. Introduction

The phase of pulses from the Crab Nebula pulsar has been followed for over a year now by radio and optical observers (Boynton *et al.*, 1969a; Richards *et al.*, 1970; Duthie and Murdin, 1970; Nelson *et al.*, 1970). This means, in effect, that the last 10^9 or so pulses from NP 0532 have been counted, and that for some of these pulse arrival times have been measured with respect to precision atomic clocks.

Since arrival times can be easily determined to better than 10^{-3} cycles ($\sim 30 \mu$s) in one night, the inherent precision of the measurement is better than 1 part in 10^{12} per year. However, in order to take full advantage of this precision, we must understand simultaneously (1) the pulsar clock mechanism, (2) any effects causing relative acceleration between the telescope and the pulsar, (3) possible effects on propagation of the pulses, and (4) relativistic effects on the clocks. The physics of (1) is poorly understood and could turn out to be very complicated at this level of precision. Most of the effects of interest in (2) are periodic, such as orbital displacements of the earth (taken out by an ephemeris) and of the pulsar (unknown). To fully utilize the current timing accuracy, we must understand those accelerations which change the Earth-pulsar distance by as little as a few kilometers.

Finally, to complicate matters still further, there was a 'sudden' decrease in the slowing-down rate of NP 0532 in late September, 1969 (Boynton *et al.*, 1969b; Richards *et al.*, 1969). Unfortunately, data is particularly sparse around that time, and progress toward understanding this 'glitch' has been slow and inconclusive to date.

---

* This research was supported in part by the Office of Naval Research and the National Science Foundation.
** Now located: Astronomy Dept., Univ. of Washington, Seattle, Wash., U.S.A.
† Now located: Haverford College, Haverford, Penn., U.S.A.
†† Alfred P. Sloan Fellow.

So, at the moment we are in the exciting, but frustrating, position of having data which appears to be considerably better than our understanding of the underlying physics. Putting it another way, there are still real effects buried in the timing data which must be recognized by their particular time dependence, and then removed. Therefore, we should not be surprised, or discouraged, if different observers present quite different interpretations of what, in the final analysis, is the same physical situation. We have different samples of a very complicated time series, and are using different approaches to understanding the data. Publishing and/or pooling raw timing data would probably lead to the resolution of most of these differences.

2. Data Analysis

At Princeton we have approached the data analysis as follows. (1) Telescope arrival times are transformed to the solar system barycenter, and known relativistic corrections (Hoffman, 1968) associated with the Earth's motion are removed. (2) A possible physical model is assumed and described by an equation involving time and some adjustable parameters. (3) The modeled phase is fitted to the measured phase by the method of least squares, giving best estimates of the model parameters and a goodness-of-fit estimate, χ^2. (4) As the model improves, the value of χ^2 approaches the number of degrees of freedom of the fit. This method of analysis assumes that the temporal behavior of the pulsar can be modeled and predicted, except for occasional small discontinuities in the rate. This assumption may be overly optimistic, but it does emphasize a physical, rather than a phenomenological, treatment of the data.

Our data span the dates March 15, 1969 to April 26, 1970 with a gap from May to August 1969, when the Crab was too near the Sun. Some of our analysis has also included data of the Arecibo radio timing groups which fills in the summertime gap. Here we will present only some qualitative results; details, including graphs of residuals and tables of data, are being prepared for publication elsewhere.

A. POLYNOMIAL BRAKING

For some time it has been recognized that NP 0532 is secularly slowing down (Richards and Comella, 1969). One explanation (Gold, 1968; Pacini, 1967; Gunn and Ostriker, 1969) has a rotating neutron star (frequency v) losing angular momentum, with the pulsation frequency equal to the rotational frequency. We model this particular behavior either as a polynomial expansion of the phase

$$\varphi(t) = \varphi_0 + v_0 t + \tfrac{1}{2}\dot{v}_0 t^2 + \tfrac{1}{6}\ddot{v}_0 t^3 + \cdots, \tag{1}$$

or as the more restrictive equation for $\varphi(t)$ which satisfies power-law radiation, $\dot{v} \propto v^n$. In the power-law fit, derivatives of φ larger than $\ddot{\varphi}$ are constrained, and the results of fitting a cubic polynomial are essentially the same as those of a power-law fit. The parameter of most interest in these fits is n – the braking index, because n

determines the braking mechanism ($n=3$ for magnetic dipole radiation, etc.). Depending on the particular piece of data fitted, we obtain values of n anywhere from 2 to 5. A simple power-law fit to all of our data gives $n=2.5$, but this is a very poor fit, partly because of the glitch, but mostly due to a large quartic-looking residual. This residual is not understood; it is two orders of magnitude larger than the quartic term expected from a power-law expansion.

The apparent instability of n may be real, if the braking mechanism is indeed complicated and time dependent, or it may be due to underlying transient or periodic effects which have not been removed from the data. This situation should become clearer as more data is obtained, and long term effects are evaluated. The braking effect is included in all of the following fits, either as a polynomial or a power-law.

B. SMALL SINE WAVE

There is a relatively quiet stretch of data lasting for about 4 months between December and March. Simple power-law fits in this region leave periodic residuals and give $\chi^2 = 500$ for 28 deg of freedom. If a sine wave is included in the fit the following best-fit parameters are obtained:

$$75 < \text{Amplitude} < 150 \ \mu\text{sec},$$
$$50 < \text{Period} < 60 \text{ days},$$
$$13 \text{ Jan.} < \text{Maximum} < 23 \text{ Jan. } 1969.$$

Including the sine wave in the fit gives a χ^2 of 50, for 25 deg of freedom. This improvement of χ^2 by a factor of 10 is a strong indication that this effect is real. It should be noted that more than 2 cycles of the sine wave are included in this data string; fitting sine waves of fewer than 2 cycles can lead to serious interference from the cubic polynomial. Unfortunately, we have no other piece of data which is long enough and clean enough to look for this sine wave. If the effect persists with stable frequency and amplitude, a planetary interpretation is favored. However, if the period and amplitude are found to change, there are other more complicated models (Ruderman, 1970).

C. TRANSIENT GLITCH FIT

As mentioned above there is a transient in pulse arrival times occurring in late September 1969, and lasting through early October. Qualitatively the feature is consistent with two-component neutron star models with sudden crust speed-up (Greenstein and Cameron, 1969; Baym et al., 1969) followed by a period of decay back to steady state rotation. We have modeled this event as a sudden change in frequency Δv, followed by an exponential decay back to the original braking curve, except that provision is made for a permanent change in v at the time of the glitch. This is to allow for a possible permanent change in the moment of inertia of the star.

The results of this fit are a function of the length of data included on either side of the glitch, again indicating that underlying effects are 'pulling' the results. For this particular model the fitted parameters usually fall within the following ranges:

0600 UT 28 Sept. < Glitch Time < 0600 UT, 29 Sept.

$$6 \times 10^{-9} < \frac{\Delta \nu}{\nu} < 8 \times 10^{-9}$$

$$0.4 \times 10^{-3} < \frac{\Delta \dot{\nu}}{\dot{\nu}} < 1.5 \times 10^{-3}$$

4 < Recovery Time Constant < 16 days,

$$-1 \times 10^{-9} < \text{Permanent } \frac{\Delta \nu}{\nu} < +1 \times 10^{-9}.$$

For the crust-cracking model (Ruderman, 1969) the values of Q obtained from these fits range from 0.85 to 1.17, where a value greater than 1.0 rules out the model (Baym *et al.*, 1969). However, the larger values of Q are obtained from longer pieces of data where the results may be more susceptible to distortion by other effects.

D. ECCENTRIC PLANET GLITCH FIT

Another model that has been proposed (Michel, 1970; Hills, 1970) to explain the 'sudden' changes in pulsar frequencies is that of a planet in a highly eccentric orbit around the pulsar. We have attempted to fit the September 1969 glitch with such a model, with surprisingly good results. The best fits to the glitch are given by a planet which moves the pulsar in an orbit, with the following elements:

200 < Period < 220 days,
0.3 < Eccentricity < 0.8,
1000 < a sin *i* < 2000 km
Oct. 1 < Periastron Passage < 10 Oct.,
170° < Longitude of Periastron < 210°.

Again the ranges are not formal fitting errors, which are much smaller, but estimates of the stability of the results for different fitting models and various pieces of the data.

It may seem surprising that the glitch data can be fit equally well by a transient model and by a relatively smooth elliptic orbit model. There are two reasons for this. First, the data are sparse; to our knowledge no one has data during the week preceding and the week following 29 September. So the detailed behavior near the glitch is not known. Secondly, the glitch in NP 0532 was small. A build-up of phase error over a period of several weeks was needed to see the effect; the total cumulative phase displacement caused by the glitch is only about 0.1 cycles (3 msec).

3. Conclusion

The high inherent precision of the Crab pulsar timing measurement is a mixed blessing at the moment, but if we are patient the rewards may be great. These include the possibility of examining (1) the pulsar clock mechanism, (2) possibly some details of the most interesting physics of neutron star interiors and crusts, or perhaps the orbits of a planetary system around a supernova remnant. Some things can already be said about this complicated system. Results to date on the slowing down rate favor a long-term average braking index between 2 and 2.5. If the glitch was caused by a sudden speed-up of the crust, it took the two components a week or two to relax back to steady state rotation, indicating a very weak coupling which points to a superfluid interior (Migdal, 1959; Ginzburg and Kirzhnits, 1964). Finally, there remains the possibility that the Crab glitch was caused by the close passage of a planet in an eccentric orbit. We should know about this within the next few months as more data allows us to explain and remove effects which are interfering with current models. Also, with the increased number of observers, data on the next glitch should be much better.

Acknowledgements

We are indebted to the Princeton Observatory for generous grants of time on their 36″ telescope. The ephemeris is supplied to us by Dr. J. Derral Mulholland and the Jet Propulsion Laboratory.

References

Baym, G., Pethick, C., Pines, D., and Ruderman, M.: 1969, *Nature* **224**, 872.
Boynton, P. E., Groth, E. J., Partridge, R. B., and Wilkinson, D. T.: 1969a, *Astrophys. J. Letters* **157**, L197.
Boynton, P. E., Groth, E. J., Partridge, R. B., and Wilkinson, D. T.: 1969b, *IAU Circ.* No. 2179.
Duthie, J. G., Murdin, P.: 1970, preprint.
Ginzburg, V. L. and Kirzhnits, D. A.: 1965, *Soviet Phys. – JETP* **20**, 1346.
Gold, T.: 1968, *Nature* **218**, 731.
Greenstein, G. S. and Cameron, A. G. W.: 1969, *Nature* **222**, 862.
Gunn, J. E. and Ostriker, J. P.: 1969, *Nature* **221**, 454.
Hills, J. G.: 1970, *Nature* **226**, 730.
Hoffman, B.: 1968, *Nature* **218**, 667.
Michel, F. C.: 1970, *Astrophys. J. Letters* **159**, L25.
Migdal, A. B.: 1960, *Soviet Phys. – JETP* **10**, 176.
Nelson, J., Hills, R., Cudaback, D., and Wampler, J.: 1970, *Astrophys. J. Letters* **161**, L235.
Pacini, F.: 1967, *Nature* **216**, 567.
Richards, D. W. and Comella, J. M.: 1969, *Nature* **222**, 552.
Richards, D. W., Pettengill, G. H., Counselman, C. C., and Rankin, J. M.: 1970, *Astrophys. J. Letters* **160**, L1.
Richards, D. W., Pettengill, G. H., Roberts, J. A., Counselman, C. C., and Rankin, J.: 1969, *IAU Circ.* No. 2181.
Ruderman, M.: 1969, *Nature* **223**, 597.
Ruderman, M.: 1970, *Nature* **225**, 838.

Discussion

F. D. Drake: Your two-planet model would predict that in April the Arecibo timing measurements would have shown a period discontinuity very similar to that seen in September 1969. Such an event did not occur.

D. T. Wilkinson: There are interactions of the two large 'planetary' effects which modify the shapes of discontinuity events. If this model makes any sense, the predicted April 1970 discontinuity could look quite different from the September 1969 one. We are currently trying to predict this shape.

Secondly it is possible to force a polynomial fit to partially smooth out a discontinuity event. It was nearly the end of October before the September discontinuity became apparent, and then only by fitting the data in a particular way. Before that it merely appeared that the data had become very noisy.

N. Visvanathan: There was a report in an *IAU Circular* in April this year indicating a change in period of the Crab Pulsar by Dr. Manchester. I understand it is now withdrawn.

2.11   TIMING OBSERVATIONS OF THE CRAB NEBULA PULSAR AT THE ARECIBO OBSERVATORY

J. A. ROBERTS* and D. W. RICHARDS**

Arecibo Observatory, Arecibo, Puerto Rico

Abstract. Timing observations of the Crab pulsar were made between 10 May 1969 to 3 July 1970. Arrival times were corrected to the barycentre of the solar system; a further correction was made for the effect of dispersion. The only jump in period occurred in September 1969, although small irregularities in the period occurred at other times. A third order polynomial fitted all the observed data to within ± 0.15 pulse periods.

Since May 1969 timing observations of the Crab Nebula pulsar have been made at the Arecibo Observatory approximately twice per week. Observations are made simultaneously at 2 or more of the frequencies 430, 318, 196.5, 111.5 and 73.8 MHz, with a sampling interval of 32 μsec. For each run of approximately 18 min (32 000 pulse periods) the corresponding samples in successive pulse periods are summed. Between 5 and 9 such runs are usually made on one observing day. Topocentric arrival times are found by cross-correlating the summed data with an expected pulse shape which includes the effects of the receiver parameters, and the pulse smearing which is presumed to be caused by interstellar multipath propagation. Further details are given by Rankin *et al.* (1970).

By using Loran C transmissions the observatory time is referenced to the U.T.C. system to better than 10 μsec. Corrections are made for the *periodic* error of a terrestrial clock caused by the non-circular orbit of the earth (Clemence and Szebehely, 1967), and arrival times are referenced to the solar system barycentre using an ephemeris kindly provided by Ash, Shapiro and Smith of the MIT Lincoln Laboratory. A ten day average of the dispersion constant deduced from the same observations (Rankin and Roberts, 1970) is used to extrapolate the 430 MHz barycentric arrival times to infinite radio frequency. Pulse numbers are assigned successively by extrapolating the known previous behaviour of the pulsar. Finally, daily mean arrival times are found from cubic fits of pulse number versus arrival time made to data from several successive observing days. For the interval 10 May 1969 to 3 July 1970 there are 112 such daily mean arrival times.

To examine irregularities in the pulsar behaviour, observations on successive observing days are used to derive a mean pulsar repetition frequency for the intervening days. The pulsar slows down so rapidly that it is necessary to add a linear function of the time to these frequencies before plotting. Figure 1 shows such plots for successive, and partly overlapping, intervals of approximately 100 days. A different multiple of the time was chosen for each section so that the residuals would form approximately symmetric curves.

* On leave from CSIRO Radiophysics Laboratory, Epping, N.S.W., 2121, Australia.
** Present address: Air Force Cambridge Research Laboratories, Bedford, Mass., U.S.A.

Davies and Smith (eds.), The Crab Nebula, 125–128. All Rights Reserved.
Copyright © 1971 by the IAU.

The frequency jump in September, 1969 (Boynton *et al.*, 1969; Richards *et al.*, 1969) is clearly seen in the second curve from the bottom. There are no other such obvious frequency jumps. There are, however, times when the pulsar behaviour is irregular and some of these irregularities may be caused by smaller frequency jumps, as for example in the interval between 12 April 1970 and 24 April 1970 (left hand side of the top curve). However the much larger irregularities in the mean frequencies seen in June and July 1970 (right hand side of upper curve) do not appear to indicate permanent frequency jumps.

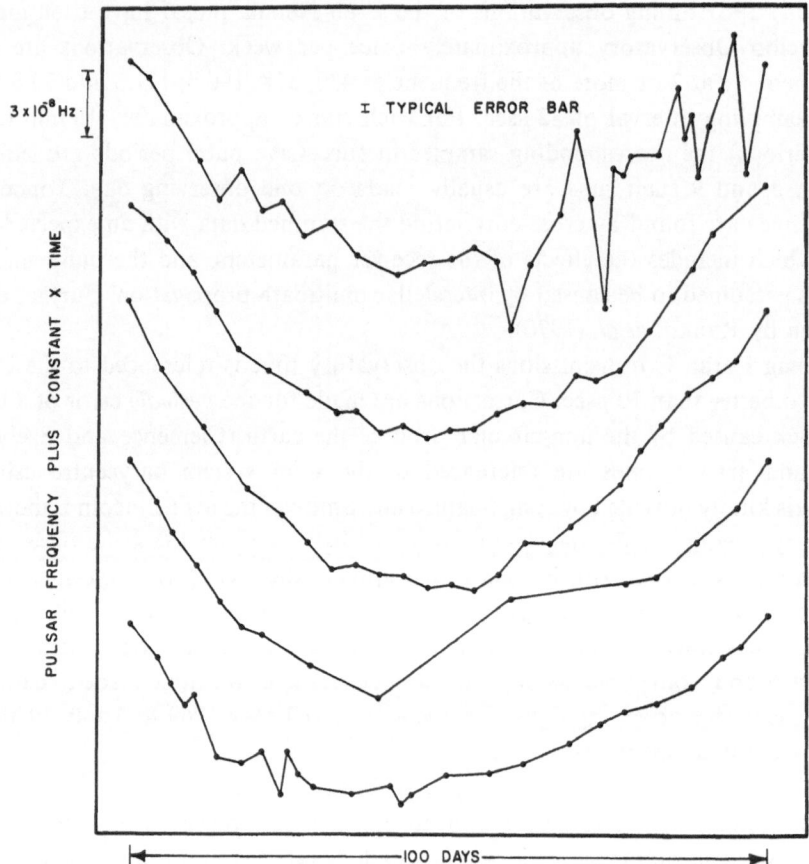

Fig. 1.   Mean pulsar repetition frequency deduced from arrival times on two successive observing days, plus a multiple of the mean of the two arrival times plotted as a function of this mean time. The 'typical error bar' corresponds to errors of 20 μsec ($\sim$ twice the rms error) at either end of a 3 day interval. The error increases for more closely spaced observations and decreases for more widely spaced observations. Bottom curve: 1969 May 10–1969 August 10; Multiplier 3.857972×10^{-10} sec^{-2}. Second bottom: 1969 July 30–1969 November 12; Multiplier 3.857096×10^{-10} sec^{-2}. Middle curve: 1969 October 20–1970 January 20; Multiplier 3.856300×10^{-10} sec^{-2}. Second top: 1969 December 30–1970 April 6; Multiplier 3.855456×10^{-10} sec^{-2}. Top curve: 1970 March 25–1970 July 3; Multiplier 3.854558×10^{-10} sec^{-2}.

It was hoped that a polynomial fitted to pulse number versus arrival time would describe the steady rotation and slowing down of the star, and leave any irregularities as residuals to the fit. However it is necessary to use at least a cubic to fit the observations, and the interpretation is not clear. Experiments show that a polynomial fitted to a set of observations is not a reliable means of predicting the future behaviour of the pulsar. Fits to the data prior to the end of September 1969 correspond to solution of

$$dv/dt = - Kv^n,$$

with $n \simeq 2$. However after the frequency jump n increased to nearly 3, and then decreased to approximately 2.5. The quasi-sinusoidal residuals evident prior to the jump (Richards *et al.*, 1970) perhaps continued until January 1970 and then disappeared rather suddenly.

If a polynomial fitted to arrival times prior to the September 1969 frequency jump is used to predict the future behaviour of the pulsar, it is found, as expected, that actual arrival times are systematically earlier. By 3 July 1970 the accumulated error is 13 full pulse periods. However, if a cubic is fitted to the whole data span from 10 May 1969 to 3 July 1970, the departures from the fit are never greater than ± 0.15 pulse periods. Hence, it is possible that the stellar rotation changed smoothly with time, approximately as described by this cubic ($n=2.6$), and that the source moved relative to this rotating frame by up to $\pm 50°$.

This note reports one aspect of a project in which J. M. Rankin of the University of Iowa, C. C. Counselman III of MIT and G. H. Pettengill of Arecibo Observatory are also involved. A fuller report of the results will be made at a later date. The project would not have been possible without the full support of the staff of Arecibo Observatory and the financial support of sponsoring agencies. Arecibo Observatory is operated by Cornell University under contract with the National Science Foundation with partial support from the Advanced Research Projects Agency. The present work is also supported in part by Air Force Office of Scientific Research contract F44620-69-C-0092 and NASA grant NGL 16-001-002.

References

Boynton, P. E., Groth, E. J., III, Partridge, R. B., and Wilkinson, D. T.: 1969, *IAU Circ.* No. 2179.
Clemence, G. M. and Szebehely, V.: 1967, *Astrophys. J.* **72**, 1324.
Rankin, J. M., Comella, J. M., Craft, H. D., Jr., Richards, D. W., Campbell, D. B., and Counselman, C. C., III: 1970, *Astrophys. J.* **162**, 707.
Rankin, J. M. and Roberts, J. A.: 1970, 'Crab Nebula Pulsar: Temporal Variation of Dispersion Measure', in preparation.
Richards, D. W., Pettengill, G. H., Roberts, J. A., Counselman, C. C., III, and Rankin, J. M.: 1969, *IAU Circ.* No. 2181.
Richards, D. W., Pettengill, G. H., Counselman, C. C., III, and Rankin, J. M.: 1970, *Astrophys. J. Letters* **160**, L1.

Discussion

F. C. Michel: What is n for all 500 days? It seems to me that one must either decide to use

$$\frac{dv}{dt} = k(t) v^{n(t)}$$

which is what one does implicitly when one talks about a 'jump in the period', or use

$$\frac{dv}{dt} = kv^n + residuals(t)$$

where k and n are constants that are increasingly better determined as more data becomes available, and the residuals represent the short term drift effects. To compromise and fit to constant k and n for different data runs is an unfortunate mixture of these two approaches.

J. A. Roberts: $n \simeq 2.6$ for the whole time interval.

R. G. Conway: You showed two periods of time when the pulse frequency was 'rough'. Am I right in thinking that they both occurred in June when the Sun and Crab are close?

J. A. Roberts: There is no real correlation between the occurrence of roughness and being near the Sun.

R. Schwarz: Could the 'rough' periods be connected with changes in the dispersion measure?

J. A. Roberts: The effects of dispersion measure changes upon arrival times have been removed.

2.12 A SEARCH FOR VARIATIONS
IN THE INTENSITY OF THE OPTICAL PULSES
FROM NP 0532

D. HEGYI*

NASA Goddard Institute for Space Studies, New York, N.Y., U.S.A.

R. NOVICK

Columbia Astrophysics Laboratory,
Physics Department, Columbia University, New York, N.Y., U.S.A.

and

P. THADDEUS

Goddard Institute for Space Studies, New York, N.Y., U.S.A.

Abstract. A search for short-time variability of NP 0532 using the 82 and 107 in. telescopes at McDonald Observatory is described. Observations were made of the mean intensity, the mean square of the intensity and the mean autocorrelation function of the pulsar light. Limits are placed on the variability of the optical pulsar.

1. Introduction

A search for variations in the optical pulses from NP 0532 in the Crab Nebula – the only known optical pulsar – is prompted by the large and erratic variations observed in the intensity of its radio pulses, (Graham *et al.*, 1970; Heiles *et al.*, 1970; Staelin and Sutton, 1970) and in the several types of irregularities found in those of other pulsars. Since our understanding of the optical emission mechanism of NP 0532 is, if anything, even more speculative than that of its radio emission, the detection of such variations is potentially of considerable interest, and should be pursued not only on the time scale of milliseconds, but on much shorter and longer scales as well. With modern photoelectric and digital techniques it is possible to do this all the way from the scale of seconds or minutes associated with conventional photometry to nanoseconds.

Some limits have already been set on optical pulse variation, but the data are fragmentary, and occasionally contradictory. Various authors, ourselves included, have shown that no extreme fluctuations occur in the pulsar light on the scale of nanoseconds or microseconds (Duthie *et al.*, 1969; Anderson *et al.*, 1969; Ögelman and Sobieski, 1969; Hegyi *et al.*, 1969; Jelley and Willstrop, 1969); observations for the most part, however, were made with small telescopes, and as the following discussion will show, this severely limits the sensitivity that can be attained on a very short time

---

* National Research Council Postdoctoral Research Associate, currently at the Department of Physics and Astronomy, Boston University.

Davies and Smith (eds.), The Crab Nebula, 129–141. All Rights Reserved.
Copyright © 1971 by the IAU.

scale. The possible existence on a scale of milliseconds or a fraction thereof of optical analogues to the 'marching subpulses' observed in several pulsars (Drake and Craft, 1968; Sutton *et al.*, 1970) or the 'giant' radio pulses of NP 0532 itself has received little discussion; presumably rather restrictive limits to such structure could be extracted from the large existing body of optical timing and polarization observations. Finally, on a scale of minutes rather extreme variations in the optical intensity of a given component of polarization have been reported (Freeman *et al.*, 1969; Cocke *et al.*, 1969), but there now seems to be widespread agreement among observers that this effect is spurious (cf. Cocke *et al.*, 1970).

Even with a large reflecting telescope only 10–50 photoelectrons are received over the 33 msec period of NP 0532, and it is not possible from the record of individual pulses to determine their shape to any great precision. The usual way to study the structure in greater detail is to coherently add together in a multichannel recording instrument the photoelectric counts received from many thousands of pulses; but this technique cannot distinguish between the case where all pulses are identical, and the more general situation where the intensity of individual pulses $I(t)$, is a *random variable* whose mean value $\langle I(\phi) \rangle$, as a function of phase ϕ, is the quantity observed.

To decide between these alternatives it is necessary to observe higher moments of the probability distributions which define a random process. Here we will describe observations of the mean intensity, the mean square of the intensity, and the mean autocorrelation function of the pulsar light which we have made with the 82 and 107 in. telescopes at McDonald Observatory. From intercomparison of these quantities it is found that little if any random variation in the pulse structure exists, or in other words that the individual optical pulses of NP 0532 appear to be identical, one to another. Limits are also imposed on gradual secular variation of the intensity of the optical pulsar, but these results are not as sensitive as those which can be obtained by standard photometry.

A few elementary quantitative considerations will make these points more precise, and will indicate the way in which the data have been analyzed. Let $W_1(I, t)$ and $W_2(I_1, t_1, I_2, t_2)$ be the first and second probability distributions which describe the postulated random intensity. These are defined in the usual way such that $W_1(I, t)\,dI$ is the probability of finding the intensity between I and $I+dI$ at time t, and $W_2(I_1, t_1, I_2, t_2)\,dI_1dI_2$ is the joint probability of finding a pair of values of I in the ranges I_1, I_1+dI_1 and I_2, I_2+dI_2 at the respective times t_1 and t_2. In the subsequent discussion it will be assumed that these two distributions are periodic with the pulsar period of 33 msec, so that the pulsar phase ϕ, rather than t, becomes the appropriate independent variable. The quantities which are yielded by the observations described below are the first and second moments of the first probability distribution, i.e., the mean intensity

$$\langle I(\phi) \rangle = \int_0^\infty I W_1(I, \phi)\, dI,\tag{1}$$

and the mean intensity squared

$$\langle I^2(\phi) \rangle = \int_0^\infty I^2 W_1(I, \phi) \, dI,$$ (2)

and in addition the mean autocorrelation function, which in terms of the second probability distribution we define to be

$$G(\tau) = \frac{1}{T} \int_0^T \int_0^\infty \int_0^\infty I_1(\phi - \tau) I_2(\phi) W_2(I_1, \phi - \tau, I_2, \phi) \, dI_1 \, dI_2 \, d\phi,$$ (3)

where $T \approx 33$ msec is the pulsar period.

It now follows directly from consideration of the variance $\langle (I - \langle I \rangle)^2 \rangle$ that

$$\langle I^2(\phi) \rangle \geqslant \langle I(\phi) \rangle^2,$$ (4)

the equality holding when all pulses are identical. This is the fundamental condition on which our most sensitive search for pulse variations will be based, the results applying to time scales from roughly microseconds to tens of minutes. Information on the pulse structure on the scale of nanoseconds will be provided by measurement of $G(\tau)$, which in the limit of identical pulses and $\tau \to 0$ reduces to

$$G_0 = \frac{1}{T} \int_0^T \langle I(\phi) \rangle^2 \, d\phi.$$ (5)

2. Apparatus

A. FOR COMPARISON OF $\langle I^2 \rangle$ WITH $\langle I \rangle^2$

A schematic illustration of the apparatus in the 'intensity squared' mode is given in Figure 1. Light from an aperture in the focal plane of the telescope was split into two beams of approximately equal intensity by front surface reflection from an aluminized right-angle prism. The beams were then detected by two fast Amperex 56DVP photomultipliers located 50 in. apart in order to reduce fast (nanosecond) cosmic ray coincidences to a negligible level. Each tube was followed by conventional stages of amplification and discrimination, but the discriminators were turned off for 10 μsec, following each photoelectron pulse, by a 'deadtime' circuit to eliminate the effects of phototube afterpulsing. The total number of counts received in each channel during an observing run was recorded by the scalers which, in Figure 1, are shown adjacent to the discriminators.

Following discrimination, pulses from phototube 1 triggered a single pulse generator ('one-shot') which in turn opened a coincidence gate for a predetermined length of time $\Delta\tau$. Pulses from tube 2 which passed through the gate were then stored in a 511 channel multichannel scaler which was synchronized to the pulsar frequency. All

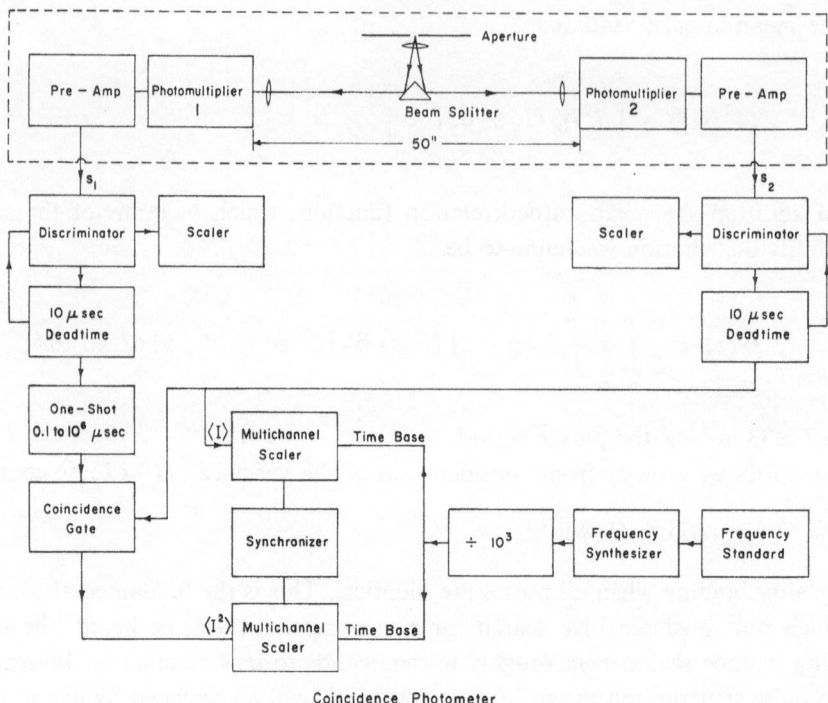

Fig. 1.   Schematic diagram of the apparatus.

pulses from tube 2, whether they passed through the gate or not, were stored in a second multichannel scaler also driven synchronously with the pulsar.

The time base for both multichannel scalers was provided by the apparatus shown in the lower right hand corner of Figure 1. A frequency synthesizer stabilized by the rubidium frequency standard at McDonald Observatory was manually adjusted at about 5 minute intervals to precisely 511 000 times the calculated* pulsar frequency. Division by 1000 then provided a sequence of timing pulses which advanced the multichannel scalers from one channel to the next. It is believed that phase slippage during the longest observations ($\sim$ 1 h) amounted to only a fraction of a channel.

To understand the essential operation of the device it is best to consider only low light intensities, such that the probabilities $s\,\Delta\tau$ and s $\Delta\tau_D$ of counts arriving within the deadtime $\Delta\tau_D = 10$ μsec, and the coincidence gate resolution time $\Delta\tau$, are infinitesimal. In fact, these probabilities while always small, were not always negligibly so, and it was necessary to make small corrections to the recorded counts in order to compare $\langle I \rangle^2$ with $\langle I^2 \rangle$. For the sake of brevity the details of these corrections will be omitted from the following discussion.

Suppose now that $s_1(t)\,dt = \alpha_1 I(t)\,dt$ and $s_2(t)\,dt = \alpha_2 I(t)\,dt$ are the probabilities, in channels 1 and 2 respectively, of the arrival of a photoelectron count in the interval

* Based on measurements of the NP 0532 frequency and its time derivatives made by Boynton *et al.* (1969).

between t and $t+dt$. These probabilities, although comparable, are not precisely equal because of the differences in the phototube efficiencies and inaccuracy in splitting the optical beam; we will only assume that their ratio, and hence the ratio of α_1 to α_2, is constant over an observing run. The probability $c(t)\,dt$ of a pulse passing through the coincidence gate during the interval $t, t+dt$ is then $s_2(t)\,dt$ times the probability P_0 that the gate is open, which in turn is

$$P_0(t) = \int_0^{\Delta\tau} s_1(t-\tau)\,d\tau. \tag{6}$$

It is now desirable to either assume that $I(t)$, and hence $s_1(t)$, possess no structure on a scale finer than $\Delta\tau$, or better still, to redefine $I(t)$ to be the instantaneous intensity averaged over a time of order $\Delta\tau$, so that Equation (6) simplifies to

$$P_0(t) = s_1(t)\,\Delta\tau, \tag{7}$$

and consequently

$$c(t) = s_1(t)\,s_2(t)\,\Delta\tau = \alpha_1\alpha_2\,\Delta\tau I^2(t). \tag{8}$$

The number of counts C_i stored in the i'th channel of the $\langle I^2 \rangle$ multichannel scaler is then simply $c(t)$ integrated over the channel width $\Delta\tau_c = T/N_c$, (where $N_c = 511$ is the total number of channels), and summed over all the NP 0532 pulses in the observing run of duration T_r. The result is

$$C_i = \frac{\alpha_2^2 N_1 \,\Delta\tau T_r}{N_2 N_c} \langle I^2(\phi_i) \rangle, \tag{9}$$

where N_1 and N_2 are the total counts recorded by the two scalers, and it has been assumed that $\alpha_1/\alpha_2 = N_1/N_2$, a relation which neglects phototube dark counts, which were very low in the phototubes used. Similarly, the number of counts stored in the i'th channel of the $\langle I \rangle$ multichannel scaler is found to be

$$S_i = \frac{\alpha_2 T_r}{N_c} \langle I(\phi_i) \rangle. \tag{10}$$

Thus from Equations (9) and (10) we see that the fundamental condition $\langle I^2(\phi) \rangle \geqslant \langle I(\phi) \rangle^2$ is equivalent to

$$C_i \geqslant \frac{N_1 \,\Delta\tau N_c}{N_2 T_r} S_i^2. \tag{11}$$

Since all the quantities occurring in Equation (11) are well defined, being either parameters of the apparatus $(\Delta\tau, N_c)$, or numbers furnished by the counters (C_i, S_i, N_1, N_2), analysis of the data is straightforward.

B. FOR COMPARISON OF $G(\tau)$ WITH G_0

The configuration of the apparatus in the mode used to determine the 'mean auto-correlation function' $G(\tau)$ has been previously described (Hegyi *et al.*, 1969). In this mode the coincidence gate shown in Figure 1 is replaced by a time-to-amplitude converter (TAC), which is started by photoelectric counts from phototube 1 and stopped by counts from phototube 2 which have been delayed ~ 30 nsec by a length of line. The TAC provides an output pulse whose height is proportional to the time elapsed between the arrival of the start and the arrival of the stop counts, but only if it is stopped before reaching the end of its range; if it runs off range no output pulse is produced. Nominal ranges of 100 nsec, 300 nsec, 1 μsec,..., 30 μsec are provided.

It is now easy to see that pulse height analysis of the TAC output yields essentially $G(\tau)$. The probability in the time interval t, $t+dt$ of receiving an output pulse from the TAC corresponding to a time delay in the range τ, $\tau+\Delta\tau$ is $s_1(t-\tau)s_2(t)dt\,d\tau = \alpha_1\alpha_2 I(t-\tau)I(t)dt\,d\tau$. Hence the total number of counts C_i received in the delay range τ, $\tau+\Delta\tau$ over a run of duration T_r, and stored in the i'th channel of the pulse height analyzer, is

$$C_i = \alpha_1\alpha_2 T_r\,\Delta\tau\langle\int_0^T I(\phi-\tau)I(\phi)\,d\phi\rangle\cdot \tag{12}$$

Since the average over many periods of NP 0532, indicated by the $\langle\ \rangle$ in Equation (12), must, by the hypothesis that the distribution functions are periodic with the pulsar, yield the same result as the ensemble average of Equation (3). Equation (12) becomes

$$C_i = \alpha_1\alpha_2 T_r\,\Delta\tau G(\tau) \tag{13}$$

If again N_1 and N_2 are the total counts recorded during a run by the two scalers, and the S_i are the total number of counts recorded in the i'th channel of the $\langle I\rangle$ multichannel recorder shown in Figure 1 (there is of course no $\langle I^2\rangle$ analyzer in this mode of operation), then by Equation (10) the limit $G(\tau)=G_0$ [Equation 5)], is equivalent to

$$C_i = \frac{N_1 N_2\,\Delta\tau\xi}{T_r}, \tag{14}$$

where the enhancement factor ξ, which is unity for steady light, is defined by

$$\xi = N_c\sum_i S_i^2/\left(\sum_i S_i\right)^2. \tag{15}$$

All the terms appearing in Equation (14) are well defined experimentally, so that analysis of the data simply consists in comparing the left with the right hand side of the equation. ξ should remain constant under ideal observing conditions for a given aperture; it was found to change slowly by up to a factor of two in the course of a night during our observing period because of a varying haze layer which, by scattering moonlight, enhanced the background light relative to that of the pulsar. Unfortunately, for these observations, only a single multichannel instrument was available (Northern Scientific NS-550 'Digital Memory Oscilloscope'), and this did double service as both a multichannel scaler and pulse height analyzer. It was therefore not possible to determine C_i and ξ simultaneously, as would have been most desirable; instead the S_i and hence ξ were determined by a brief (5 or 10 min) intensity run just before and, when possible, immediately after the generally longer observation of the C_i.

3. Results

A. $\langle I^2 \rangle$ VS. $\langle I \rangle^2$

Observations of the mean intensity squared of NP 0532 were made at the Cassegrain focus of the 82 inch telescope on the nights of 13, 14, and 15 February 1970 UT. The moon passed near the Crab Nebula toward the end of this period, and the observations were occasionally interrupted by clouds, but useful data were obtained on all three nights.

A typical observation lasted from 5 min to 1 h. In order to exclude scattered moonlight and background nebular light, observations were made with the smallest aperture that the seeing allowed; on a few exceptional occasions a 3.5″ aperture was tried with some success, but generally a 4″ one was the smallest used under good seeing conditions. Since slow modulation of the pulsar light, owing for example to guiding irregularities or clouds, can be as effective as short term fluctuations in enhancing $\langle I^2 \rangle$ with respect to $\langle I \rangle^2$, the mean counting rate (time average about 1 sec) of one of the phototubes was continuously monitored by a chart recorder. In this way it was established for the best runs that long term atmospheric or guiding effects probably caused an enhancement of $\langle I^2 \rangle$ of less than 1%, and can therefore be safely neglected in the data analysis. Observation of field stars confirmed this conclusion, and showed that short term seeing fluctuations were negligible as well.

The results of a good (but short) observation, a 5 min run starting at 3:39, 13 February 1970 UT, for which the coincidence gate resolution $\Delta\tau$ was 30 μsec, are given in Figure 2; in the units of Equation (11) the lightly plotted line is $\langle I^2 \rangle$, and the heavily plotted one is $\langle I \rangle^2$. Data from only about 30% of the instrumental channels – corresponding to the 30% of the NP 0532 period centered on the main pulse – are shown in the figure; the omitted data show comparable noise and agreement of the two curves for the interpulse.

Neither in Figure 2 nor in any other observation at whatever $\Delta\tau$ was any overall enhancement of $\langle I^2 \rangle$ with respect to $\langle I \rangle^2$ discernible. But the question remains as to whether the actual complexion of the noise in $\langle I^2 \rangle$ can be attributed to purely statistical fluctuations in the number of received photoelectrons, or whether structure is present which might result from phase-dependent rapid fluctuation in the pulsar light. For many runs there was a temptation for the eye to pick out structure on the leading edge of the main pulse, but it has been concluded that the noise is always of a purely statistical origin, since it was found that the records in question could

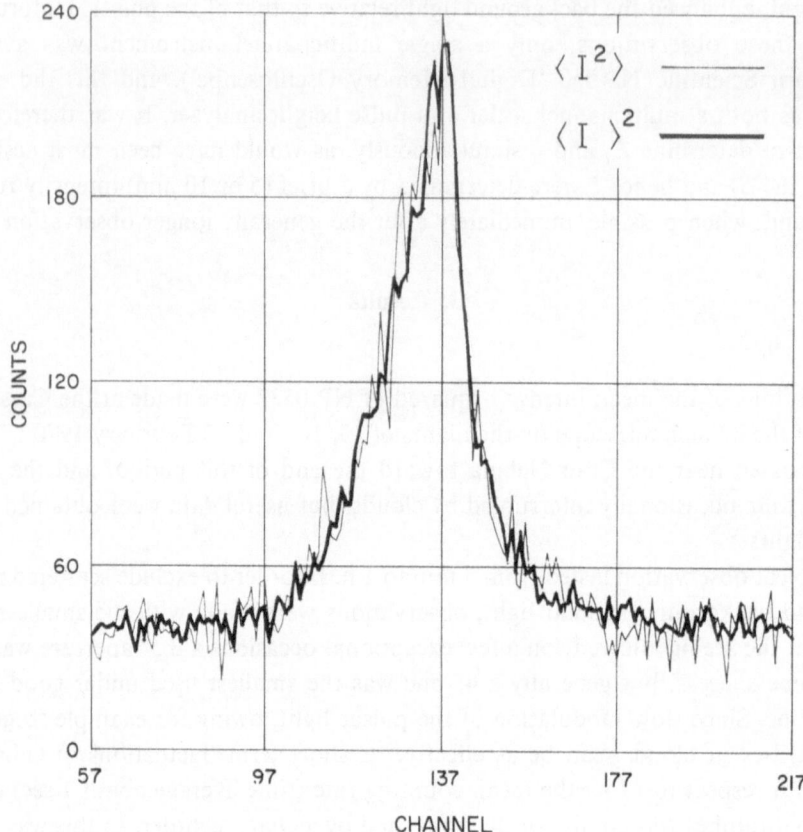

Fig. 2.   Comparison of $\langle I^2 \rangle$ with $\langle I \rangle^2$, – a single 5 min observation. Only the main
NP 0532 pulse is shown.

never be distinguished with any confidence from fictitious ones whose noise was
generated by Monte Carlo technique on a digital computer. Also, none of the supposed structure remained when the best observations taken at a given $\Delta\tau$ were added
together coherently in pulsar phase*.

This is shown in Figures 3 and 4, which are the result of synthesizing the best
observations taken at $\Delta\tau = 30$ and 3 μsec, respectively, and where the excellent agreement of $\langle I^2 \rangle$ with $\langle I \rangle^2$ is manifest. When the counts due to background are subtracted off, the limits which these observations provide on the intrinsic enhancement
factor $\xi = \langle I^2 \rangle / \langle I \rangle^2$ of the pulsar light are then summarized in Table I. These limits
pertain to the 21 channels centered on the peak of the main NP 0532 pulse; the first
entry ($\Delta\phi = 65$ μsec) for each $\Delta\tau$ is the maximum value of ξ permitted for any channel
within this range; the second results when all channels are summed together and

* Absolute phase was not maintained from one run to the next by the apparatus shown in Figure 1,
but it proved possible to reconstruct it to an accuracy of about one-third of a channel – corresponding
to 21 microseconds – by cross correlation of different runs over the main pulse.

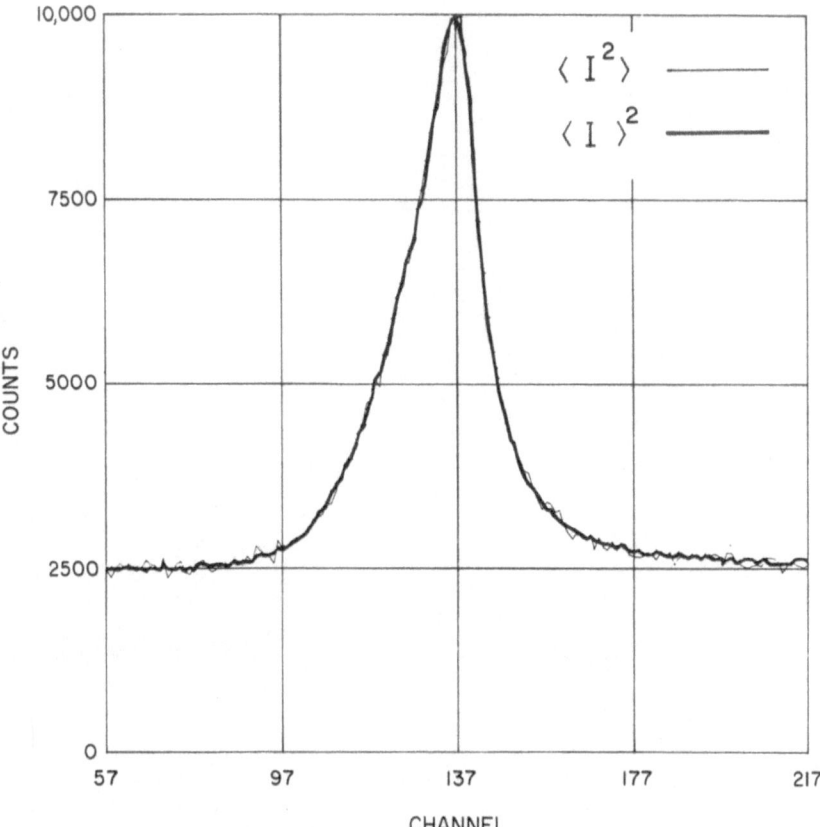

Fig. 3.   Synthesis of $\Delta\tau = 30$ μsec observations. Here, as in Figures 2 and 4, the two curves have not been fit but result from analysis of wholly independent data.

TABLE I

Upper limits to the enhancement factor

Number of Observations	Total obs. time (min)	$\Delta\tau$ (μsec)	$\Delta\phi$ (μsec)	ξ
6	192	30	65	< 1.04
			1356	< 1.02
2	61	3	65	< 1.14
			1356	< 1.04

treated as a single channel of width $\Delta\phi = 1356$ μsec. Naturally, somewhat less constrictive limits than those listed in the table apply to the wings of the main pulse or the interpulse of NP 0532.

It is now of interest to briefly consider the restrictions on specific kinds of pulse variation set by the data in Table I.

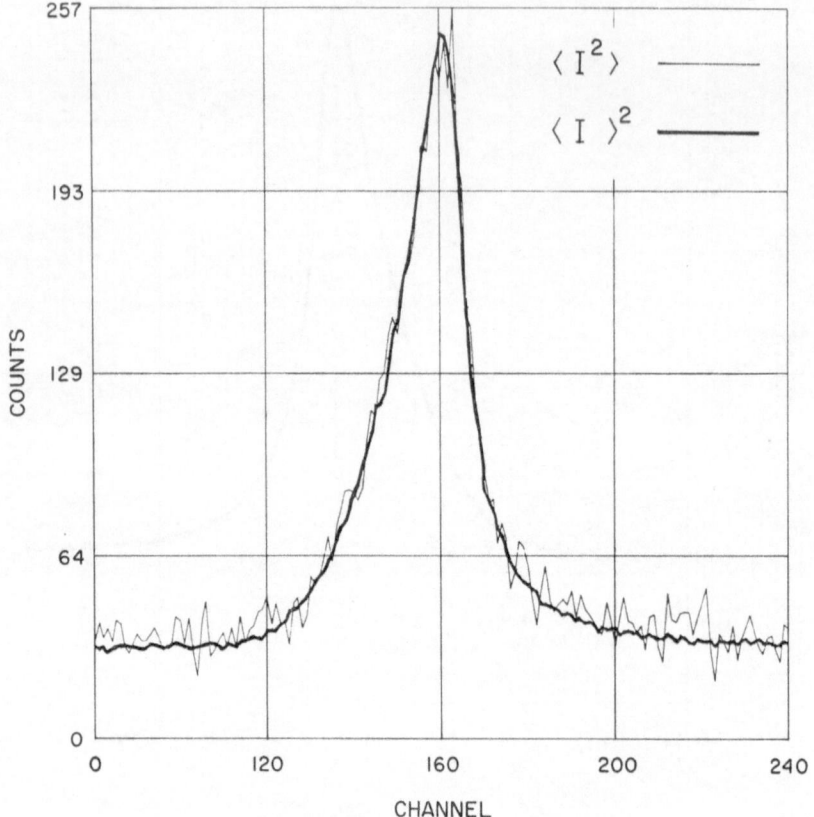

Fig. 4.   Synthesis of $\Delta \tau = 3$ μsec observations.

1. *Slow Secular Variation*

Suppose that the intensity of the pulsar varies linearly with time during the course of an observation. If the total change in intensity is a fraction f of the mean, then one readily calculates that $\xi = (1 + f^2/12)$. The most restrictive limit in Table I, $\xi < 1.02$, then implies that $f < 0.33$, which is equivalent to a change in intensity of 0.44 mag. It is evident from this that a comparison of $\langle I^2 \rangle$ with $\langle I \rangle^2$ is not as sensitive as conventional photometry in detecting fluctuations of this kind.

2. *Sudden Enhancement or 'Giant' Pulses*

Suppose that one or more pulses, comprising a small fraction g of the total number, are more intense than the rest by a factor F. As long as F is less than about 20 (see below), the enhancement factor is

$$\xi = 1 + Fg^2. \tag{16}$$

Thus the limit $\xi = 1.02$ implies that for g equal to 10^{-1}, 10^{-2}, 10^{-3}, and 10^{-4}, F is, respectively, less than 0.5, 1.4, 5, and 14. These figures clearly show that the present technique has a real edge over conventional photometry in the detection of fluctuations of this kind, since a 14-fold increase in the intensity of a single pulse would, for example, produce only a 0.05 magnitude increase in the mean intensity of NP 0532 for photometric observation over an interval as short as 10 sec.

For 'giant' pulses, with F much greater than 20, the instrument saturates, either because of the 10 μsec deadtime $\Delta\tau_D$, or the coincidence gate resolution time $\Delta\tau$, whichever is greater: a maximum of $\Delta\tau_c/\Delta\tau_D \sim 6$ counts per channel in the $\langle I^2 \rangle$ multichannel counter will result from a very intense pulse, and the limits which can be set on infrequent giant pulses are therefore not as restrictive as Equation (16) would indicate.

Referring to the counts shown in Figure 4, we now note that two such giant pulses occurring during the course of the $\Delta\tau = 3$ μsec observations would have produced an enhancement factor of $(250 + 12)/250 = 1.05$, greater, according to Table I, than the upper limit of 1.04 obtained from the twenty-one channels centered on the pulsar main peak, and therefore marginally detectable. It is thus possible to conclude that giant optical pulses (if of the same width as the usual ones) occurred at a rate of less than about 2/h during the period of our observations.

3. Marching Subpulses

The limits just set on sudden enhancement of entire pulses apply to marching subpulses as well, if g now denotes the ratio of the subpulse length to the subpulse repetition period (i.e. g is the fraction of the time that the subpulse is on).

B. $G(\tau)$ VS. G_0

Observations of $G(\tau)$ were made at the Cassegrain focus of the 107 inch telescope on the nights of 12, 13, and 14 December 1969 UT, for the most part under photometric sky conditions with seeing in the range 1–2". Figure 5a shows the result of a 55 min observation begun at 8:21 UT on 14 December, with a 4" aperture and the time-to-amplitude converter set on a full range of ~ 100 nsec. Counts have been summed over successive blocs of 20 channels of the pulse height analyzer to reduce statistical fluctuations. The enhancement factor ξ used to calculate G_0 via Equations (14) and (15) was derived from a 5 min determination of $\langle I \rangle$ taken immediately before the run; because of the small aperture, the good seeing, and the absence of moonlight, the background counts were low, and the resulting enhancement factor $\xi = 2.10$ was about as large as was ever obtained. Consequently, most of the counts contributing to $G(\tau)$ in Figure 5a come from the pulsar itself and not the background.

Figures 5b and 5c are similarly the result of observing runs taken again with a 4" aperture and with the TAC on nominal full ranges of, respectively, 3 μsec and 30 μsec. The 3 μsec full-range observations began at 7:30 UT, 14 December 1969 and lasted 30 min; the counts as before have been summed over many channels – 10 in this case – to reduce statistical fluctuations. The 30 μsec full-range observations began at 9:51 UT, 14 December and lasted 50 min; there has been no attempt to reduce

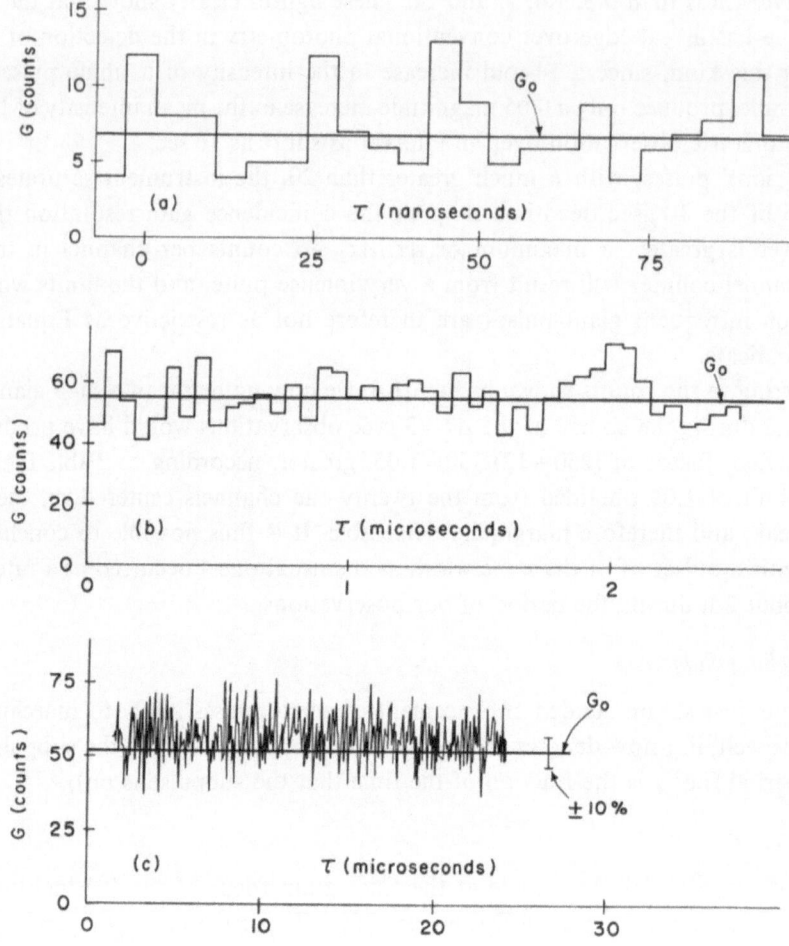

Fig. 5.   Comparison of $G(\tau)$ with G_0, for three ranges of the time-to-amplitude converter.

statistical fluctuations at the expense of resolution in delay time τ: counts received in each channel of the pulse height analyzer are shown.

The most important conclusion which we draw from Figure 5 and similar data is that $G(\tau)$ is essentially flat over the interval of delay times τ from about 1 nsec to 30 μsec, the observed noise being purely statistical in character; therefore little if any variation in the pulsar light occurs on a time scale too short to have been detected by the observations described in Section 3A. Fluctuations on a much longer time scale will of course enhance $G(\tau)$ over G_0 in the way that they enhance $\langle I^2 \rangle$ with respect to $\langle I \rangle^2$, but the data at hand adds nothing in this respect to the limits set in Section 3A – entirely owing to the uncertainty of roughly 10% introduced into the determination of ξ by the sequential observation of the C_i and S_i. To within this uncertainty, as Figure 5 shows, $G(\tau)$ and G_0 are in good agreement.

Acknowledgements

We wish to thank Dr. Harlan Smith, Director of McDonald Observatory, for the use of the 82 in. and 107 in. telescopes, and Dr. Brian Warner for assistance in their use. It is also a pleasure to acknowledge the help of Messrs. I. Beller, John Grange, Edward M. Strong, and the staff of the Columbia Physics Department Shop, in the construction of the apparatus. One of us (D. H.) would like to thank Dr. Robert Jastrow for the hospitality of the Institute for Space Studies during the course of this work. This work was supported by the National Aeronautics and Space Administration under grant NGR-33-008-012; it is Contribution No. 32 from the Columbia Astrophysics Laboratory.

References

Anderson, J. A., Crawford, F. S., and Cudaback, D. D.: 1969, *Nature* **222**, 861.
Boynton, P. E., Groth, E. J., Partridge, R. B., and Wilkinson, D. T.: 1969, *Astrophys. J. Letters* **157**, L197.
Cocke, W. J., Disney, M. J., and Gehrels, T.: 1969, *Nature* **223**, 576.
Cocke, W. J., Disney, M. J., Muncaster, G. W., and Gehrels, T.: 1970, *Nature* **227**, 1327.
Drake, F. D. and Craft, H. D.: 1968, *Nature* **220**, 231.
Duthie, J. G., Sturch, C., Richer, H. B., and Rodney, P.: 1969, *Science* **165**, 1320.
Freeman, K. C., Rodgers, A. W., Rudge, P. T., and Lynga, G.: 1969, *Nature* **222**, 459.
Graham, D. A., Lyne, A. G., and Smith, F. G.: 1970, *Nature* **225**, 525.
Hegyi, D., Novick, R., and Thaddeus, P.: 1969, *Astrophys. J. Letters* **158**, L77.
Heiles, C., Campbell, D. B., and Rankin, J. M.: 1970, *Nature* **226**, 529.
Jelley, J. V. and Willstrop, R. V.: 1969, *Nature* **224**, 568.
Ögelman, H. and Sobieski, S.: 1969, *Nature* **223**, 47.
Staelin, D. H. and Sutton, J. M.: 1970, *Nature* **226**, 69.
Sutton, J. M., Staelin, D. H., Price, R. M., and Weimer, R.: 1970, *Astrophys. J. Letters* **159**, L89.

Discussion

J. Nelson: Concerning the shape of the light curve: we find the phase angle between the main pulse and the interpulse is constant to within our errors of $\pm 0.03°$ (or ± 2.5 μsec). Also the shape of the light curve (averaged over 15 minutes) does not change during our observation period.

2.13 RESULTS OF OPTICAL TIMING MEASUREMENTS
OF THE CRAB NEBULA PULSAR

C. PAPALIOLIOS and N. P. CARLETON

Smithsonian Astrophysical Observatory, and Harvard University

and

P. HOROWITZ

Harvard University

Abstract. Absolute time of arrival measurements were made of the optical pulses from the Crab pulsar between September 1969 and April 1970; they were corrected to the solar system barycentre. The fit to the timing data indicates that the slowdown is due to magnetic dipole radiation, but there are significant deviations indicating the presence of small fluctuations and major jumps. There is no evidence of the quasi-sinusoidal behaviour reported by Arecibo. These measurements allowed an integrated light curve to be constructed with high precision.

1. Introduction

Absolute time of arrival measurements of the optical pulses from the Crab pulsar NP 0531 have been made from September 1969 to April 1970, using the 61-in. reflector at Harvard's Agassiz Station. The timing accuracy, limited only by the available signal strength and the sky background level, is typically 3–8 μsec after an hour of observation.

The signal was accumulated in the usual way (Papaliolios *et al.*, 1968, Horowitz 1969) with some modifications made to allow for accurate absolute timing information (UTC to nearest μsec) available from East Coast Loran-C to be inserted into the multichannel analyzer. A system similar to ours has been previously reported (Boynton *et al.* 1969a).

2. Analysis and Results

Each night's data consisting of 4 to 10 individual 10-minute runs are combined to give a precise arrival time at the telescope, for a single optical pulse within the observing period. This site arrival time is reduced to a solar-system-barycenter arrival time (Richards *et al.*, 1970). The secular behavior of the pulsar period between successive observations is sufficiently stable so that each of the above pulses can be numbered unambiguously. The result of this data reduction process is shown in Table I. Included in these calculations are the second order general relativistic effects arising from the earth's eccentric orbit around the sun (Hoffman, 1968; Counselman and Shapiro, 1968).

We next attempt to reduce the contents of Table I to some analytic form in order 1) to display in a more transparent way the behavior of the pulsar, and 2) to gain some insights into the physical processes that determine the observed behavior.

TABLE I

Pulse arrival times at site and solar system barycenter, and corresponding pulse number

• •

CRAB PULSAR TIMING DATA

• •

DATE MO DA YEAR	SITE ARRIVAL TIME HR MIN SEC	SIGMA μSEC	BARYCENTRIC ARRIVAL TIME JUL. DAY SECONDS	PULSE NUMBER
3 17 1969	01 47 39.983604	50.0	2440297.5 6437.513579	0
9 12 1969	08 57 06.023401	10.6	2440476.5 32189.666495	468033096
9 13 1969	08 11 02.024519	11.5	2440477.5 29433.917910	470559950
9 16 1969	08 39 46.009152	7.2	2440480.5 31183.686258	478443104
10 9 1969	08 19 57.002321	5.1	2440503.5 30186.164151	538444366
10 10 1969	07 56 09.984035	4.1	2440504.5 28766.894782	541011518
10 17 1969	09 09 29.992356	10.6	2440511.5 33220.661799	559416162
10 18 1969	08 02 44.990390	4.6	2440512.5 29222.660511	561905392
10 19 1969	08 37 14.983562	3.4	2440513.5 31300.091969	564578148
11 16 1969	07 11 31.980315	5.9	2440541.5 26320.609244	637506568
12 5 1969	07 45 55.005933	4.4	2440560.5 28439.187459	687158492
12 12 1969	06 21 25.005616	8.1	2440567.5 23376.294043	705274534
12 18 1969	08 19 56.985649	4.4	2440573.5 30488.458890	721148376
1 1 1970	02 05 24.999442	11.8	2440587.5 7996.335239	757006265
1 3 1970	03 18 44.986356	6.8	2440589.5 12390.913015	762358574
1 11 1970	02 22 39.980689	7.1	2440597.5 8999.188371	783134278
1 14 1970	06 03 42.002578	5.3	2440600.5 22248.193239	791363730
1 16 1970	06 13 37.008305	10.6	2440602.5 22834.231771	796600923
1 31 1970	03 39 17.008171	6.1	2440617.5 13492.359294	835464572
2 5 1970	01 38 04.982081	7.3	2440622.5 6187.966638	848292409
2 8 1970	04 44 46.993842	9.7	2440625.5 17368.167257	856459147
2 24 1970	00 20 01.998223	12.2	2440641.5 1361.581707	897730144
2 27 1970	01 05 55.000236	4.5	2440644.5 4089.601008	905641422
3 4 1970	00 18 26.978656	4.1	2440649.5 1199.865754	918602218
3 7 1970	01 23 11.986946	11.7	2440652.5 5058.931412	926547588
3 9 1970	01 12 56.984122	7.4	2440654.5 4426.844686	931747690
3 28 1970	01 01 14.993929	7.2	2440673.5 3562.137620	981303336
4 4 1970	01 13 26.979862	7.7	2440680.5 4236.229829	999590399
4 6 1970	02 11 46.998164	9.7	2440682.5 7719.818813	1004914645
4 8 1970	01 34 24.999011	7.0	2440684.5 5462.152386	1010065478
4 16 1970	01 22 27.006979	14.2	2440692.5 4683.327070	1030917923
4 26 1970	01 02 21.990707	49.0	2440702.5 3409.390075	1056974148

The first relationship between t, the barycentric arrival time, and N, the pulse number that we try is the truncated power series

$$\phi = \phi_0 + f_0 t + \tfrac{1}{2}\dot{f}_0 t^2 + \tfrac{1}{6}\ddot{f}_0 t^3 \qquad (1)$$

where we have replaced the integer N by ϕ, a continuous variable. The parameters $\phi_0, f_0 \dot{f}_0, \ddot{f}_0$ are adjusted by the method of least squares to minimize the residuals defined by $r = N - \phi$. If such a fit is performed over the data from March 1969 to the end of September 1969, we have the residuals (converted to units of time) shown in Figure 1. It is clear that for the days not involved in the fit, i.e., past September 1969, the residuals diverge quite markedly. This is not because of the fitting procedure, but because of a sudden increase in the frequency of the pulsar, as others have announced

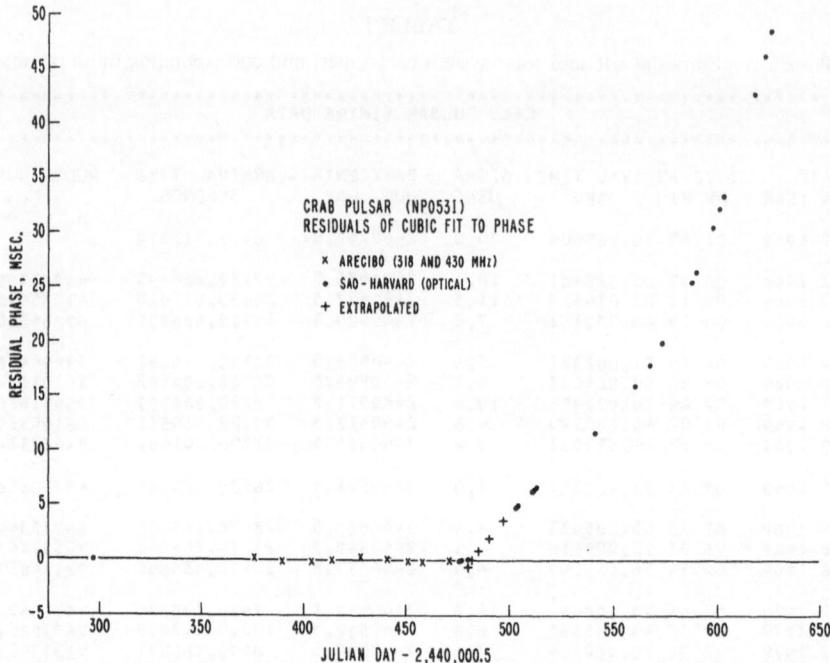

Fig. 1. Residuals *vs.* Julian day, after weighted arrival times prior to 22 September 1969 (JD 2440484.5) are fitted by equation (1). The phase residuals resulting from this calculation have been converted to units of time.

(Boynton *et al.*, 1969b). A cubic fit to the data past the jump allows us to give the following values to the parameters before the jump, and their changes (final-initial) at the jump,

$$f = 30.209\,297\,624 \text{ Hz} \qquad \Delta f = 9.3 \pm 1 \times 10^{-8} \text{ Hz}$$
$$\dot{f} = -3.857212 \times 10^{-10} \text{ Hz/sec} \qquad \Delta \dot{f} = -1.8 \pm .1 \times 10^{-14} \text{ Hz/sec}$$
$$\ddot{f} = 1.04 \times 10^{-20} \text{ Hz/sec}^2 \qquad \Delta \ddot{f} = 5.9 \pm 1 \times 10^{-21} \text{ Hz/sec}^2$$
$$t = \text{JD } 2440484.5 \pm 4 = 20 \text{ September } 1969 \pm 4 \text{ days}$$

If this jump is the result of a sudden readjustment of the equatorial radius of the outside crust, then this radius changed by only 15 μ. It can easily be shown that changes of this size should be happening about once a week, but no other has been observed. Another explanation for the jump could lie in Scargle's observation (Scargle and Harlan, 1970) of changes in nebulosity following the jump. A cubic fit to all the data following the jump up to 26 April 1970 results in residuals shown in Figure 2. These residuals, particularly the early ones, are far greater than what one would expect from the measuring process. There is some evidence of a post-jump relaxation lasting up to 1 January 1970. Using the data past 1 January 1970 results in residuals shown by the solid circles of Figure 2.

Since the residuals are considerably larger than the measurement errors (which are too small to be shown in Figure 2 but are listed in Table I) we conclude for the present

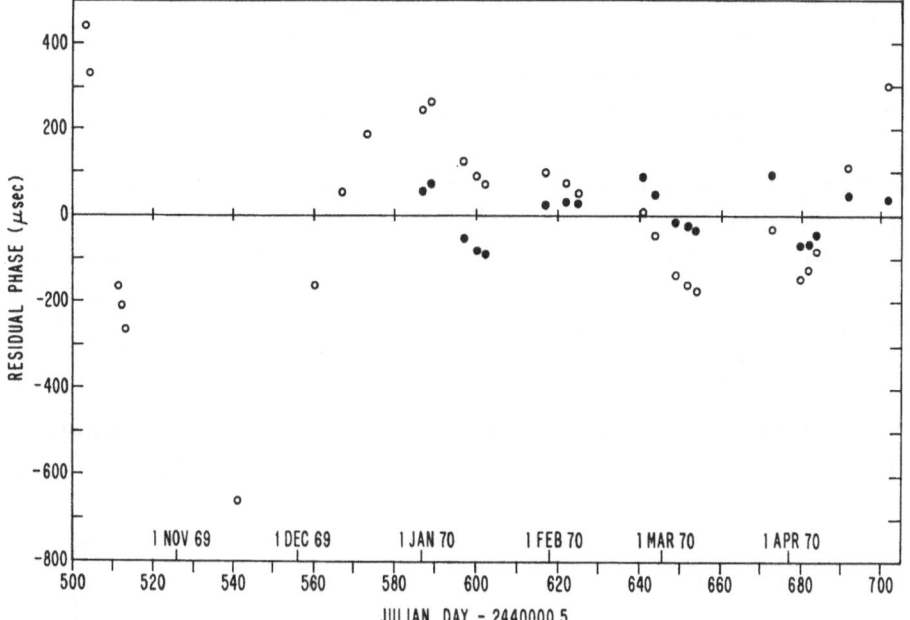

Fig. 2. Residuals *vs.* Julian day; $\bigcirc$ = residuals of a cubic fit from 1 October 1969 to 26 April 1970. $\bullet$ = residuals of a cubic fit after 1 January 1970. The braking parameter n is 2.70 and 2.61, respectively, with an uncertainty of about 5% due to possible errors of a second of arc in the alignment of solar system coordinates relative to NP 0531. The measurement errors are too small to show in this figure but can be obtained from Table I.

that Equation (1) is not the appropriate relationship between t and ϕ. Modifying Equation (1) by adding a quartic term reduces the residuals somewhat, but this modification is rejected because of the anomalously large contribution to ϕ from the additional term.

Another useful relationship is obtained from

$$\dot{f} = -af^n \qquad (2)$$

The braking parameter n defined by Equation (2) has the value 3 if the slowdown of a spinning rigid neutron star is due to solely dipole radiation, and has the value 5 for quadrupole radiation. The solution of (2) is

$$\phi = \phi_1 + \alpha (1 + t/L)^k \qquad (3)$$

where $k = (n-2)/(n-1)$, and L is a 'lifetime' parameter.

Least-squares adjustment of the four parameters in equation (3) results in residuals that are essentially *identical* to the ones obtained from the use of Equation (1). Even such diverse forms as

$$\phi = \phi_1 + \alpha \{(t + L) + \beta [\ln (t + L) + 2]\}^{1/2} \qquad (4)$$

and

$$\phi = \phi_1 + \alpha (t + L)^{3/4} + \beta (t + L)^{5/4} \qquad (5)$$

which are the solutions expected if both dipole and quadrupole radiations are present where dipole radiation is dominant in (4) and quadrupole is dominant in (5), result in residuals *identical* to those previously obtained.

These results indicate that it is very difficult to compare theoretical explanations with one another, since only the number of adjustable parameters seems to matter. The underlying reason for this unfortunate property appears to be noise, so that *all simple relationships between ϕ and t will result in excessive residuals.*

The braking parameter n varies between 2.6 and 3.7 depending on the string of data one uses in the least-squares fitting process. If we set $n=3$ (the case for pure dipole slowdown) then we see from equation (3) that

$$t = t_0 + P_0\phi + \tfrac{1}{2}P_0'\phi^2 \qquad (6)$$

This leads to another means of data analysis, the method of divided differences (Buckingham, 1957) which allows us to recognize the existence of small fluctuations

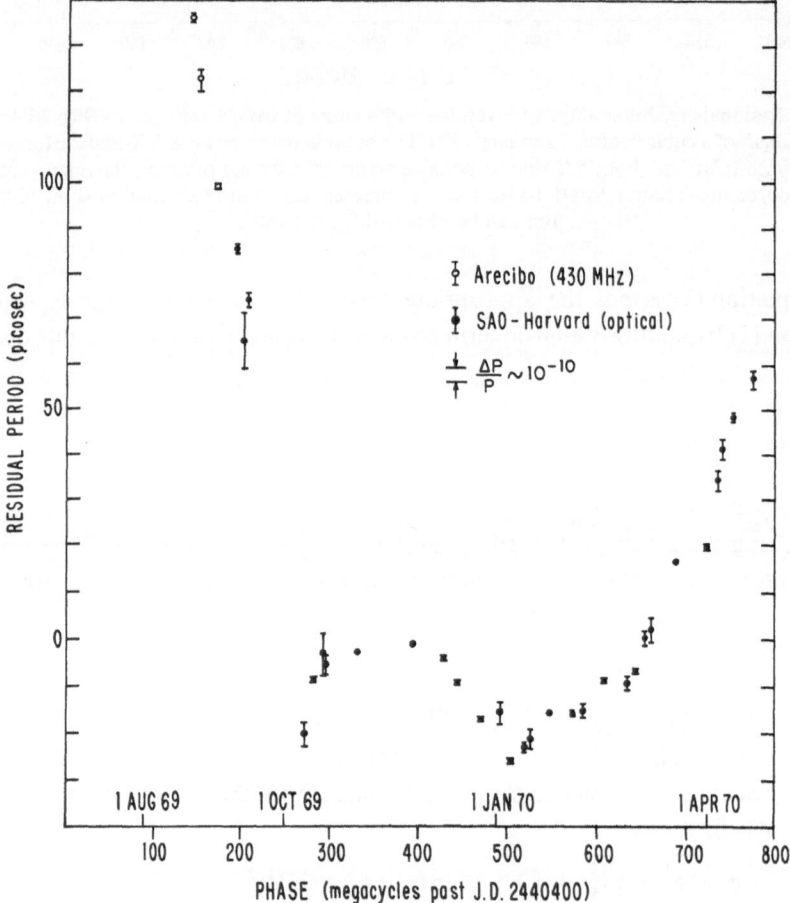

Fig. 3.　Residual period *vs.* phase, calculated by divided differences, with a linear term subtracted. Error bars are omitted when less than 0.35 picosec.

in the arrival times as well as the major ones such as the jump previously noted. In addition this method is able to suggest interpretations that might not otherwise have been suspected. If we assume that Equation (6) is valid between every pair of measured points (t, N) then at $N = \frac{1}{2}(N_1 + N_2)$ the period P is *exactly* $(t_2 - t_1)/(N_2 - N_1)$. Equation (6) also implies that the period P is a linear function of N (or ϕ). We perform these calculations, subtract out a linear dependence of P on ϕ, and plot the remaining residual periods vs. ϕ in Figure 3. This graph shows clearly the jump in period that occurred on 20 September 1969, the relaxation following the jump, and the small fluctuations throughout the year. These fluctuations in period of about one part in 10^{10} make it difficult to extract from the present timing measurements any useful information on the gravitational red shift, and on the solar system ephemeris. There is also no evidence of a continuation of the sinusoidal oscillation that was reported to exist before the jump by Richards *et al.* (1970).

If we perform the above analysis on the Arecibo data, kindly supplied to us by the Arecibo group, then the results are as shown in Figure 4. The two straight lines suggest a pure-dipolar slowdown with a single jump in dipole moment. The phase residuals calculated on the basis of this interpretation, which incidentally contains only five adjustable parameters instead of Arecibo's seven, are shown in Figure 5 and are as small as those of the Arecibo group. If our interpretation is correct then there is no sinusoidal oscillation, but instead the dipole moment increased by 3.7 parts in 10^5 on JD 2440410.4.

Finally, as a simple byproduct of our timing measurements, we can superpose the light curves taken throughout the year, and still maintain the 32-μsec resolution. The

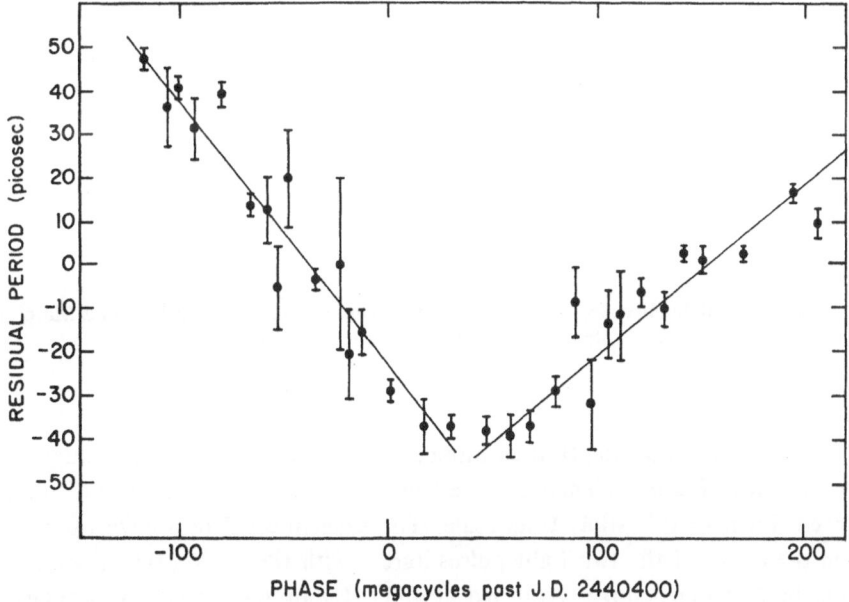

Fig. 4.   Residual period vs phase of the summer Arecibo data, with a linear term subtracted.

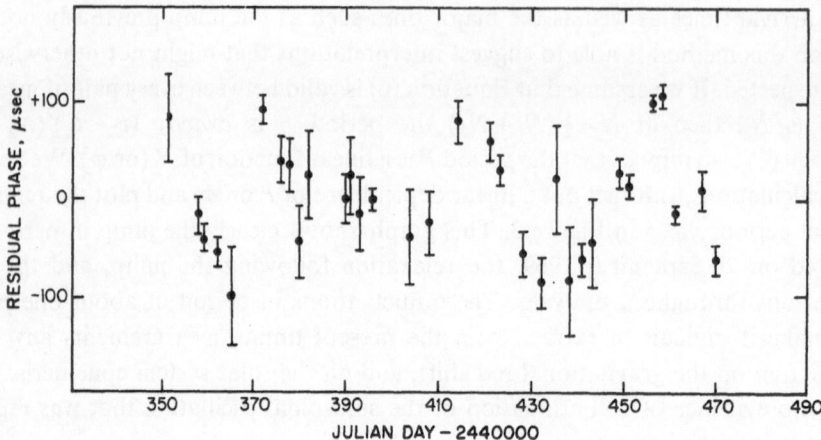

Fig. 5.   Residuals *vs.* Julian day, after fitting a dipole braking law with a single jump in the dipole moment to the radiofrequency arrival times obtained during the summer of 1969 by the Arecibo group.

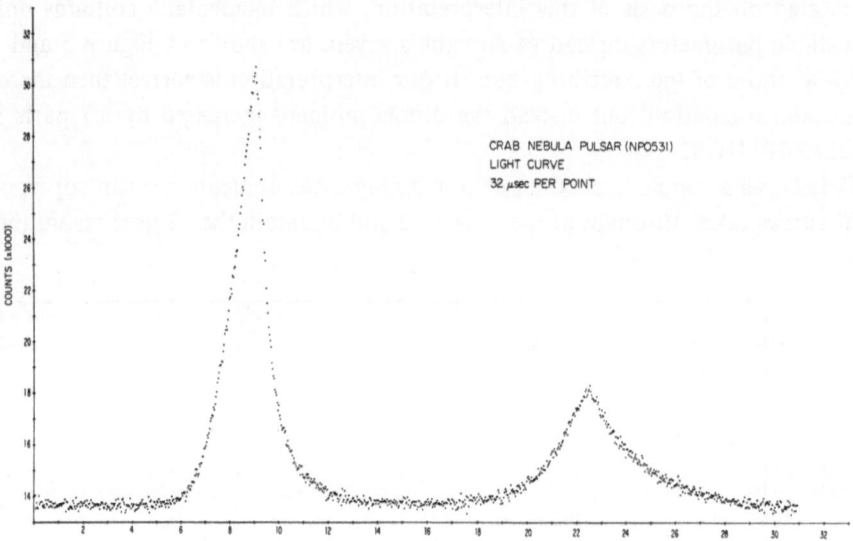

Fig. 6.   Superposed light curve from ten nights of observing between October 1969 and March 1970. Total observing time is about ten hours.

result of such a superposition (representing about 10 h of observation throughout the year) is shown in Figure 6. There is no evidence of rounding at the top of the main peak even down to this short time scale. The time interval of 13.376±0.032 msec between the peaks of the two light pulses agrees with the 13.36±0.03 msec interval between the two sharp spikes recorded at 430 MHz by the Arecibo group (Rankin *et al.*, 1970).

Acknowledgements

We are very grateful to I. I. Shapiro and C. C. Counselman, III, of MIT, and to F. D. Drake and D. W. Richards of Cornell University for their generous sharing of information, and for many interesting and stimulating conversations. We are similarly grateful to our colleagues G. G. Fazio, L. Goldberg, D. Hearn, B. Kaplan, R. V. Pound, and E. M. Purcell, and members of the Harvard Bubble Chamber Group. Messrs. C. R. H. Tsiang, J. T. West, and W. Wright of the Smithsonian Observatory aided us greatly in our measurements of absolute time, and M. Mattei of Agassiz Station was a valuable assistant in the observing. One of us (P. H.) is grateful to the Society of Fellows for support.

References

Boynton, P. E., Groth, E. J., III, Partridge, R. B., and Wilkinson, D. T.: 1969a, *Astrophys. J. Letters* **157**, L197.
Boynton, P. E., Groth, E. J., III, Partridge, R. B., and Wilkinson, D. T.: 1969b, *IAU Circ.* No. 2179.
Buckingham, R. A.: 1957, *Numerical Methods*, Pitman, London, p. 92.
Counselman, C. C., III, and Shapiro, I. I.: 1968, *Science* **162**, 352.
Hoffman, B.: 1968, *Nature* **218**, 667.
Horowitz, P.: 1969, *Rev. Sci. Instr.* **40**, 369.
Papaliolios, C., Carleton, N. P., Horowitz, P., and Liller, W.: 1968, *Science* **160**, 1104.
Rankin, J. M., Comella, J. M., Craft, H. D. Jr., Richards, D. W., Campbell, D. B., and Counselman, C. C., III: 1970, *Astrophys. J. Letters* **160**, L1.
Richards, D. W., Pettengill, G. H., Counselman, C. C., III, and Rankin, J. M.: 1970, *Astrophys. J. Letters* **160**, L1.
Scargle, J. D. and Harlan, E. A.: 1970, *Astrophys. J. Letters* **159**, L143.

Discussion

D. T. Wilkinson: It now is clear that the behaviour of residuals is a strong function of the model chosen and the distribution of the data. We disagree with several of the results of the Harvard group. Our data indicate that the exponential decay of the transient part of the discontinuity has a time constant of about 2 weeks; its effect is essentially gone by 1 December. The value of Q, as defined by Pines *et al.*, is about 1 or larger in our fits. The changes in n reported are modelled in our fits as quartic or periodic terms. Finally, we also see the bumps in March and April of 1970, seen by the Lick group. They are unexplained in any models discussed here.

C. Papaliolios: We do observe a different behaviour for the pulsar before about 1 January 1970, and then after that date. This difference is not an artifact of the fit but is clearly indicated by the method of divided differences.

Hills: As a member of the Berkeley group I would like to comment on the apparent differences in the results of the pulse timing presented by the different groups. As far as we know there are no actual contradictions in the arrival times measured but clearly the phenomena which show up depend very much on the details of the observations made. Thus our measurements which have greater timing accuracy (< 2 μsec for a good night) and more frequent observations, but a shorter time base than the others, show only the structure on a scale which is very small (some might say insignificant), whereas the measurements over longer periods with lower resolution show the more dramatic changes. We look forward to combining our data with that of other groups to check our measurements and to try to understand whether the phenomena described can all be considered part of the same behaviour or whether a more complex model must be used.

C. Papaliolios: In studying the slow-down of the pulsar, i.e., the braking parameter, it is not sufficient to consider just the radiation torques. The changes in moment of inertia can also play an important role; after all, we know that the pulsar is slowing down, therefore, its oblateness is also changing. Although this effect is not in itself sufficient to account for the total discrepancy between $n = 3$ (magnetic dipole slowdown) and the observed average value of about 2.6, it is in the proper direction. Other effects which should be included in a detailed calculation are the changing dipole moment that comes from the shape changes and the additional effects due to the compressibility of the pulsar.

2.14 OPTICAL TIMING
OF THE CRAB NEBULA PULSAR NP 0532*

JERRY NELSON

Lawrence Radiation Laboratory,
University of California, Berkeley, Calif., U.S.A.

RICHARD HILLS and DAVID CUDABACK

Astronomy Department and Radio Astronomy Laboratory,
University of California, Berkeley, Calif., U.S.A.

and

JOSEPH WAMPLER

Lick Observatory, Board of Studies in Astronomy and Astrophysics,
University of California, Santa Cruz, Calif., U.S.A.

Abstract. Accurate pulse arrival times have been measured for NP 0532 during the period 15 December 1969 to 3 May 1970, and have been fitted to simple models of the pulsar braking mechanism. A good fit could not be obtained to all the data at once, because of deviations on a time scale of several days. However it was possible to divide the observing period into four shorter intervals in such a way that the data within each deviated only slightly from smoothly varying functions. The difference in the parameters of these four functions may indicate sudden events in the pulsar producing changes of order of 1 part in 10^9 in the pulsar frequency and 4 parts in 10^5 in the rate of change of frequency. In each case the difference in frequency from one interval to the next implies a slowdown of the pulsar.

We found that the average value of the 'braking parameter' n in the equation $dE/dt = -A\omega^n$ was 3.63, but dividing the data into shorter intervals gave values between 0 and 5. We found no changes in the mean shape of the pulses, or the phase of the interpulse relative to the main pulse.

* This is the abstract of an article published in *Astrophys. J.* **161**, L235.

2.15   PHOTOMETRY AND POLARIMETRY OF THE CRAB PULSAR

N. VISVANATHAN

Harvard College Observatory, Cambridge, Mass., U.S.A.

Abstract. Observational results are presented for the pulse component, the nebula component and the wisp component of the Crab Nebula. The pulsar polarization is separated from the background nebular polarization. The relation between the pulsar and the adjacent nebula is discussed.

1.   The discovery that the central star of the Crab Nebula pulsates with the shortest period known, 33 msec, played a decisive role to the idea put forward by Gold, that pulsars are rotating neutron stars. Scargle (1969) has studied the semi-periodic nature of wisps around the pulsar and recently he and Harlan (1970) found evidence that the recent spin up of the pulsar in September 1969 was followed by a change of structure in the wisps. Further Arecibo observations (reported in this Conference) show an increase in the dispersion measure in the direction of the Crab. All these show an intimate electrodynamic link between the Crab Nebula and pulsar in the center.

In this paper, observational results obtained for the pulse component, nebula component and wisp component are presented. These observations were made at the prime focus of 200-in. telescope in collaboration with Dr. Kristian. For a description of instrumentation used, kindly refer to Kristian *et al.* (1970). Nebula observations were made at the Cassegrain focus of 84-inch at Kitt Peak National Observatory with an automatic photometer-polarimeter.

Figure 1 is the 50 μsec light curve of the pulsar obtained on 18 October 1969 with a total integration time of 1 h. The following features are to be noted in the figure:

1. (a) The asymmetrical shape of the main and sub pulses with respect to the peak. The falling edge is steeper than the rising edge in the case of the main pulse while it is the other way around in the case of sub-pulse.

(b) The peak is sharp and the intensity reaches peak within an interval of time 40 μsec and the width at 90% of maximum is of the order of 100 μsec. This means that the core of the pulse, if it is a sweeping beam, has an angular size of a degree.

(c) In between the main and sub pulses the intensity is more than that in the other side of the pulse. This excess is about 1.5% of the peak intensity of the main pulse. This implies that the main pulse has a long tail.

(d) The peaks of the main and sub pulses are separated by 13.5 msec which is 0.41 of the phase.

(e) The area of the main pulse is twice that of the sub pulse.

2. Figure 2 is a comparison of the light curve of the pulsar in two filters – visual (5500 Å) and ultraviolet (3600 Å). Each point in the pulse corresponds to 40 μsec. To make the comparison easier the intensities are plotted on a log scale and shifted vertically. It can be seen that the differences between two filters at different points in the phase is about a few percent. Thus the spectrum at different points at the pulse is exactly the same within a few percent.

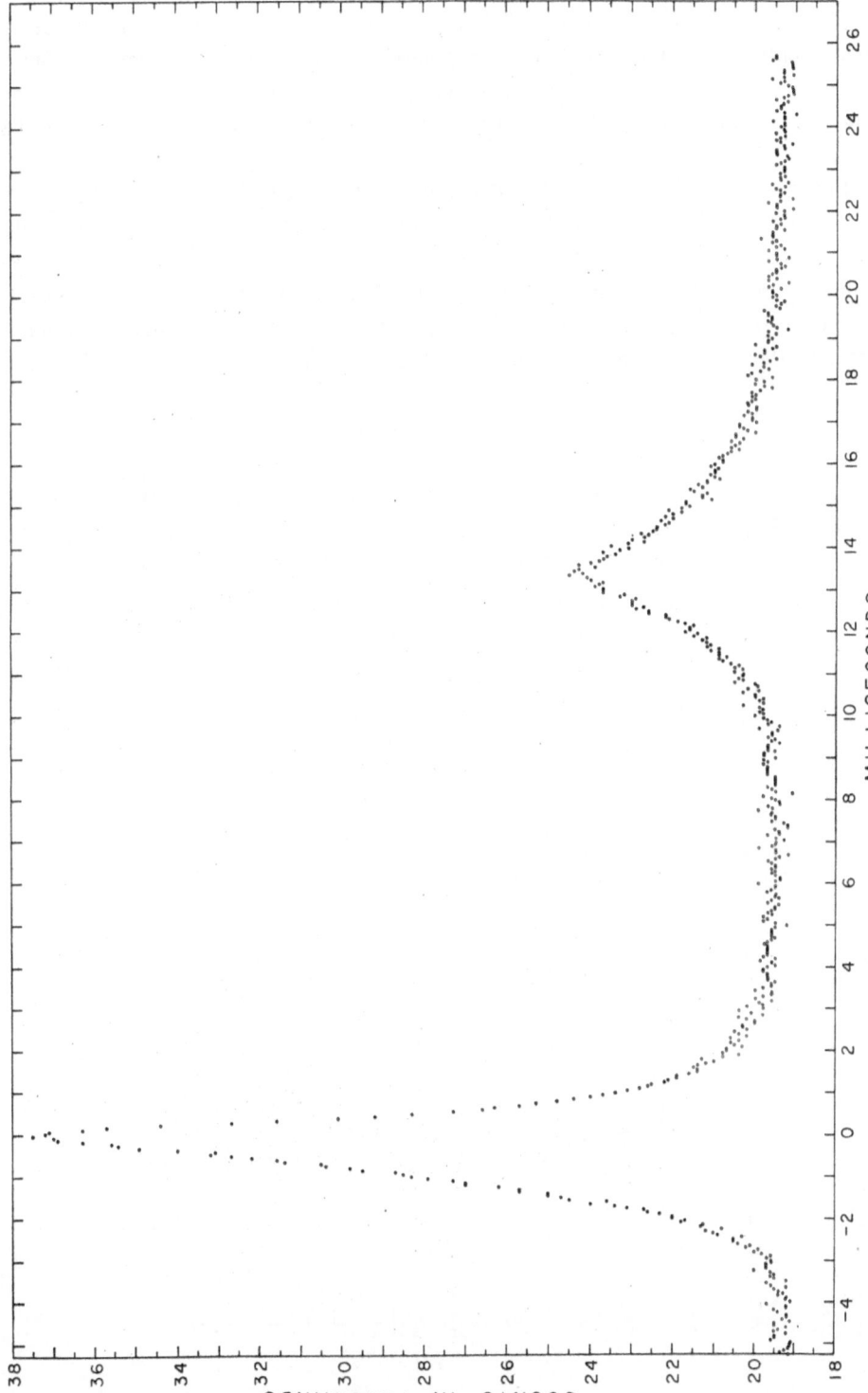

Fig. 1. Light curve of the Crab pulsar taken on 18 October 1969, at the prime focus of the 200-in. telescope.

3. The linear polarization of the pulsar was measured on the nights of 8 February, 18 and 19 August, 19 October and 3 December 1969. Most of the polarization measurements were made by rotating the polaroid filter in 30° steps from 0° to 360°. The starting time of each run was also known to within 50 microsec so that the detailed time behaviour of the polarization within the averaged pulses could be studied.

4. Figure 3 is the plot of **P** and **θ** versus phase of the pulse. Each point corresponds to the integrated **P** and **θ** for 4 channels. It can be seen that **P** and **θ** are different only where the pulses occur, otherwise they are constant. The constant polarization represents the polarization of nebular light around the pulsar. The polarization at each point in the pulse represents the vectorial addition of the constant nebular polarization and the polarization of the pulse at that point. The fact that **P** and **θ** of the

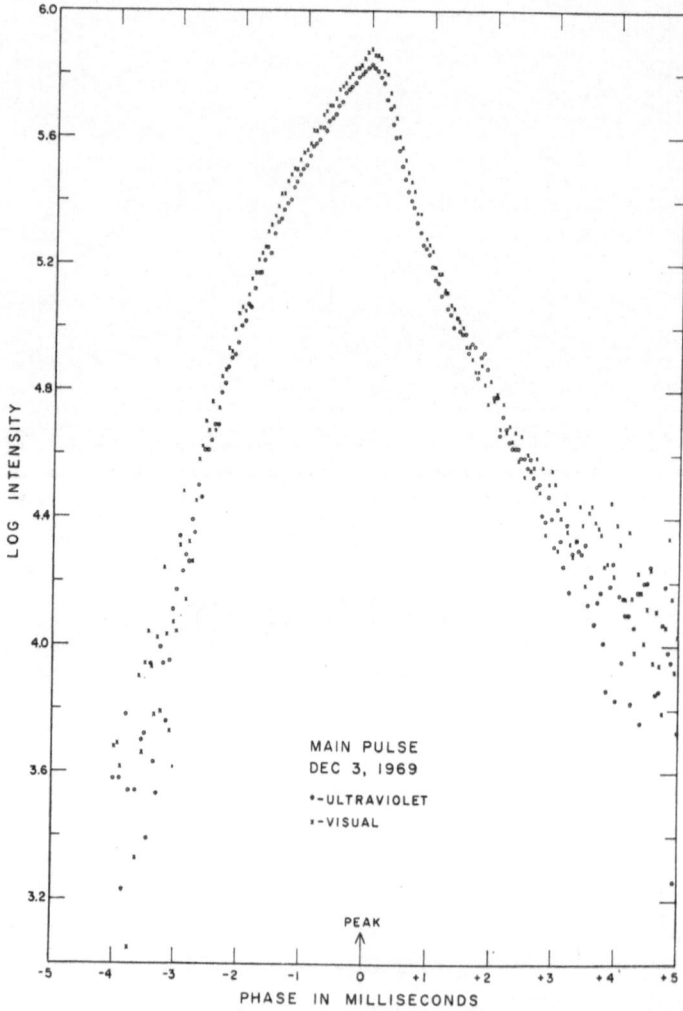

Fig. 2.   Comparison of main pulse at visual and ultraviolet wavelengths.

pulses are different from nebular polarization indicates the pulses are linearly polarized.

2. After subtracting the background polarization at each point, the runs of **P** and **θ** within the main and sub pulses have been constructed. The results in both the pulses show that **P** drops smoothly from about 25% in the leading edge to a minimum of about two percent just after the peak of the light curve, then increases in a roughly symmetric fashion to at least 10% in the trailing edge. The position angle of the electric vector **θ** varies continuously from the rising edge of the pulse to the falling side of the pulse from 80° to 50°. The rotation of the electric vector within the pulses can be explained by a general geometrical model by assuming that the pulses are due to a

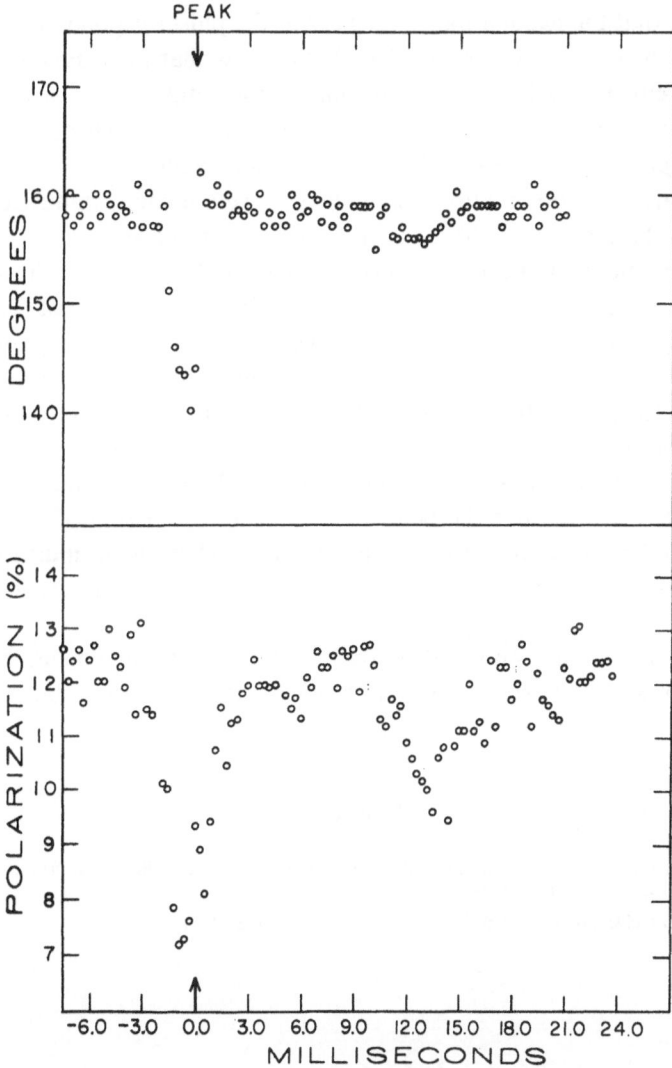

Fig. 3. Run of P and θ with phase of the Crab pulsar along with background nebula, observed with 7 sec of arc aperture, at the prime focus of the 200-in. telescope, on 18 October 1969.

polarized radiation pattern fixed in the pulsar which is scanned as the pulsar rotates
For each of the pulses we assume a unit vector **p** fixed in and rotating with pulsar. At any
instant, the position angle of electric vector θ we measure is the angle which the
projection of the vector **P** makes in the plane of the sky. The observed run of θ with
phase suggests that the observer must lie within 35° of the equator and the emission
pattern in the pulsar must lie within 25° of the equator.

Another feature of this model is the run of θ with phase and it will show inversion
symmetry about $\omega t = 0$ (when the observer is at the longitude of the emission pattern)
when the vector **P** lies in the plane defined by rotation axis and the observer. At this
time the electric vector θ will be either parallel or perpendicular to the projection of
the rotation axis on the plane of the sky. This angle comes out to be 160°. It is interesting to note that the nebular light very near to the pulsar is polarized and the position angle of the electric vector θ is 160°. These show that the magnetic field around
the pulsar is either parallel or perpendicular to the rotation axis of the pulsar. We
think that this relationship between rotation axis and the outside field suggests a
direct causal connection between the pulsar and the nebula.

Linear polarization observations of the nebula were made with different apertures
ranging from 2 sec of arc to 20 sec of arc around the pulsar.

These observations show a region of constant **P** and θ asymmetrically situated with
respect to the pulsar, extending 16 sec in the west and 8 sec in the east. These indicate
that the magnetic field structure around the pulsar is uniform in that region and the fine
scale structure is smaller than 0.02 parsec. The uniform field in this region is at an
angle of 69°. In the northwest side of the pulsar there are many wisps seen in the
continuum photographs. The axes of these wisps are at an angle of 30° to the general
magnetic field in this region. Scargle and Harlan (1970) found that these wisps had
changed their structure after the September jump in the period of the pulsar, thus
suggesting a relation between the two phenomena. There is an indication that the
magnetic fields in the wisps are at an angle 30° from the general magnetic field around
the pulsar.

Thus the Crab nebula shows a linearly polarized pulse component within which
electric vector θ rotates, surrounded by a constantly polarized nebular component
and slowly variable polarized wisp component.

References

Kristian, J., Visvanathan, N., Westphal, J. A., and Snellen, G. H.: 1970, *Astrophys. J.* **162**, 475.
Scargle, J. D.: 1969, *Astrophys. J.* **156**, 401.
Scargle, J. D. and Harlan, E.: 1970, *Astrophys. J. Letters* **159**, L143.

2.16 MEASUREMENT OF THE CIRCULAR POLARIZATION
OF PULSAR NP 0532

J. R. P. ANGEL*

Columbia Astrophysics Laboratory, Columbia University, New York, N.Y., U.S.A.

D. HEGYI**

NASA Goddard Institute for Space Studies N.Y,, New York, U.S.A.

J. D. LANDSTREET†

Department of Astronomy, Columbia University, New York, N.Y. U.S.A.

Abstract. The polarizer is briefly described. The polarization of the whole pulse is not significantly different from zero (-0.15 ± 0.70 per cent). Polarization measurements of the leading edge and trailing edge of the main pulse are also given.

1. Observations and Instrumentation

We present below the results of measurement of the optical circular polarization of the pulsar NP 0532. The observations were made on 11/12 February 1970 with the 82-in. telescope of McDonald Observatory.

The instrument used was the Cassegrain photoelectric polarimeter described by Angel and Landstreet (1970) and designed specifically for measuring circular polarization, together with a data handling system assembled for the occasion. The basic polarimeter consists of a KD*P electro-optic crystal (a Pockels cell) used as a reversible quarterwave plate, followed by a Wollaston prism which directs the light into two Channeltron photomultipliers. A periodically reversing square wave voltage is applied to the Pockels cell which causes one photomultiplier to measure alternatively the intensity of right and left circularly polarized light. The polarization of a star can thus be determined from the output of either photomultiplier alone.

The Pockels cell was set for quarter-wave retardation at about 5000 Å. The 1-mm S-20 photocathodes of the Channeltrons were unfiltered, but the dispersion of the beam by the Wollaston prism limits the light falling on the photocathodes to approximately the 4000–7000 Å range. The error in retardation in the Pockels cell for wavelengths other than 5000 Å lowers the sensitivity of the system to circular polarization only a few percent.

The output signals from the photomultipliers were fed into two Northern 512 channel analyzers, one for each photomultiplier. The analyzers were used in a multiscaling mode; the channel address was advanced by a signal from a General Radio frequency synthesizer which was set to scan through the 511 data channels in exactly two pulsar periods. The voltage on the Pockels cell was reversed at the first and at

---

* Alfred P. Sloan Research Fellow.
** Present address: Department of Astronomy, Boston University, Boston, U.S.A.
 † Present address: Department of Astronomy, University of Western Ontario, London, Canada.

Davies and Smith (eds.), The Crab Nebula, 157–159. All Rights Reserved.
Copyright © 1971 by the IAU.

the 257th channels. Each multichannel analyzer thus recorded a full pulsar period in one circular polarization followed by a full period in the other polarization. It was not necessary for our purposes to go to great lengths to insure exact registration of the superposed data, but we may estimate from the data that even during the longest run (40 minutes) the phase error was no more than 2 or 3 channels, or less than 2%.

Because this is a two-channel instrument, with measurements of both handednesses of circular polarization being done essentially simultaneously and through the same optical and electronic train, the systematic zero-point errors are extremely low (less than 0.05%) and the measurement is not adversely affected by variations in transparency or background light. Our accuracy is thus limited mainly by statistical fluctuations in the signal and in the background light, which was primarily moonlight and whose amplitude equalled the peak height of the pulse.

The system was checked periodically by putting a circular polarizer (a piece of polaroid followed by a quarter-wave plate) in the beam before the Pockels cell. We also measured the circular polarization of comparison stars believed to be unpolarized. With the circular polarizer in the beam, we generally measure a polarization of more than 90%, while the measured circular polarization of an unpolarized star is always less than 0.05%.

TABLE I

Polarization of primary pulse of NP 0532

Component of pulse	Measured polarization (%)
Leading edge	-0.19 ± 0.89
Trailing edge	-0.09 ± 1.10
Whole pulse	-0.15 ± 0.70
Leading minus trailing edge	-0.08 ± 0.70

2. Analysis and Results

The fractional circular polarization was determined for the leading and trailing edges of the pulse separately by the following formula. Let the total number of counts in the leading edge of the pulse (which was taken to be the 15 channels (1.9 msec) before the peak) be $N(l, R)$ and $N(l, L)$ in right and left circularly polarized light, and $N(b, R)$ and $N(b, L)$ be the corresponding numbers of counts of background light as given by the same number of channels, centered about 6 msec before the main pulse. Then the fractional polarization is

$$V(l) = \frac{[N(l, R) - N(b, R)] - [N(l, L) - N(b, L)]}{[N(l, R) - N(b, R)] + [N(l, L) - N(b, L)]}$$

with a standard deviation given by

$$\sigma(l) = \frac{\sqrt{N(l, R) + N(b, R) + N(l, L) + N(b, L)}}{[N(l, R) - N(b, R)] + [N(l, L) - N(b, L)]}$$

Similar expressions apply for the trailing edge of the pulse, which is taken to be the 10 channels following the peak.

The resulting circular polarization from all runs summed together is shown in Table 1 for the leading and trailing edges of the primary pulse separately, for the whole pulse, and for the leading minus the trailing edge (this would be non-zero if the circular polarization changed sign at the peak of the pulse). It is seen that the circular polarization is zero to within the errors of measurement. This is in agreement with the previous results of Wampler *et al.* (1969), who find $V=3.9\pm4.9\%$, and with the results of Cocke *et al.* (1969), who measure $V=2\pm1\%$. The background nebula light is also found to be unpolarized.

This measurement was greatly reduced in accuracy by the fact that we made our observations on a moonlit night. We intend to repeat the measurement in the fall and expect to be able to reach a limit of no more than 0.1% uncertainty.

Acknowledgements

We are grateful to the Director of McDonald Observatory for the hospitality of the Observatory, and to Dr. Brian Warner and Dr. Patrick Thaddeus for their help and cooperation. This work was supported in part by the National Aeronautics and Space Administration, the Air Force Office of Scientific Research, and the Research Corporation. It is Columbia Astrophysics Laboratory Contribution No. 27.

References

Angel, J. R. P. and Landstreet, J. D.: 1970, *Astrophys. J. Letters* **160**, L147.
Cocke, W. J., Disney, M. W., and Gehrels, T.: 1969, *Nature* **223**, 576.
Wampler, E. J., Scargle, J. D., and Miller, J. S.: 1969, *Astrophys. J. Letters* **157**, L1.

2.17 SEARCH FOR

PULSED HIGH ENERGY GAMMA RADIATION FROM NP 0532

G. FAZIO, D. HEARN, H. HELMKEN, G. RIEKE,
T. WEEKES, and F. CHAFFE

Smithsonian Astrophysical Observatory, Cambridge, Mass., U.S.A.

Abstract. During winter 1969 the 10 m optical reflector at Mt. Hopkins, Arizona, was used to search for periodic gamma ray emission above 10^{11} eV from NP 0532. Based on predicted optical period and phase, approximately 57 h of data were summed together. No evidence of pulsed radiation was found.

1. Introduction

A search with long integration times was undertaken from 11 December 1969 to 14 January 1970 for periodic gamma ray emission from NP 0532. Theoretical predictions for the existence of this radiation are based on inverse Compton scattering by electrons and high energy proton interactions. Above 10^{11} eV, gamma rays are detected from the Čerenkov light pulse emitted by their atmospheric electromagnetic cascade. Cosmic ray protons also give rise to many, more numerous cascades with similar light pulses. For pulsar search, the time of arrival of each Čerenkov light pulse is recorded and sorted into bins according to the period of the object under investigation. Best previous upper limits (Fazio *et al.*, 1970) to the gamma ray flux were based on 1 hour runs.

2. Techniques

The 10 m optical reflector at Mt. Hopkins, Arizona, was used for the collection of the Čerenkov light. The reflector has a collecting area of 1.3×10^4 m^2 and a zenith energy threshold of 9×10^{10} eV. Two 5″ RCA 4522 phototubes separated by approximately 2.4° were placed at the prime focus of the reflector. Each phototube successively tracked 'on' the pulsar direction for 9 min. One additional minute was allowed for slewing the reflector to interchange the 'on' phototube. Each phototube was light servoed to give a constant current response.

Each event from the 'on' phototube was recorded on one track of a stereo tape recorder for periods up to 115 min. On the adjacent stereo track, a 5 kHz timing signal from the clock at the satellite-tracking station on Mt. Hopkins was recorded. Frequency stability of the clock was 1 part in 10^{10}. Preceding each event record were two known minute time marks. Absolute phase of the pulsar was referenced to these time marks. During the playback of each tape the time of each event, to the 5 kHz clock accuracy, was recorded on a digital stepping recorder. A 2 period analysis into 100 channels was made with the CDC 6400 computer and the predicted phase normalized the 'peak' to channels 10 and 60. Period and phase predictions were based on the optical ephemeris compiled at the Harvard Agassiz Station by Horowitz and Papaliolios (1971). Verification of the predicted phase was achieved at Mt. Hopkins by observing the optical pulsar with the No. 2 Kitt Peak 36″ reflector and telemetering the individual pulses to Mt. Hopkins, some 40 miles distant. The telemetry system had a 4 MHz bandwidth to ensure good pulse response.

In the digital analysis, a free running 5 kHz clock was phase locked with the recorded 5 kHz signal. Thus a missing pulse due to tape dropout would be correctly reinserted. The recorded tapes had an additional set of minute marks at the end of a run and thus verified that the synthesized clock neither added nor deleted clock pulses.

3. Results

Three sums for each month's operations were made. The first summed the entire month's data, the second only those runs in which the average zenith angle of NP 0532 was less than 30°, the third for those with average zenith angle greater than 30°. Finally, the data from both months' observations were summed together.

In 57 hours of combined data, no excess was seen at the expected channel or at any other channel. The high energy pulsed gamma ray emission was assumed to occur over just one channel width, i.e., 0.6 msec. The flux that could have been detected was taken as the effect in the predicted channel plus three standard deviations. Thus the upper limits to the pulsed gamma radiation emitted in phase with the optical pulsations for the period December 1969 to January 1970 are:

Data	Energy threshold (eV)	Upper limit photons cm^{-2} sec^{-1}
All zenith angles	$2.2 \times 10^{+11}$	6.0×10^{-12}
Zenith angles less than 30°	$1.2 \times 10^{+11}$	2.3×10^{-11}
Zenith angles greater than 30°	$3.2 \times 10^{+11}$	3.5×10^{-12}

References

Fazio, G. G., Hearn, D. R., Helmken, H. F., Rieke, G. H., and Weekes, T. C.: 1970, in L. Gratton (ed.), 'Non-Solar X- and Gamma-Ray Astronomy', *IAU Symp.* **37**, 192.
Horowitz, P. and Papaliolios, C.: 1971, to be published in *Nature*.

Discussion

F. C. Michel: Is the A. C. flux limit a mean flux or a peak flux?

H. Helmken: The periodic result is the upper limit to the flux per second at the expected position of the pulse for a mean pulse width of 0.6 msec.

N. Visvanathan: You reported 113 hours observing but have only reported results from 57 hours.

H. Helmken: The analysis is incomplete.

SESSION 3

OBSERVATIONS OF OTHER PULSARS

3.1   THE GALACTIC POPULATION OF PULSARS

M. I. LARGE*

University of Manchester, Nuffield Radio Astronomy Laboratory, Jodrell Bank, U.K.

Abstract. This paper examines the distribution of pulsars in the Galaxy and compares it with the distribution of supernova remnants. The dispersion measure has been used as a distance indicator to obtain the distribution in luminosity, z-distance and period of the pulsars. The z-distance distribution is similar within the uncertainties to that of the supernova remnants. An estimate of the total number of pulsars in the Galaxy with a peak luminosity $\geqslant 1$ fu$(dm)^2$ gives 5×10^5 within a factor of 10. The data is consistent with the hypothesis that pulsars are formed at the same rate as supernovae, if the pulsars are assumed to have a timescale of 10^7yr.

1. Introduction

The extent to which pulsars and supernovae are physically associated is uncertain. It is now generally accepted that pulsars are beamed radiation from neutron stars, and a supernova explosion has been thought to provide suitable conditions for the formation of neutron stars (Colgate and White, 1966) although there are theoretical difficulties (Arnett, 1969). It is therefore of considerable interest to determine whether pulsars and supernovae do have a common origin. The direct observational evidence is unsatisfactory: of 55 pulsars** and about 90 known or suspected supernova remnants (see for example, Milne, 1970) there is one unequivocal association (NP 0532 with the Crab Nebula) and one probable but by no means certain association (PSR 0833-45 with the Vela remnant).

Several factors could contribute to the lack of obvious correlation between pulsars and supernovae. The detectable traces of a supernova explosion seem to last for about 10^5 yr, whereas most pulsar ages derived from the rate of change of period are at least an order of magnitude greater than this (Ostriker and Gunn, 1969). It is quite possible that the comparatively young pulsars in known remnants have not been found simply because adequate sensitivity has not been achieved for short period pulsars (Large, 1970a). Furthermore a large proportion of all pulsars may be missed on account of their orientation if the beam of the pulsar is restricted in the plane containing the rotation axis. It is also possible that pulsars are formed with high velocity so that they appear coincident with the parent supernova for a short fraction of their lifetime only. Using this idea Prentice (1970) has suggested that several pulsars might be associated with near (but not coincident) supernovae. It is interesting that the two longest period pulsars (NP 0527, JP 2319) are fairly near the two most spectacular supernovae, although physical association in both cases seems improbable.

From the foregoing remarks, it appears that despite the paucity of individual associations, supernovae could well be the origin of some or all pulsars. A phe-

---

* On leave of absence from Department of Physics, University of Sydney, Australia.
** A recent compilation by Maran and Modali (1970) includes 50 pulsars. A further 5 were reported by Davies *et al.* (1970).

nomenological analysis of radio evolution of pulsars by Aizu and Taketani (1970) supports this hypothesis. The main objective of this talk is to examine the distribution and number of pulsars in the Galaxy to find out whether they are consistent with the probable rate of production of supernovae in the Galaxy. With the comparatively small sample of pulsars so far discovered (55) it is not possible to deduce the finer details of the galactic distribution. I have therefore chosen three parameters which seem to me of particular importance namely the pulsar period (P), the absolute radio luminosity (L) and the height above the plane of the Galaxy (z) and attempted to derive the spatial density of pulsars $\varrho(P, L, z)$ in terms of these three quantities. The first part of the talk is concerned with the analysis of the available data and discussion of the likely sources of error in the derived distributions. The second half is more speculative in nature, being devoted to astrophysical interpretation of these results and comparison with the supernova data. In a series of papers on the nature of pulsars Gunn and Ostriker (1970) formulated statistical methods for the analysis of pulsar observations based on their magnetic dipole model, and applied this formalism to the extant observations. It is interesting that many of the results that I derive without assuming a theoretical model of pulsars are in broad agreement with Gunn and Ostriker's conclusions.

2. The Data

In order to derive the galactic density of pulsars from the observed numbers, it is necessary to have a means of determining pulsar distances, and to know the instrumental selection effects in terms of such quantities as luminosity, period, dispersion measure and galactic z-distance.

The distance to each pulsar, in terms of the integrated electron density in the line-of-sight, is known from the dispersion measure, since it is unlikely that an appreciable fraction of the dispersion arises in the source (Rickett, 1969). Many workers have estimated the mean electron density in the Galaxy in order to obtain a calibration of the dispersion measure distance scale. Prentice and ter Haar (1970) in particular have studied the distribution of H II regions within 1 kpc of the Sun in order to determine their individual contribution to the dispersion of pulsars, and this type of approach is undoubtedly valuable in determining the distance to particular pulsars. However their work shows that a single H II region rarely contributes as much as 50% of the total pulsar dispersion, suggesting that the dispersion measure is quite a good statistical measure of pulsar distances. In the work described in this paper I have used the dispersion measure as the distance measure, deferring to the later sections the question of relating these distances to true geometrical distances.

The assessment of search selection effects is more difficult. Most pulsars have been discovered as individual spikes on chart records, with a sensitivity which depends in a complicated way on the response of the pen, the statistics of the pulse amplitudes and the quickness of the observer's eye. Furthermore many pulsars in the northern sky have been found at relatively low radio frequencies (~ 100 MHz) where dispersion broadening of the pulses in the receiver bandwidth greatly reduces the sensitivity to highly dispersed pulsars. It seems impossible to make proper allowance for these complicated selection effects in the various searches, and I have made no attempt to do so. Instead I have used the relatively homogeneous sample of 29 pulsars discovered at the Molonglo Radio Observatory. The Molonglo pulsars were discovered in uniform and well calibrated searches of about 7 ster of the southern sky. The Molonglo Cross operates at 408 MHz, and most of the searches were made using the full 4 MHz bandwidth and split beams of the E–W arm. Most of the accessible longitudes of the galactic plane were further searched between $\pm 10°$ in latitude using 'dispersion removers' to extend the sensitivity to dispersion measures of about 400 cm^{-3} pc. The confining of this latter search to the Galaxy does not involve a selection, since pulsars of dispersion measure greater than 50 cm^{-3} pc are not found more than 10° from the galactic plane. The Molonglo searches have been calibrated by generating artificial pulsar signals of known intensity and assessing the limiting mean flux density by examination of chart records using the same procedures used in finding real pulsars. Despite the subjective aspects of this approach, the calibration is self-consistent within about 1 dB since independent observers were in agreement in determining the limiting flux densities. Data are thus available on the sensitivity of the Molonglo searches in terms of period and dispersion measure. The main results of this calibration have been published in graphical form (Large, 1970b) and further details will be given in due course (Large and Vaughan, 1971).

The observed distributions of the 29 Molonglo pulsars in period, luminosity, and z-distance are shown in Figures 1, 2 and 3. As explained in the introduction, dispersion measure has been used as a measure of distance so that z-distance is plotted in units

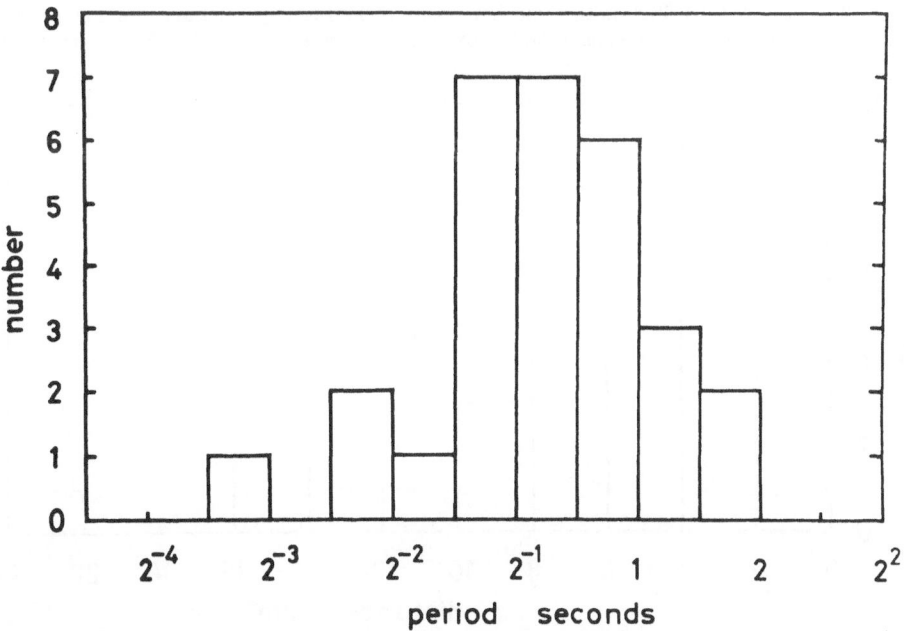

Fig. 1.   The observed distribution in period of 29 Molonglo pulsars.

of cm^{-3} pc. The luminosity is defined as the flux density at 408 MHz averaged over the entire period multiplied by the square of the dispersion measure. The unit of luminosity is therefore 10^{-26} Wm^{-2} Hz^{-1} (cm^{-3} pc)^2 which is abbreviated to fu (dm)^2 throughout this paper. In drawing the histograms of Figures 1, 2 and 3 the intervals of 0.5 in $\log_2 P$, 0.5 in $\log_{10} L$, and 2 cm^{-3} pc in z were chosen to divide the pulsars into about 5 groups in each distribution. The distributions represent the true galactic distributions of the pulsars as modified by the selection effects in the searches, with large statistical errors.

Several workers have drawn attention to correlations between pulsar parameters. In particular there is apparently a weak tendency for higher latitude pulsars to have

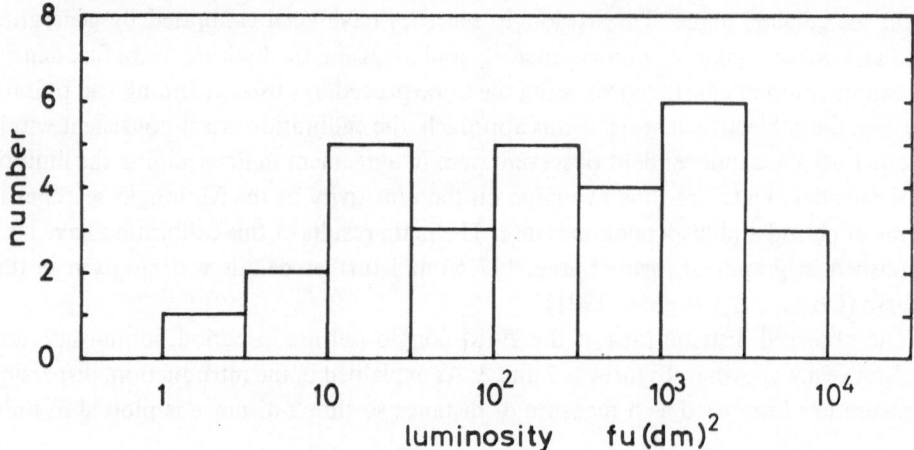

Fig. 2.   The observed distribution in radio luminosity of 29 Molonglo pulsars. The luminosity is based on the pulsar mean flux density and is expressed in units of 10^{-26} Wm^{-2} Hz^{-1} (cm^{-3} pc)^2.

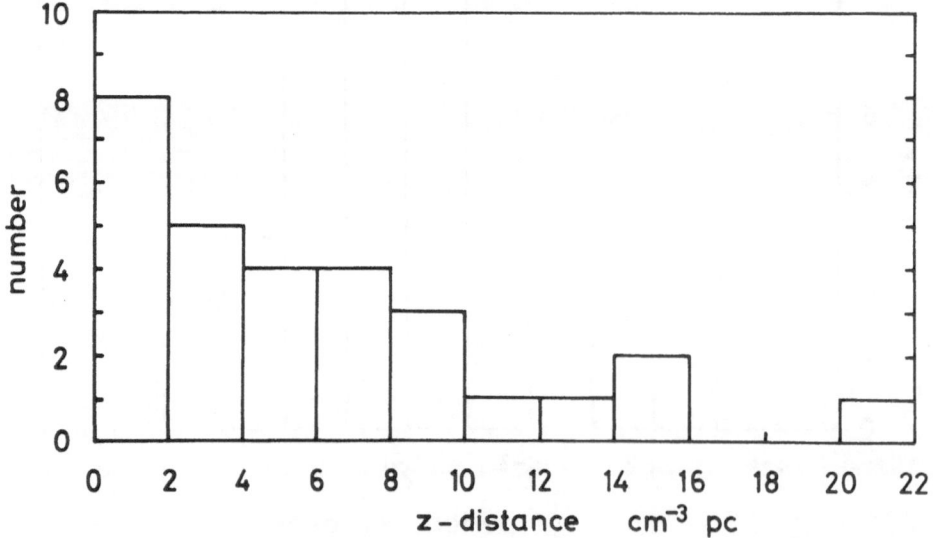

Fig. 3.   The observed distribution in galactic z-distance of 29 Molonglo pulsars.

longer periods, and this has been interpreted as indicating that longer period pulsars have lower absolute luminosities (Gold, 1970; Notni *et al.*, 1970). Certainly such correlations are to be expected if pulsars typically become weaker and increase in period as they age. However, the observed correlations are weak, subject to selection effects and barely detectable among the Molonglo pulsars. For example, in Figure 4, the periods and luminosities of the Molonglo pulsars are plotted as a scatter diagram. The correlation coefficient between $\log P$ and $\log L$ is 0.30 which is significant at the

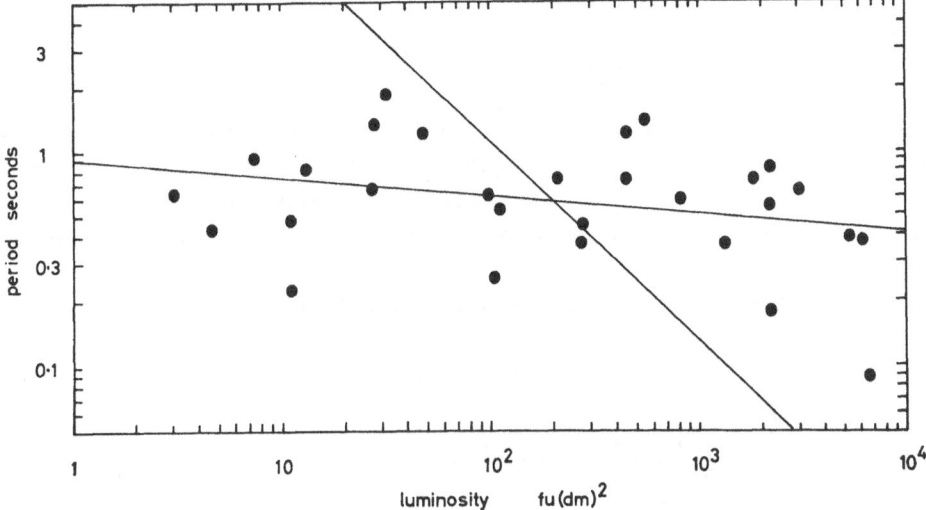

Fig. 4.   Diagram showing the period and luminosity of 29 Molonglo pulsars. The correlation between $\log P$ and $\log L$ is barely significant, and is largely produced by one pulsar. The two regression lines are shown.

10% level only. The two regression lines of $\log P$ on $\log L$ and *vice versa* are shown on the diagram. Most of the correlation is produced by one pulsar (PSR 0833-45, period 0.089 sec). The regression lines show that pulsars of period ~ 0.3 sec perhaps have luminosities about 3 times that for periods of about 1 sec. Similarly, over the full range of luminosity plotted, the periods change by perhaps a factor of 2. However, over the observed ranges of period and luminosity, the systematic variation of period with luminosity is considerably less than the intrinsic scatter of the observed periods and luminosities.

There is a systematic difference between the Molonglo pulsars and the rest in this respect. Molonglo pulsars are of shorter period and lower latitude than the others. This difference accounts for most of the correlation between latitude and period, and hence needs examining carefully for instrumental selection effects.

3. Method of Analysis

The number of pulsars observed in ranges dP, dL and dz of period, luminosity and z-distance can be written

$$N(P, L, z)\, dP\, dL\, dz = A(P, L, z)\, \varrho(P, L, z)\, dP\, dL\, dz \tag{1}$$

where $A(P, L, z)\, dz$ is the volume of space effectively explored at period P, luminosity L and z distance between z and $z + dz$. In principle the required spatial density $\varrho(P, L, z)$ can be derived from the observed distributions and the sensitivity factor $A(P, L, z)$. It is clear however that without some simplifying assumptions this cannot be done adequately on the basis of 29 (nor even 55) pulsars.

As a first approximation, the density is written as the product of three independent functions,

$$\varrho(P, L, z) = \varrho_0 \varrho_1(P)\, \varrho_2(L)\, \varrho_3(z) \tag{2}$$

Separation of the variables in this way makes it possible to derive estimates of the galactic distribution. It is justified only in so far as the respective distributions are uncorrelated. It will be shown that the small degree of correlation indicated by Figure 4 introduces negligible errors in the derived distributions compared with the statistical errors.

From Equations 1 and 2 the distribution in period can now be written

$$\varrho_1(P) \propto \frac{\displaystyle\int\int N(P, L, z)\, dL\, dz}{\displaystyle\int\int A(P, L, z)\, \varrho_2(L)\, \varrho_3(z)\, dL\, dz} \tag{3}$$

with similar expressions for $\varrho_2(L)$ and $\varrho_3(z)$. These equations have been solved iteratively using the observed distributions of Molonglo pulsars (Figures 1, 2 and 3) and the data on the search calibrations to compute the sensitivity function $A(P, L, z)$. The resulting derived distributions $\varrho_1(P)$, $\varrho_2(L)$ and $\varrho_3(z)$ are shown in Figures 5, 6 and 7, with distance measured in dispersion units of $(cm^{-3}\ pc)$ and luminosity in flux units (dispersion measure)^2 as explained previously.

4. Result and Discussion of Errors

It will be seen from the previous section that each distribution, $\varrho_1(P)$, $\varrho_2(L)$ and $\varrho_3(z)$ is derived from the corresponding observed distribution by dividing by a scaling function. For example Equation 3 can be written

$$\varrho_1(P) = \frac{1}{D(P)}\, N(P) \tag{4}$$

where $N(P)$ is the observed distribution in period and $D(P)$ is the scaling function computed from the known sensitivity and the other two distributions $\varrho_2(L)$ and $\varrho_3(z)$.

The observed distributions $N(P)$, $N(L)$ and $N(z)$ contain large random errors as they are based on only 29 pulsars. The random errors in the derived distributions are largely caused by these random errors in the corresponding observed distribution.

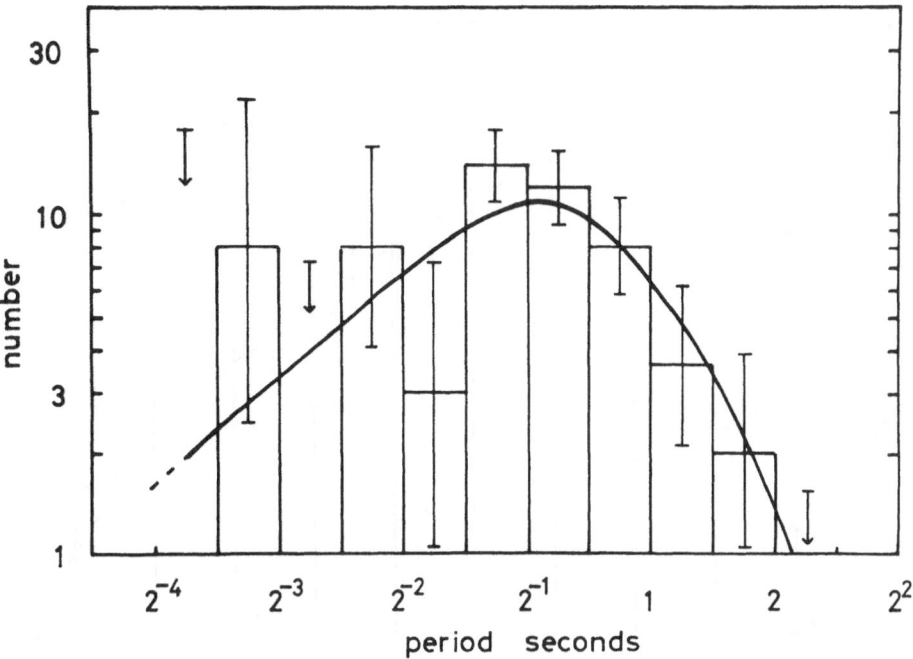

Fig. 5. The distribution, $\varrho_1(P)$, of pulsars in period derived from the observed distribution of the Molonglo pulsars (Figure 1) and a scaling function calculated from the measured sensitivity data and the distributions of pulsars in luminosity and z-distance. The probable error limits reflect sampling errors in the observed distribution, and are calculated from cumulative sums of the Poisson distribution.

The random errors in the scaling denominators are relatively small as the denominators are relatively insensitive to changes in the assumed distributions. The random errors indicated in Figures 5 and 6 are 'probable error' points determined by assuming that the number of pulsars observed in each class is Poisson distributed. Systematic errors are discussed for each distribution separately.

A. THE DISTRIBUTION IN PERIOD

The derived distribution in period differs somewhat from the observed distribution. In particular the relative abundance of short period pulsars is considerably greater than the observed distribution indicates. This conclusion supports a less complete analysis which showed (Large, 1970b) that for periods of ~0.2 sec the Molonglo search might be missing more than half of the pulsars within the theoretical sensitivity limit. At the long period end of the distribution the sharp cut-off appears to be real, as the Molonglo search maintained good sensitivity to well beyond 4s period.

The assumption that period is independent of luminosity might have introduced systematic distortion into the derived period distribution. If in fact the period changes by a factor of about 2 over the observed range of luminosity as suggested by Figure 4, then the true period distribution would be shifted towards longer periods for low luminosities, and towards shorter periods for high luminosities, with the mean curve

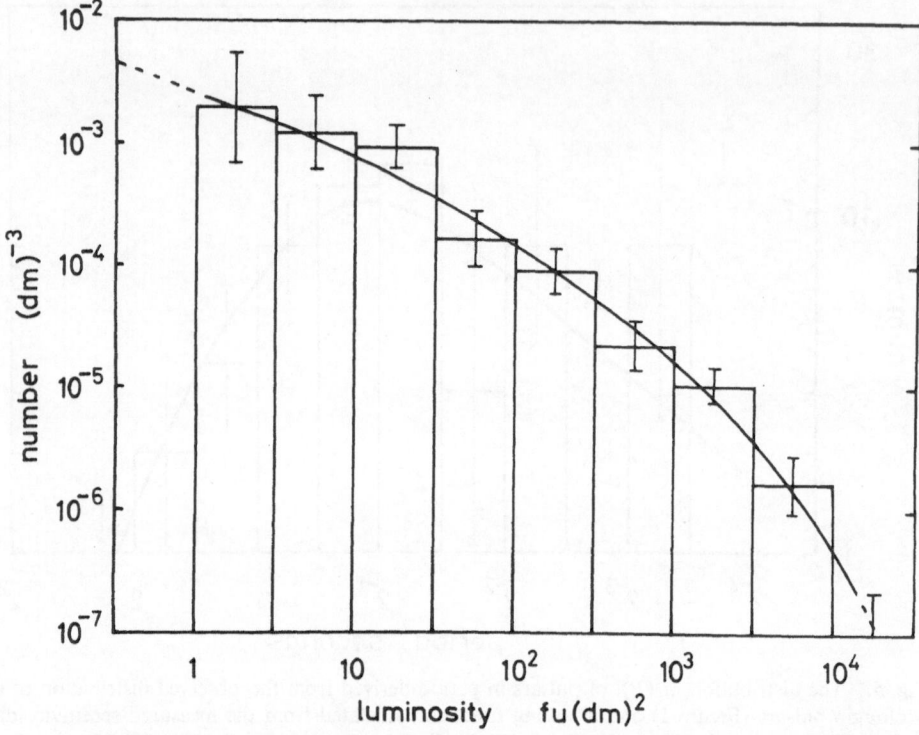

Fig. 6.   The distribution $\varrho_2(L)$, of pulsars in absolute luminosity derived from the observed distribution of Molonglo pulsars (Figure 2). The histogram shows the number of pulsars per unit volume in each range of luminosity, with the probable sampling errors. As elsewhere, 'distance' is measured in terms of dispersion measure in units of cm^{-3} pc.

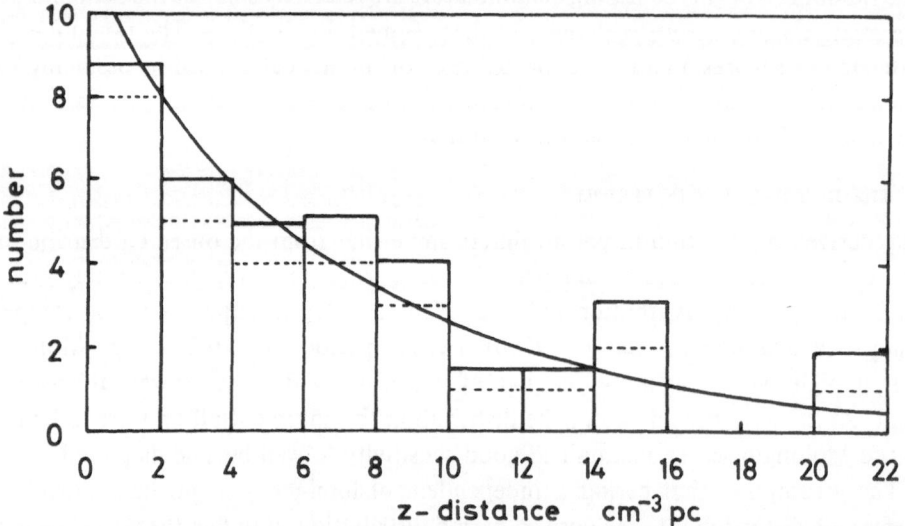

Fig. 7.   The distribution, $\varrho_3(z)$, of pulsars in galactic z-distance derived from the observed distribution of Molonglo pulsars (shown dotted) and calibration data. Distance is expressed in terms of dispersion measure, in units of cm^{-3} pc. The distribution is adequately represented by an exponential of scale height 7 cm^{-3} pc.

shifted slightly towards longer periods corresponding to the more numerous low luminosity pulsars.

B. THE DISTRIBUTION IN LUMINOSITY

It is apparent from Figure 6 that the numbers of pulsars per unit volume of space in each logarithmic interval of luminosity falls rapidly with luminosity. A power law of the form

$$\varrho(L)\, dL \propto L^{-x}\, dL \tag{5}$$

represents the data moderately well with $x=2$. However the errors derived from assuming a Poisson distribution in each class of the observed distribution show quite clearly that the luminosity function is curved, with x increasing from about 1.5 at low luminosity ($L=1$ fu(dm)^2) to about 3 at high luminosity ($L=10^4$ fu(dm)^2).

The form of this luminosity curve is very little affected by the assumption that period and luminosity are unrelated. If the degree of correlation indicated in Figure 4 is allowed for, the change in the derived luminosity curve is within the random error limits, with marginally fewer low luminosity pulsars, and more high luminosity ones.

The overall scaling and general form of the luminosity curve have been experimentally confirmed in a recent pulsar search at Jodrell Bank (Davies *et al.*, 1970). A search was made of 431 points within 1° of the galactic plane at 408 MHz, analysing for periodicities in 13 minutes data from each point. The luminosities of the seven pulsars detected (of which 5 were new discoveries) are plotted as a (cross-hatched) histogram in Figure 8. The plain histogram represents the predicted numbers and luminosities expected using the Molonglo luminosity curve (Figure 6) and the distribution in z (Figure 7). The dotted histogram shows the predicted numbers supposing all 431 points of observation had been exactly on the galactic plane. The excellent agreement between the predicted and observed distributions provides independent confirmation that the pulsar luminosity curve is reasonably well established at 408 MHz.

C. THE DISTRIBUTION IN z-DISTANCE

Comparison of the derived distribution in z with the observed distribution shows that they are almost identical. The distribution is adequately described by an exponential function of scale height 7 cm^{-3} pc in dispersion measure units, and this result is very little affected by selection effects. It cannot therefore be appreciably in error as a result of assuming the distributions in P, L and z to be independent. If $\bar{n}_e$ is taken to be 0.05 cm^{-3}, the scale height is 140 pc.

5. Interpretation

A. THE NUMBER OF PULSARS IN THE GALAXY

The pulsar luminosity function shown in Figure 6 gives the number of pulsars per unit volume in each logarithmic interval of apparent luminosity. It is probable that pulsars have a finite beamwidth in latitude (i.e. in the plane containing the rotation axis) so that only a fraction are seen with their peak luminosity. The remainder will be seen in weak side lobe radiation or not at all. In order to make some estimate of how such 'latitude beaming' might affect the observed luminosity function, I have considered two models of the beam in latitude. One model, defined by Equations 6,

$$L = L(0)\left(1 + (\theta/\theta_0)^2\right)^{-1} \tag{6}$$
$$\theta_0 = 5°$$

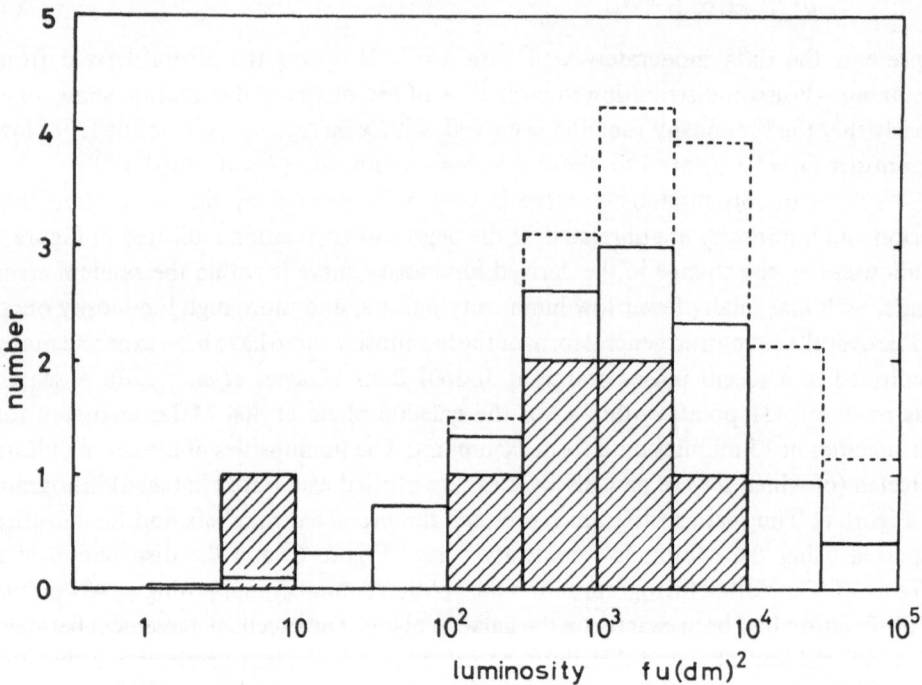

Fig. 8.   The cross-hatched histogram shows the luminosity distribution of 7 pulsars observed in a recent 408 MHz search at Jodrell Bank. The plain histogram shows numbers predicted for this search using the distributions derived from the Molonglo searches. The agreement confirms the general features and scaling of the derived distributions. The dotted histogram shows the predicted numbers that would have been observed if all the Jodrell observations had been confined to $b = 0°$.

is intended to represent a radiation pattern that remains reasonably strong in the side lobes. The second model defined by the gaussian function

$$L = L(0) \exp -\tfrac{1}{2}(\theta/\theta_0)^2 \tag{7}$$
$$\theta_0 = 5°$$

represents a beam with very little side lobe radiation.

   For both models I have computed a luminosity function in terms of $L(0)$ that would produce the distribution of apparent luminosities shown in Figure 6, supposing pulsar axes to be randomly orientated. The results are plotted in Figure 9 which shows the apparent luminosity curve as well as those derived in terms of $L(0)$ for the two models. The number scale is now expressed in terms of number per cubic kpc in each interval of 0.5 in $\log_{10} L$ based somewhat arbitrarily on taking the mean electron density in the galaxy to be $\bar{n}_e = 0.05$ cm^{-3}. In terms of $L(0)$ there are several times more pulsars in each luminosity interval, and the curvature of the luminosity curve is increased.

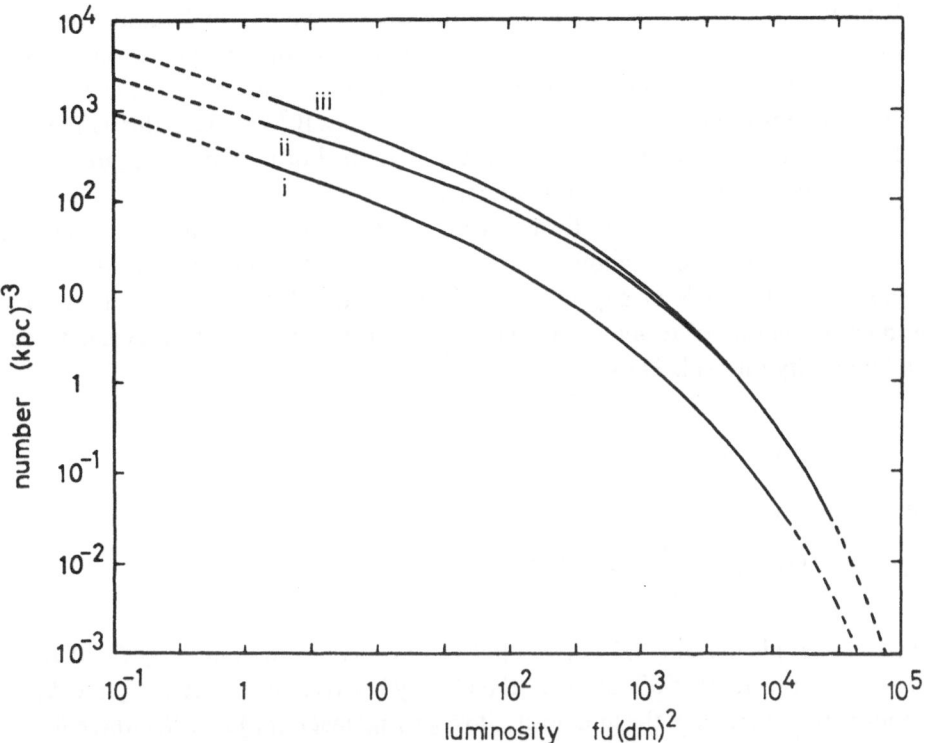

Fig. 9. Luminosity functions showing the number of pulsars in each interval of 0.5 in $\log_{10} L$ per cubic kiloparsec of the galaxy. The scaling is based on assuming that 1 dispersion measure unit is 20 parsecs, i.e. $\bar{n}_e = 0.05$ cm^{-3}. Curve (i) is the distribution of apparent luminosities. Curve (ii) represents the distribution of peak luminosities assuming a beam in latitude given by $L = L(0) (1 + (\theta/\theta_0)^2)^{-1}$ with $\theta_0 = 5°$. Curve (iii) is similar to (ii) but calculated for a gaussian beam $L = L(0) \exp -\tfrac{1}{2} (\theta/\theta_0)^2$, again with $\theta_0 = 5°$.

It is now possible to estimate the total number of pulsars in the Galaxy with luminosity greater than a suitably chosen lower limit. Using a value of mean electron density $\bar{n}_e = 0.05$ cm^{-3} and the gaussian model of the beam in latitude (Equation 7) the luminosity curve Figure 9 (iii) indicates that there are 2500 pulsars of luminosity greater than $L(0) = 1$ fu (dm)^2 per cubic kpc. Taking the scale height to be 140 pc (7 cm^{-3} pc) in galactic z-distance and a value of 15 kpc for the radius of the Galaxy, we obtain the result that there are 5×10^5 pulsars of peak luminosity exceeding 1 fu (dm)^2 in the Galaxy. This estimated number depends inversely on the assumed width of the beam in latitude, and directly on the square of the assumed mean electron density (not on $\bar{n}_e^3$ since the scale height in z-distance depends inversely on $\bar{n}_e$), and must as a result be regarded a good to within a factor of 10 only.

B. COMPARISON WITH RATE OF OCCURRENCE OF SUPERNOVAE

If it is assumed that there is a one-to-one correspondence between pulsars and supernovae, then the typical life-time of pulsars can be estimated from knowledge of the rate of occurrence of supernovae in the Galaxy and the total number of pulsars in the Galaxy. Taking the rate of occurrence of supernovae to be one every 50 years and the total number of pulsars to be 5×10^5 indicates that pulsars typically radiate with a luminosity in excess of 1 fu $(\mathrm{dm})^2$ for $\sim 2.5 \times 10^7$ yr. Since this is the order of life-time suggested by the observed rates of change of period of pulsars, it seems worth pursuing this argument in more detail.

Suppose that all pulsars are formed with a luminosity exceeding some value $L_{\min}$ and then evolve with continuously decreasing luminosity. Then in a steady state the number of pulsars dN in any luminosity interval dL (below $L_{\min}$) is given by the product of the rate of formation of pulsars n_0 and the time taken to evolve through the luminosity interval. Thus

$$dN = n_0 \frac{dt}{dL} dL = n_0 \, dt. \tag{8}$$

Integration gives

$$(t_2 - t_1) = \frac{1}{n_0} (N_2 - N_1) \tag{9}$$

where $(t_2 - t_1)$ is the time taken for a pulsar to evolve from luminosity L_2 to L_1 and $(N_2 - N_1)$ is the number of pulsars in the Galaxy between luminosities L_2 and L_1.

Figure 10 is a graph of the luminosity $(\log_{10} L)$ against $(1/n_0) N(>L)$ where $N(>L)$ is the number of pulsars of luminosity greater than L. In view of Equation 9, the graph can be regarded as representing the evolution of the luminosity of a typical pulsar below some (undetermined) luminosity $L_{\min}$. The abscissa of Figure 10 is a linear time scale in arbitrary units, t_0. The unit is 10^7 years for the particular set of parameters we have been using, namely $\bar{n}_e = 0.05$ cm^{-3}, $\theta_0 = 5°$ and $n_0 = \frac{1}{50}$ yr^{-1} and is given more generally by

$$t_0 = \text{time unit} = 10^7 \left(\frac{\bar{n}_e}{0.05}\right)^2 \left(\frac{5}{\theta_0}\right) \frac{1}{n_0} \text{ years}$$

It is not possible to determine the value of $L_{\min}$ from the present data, and the interpretation of the abscissa of Figure 10 as a time scale may be meaningless. However, inspection of the graph and also the luminosity curves (Figure 9) suggests to me that above luminosities of order 10^1 to 10^2 fu $(\mathrm{dm})^2$ we are seeing the creation and/or rapid evolution of pulsars, and below luminosities of 10 fu $(\mathrm{dm})^2$, the steady evolution of luminosity with a time scale of a few units, i.e. $\sim 10^7$ yr.

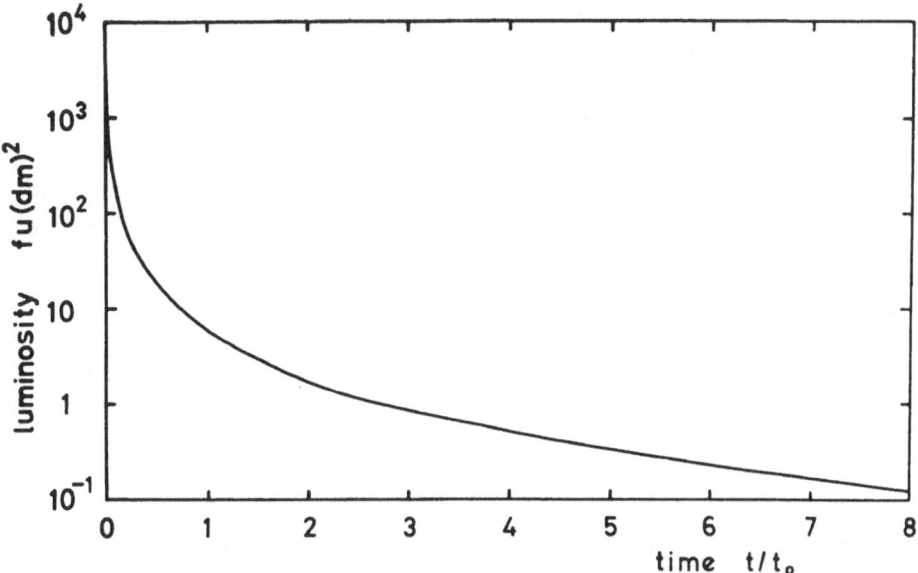

Fig. 10.   An interpretation of the pulsar luminosity function (Figure 9 (iii)) in terms of the evolution in luminosity of individual pulsars. Assuming all pulsars are initially formed with luminosity in excess of L_{min}, the curve represents the decay in luminosity below that luminosity. If pulsars are created with supernovae at the rate of about 1 every 50 years, the unit, t_0, of the time scale is $\sim 10^7$ yr. It is tentatively suggested that pulsars are generally created with luminosities in excess of 10 fu (dm)^2 and that below this luminosity, the curve indicates the time scale of their evolution in luminosity.

C. COMPARISON OF THE z-DISTRIBUTIONS OF PULSARS AND SUPERNOVAE

The supernovae appear to be concentrated in a thin layer in the galaxy. According to Poveda and Woltjer (1968) the mean z-distance $|\bar{z}|$ for nearby supernovae is 64 pc. Milne (1970) finds that more than half the supernovae in his compilation lie within 50 pc of the galactic plane. The mean height of the pulsars above the galactic plane (see Figure 7) is 7 cm^{-3} pc in dispersion measure units. Accordingly, if we are to match the two distributions it is necessary to assume that 1 dispersion measure unit corresponds with about 8 pc, implying a mean electron density of 0.12 cm^{-3}. This value is slightly high compared with that deduced from the known dispersion and distance of the Crab pulsar (0.03 cm^{-3}) but is consistent with the data for the Vela pulsar. In a recent note, Gold and Newman (1970) also note that a value of $\bar{n}_e \sim 0.1$ cm^{-3} is required to match the thickness of the pulsar distribution to other galactic populations. Alternatively, the mean height of the pulsar distribution in z-distance may be significantly greater than that of the supernova remnants, either because they were formed at high velocities as Gunn and Ostriker (1970) and others have suggested, or because they are a different class of object.

6. Conclusions

The distributions of pulsars at 408 MHz in period, luminosity and galactic z-distance are reasonably well represented by Figures 5, 6 and 7. The principal difficulties in comparing these distributions with supernova data lie in the conversion from a dispersion measure distance scale to a geometric distance scale (involving knowledge of the mean electron density in the Galaxy) and in assessing the extent to which the pulsar radiation is concentrated in the plane containing the rotation axis. However by making reasonable estimates of these factors it is shown that the total number of pulsars in the Galaxy is consistent with the hypothesis that pulsars are formed at the same rate as supernovae, and that the pulsar luminosity decays with a time scale of order of 10^7 yr. Uncertainties in the available data on the distribution of electrons in the Galaxy also limit the precision with which the height of pulsars above the galactic plane can be measured. As far as one can tell, the pulsars and supernovae extend over a similar range of z-distance, but it is possible that the pulsar distribution is slightly wider.

7. Acknowledgements

I thank Professor Sir Bernard Lovell and the University of Manchester for a Research Fellowship which enabled me to spend my year's sabbatical leave from the University of Sydney at Jodrell Bank.

References

Aizu, K. and Taketani, M.: 1970, Preprint, Rikkyo University, Tokyo.
Arnett, W. D.: 1969, *Nature* **222**, 359.
Colgate, S. A. and White, R. H.: 1966, *Astrophys. J.* **143**, 626.
Davies, J. G., Large, M. I., and Pickwick, A. C.: 1970, *Nature* **227**, 1123.
Gold, T. and Newman, H. M.: 1970, *Nature* **227**, 151.
Gunn, J. E. and Ostriker, J. P.: 1970, *Astrophys. J.* **160**, 979.
Large, M. I.: 1970a, *Astrophys. Letters* **5**, 11.
Large, M. I.: 1970b, *Proceedings of Rome Conference on Pulsars and High Energy Activity in Super-novae*.
Large, M. I. and Vaughan, A. E.: 1971, *Monthly Notices Roy. Astron. Soc.* **151**, 277.
Maran, S. P. and Modali, S. B.: 1970, *Earth Extraterest. Sci.* (in press).
Milne, D. K.: 1970, *Australian J. Phys.* **23**, 425.
Notni, P., Oleak, H., and Schmidt, K.-H.: 1970, *Astrophys. Letters* **6**, 61.
Ostriker, J. P. and Gunn, J. E.: 1969, *Nature* **223**, 813.
Poveda, A. and Woltjer, L.: 1968, *Astron. J.* **73**, 65.
Prentice, A. J. R.: 1970, *Nature* **225**, 438.
Prentice, A. J. R. and ter Haar, D.: 1969, *Monthly Notices Roy. Astron. Soc.* **146**, 423.
Rickett, B. J.: 1969, *Nature* **221**, 158.

Discussion

J. P. Ostriker: While I agree with many of your conclusions, I have some quibbles with your methodology. Why do you assume

$$N(L, P, z) = A(L, P, z)\,\varrho\,(L, P, z)?$$

Must not the relation be

$$N(L, P, z) = \iiint A(L', P', z'; L, P, z)(L,' P', z')\,\mathrm{d}L'\,\mathrm{d}P'\,\mathrm{d}z'?$$

M. I. Large: The degree of dependence of P on L is weak in the data available, and does not affect the analysis much.

D. H. Staelin: A critical assumption is the relationship between distance and dispersion measure. What is the effect of H$_{II}$ regions upon the inferred distribution luminosities?

M. I. Large: I think the dispersion measure is a reasonable statistical indicator of distance and is justified by the large range in luminosity in my sample over a factor of 10^4 which corresponds to 10^2 in distance. Prentice and ter Haar showed that an H$_{II}$ region may increase the dispersion measure of a pulsar by a factor of 2 at the most.

J. A. Roberts: I would like to know whether there have been properly designed searches for radio pulsars in supernova remnants, which allowed for the expected large dispersion and short period by some form of dedispersing. Also, have there been searches for optical pulsars in the remnants, since optical searches are not subject to the problems of large dispersion coupled with short periods?

M. I. Large: In my opinion searches for pulsars in known supernovae have not been made with adequate sensitivity for periods < 0.25 sec.

D. Horowitz: In answer to Robert's question suggesting an optical search for pulsars near super-nova remnants, I would like to say that our group at Harvard has been doing this for the past year and a half, and I will report on this later in the session.

J. E. Baldwin: What do you believe to be the relative z distributions of the pulsars and the ionised material causing dispersion?

M. I. Large: I have assumed the pulsar z distribution is narrower than the ionized layer. If the pulsar distribution is wider than that of the ionized layer then the real width of the pulsar distribution is greater than I have given in my paper. Already the pulsar distribution is wider than the supernovae.

J. V. Jelley: You mentioned there were $\sim 5.10^5$ pulsars in our own Galaxy. I would like to ask, if there were a comparable number in M31, would they contribute significantly to the total radio yield of this or other nearby galaxies?

M. I. Large: If there are a similar number of pulsars in M31, they would contribute negligibly to the continuous radio emissivity. Furthermore my luminosity curve is very steep at the high luminosity end. Extrapolation indicates that the most luminous pulsar in M31 would be quite undetectable with present techniques.

F. D. Drake: There have been discussions, particularly by Gold and Newman, of the apparent greater concentration towards the galactic plane of the shorter period pulsars. This seems to imply greater luminosity for the short period pulsars. If there is no dependence of luminosity on period, perhaps the long period pulsars have a greater mean z. What is your opinion?

M. I. Large: The effect described by Gold and Newman is plotted in Figure 11. I feel that the

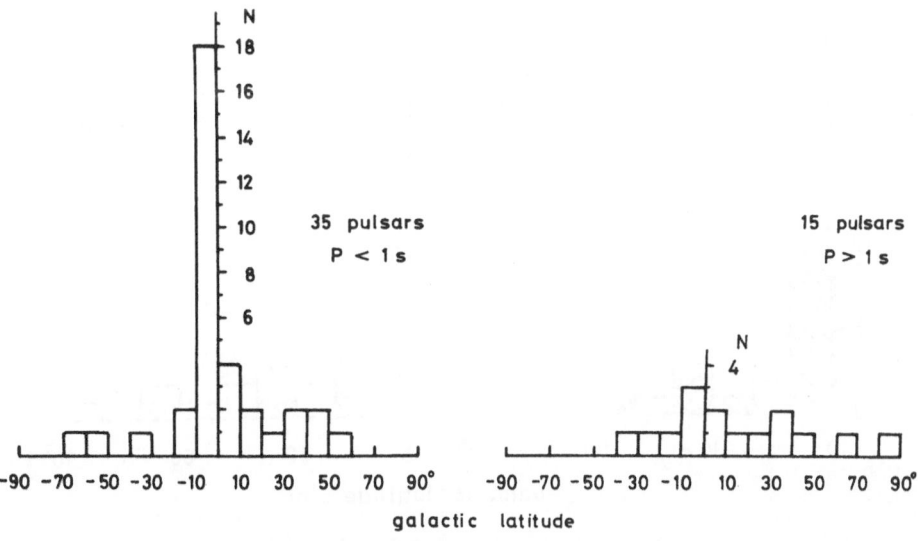

Fig. 11.   The latitude distribution of known pulsars divided into (a) those with a period < 1 sec and (b) those with a period > 1 sec.

difference in the distribution of pulsars with $P < 1s$ and those with $P > 1s$ is not well established. An alternative presentation of the data is given in the top section of Figure 12 in which the Molonglo pulsars are shaded. The distribution for $P > 1s$ is multiplied by $\frac{35}{15}$ to make it comparable with the distribution for $P < 1s$ and it is seen that the distributions are not significantly different. If the range of periods is split at $P = 0.7$ sec so that there are 25 in each class then there is no significant difference between the two distributions as shown in the bottom half of Figure 12.

Notni, Oleak and Schmidt conclude there is a correlation between period and galactic latitude at the 5% level. The data is given in Figure 13 where the Molonglo pulsars are shown as open circles and the Rest of the World as filled circles. Because of the observing technique (particularly the frequency) used the Molonglo pulsars tend to have a shorter period and lower latitude than the others. This

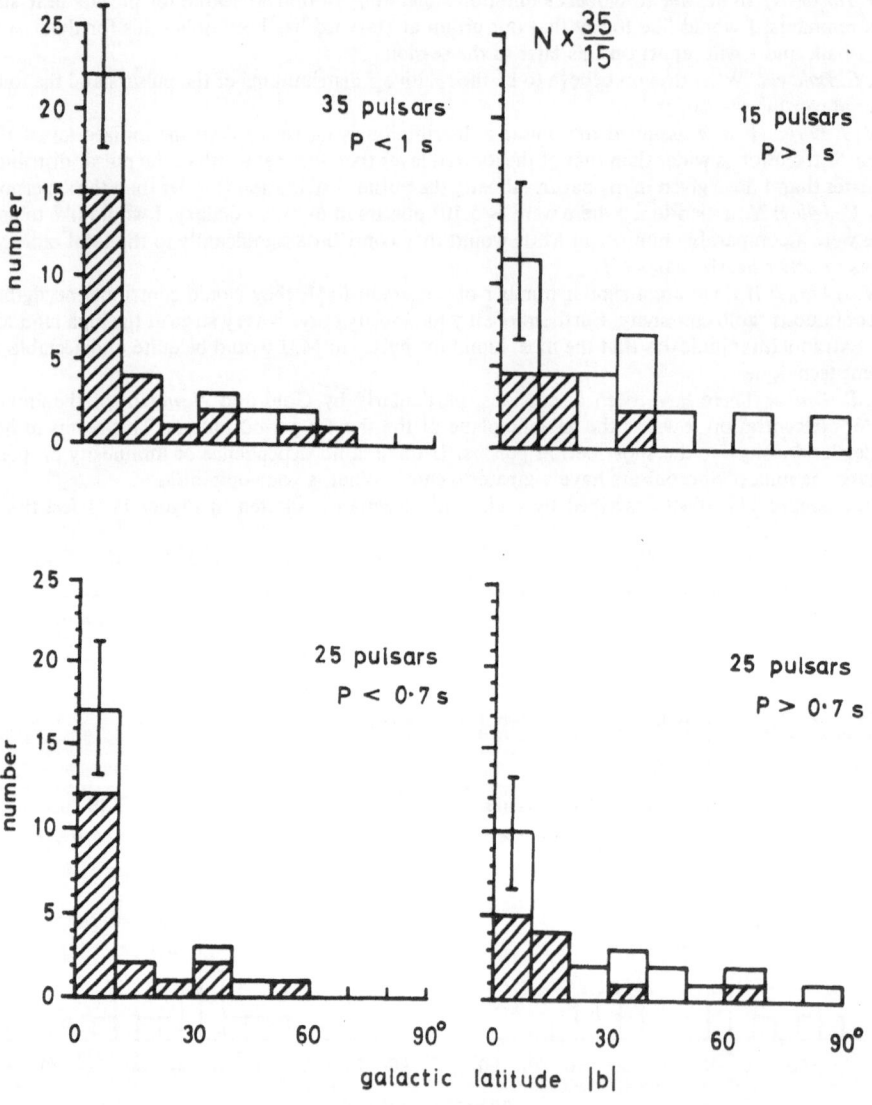

Fig. 12. The same data as in Figure 11, plotted as a function of $|b|$. The top half of the figure shows the normalized distributions. Error bars indicate the statistical uncertainty. The bottom half shows the data divided into equal groups with the division at a period of 0.7 sec.

alone is responsible for some of the systematic trends in the statistics. A recent pulsar search at 408 HMz using the Mk I telescope at Jodrell Bank discovered a number of low latitude, long period pulsars which when plotted on Figure 13 remove any correlation between period and latitude which may be present.

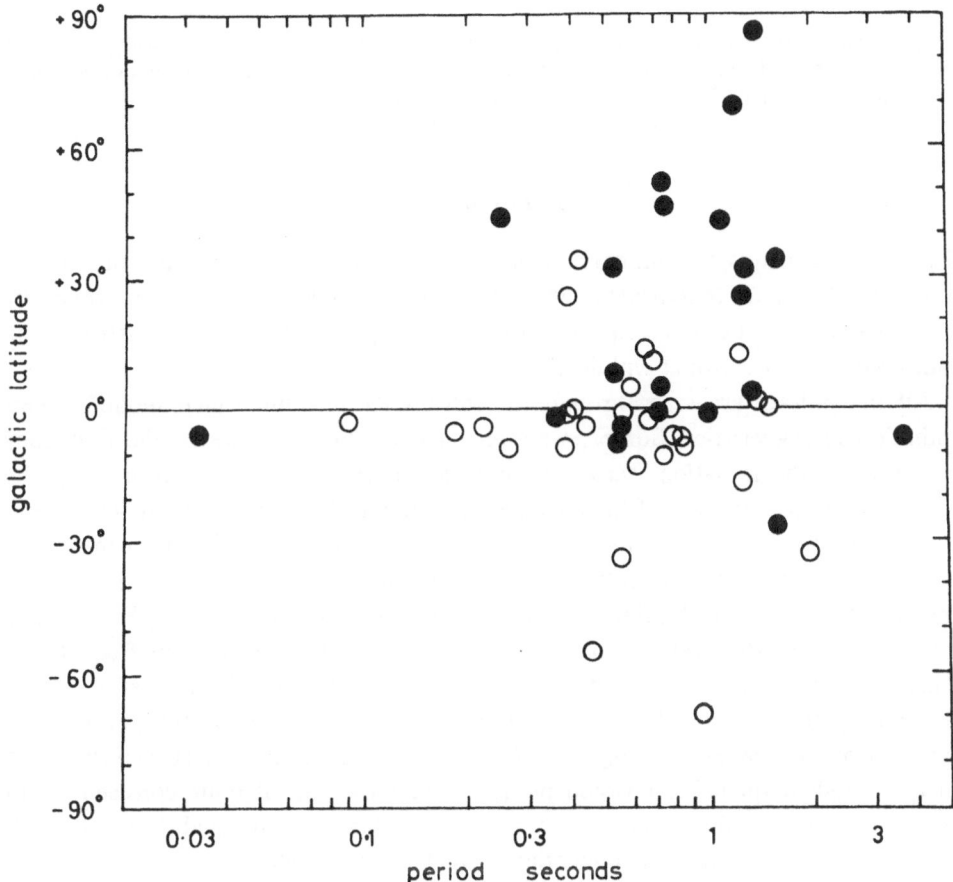

Fig. 13.   A plot of the distribution of pulsars in period and galactic latitude. Molonglo pulsars are shown as open circles, the Rest of the World as filled circles.

3.2 THE COMPARATIVE PROPERTIES OF THE PULSARS

A. G. LYNE

University of Manchester, Nuffield Radio Astronomy Laboratories, Jodrell Bank, U.K.

Abstract. This paper reviews the outstanding characteristics of the pulsars and, in particular, compares the properties of the Crab Nebula pulsar with those of the other pulsars. The Crab pulsar is unique in that it has a remarkably short period, emission outside the radio band, some strange amplitude variations and an abnormally large interpulse.

1. Introduction

This paper attempts to summarise some of the basic properties of the pulsars. In particular, I shall try to indicate in what ways the Crab pulsar fits into the morphology of the less endowed objects. I shall not discuss the polarisation characteristics of the radio pulses as Dr. Moffet will do that.

All the pulsars are seen through the interstellar medium which modifies their radiation characteristics. Some properties such as fine structure in the frequency spectrum of the radiation, certain time scales in the intensity variations and the dispersion in arrival times of the pulses are for the most part understood. The effects can usually be identified or removed from the observations so that we can deduce what the radiation would look like in the absence of the haze of the interstellar medium. One effect which cannot be removed was mentioned yesterday by Prof. Drake who showed that the Crab pulsar pulses are lengthened drastically at low frequencies. This effect is attributed to multiple path transmission in the interstellar medium; it has been studied recently at Jodrell Bank, using recently discovered pulsars with large dispersion measures. Figure 1 shows how drastic the effect can be even at quite high frequencies. I would point out that the data is quite consistent with the convolution of the natural pulse width indicated (but still slightly lengthened) at 610 MHz, with a simple exponential of the form $e^{-t/\tau}$, where

$$\tau \sim \tau_0 \left(\frac{f_0}{f}\right)^n \quad \text{and} \quad n \sim 3\text{--}4.$$

I would point out that there is no clear cut dependence of τ_0 on dispersion measure (DM) as was suggested by Prof. Drake. The effect is not detectable in some pulsars with large dispersion measures such as JP 2003 ($\int N \, dl = 200$ pc cm^{-3}) but shows very clearly in JP 1946 ($\int N \, dl = 130$) in Figure 1. The data is nevertheless consistent with a dependence upon the square of the DM as expected from some theoretical considerations, but there is a wide scatter among pulsars of similar dispersion measure.

I shall now talk about the properties of the radiation we would see in the absence of the intervening medium.

2. Periodicity

The most notable property of the pulsars is the periodicity in arrival times. The periods range from 0.03 sec for NP 0532 to 4 sec for NP 0527. The stability of the pulsations are such that the arrival times of pulses from a number of pulsars can now be confidently predicted over periods of a year to an accuracy of a few milliseconds or better. Reichley, Downs and Morris (1970) and Hunt (1969) have provided rates of change of period for more than a dozen long period pulsars.

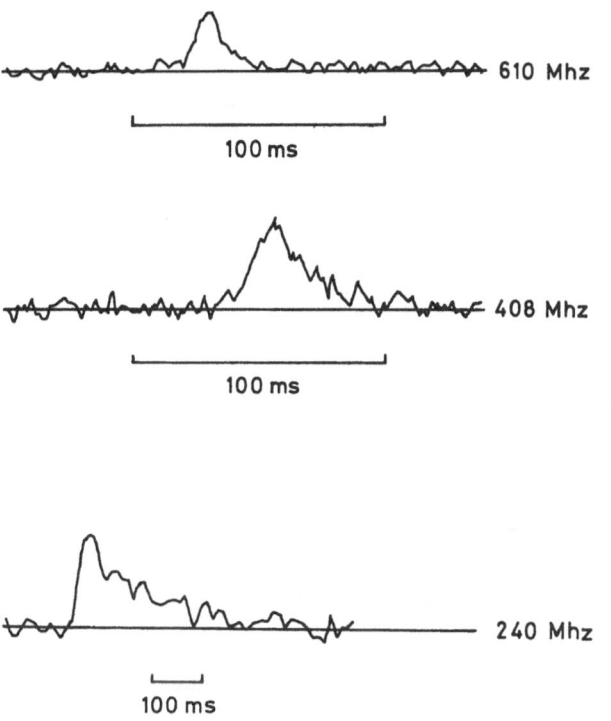

Fig. 1.   Pulse broadening in JP 1946 due to multipath interstellar propogation.

Deviations from a secular slow down rate are the exception rather than the rule and the dramatic changes seen in PSR 0833-45 last year (Reichley and Downs, 1969; Radhakrishnan and Manchester, 1969) and the complicated perturbations in NP 0532 that we heard about yesterday have not been observed in other pulsars.

Figure 2 shows a plot of $\dot{P}$ against P for all pulsars with known $\dot{P}$. Apart from the Crab and Vela pulsars there is no clear relationship.

Figure 3 shows a slightly clearer relationship between $\dot{P}$ and the luminosity. Again, the Crab and Vela pulsars in the top of the diagram are chiefly responsible for the correlation. It has been suggested that this is due to a dependence of both variables upon the magnetic field in the source (Hunt, 1969).

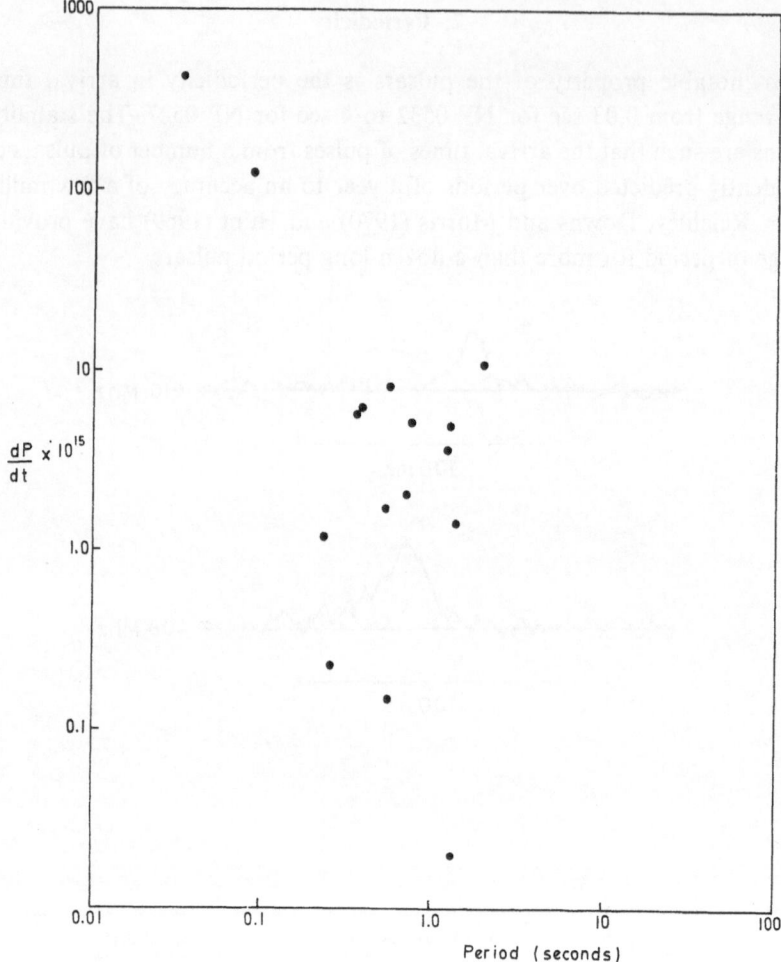

Fig. 2.   Rate of change of period plotted against period for 16 pulsars.

3. Mean Profiles

One other well defined characteristic of a pulsar is the mean pulse profile when integrated over several hundred pulses. Such a profile is stable and is quite unique to a particular pulsar. Figure 4 shows the wide range of profiles encountered at 408 MHz.

Not immediately obvious from these profiles is a very close relationship between the pulse duration and the period. Figure 5 shows the equivalent width of the pulses, defined as the total impulsive energy per period divided by the peak impulsive power, plotted as a function of period. The straight line corresponds to pulses having a width, w_p, of 2.6% of the pulse repetition period. The most striking point about this is that all but one or two pulsars lie within a factor of two of this line.

If the pulses are interpreted as being due to the sweeping of a beam across the Earth, then the angular width of this beam is 9° of longitude, with a standard deviation of about 3.5°. Any proposed emission mechanism must be able to account for this narrow range of beam-widths.

It is also clear that there is little dependence of the duty ratio upon the period. The mean duty ratio of the 14 pulsars with the shortest periods is 2.8% while that of the other 14 is 2.6%. In this respect, therefore, the long period pulsars look just like

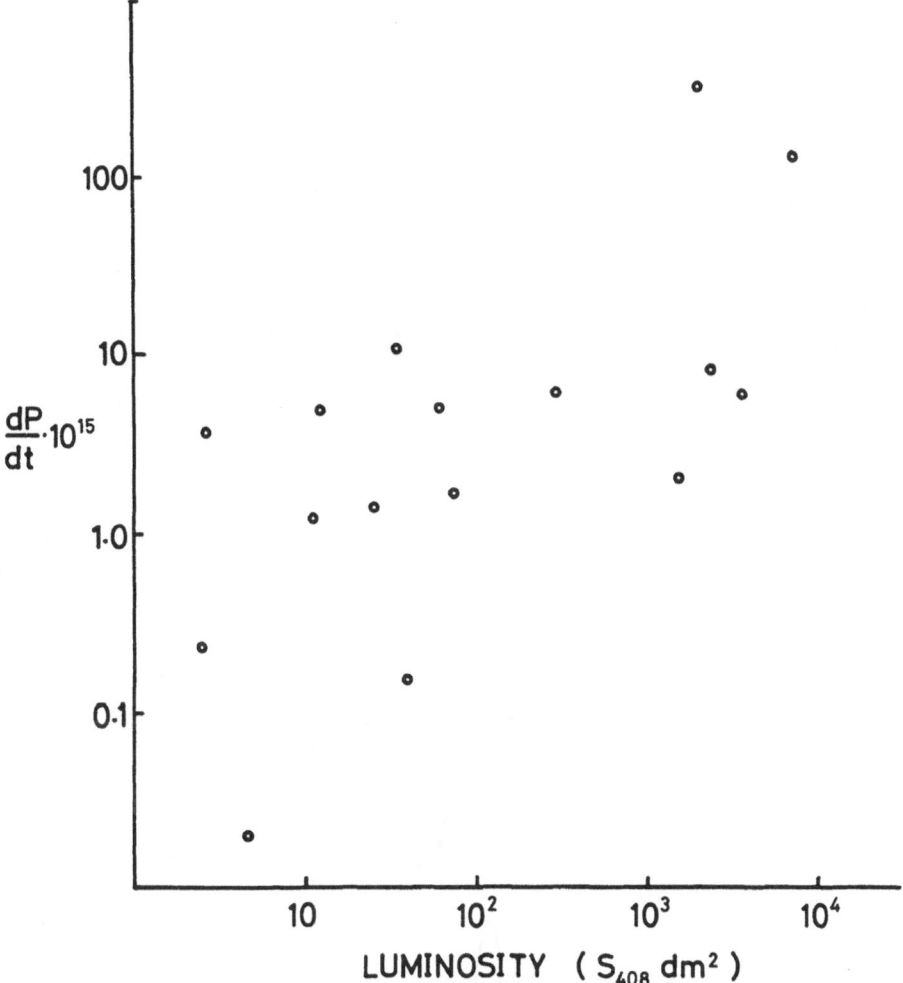

Fig. 3. Rate of change of period plotted against luminosity.

slow, short period ones. If period is any measure of age, then we can clearly say that the angular width of the beam does not evolve significantly with it.

Both the Crab and Vela pulsars have very typical duty ratios and are very similar to the others in this respect.

Figure 6 shows a weak dependence of the duty ratio upon luminosity, measured as the product of the flux at 408 MHz and the square of the dispersion measure. The

A.G. LYNE

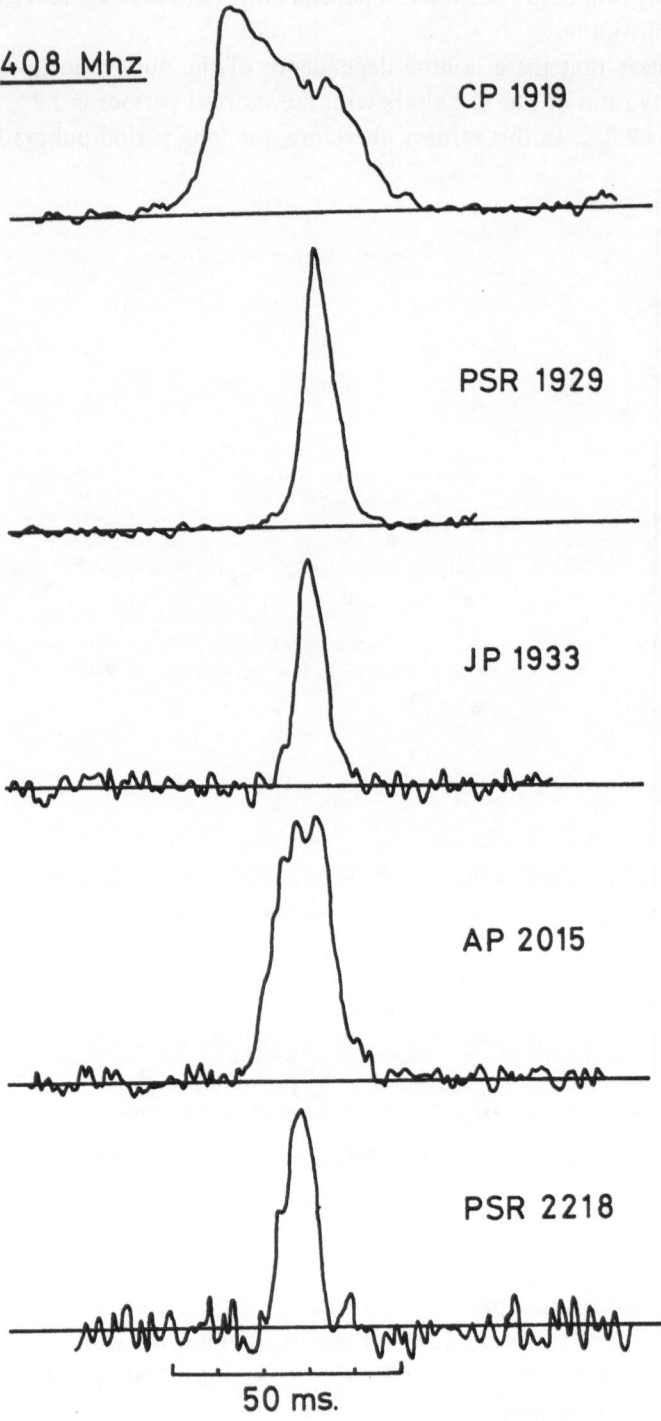

Fig. 4a.

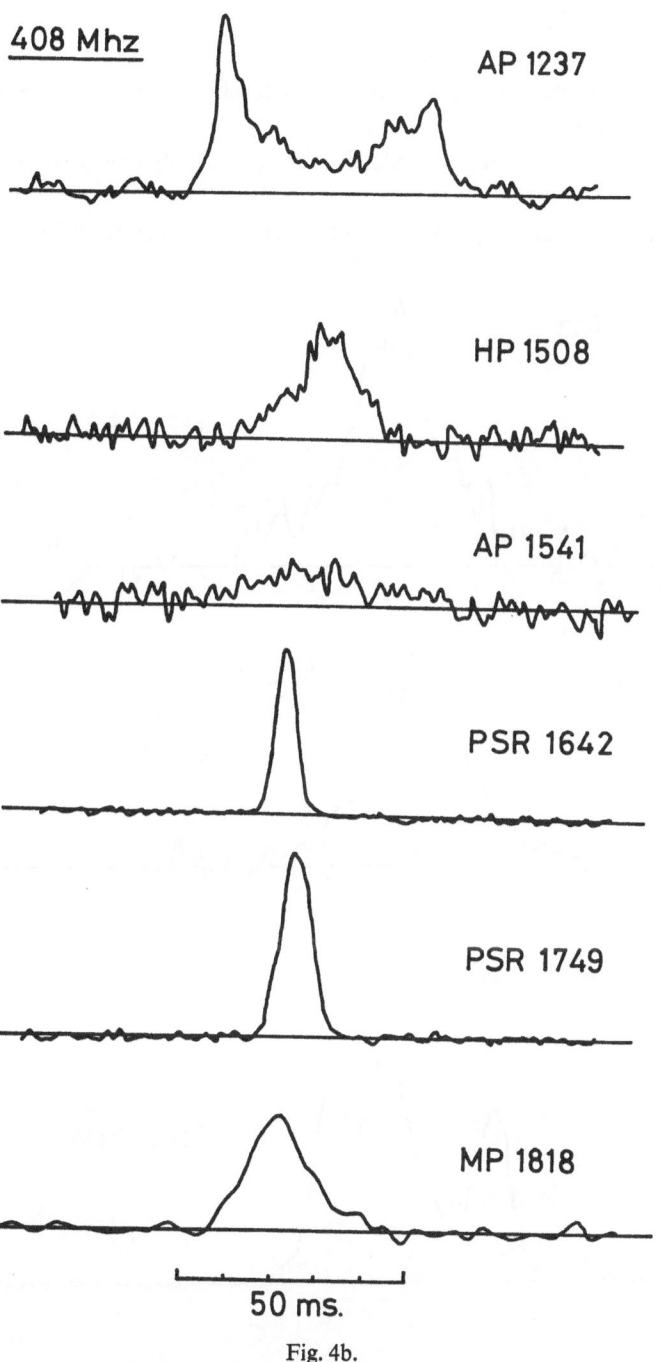

Fig. 4b.

correlation is significant and the dependence takes the form

$$\frac{w_p}{P} \propto L^{-0.1}.$$

Thus the most luminous objects tend to have slightly narrower beams. For this I offer no explanation.

I might add that there are no selection effects which can seriously affect these relationships.

The mean profiles, although well defined, often change slightly with frequency.

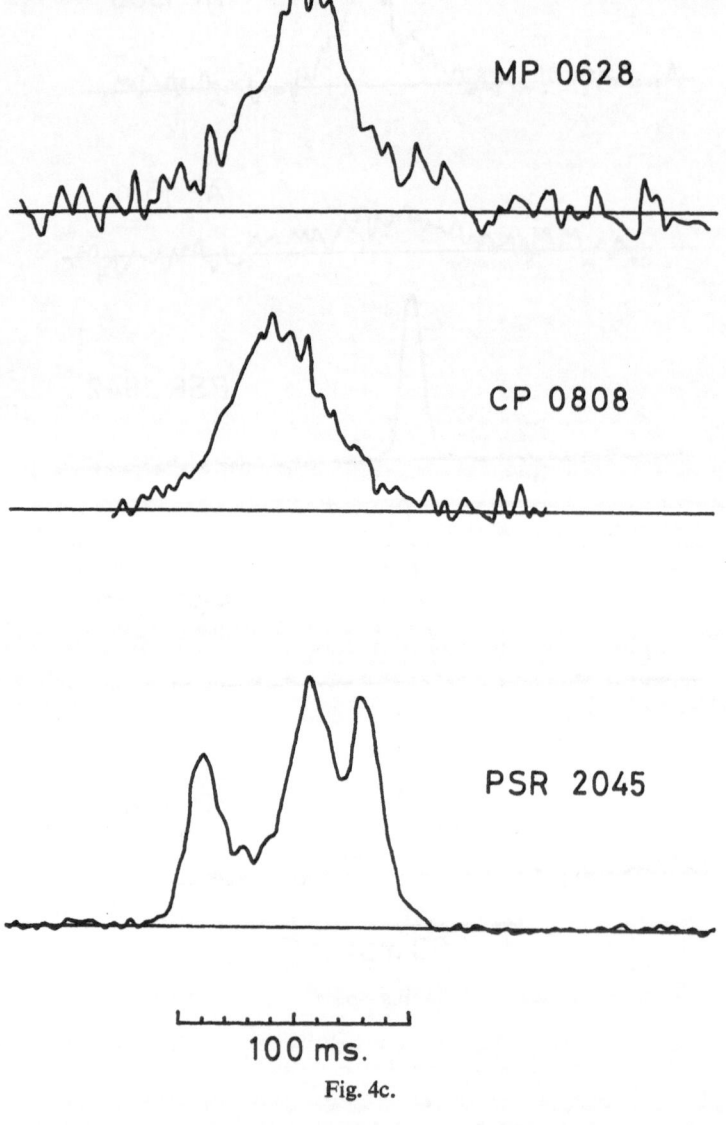

Fig. 4c.

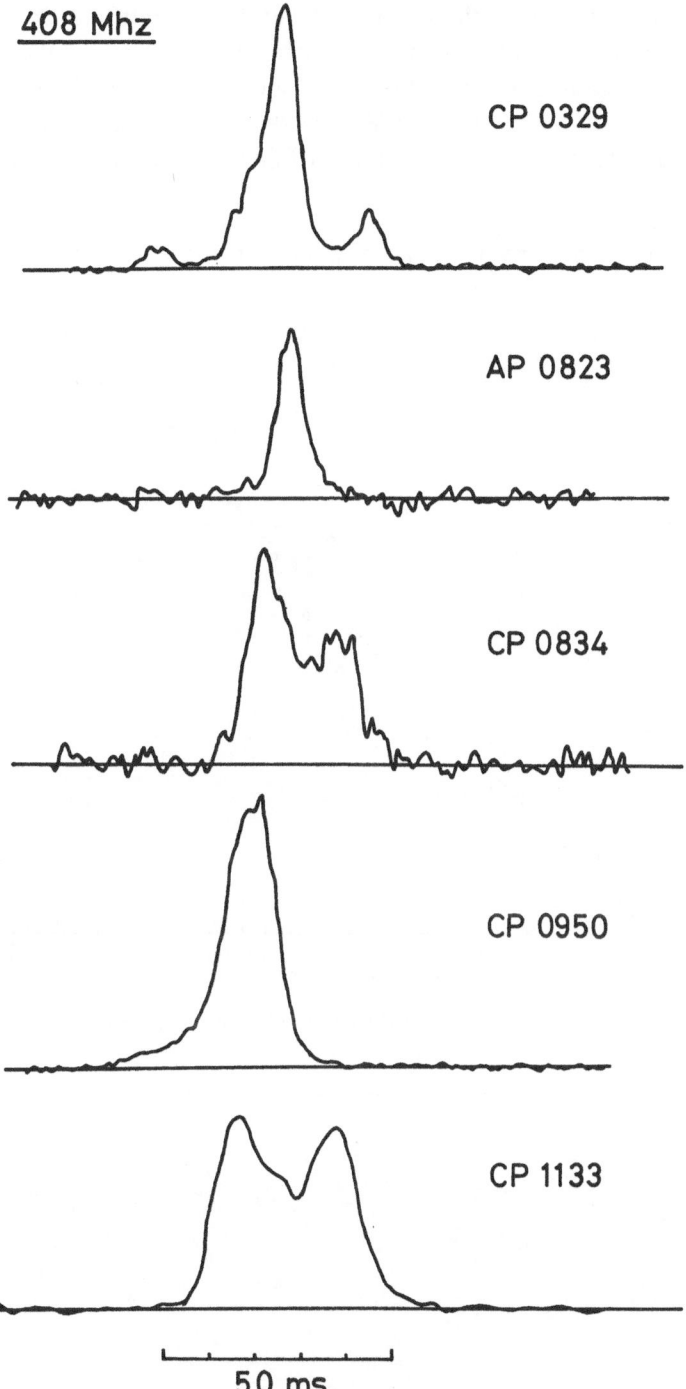

Fig. 4d.

Fig. 4a–d.   Mean pulse profiles averaged over about 1000 pulses for a number of pulsars.

The best example is shown in the Figure 7, taken from Craft and Comella (1968). Here, the separation of the two components increases as $f^{-0.25}$ and the width of the first component increases at a similar rate. The width of the second is nearly constant. Other pulsars, such as NP 0527 and PSR 2045 show similar behaviour with similar or smaller exponents. Many of the profiles in Figure 4 were obtained by Dr. B. J. Rickett.

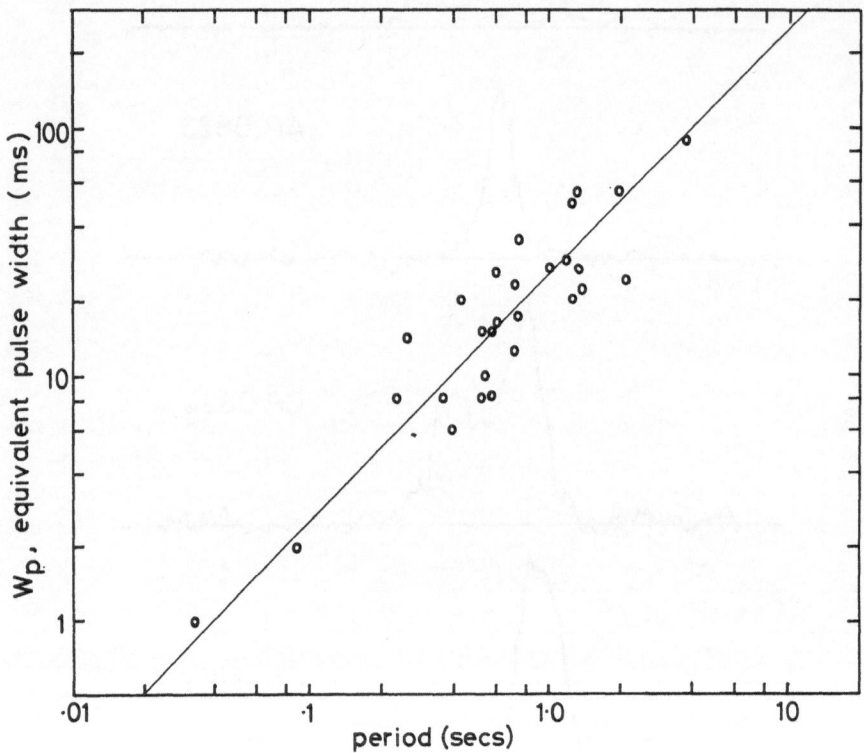

Fig. 5.   Equivalent pulse width plotted against period.

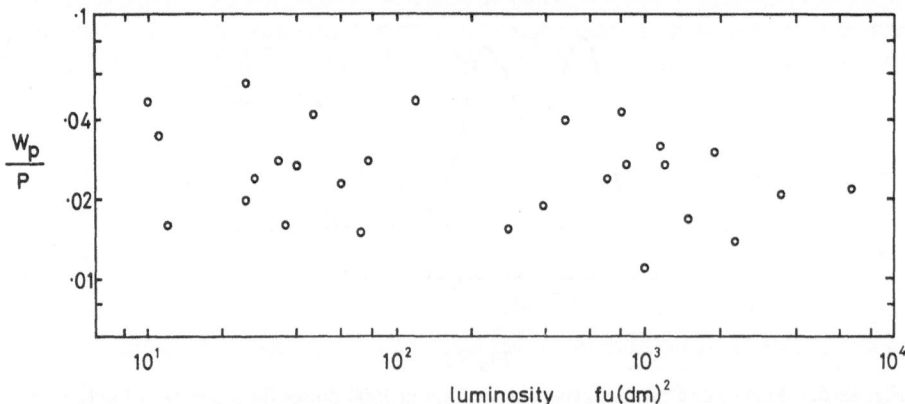

Fig. 6.   The duty ratio of a number of pulsars plotted against their luminosity.

4. Intensity Variations

Although mean pulse-shapes are very stable, individual pulses vary dramatically in both intensity and in shape. The nature of the intensity variations changes considerably from one pulsar to the next. Figure 8 shows the autocorrelation functions of pulse trains from three different pulsars. CP 1919 shows a high amplitude correlation over about 50 pulses. CP 1133 shows each pulse to be uncorrelated with its neighbours.

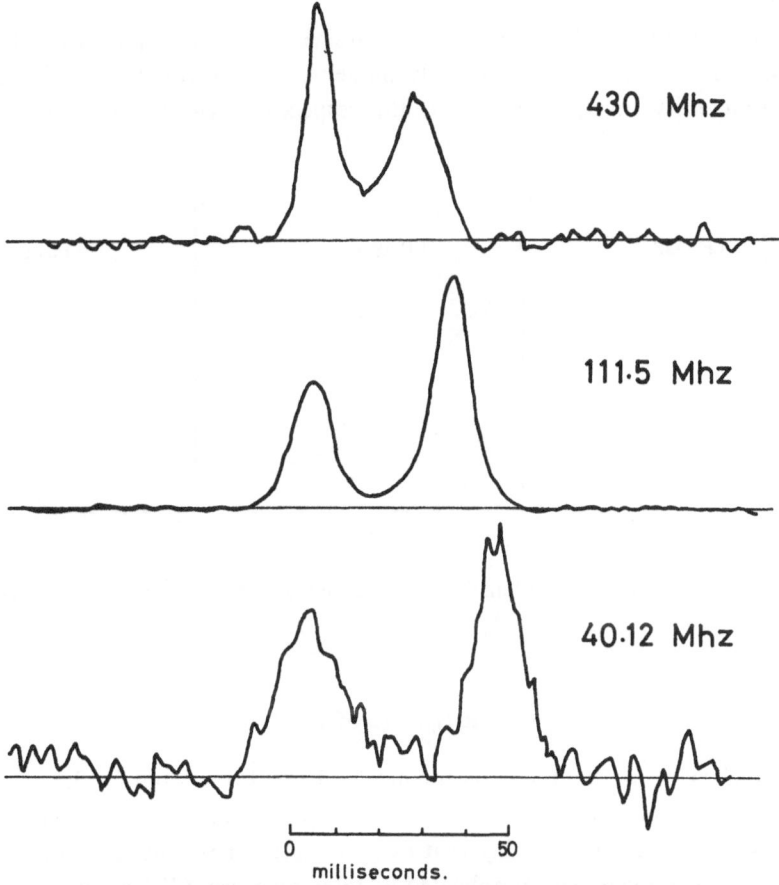

Fig. 7. The variation of the profile of CP 1133 with frequency (Craft and Comella).

CP 0328 has amplitude correlation over about three pulses followed by a negative correlation. In this case, therefore, clusters of strong pulses are usually associated with neighbouring regions of weak pulses.

The pulse energy distributions mostly show an exponential form. NP 0532 is apparently rather strange in this respect in that it appears to have at least two distinct distributions, each apparently associated with different parts of the pulse.

5. Sub-Pulse Structure

The variations in pulse profile from one pulse to the next are considerable in most pulsars, and the structure may be very complicated.

A marked regularity in the pulses from CP 1919 and AP 2015 was noticed by Drake and Craft (1968) in which features appeared to persist from one pulse to the next, drifting slowly earlier through the mean profile. In CP 0328 the features may drift either earlier or later. Typically they take ten pulse periods to drift through the pulse.

This effect can only be clearly observed when several successive pulses are observable. In the case of the Crab which gives only single isolated strong pulses it has not been possible to identify any such drifting. In this respect it is with the majority of pulsars.

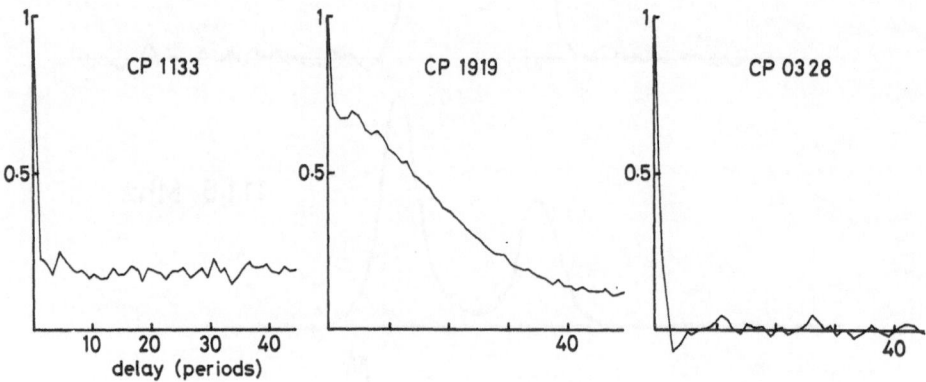

Fig. 8.   The autocorrelation functions of the amplitudes of pulses from three pulsars.

6. The Spectra of Pulsars

There is much evidence that the radiation mechanism is broadband. With most pulsars, a large pulse on one frequency occurs as a large one on another frequency some octaves away. We heard yesterday that this was not strictly the case with NP 0532.

Spectral measurements are complicated by interstellar scintillation but can be made by integration over large time or frequency intervals. The evidence is that for frequencies over 300 MHz most pulsar intensities decrease as f^{-2} or faster. There are a few notable exceptions which have much flatter spectra and are strong at high frequencies. JP 1933 and PSR 0833-45 are two such pulsars. CP 0328 has a peak in its spectrum between 400 MHz and 1000 MHz. The Crab pulsar, NP 0532, has a spectral index of about -3.0 over the radio band which is larger than most other pulsars.

Over several months, the intensities of some pulsars change by considerable factors (Cole *et al.*, 1970). As far as I know, there is no indication if this is due to a change in shape of the spectrum.

7. Interpulses

Three pulsars are known to have interpulses – regions of impulsive emission roughly midway between the main pulses. CP 0950 and PSR 1929 have interpulses containing one or two percent of the power in the main pulse. The interpulse in NP 0532 is only slightly smaller than the main pulse, and in this respect is unlike any other pulsar.

8. Conclusion

This short review summarises most of the more obvious properties of pulsars. NP 0532 is unique in only a few respects. It has a remarkably short period, emission outside the radio band, strange amplitude variations and an abnormally large interpulse.

References

Cole, T. W., Hesse, H. K., and Page, C. G.: 1970, *Nature* **225**, 712.
Craft, H. D. and Comella, J. M.: 1968, *Nature* **220**, 676.
Drake, F. D. and Craft, H. D.: 1968, *Nature* **220**, 231.
Hunt, G. C.: 1969, *Nature* **224**, 1005.
Radhakrishnan, V. and Manchester, R. N.: 1969, *Nature* **222**, 228.
Reichley, P. E. and Downs, G. S.: 1969, *Nature* **222**, 229.
Reichley, P. E., Downs, G. S., and Morris, G. A.: 1970, *Astrophys. J.* **159**, L35.

Discussion

C. Heiles: What is responsible for the differences between your relation for pulse width *vs* pulse period and the one Dr. Drake showed to us yesterday, specifically in the cases of AP 1541 and NP 0532?

A. G. Lyne: The difference comes I think, from the method of measuring pulse width. I have used the area under the pulse divided by the peak intensity as a measure of effective pulse width for all pulsars even if they have two or more peaks.

J. Rankin: Although pulse broadening has now been observed in several other pulsars which is well described by an exponential (e^{-x}, $x \geqslant 0$) broadening function, we wish to point out that this result is not perfectly general. Three observational properties of the pulsar, its low frequency pulse shapes, its spectral cut off at about 100 MHz, and its dispersion relation all indicate that a broadening function of the form xe^{-x} is more appropriate than the simple exponential.

Arguments relative to the broadening based on the above three pulsar properties are necessarily a little indirect; however, recent observation of strong single pulses at 111.5 MHz (with time bandwith product greater than 20) with good time resolution (about 3 m sec) allow observation of the broadening function direct. Such single pulses have widths of the order of 17 m sec with the peak falling approximately 11 m sec after the start of the pulses. Further, if the f^{-2} component of the delay is extracted, it is found that the start of the strong pulse follows the intrinsic main pulse by about 2 m sec. Thus, although we cannot argue at this time that the strong pulse observation precisely support the xe^{-x} broadening function hypothesis, we do argue that it is incompatible with an exponential broadening function.

F. G. Smith: Is it possible that the true pulse shape varies so much with frequency that part of your apparent smearing shape is really intrinsic? – Of course I realise that it must be possible – I ask is it reasonable?

J. Sutton: In reply to Dr. Rankin's comment concerning shapes of individual pulses from NP 0532 at 111 MHz, our observations (as described in Paper 2.5, p. 97) are inconsistent with pulse shapes of the form $te^{-at}(t>0)$ but are better described by a shape $e^{-at}(t>0)$ as consistent with electron scattering between the pulsar and us.

M. M. Komesaroff: The form of the scattering medium may determine the form of the scattering function. If it occupies a very much smaller depth than the distance between pulsar and observer I believe the scattering function will have an approximately exponential decay.

3.3 POLARIZATION OF PULSARS

ALAN T. MOFFET

Owens Valley Radio Observatory, California Institute of Technology

Abstract. A survey is given of the polarization properties of pulsars. Many pulsars, although showing pulse-to-pulse variations in polarization, have a stable mean polarization characteristic. Several pulsars like 0833–45 show changes in the position angle of linear polarization across the pulse which can be interpreted as due to the changing direction of the magnetic field as envisaged in the oblique magnetic dipole model of pulsars. Faraday rotation measurements when taken with dispersion measures yield values for the interstellar magnetic field of several microgauss.

In the first few months following the discovery of pulsars, observations made simultaneously with orthogonally polarized antennas showed that, at least at long wavelengths, individual pulses may be quite strongly polarized (Lyne and Smith, 1968). This polarization is in general elliptic, and the sense and degree of polarization vary strongly from pulse to pulse (Taylor, 1968). From these early observations it was by no means clear whether there would be a characteristic average profile for the polarization of the pulsar emission. Stated in another way, early observations had shown that each pulsar possesses a characteristic average intensity profile (the I Stokes parameter, or something closely related to it), but it was not clear whether similar non-zero profiles existed for the Q, U and V Stokes parameters.

This question was clarified by the discovery of the strongly polarized emission from PSR $0833-45$. At decimeter wavelengths this object shows very much less pulse-to-pulse intensity variation than do other pulsars; thus it may be studied by the conventional technique of rotating a single linearly polarized antenna. With other pulsars there is always some doubt whether successive averages taken in this way may be meaningfully compared, although the technique has been used successfully for several objects by Komesaroff *et al.* (1970). It is far more convincing to use a polarimeter in which all four Stokes parameters are recorded simultaneously, and such devices have been used for observations at Green Bank, Goldstone, Jodrell Bank, Parkes, and Arecibo.

The radiation from PSR 0833–45 at decimeter wavelengths has almost 100% linear polarization, and the angle of polarization rotates by nearly 180 degrees during each pulse. This rather startling discovery was made independently by Ekers and Moffet at Goldstone (Ekers *et al.*, 1969; Ekers and Moffet, 1969) and by Radhakrishnan *et al.* (1969) at Parkes. The latter authors suggested a compelling explanation for the rotation of the polarization angle as a consequence of the changing direction of the magnetic field in the oblique magnetic dipole model of a pulsar. This explanation was elaborated in a subsequent paper (Radhakrishnan and Cooke, 1969). The emission is presumed to be linearly polarized with the polarization parallel to the projection of the magnetic field normal to the line of sight to the emitting region. As the magnetic polar region passes the line of sight, this direction rotates through as much as 180°. A quantitative model for such emission is given by Komesaroff (1970). The concentration of emission in the vicinity of a magnetic pole of a rotating neutron star follows from the electro-dynamic model calculated by Goldreich and Julian (1969) in which they show that a stream of energetic particles will be ejected from such a region.

Even after the discovery of the polarized emission from PSR 0833−45 it was not clear whether other, more variable pulsars would have stable polarization averages, but this is in fact the case, as has been shown by Ekers and Moffet (1969); Komesaroff et al. (1970); Morris et al. (1970); and Manchester (1970). Because of the pulse-to-pulse variation, the degree of polarization is necessarily less than 100%, at least in the strong parts of the pulses. Manchester (1970) finds that in the wings of the pulses the average degree of polarization often approaches 100%. A uniform rotation of position angle during the pulse is often seen, although sharp discontinuities of position angle do occur. In PSR 2045−16 (Morris et al., 1970) and in PSR 1237+25 (Manchester, 1970) these are interpreted as evidence that a single magnetic pole passes very nearly across the line of sight. Ekers and Moffet (1969) observed a jump in angle of 90° during a part of the pulse from CP 1133. This was interpreted in terms of a model with two magnetic polar regions contributing to the pulse. The reality of this phenomenon is in doubt, since it was not confirmed in the more sensitive observations of Manchester (1970), although these were at a very different wavelength.

Polarization observations may help to explain what is happening in pulsars with complex intensity profiles. In both PSR 2045−16 and PSR 1237+25 the change in the polarization angle is much more regular than the change in the intensity, with the discontinuity in position angle coming near the most prominent minimum in the intensity average. Morris et al. (1970) interpret this as a relative minimum of the intensity in the immediate vicinity of the magnetic pole with more intense emission in a hollow cone around the projected direction of the pole (Komesaroff et al., 1970). Other complex pulses such as CP 0328 (Clark and Smith, 1969; Ekers and Moffet, 1969) do not appear to be so simple.

While an appreciable degree of circular polarization is often seen in individual pulses, the average circular polarization is usually quite small, although Ekers and Moffet (1969) have reported an appreciable average circular polarization in PSR 1749−28.

We have heard some discussion at this conference about whether the emission from the Crab pulsar shows a variation of position angle during the pulse. Campbell et al. (1970) find that at 430 MHz no rotation takes place. The precursor is 100% polarized while the main pulse and interpulse are about 30% polarized. Observations made at Jodrell at 408 MHz and presented at this conference by Schönhardt (1970) indicate a rotation of about 60° within the main pulse. The matter is clearly an important one and should be cleared up by observations with higher time resolution.

There is also some dispute about the polarization of individual giant pulses from the Crab. Graham et al. (1970) found about 70% linear polarization for these, while Rankin and Heiles (1970) found about 25% linear and 10% circular to be typical. Heiles et al. (1970) found that the sense of circular polarization is different for giant pulses which occur before or after the center of the average main pulse.

Where linear polarization is present, Faraday rotation can be measured. In reporting their detection of polarized emission from pulsars, Lyne and Smith (1968) pointed out the possibility that measurements of Faraday rotation and dispersion could be combined to give a measure of the mean galactic magnetic field component along the line of sight to the pulsar. The rotation measure is $RM \propto \int n_e \mathbf{B} \cdot \mathbf{ds}$, while the dispersion measure is $DM = \int n_e \, ds$. Thus $RM/DM = \langle B_{\parallel} \rangle$, the average component of B parallel to the line of sight and weighted by the electron density.

Such a measure was first obtained by Smith (1968a, b) for CP 0950, CP 0328 and AP 2015 + 28. For nearby pulsars the rotation measure will be small, and the effects of the ionosphere are not negligible (Roger and Shuter, 1968). Faraday rotations measured over narrow frequency intervals seem to be subject to errors, perhaps introduced by the interstellar scintillation. Thus Smith's first value for CP 0328 was 50% high (Goldstein and Meisel, 1969; Staelin and Reifenstein, 1969). Also Morris *et al.* (1970) find a rotation measure for PSR 0628−28 between 2650 and 1720 MHz of $+22 \, \text{rad m}^{-2}$, which differs from Vitkevitch and Shitov's (1970b) value of $45 \, \text{rad m}^{-2}$, which was measured over narrow intervals near 86 and 110 MHz.

TABLE I

Magnetic Fields Determined from Pulsar Faraday Rotations

Object	l	b	RM rad m^{-2}	DM cm^{-3} pc	$\langle B_{\parallel} \rangle$ μ G	Ref.*
CP0328	145	− 1	− 63	26.8	+2.8	1,9
NP0527	184	− 7	|36|	51.2	|0.9|	1
NP0532	185	− 6	(− 25)	56.8	(+0.55)	2
PSR0628–28	237	− 17	{+ 22	35	− 0.6	3
			(|45|	34.4	|1.6|	4
CP0808	140	+ 32	|12|	5.8	|2.5|	5
PSR0833–45	264	− 3	+ 33	53	− 0.73	6, 7
CP0950	229	+ 44	< |0.5|	3.0	< |0.2|	8
AP2015 + 28	68	− 4	− 30	14.2	+ 2.0	9

* 1. Staelin and Reifenstein (1969); 2. Verschuur (1969); 3. Morris *et al.* (1970); 4. Vitkevitch and Shitov (1970b); 5. Vitkevitch and Shitov (1970a); 6. Ekers *et al.* (1969); 7. Radhakrishnan *et al.* (1969); 8. Smith (1968a), 9. Smith (1968b).

This method has since been applied by various authors to determine the magnetic field in the directions of eight pulsars, as shown in Table I. Also given are the galactic coordinates, the rotation measure and the dispersion measure. Where the sign of the Faraday rotation is known it is given; a positive sign for the magnetic field implies $\mathbf{B}$ directed from the source towards the observer. Exempting the low value for CP 0950 because of doubt about the ionospheric contribution, the fields range from 0.55 to 2.8 μG. This seems likely to represent the typical interstellar magnetic field better than the higher values measured from the Zeeman splitting of the 21 cm absorption features produced by cold, dense clouds of hydrogen. There is reason to believe that the fields may have been amplified as these clouds condensed (Verschuur, 1969). The field given in Table I for NP 0532 assumes that the Faraday rotation observed for the Crab Nebula as a whole is applicable to the pulsar, since the rotation of the pulsar emission has not yet been measured. The field for CP 0808 comes from the observations of Faraday fading given by Vitkevitch and Shitov (1970a) and is fairly uncertain since the implied rotation measure is only 12 rad m^{-2}. For PSR 0628 – 28, values of $\langle B \rangle$ are given corresponding to the two conflicting observations of RM as mentioned before.

Observations of the polarization of the pulsar emission have given important clues about the nature of these objects. Further work will surely assist in constructing detailed models of the emission process. It will be important to compare changes in the pulse shape with changes in the polarization at different frequencies and to examine the relation between the polarization and the short time scale structure of individual pulses. In addition the measurement of Faraday rotation gives information about the interstellar magnetic field which cannot be obtained in any other way.

References

Campbell, D. B., Heiles, C., and Rankin, J. M.: 1970, *Nature* **225**, 527.
Clark, R. R. and Smith, F. G.: 1969, *Nature* **221**, 724.
Ekers, R. D., Lequeux, J., Moffet, A. T., and Seielstad, G. A.: 1969, *Astrophys. J.* **156**, L21.
Ekers, R. D. and Moffet, A. T.: 1969, *Astrophys. J.* **158**. L1.
Goldreich, P. and Julian, W. H.: 1969, *Astrophys. J.* **157**, 869.
Goldstein, S. J. and Meisel, D. D.: 1969, *Science* **163**, 810.
Graham, D. A., Lyne, A. G., and Smith, F. G.: 1970, *Nature* **225**, 526.
Heiles, C., Campbell, D. B., and Rankin, J. M.: 1970, *Nature* **226**, 529.
Komesaroff, M. M.: 1970, *Nature* **225**, 612.
Komesaroff, M. M., Morris, D., and Cooke, D. J.: 1970, *Astrophys. Letters* **5**, 37.
Lyne, A. G. and Smith, F. G.: 1968, *Nature* **218**, 124.
Manchester, R. N.: 1970, *Nature* **228**, 264.
Morris, D., Schwarz, U., and Cooke, D. J.: 1970, *Astrophys. Letters* **5**, 181.
Radhakrishnan, V., Cooke, D. J., Komesaroff, M. M., and Morris, D.: 1969, *Nature* **221**, 443.
Radhakrishnan, V. and Cooke, D. J.: 1969, *Astrophys. Letters* **3**, 225.
Rankin, J. M. and Heiles, C.: 1970, *Nature* **227**, 1330.
Roger, R. S. and Shuter, W. L. H.: 1968, *Nature* **218**, 1036.
Schönhardt, R.: 1970, this symposium, Paper 2.7, p. 110.
Smith, F. G.: 1968a, *Nature* **218**, 325.
Smith, F. G.: 1968b, *Nature* **220**, 891.
Staelin, D. H. and Reifenstein, E. C.: 1969, *Astrophys. J.* **156**, L121.
Taylor, J. H.: 1968, paper presented at Conference on Pulsars, Goddard Institute for Space Studies, New York.
Verschuur, G. L.: 1969, *Nature* **223**, 140.
Vitkevitch, V. V. and Shitov, Yu. P.: 1970a, *Nature* **225**, 248.
Vitkevitch, V. V. and Shitov, Yu. P.: 1970b, *Nature* **226**, 1235.

Discussion

L. Mestel: Can I underline Dr. Moffet's remark about the comparison between the magnetic field estimates from the ratio of rotation measure and dispersion measure, and the Zeeman measurements? A magnetic field strength is significant only when one knows the associated matter density. Verschuur's Zeeman measurements refer to moderately dense H I clouds, with the magnetic field appropriately amplified. The self-gravitation of the clouds may be sufficient to balance the outward force exerted by the distorted magnetic field. So far from there being any glaring contradiction, I feel that the evidence is at least qualitatively in agreement with what one expects from hydromagnetics.

F. C. Michel: I think it unprofitable to impose upon theoretical models morphological features exhibited by only a small fraction of pulsars. You have shown polarisation sweeps by about 10% of the pulsars, but what about the rest – is the sweep really a general effect?

A. T. Moffet: I think that in all cases where good data have been obtained on pulsars with fairly simple pulse shape a regular sweep of position angle has been found.

R. N. Manchester: Approximately 60% of pulsars observed show a continuous swing of position angle across the pulse.

M. M. Komesaroff: I would like to point out that the observations of Morris, Schwarz and Cooke made at 11 cm showed an identical sweep in position angle (except for a constant shift) with those of Manchester at 75 cm.

R. N. Manchester: This result shows that differential Faraday rotation between components of pulse is essentially zero, the $\Delta RM = 0.10 \pm 0.23$ rad/m².

J. Sutton: A search for short period pulsars in Cas A, Tycho's supernova and 3C 58 was conducted at 408 MHz using the 300 ft radio telescope at Green Bank. The beamwidth was 34′. Data were taken with bandwidths of 100 and 500 kHz, and recorded digitally at 1000 samples per second. The search involved Fourier analysis of groups of 16000 points. The data were analysed 4 times, smoothing to effective sample intervals of 1, 4, 16 and 64 msec. The dispersion limit due to bandwidth broadening was approximately $40\,P$ electrons cm⁻³, where P is the period in msec. Within the period and dispersion limits, upper limits on the fraction of energy from Cas A in pulsed radiation is $\sim 0.2\%$. Similar limits for Tycho and 3C 58 are $\sim 1\%$.

3.4 OBSERVATIONS OF PULSAR SPECTRA

M. S. EWING, R. A. BATCHELOR*, R. FRIEFELD, R. M. PRICE,
and D. H. STAELIN

Research Laboratory of Electronics, Massachusetts Institute of Technology, Cambridge, Mass., U.S.A.

Abstract. The observed width of spectral features in CP 0328, CP 0834, and CP 1919 are approximately proportional to the fourth power of frequency, thus supporting the hypothesis that the slow spectral variations of these pulsars are due to interstellar scintillation. The spectral features in CP 0834, CP 1133, and CP 1919 are observed to drift systematically at rates compatible with a simple interstellar scintillation model. Pulsar velocities of ~ 100 km sec^{-1} are inferred from these spectral drift rates.

1. Introduction

Observations of pulsar spectra yield information about the intrinsic radiation properties of pulsars, and about the intervening interstellar medium. In an effort to separate and study these two aspects of pulsar spectra, we have observed since 1968 the spectra of those pulsars within the range of the National Radio Astronomy Observatory 300-ft transit telescope in Green Bank, West Virginia. The spectrometers each had 50 channels with bandwidths of 100, 30, or 10 kHz, or had 40 channels with bandwidths of 1 MHz. The center frequencies of the filter banks ranged from 110 to 550 MHz, and all 40 or 50 channels were sampled with 8-bit accuracy every 30 msec. Subsequent computer analysis yielded the spectra of individual and average pulses. Additionally, in May, 1970, both the 300-ft transit telescope and the 140-ft fully steerable telescope in Green Bank were used in conjunction with the 384-channel autocorrelation receiver to achieve frequency resolution as high as 1 kHz.

The present report describes the results of manual analyses of the spectra of four pulsars, CP 0328, CP 0834, CP 1133, and CP 1919. A more complete analysis of all the spectral data is in preparation.

2. Widths of Spectral Features

Typical data obtained with the multichannel filter systems are presented in Figure 1. These data were processed by computer, displayed on a cathode-ray-tube, and then photographed. Each resolution element in these photographs represents the average of several pulses within a single channel. The three light levels represent relative pulse energy thresholds of 1, 2, and 4. The figure illustrates how the spectral features in each pulsar develop and change.

One parameter of interest in these spectra is the frequency dependence of spectral feature width. Feature widths were determined by averaging the visually determined instantaneous full widths at half-maximum, B, for a large number of spectral features at each frequency. Typical feature widths observed using the autocorrelator for these

---

* On leave from Division of Radiophysics, CSIRO, Sydney, Australia.

Davies and Smith (eds.), The Crab Nebula, 200–205. All Rights Reserved.

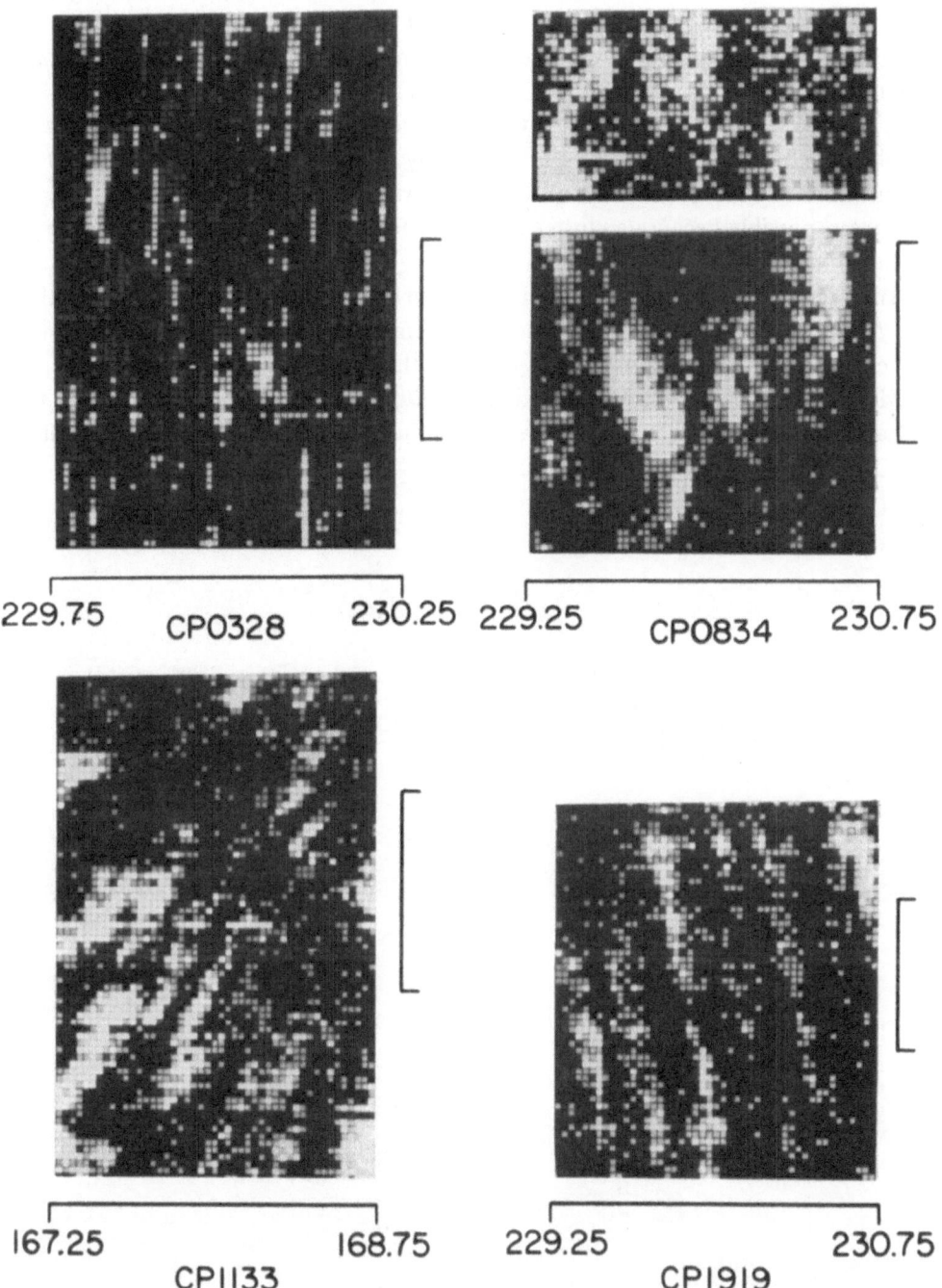

229.75   CPO328   230.25   229.25   CPO834   230.75

167.25   CPII33   168.75   229.25   CPI9I9   230.75

Fig. 1.   Representative dynamic spectra for CP 0328, CP 0834, CP 1133, and CP 1919 as observed
with 50-channel spectrometers. The light levels represent relative power thresholds of 1, 2, and 4, in
order of increasing brightness. Time increases from top to bottom, and the scale markers
indicate 10 minutes.

four pulsars are listed in Table I as a function of frequency and are plotted in Figure 2. The feature widths B are assumed to vary as f^{α}. Least squares fits for α yield the numbers in Table I. The errors quoted are estimated maximum deviations from the mean fit. The three pulsars excluding CP 1133 are fairly consistent with a value of $\alpha = 4$. An f^4 dependence is generally consistent with the results reported by Rickett (1969), Staelin (1969), and Lang (1970), and apparently inconsistent with those reported by Huguenin *et al.* (1969). Some of the widths reported here are less than those reported by others, possibly because a cluster of narrow features can resemble a single broad feature.

An f^4 dependence of feature width is predicted by scintillation theory. In particular, the f^4 dependence follows if we assume the antenna intercepts rays which have traversed different paths, and that the path lengths differ by a nominal value δ, where δ arises geometrically, and is approximately $R\theta_s^2/2$, where R is the distance to the pulsar, and θ_s is the rms angle of ray arrival at the antenna. This same model predicts

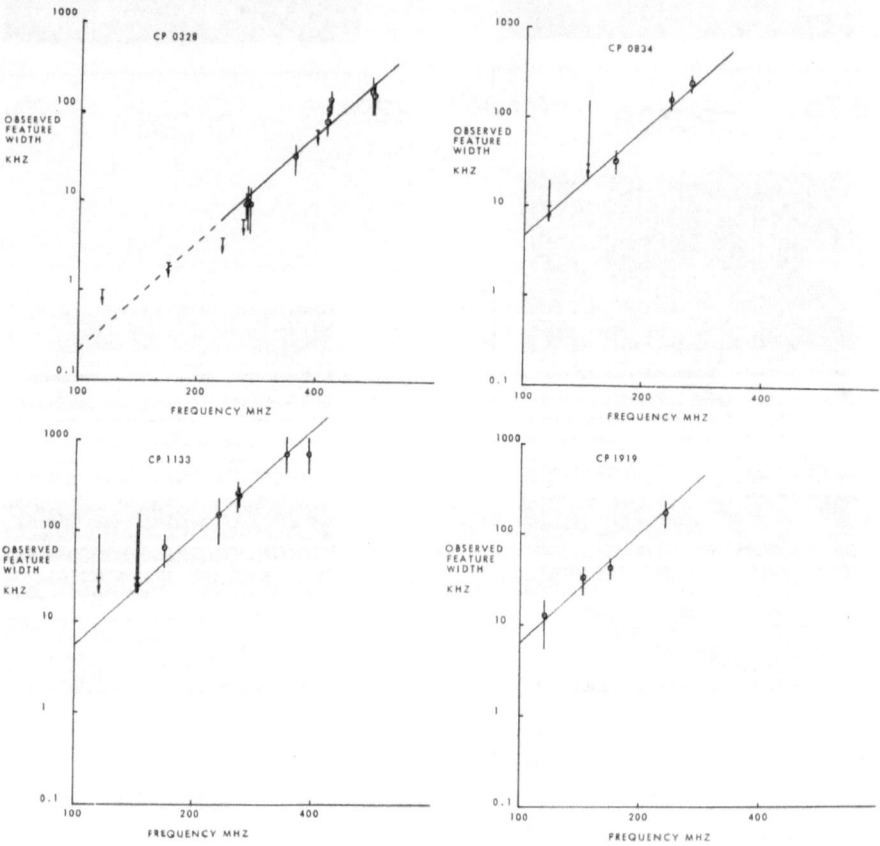

Fig. 2. Wavelength dependence of spectral feature widths. The widths represent averages of many spectral features, and the error bars indicate the total range of observed feature widths. Arrows and cross-bars represent the limitations of spectral resolution. The sloping lines correspond to a f^4 dependence of feature width.

TABLE I

Pulsar spectral feature widths

Source	Freq (MHz)	B (kHz)	α	Source	Freq (MHz)	B (kHz)	α
CP 0328	112	$< 1?$		CP 1133	112	< 90	
	168	$\leqslant 2$			142	< 150	
	267	10 ± 5			168	65 ± 25	
	350	33 ± 13			230	150 ± 80	
	405	111 ± 50			350	417 ± 70	
	560	170 ± 70	3.9 ± 0.8		405	692 ± 250	2.6 ± 0.6
CP 0834	112	< 20		CP 1919	112	13 ± 7	
	142	< 150			142	34 ± 10	
	168	34 ± 9			168	38 ± 20	
	230	162 ± 35			230	178 ± 60	3.5 ± 1.1
	258	245 ± 40	4.6 ± 0.5				

that the spectral feature widths will also be approximately inversely proportional to the square of the dispersion measure, a result compatible with the present observations. Such scintillation models have been discussed by Scheuer (1968), Salpeter (1969), and Uscinski (1968a, b).

In the case of the Crab Nebula pulsar, NP 0531, the value of δ and its f^{-4} dependence has been measured directly by Staelin and Sutton (1970), and Rankin *et al.* (1970). Using the observed value of 13 msec for the multipath delay at 115 MHz, this model predicts feature widths for NP 0531 of ~ 2 kHz near 300 MHz. This is too narrow to be observed readily, and may account for the apparent lack of spectral features in the Crab Nebula pulsar. This feature width is narrower than those of other pulsars scaled by the inverse square of the dispersion measure, but is within a factor of 8 of the expected value, and within a factor of 2 of values extrapolated from CP 0328 and CP 1133. It is therefore possible that the general interstellar medium, rather than the pulsar environment, is responsible for the multipath effects and lack of observable scintillation in NP 0531.

3. Drifting of Spectral Features

A very interesting property of some pulsars is systematic drifting of spectral features, as illustrated for CP 1919 and CP 1133 in Figure 1. Drifting has been observed in each of these two pulsars on several occasions, although the phenomenon is readily overlooked if the spectrometer resolution is not appropriate, or if the operating frequency is such that the drift rate is too slow or obscured by variations in intrinsic pulsar intensity. Drifting has also probably been observed in CP 0834. Representative observed drift rates for CP 0834, CP 1133, and CP 1919 are 130 Hz sec^{-1} at 168 MHz, 1 kHz sec^{-1} at 230 MHz, and 500 Hz sec^{-1} at 230 MHz, respectively. The drift rates appear to vary from month to month, and may vary on shorter time scales. The drifts of both CP 1919 and CP 1133 have changed directions, and at times different simultaneous spectral features may have different drift rates.

A simple model for interstellar scintillation yields an understanding of spectral drifting. We may assume that the radiation propagating from the pulsar to Earth is composed of rays, each executing a random walk characterized by θ_s, the rms angle between any ray segment and the direct path. These rays converge at the Earth with different arrival angles and different delays. If many scattering events occur for each ray, then the propagation delay and the angle of arrival may be weakly correlated, in contrast to the single thin-screen model. Consider the case where the radiation incident upon the Earth is dominated by two rays. The interference of the rays produces an interference pattern through which the Earth moves at velocity v_0. In this case the lifetime Δt of a single spectral feature, i.e. the time between half-power points at a single frequency, is the time required for the Earth to move past one lobe of the interference pattern, i.e.

$$\Delta t \cong \frac{\lambda}{\sqrt{8}\,\theta_s v_0}$$

where λ is the wavelength, and $\sqrt{8}$ is an approximate geometrical factor. A frequency drift can result if the propagation delays for these two rays are different. The phenomenon is analogous to the movement of an observer through the frequency-dependent lobes of a transmitting interferometer. If the delays differ by the reasonable value $R\theta_s^2/2$, where R is the pulsar-Earth separation, then the feature drift rate $\dot{f}$ is

$$\dot{f} \cong f v_0 \sqrt{\frac{8B}{Rc}}.$$

Since the nominal width B is proportional to $f^4 R^{-2}$, it follows that the drift rate $\dot{f}$ should be proportional to $f^3 R^{-3/2}$. Consistency of these expressions requires that $\dot{f} \cong B/\Delta t$, where $B \cong c/R\theta_s^2$.

By averaging the drift rates of several spectral features it is possible to estimate the magnitude and frequency dependence of the drift rates. Since rays of different frequencies have different interstellar propagation paths, and since the paths are time varying, many observations will be required. For the present limited data the drift rate is proportional to $f^{3\pm1.5}$. Although the observations are consistent with the $R^{-3/2}$ dependence upon pulsar distance, the results are not definitive. The drift rate predicted for CP 0328 is approximately 70 Hz sec^{-1} at 200 MHz, which is too small to be evident in the spectra of Figure 1.

The observations of $\dot{f}$ can yield an independent estimate of the transverse pulsar velocity v. We deduce approximate velocities of 100 km sec^{-1} for all three pulsars using the drift rate, bandwidths, and equations cited earlier, and assuming the interstellar electron density is 0.03 cm^{-3} (Staelin and Reifenstein, 1968). This equals the velocities deduced from the feature lifetimes Δt. For example, for the data of Figure 1 and Table I, the approximate velocity formula

$$v \cong \sqrt{\frac{cRB}{8}}\,\frac{1}{v\,\Delta t}$$

yields for CP 0328, CP 0834, CP 1133, and CP 1919 velocities of 100 km sec^{-1}, within a factor of 2. Rickett (1969) and Lang (1970) have deduced similar velocities from their feature-lifetime data. Further corroboration follows from the measurements by Lang and Rickett (1970) of scintillation delay between spectral features observed at Arecibo and Jodrell Bank, which yielded velocities for CP 1133 of approximately 100 km sec^{-1}.

Since the velocity of the Earth with respect to the interstellar medium within 2 kpc is generally less than 50 km sec^{-1}, these data suggest that the average transverse velocity of these three pulsars with respect to the interstellar medium may be of the order of 100 km sec^{-1}, which is consistent with the velocities of runaway stars (Gott *et al.*, 1970; Gunn and Ostriker, 1970; and Prentice, 1970) and of NP 0531. More extensive drift rate data and more accurate theoretical analysis could further strengthen this conclusion.

The consistency of these spectral observations with theoretical models of interstellar scintillation further supports suggestions of Rickett (1969) and others that the observed slow spectral changes originate in the interstellar medium. Our observations have extended this conclusion to spectral features of widths ranging from 2 kHz to several MHz.

Acknowledgements

We thank J. Sutton for helpful conversations and we acknowledge the cooperation and assistance of W. Brundage, J. Greenhalgh, and other staff of the National Radio Astronomy Observatory, which is operated by Associated Universities, Inc., under contract with the National Science Foundation.

The work of M. S. Ewing and R. M. Price was supported principally by National Science Foundation grant GP13056, and that of D. H. Staelin was supported principally by grant GP14854.

References

Gott, J. R., III, Gunn, J. E., and Ostriker, J. P.: 1970, *Astrophys. J.* **160**, L91.
Gunn, J. E. and Ostriker, J. P.: 1970, *Astrophys. J.* **160**, 979.
Huguenin, G. R., Taylor, J. H., and Jura, M.: 1969, *Astrophys. Letters* **4**, 71.
Lang, K. R.: 1970, 'Interstellar Scintillation of Pulsar Radiation', preprint.
Lang, K. R. and Rickett, B. J.: 1970, *Nature* **225**, 528.
Prentice, A. J. R.: 1970, *Nature* **225**, 438.
Rankin, J. M., Comella, J. M., Craft, H. D., Jr., Richards, D. W., Campbell, D. B., and Counselman, C. C., III: 1970, *Astrophys. J.* **162**, 707.
Rickett, B. J.: 1969, *Nature* **221**, 158.
Rickett, B. J.: 1970, *Monthly Notices Roy. Astron. Soc.* **150**, 67.
Salpeter, E. E.: 1969, *Nature* **221**, 31.
Scheuer, P. A. G.: 1968, *Nature* **218**, 920.
Staelin, D. H.: 1969, contribution to Accademia Nazionale Dei Lincei meeting on 'Pulsars and High Energy Activity in Supernovae Remnants'.
Staelin, D. H. and Reifenstein, E. C., III: 1968, *Science* **162**, 1481.
Staelin, D. H. and Sutton, J. M.: 1970, *Nature* **226**, 69.
Uscinski, B. J.: 1968a, *Phil. Trans.* **A262**, 609.
Uscinski, B. J.: 1968b, *Proc. Roy. Soc.* **A307**, 471.

3.5 POLARIZATION MEASUREMENTS OF OTHER PULSARS AT JODRELL BANK

D. A. GRAHAM

University of Manchester, Nuffield Radio Astronomy Laboratories, Jodrell Bank, U.K.

The polarization properties of a number of pulsars have been studied at Jodrell Bank using the 76 m Mark I radio telescope.

The feed system has outputs responding to the two senses of circular polarization, which are combined in the receiver to give the four Stokes parameters I, V, Q and U. These are either displayed directly on an oscilloscope triggered in synchronism with the pulsar period, or integrated by an on-line computer. Some of these results are presented for comparison with the Crab Nebula pulsar.

Complex structure is frequently observed in single pulses, particularly in circular polarization. CP 0328, however, shows a rather regular reversal of circular polarization in single pulses (Clark and Smith, 1969), still seen to a lesser degree in the average picture. The Stokes parameters I and V are shown in Figure 1 for two pulses recorded at 408 MHz, and also shows I and V for the same pulses at 240 MHz, having allowed for the dispersion delay of several periods. The similarity seen here between results at two different frequencies is mirrored by the integrated pulse results. Figure 2 shows integrated linear polarization for PSR 2045 at 408 and 240 MHz. Position angles at the lower frequency are displaced by a constant angle to emphasize the similarity. The swing of position angle in this source may be closely fitted by the geometrical model described by Radhakrishnan and Cooke (1969).

The variation of position angle through the pulse for six pulsars may be seen to better advantage in Figure 3, in which the intensity of linear polarization is shown by the length of a line, whose inclination from a constant but arbitrary zero is the position angle of linear polarization.

PSR 1929 is 100% ($\pm 5\%$) polarized in the early part of the pulse, and about 70% in the latter half. Its position angle is seen to swing at a rate of about 1.3° per degree of assumed pulsar rotation.

The net swing of position angle over the whole pulse for AP 1237 is seen to be almost zero, although some swing occurs in each of the two apparently separate double components.

The linear polarization picture for 200 ms of the period of CP 0950 at 408 MHz clearly shows the interpulse, since, unlike the interpulse of NP 0532, it is 100% ($\pm 20\%$) polarized, and the average polarization of the main pulse is some 40%.

Both early and late 'interpulses' are seen for CP 0328 at 408 MHz.

References

Clark, R. R. and Smith, F. G.: 1969, *Nature* **221**, 747.
Radhakrishnan, V. and Cooke, D. J.: 1969, *Astrophys. Letters* **3**, 223.

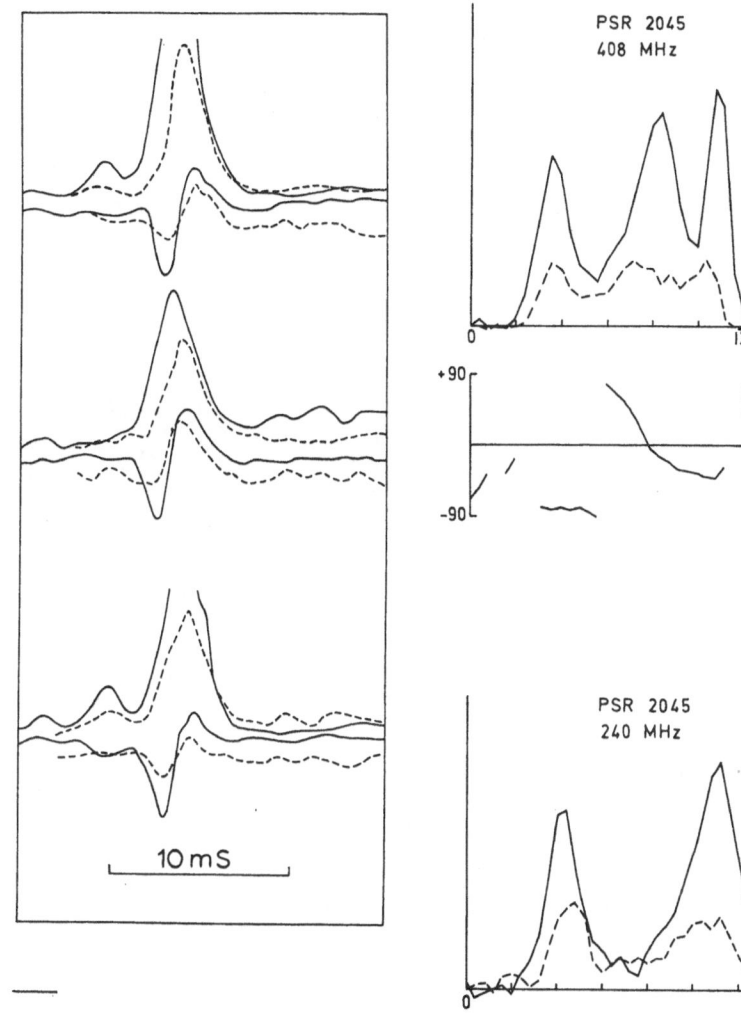

Fig. 1. Three pulses of CP 0328 in May 1970 recorded at 408 MHz (——), $\Delta f = 0.33$ MHz and 240 MHz (-----), $\Delta f = 0.1$ MHz. In each pulse the upper traces are Stokes parameter I (total intensity) and the lower traces are Stokes parameter V (circular).

Fig. 2. The linear polarization of PSR 2045 at 408 MHz and 240 MHz integrated over many pulses.

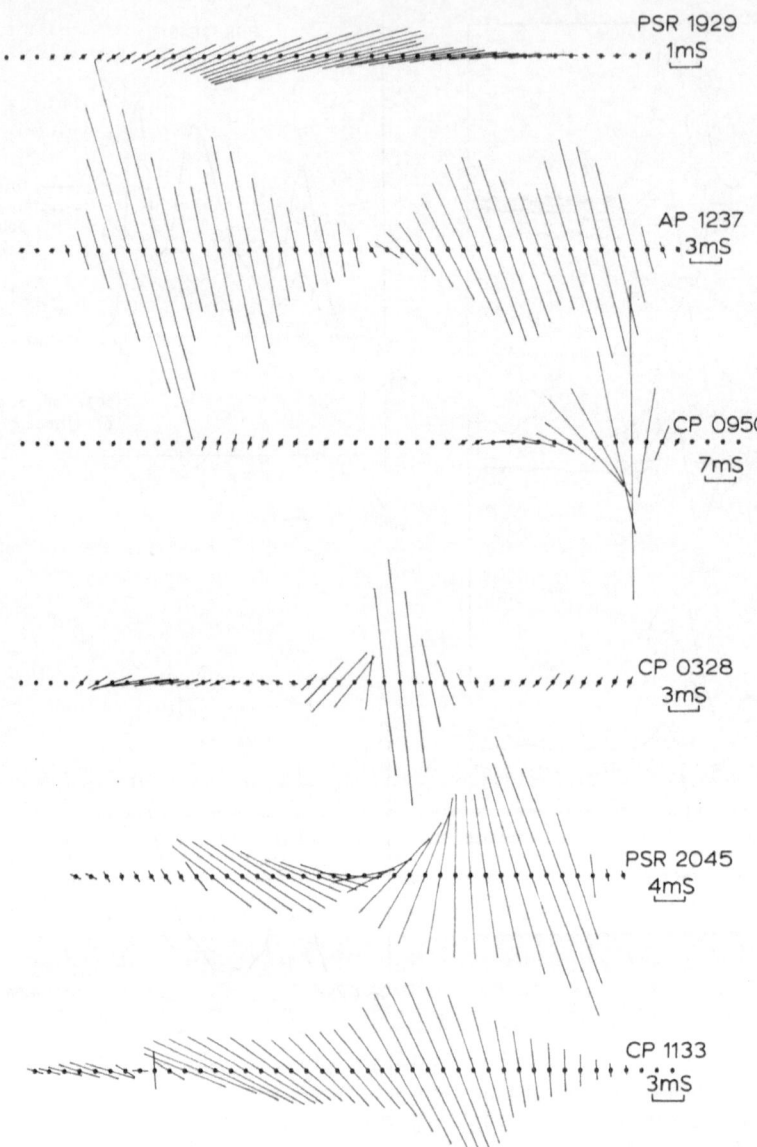

Fig. 3. The variation of the position angle of the linear polarization through the average pulse of six pulsars.

3.6 CRAB PULSAR RADIATION CHARACTERISTICS

R. N. MANCHESTER

National Radio Astronomy Observatory*, Green Bank, W. V., U.S.A.

It has been suggested that for the Crab pulsar the radio radiation mechanism may be different from that in other pulsars. The principal observations leading to this suggestion are, firstly, the essentially constant position angle of the highly linearly polarised precursor, and secondly, the occasional large increases in intensity of the main pulse.

I would like briefly to point out that similar characteristics have been observed in other pulsars, and therefore to suggest that the Crab pulsar is not unique in this respect. Highly polarised precursor pulses have been observed in a number of pulsars, particularly PSR 0329 + 54 and PSR 0950 + 08. In each case the variation of position angle through this precursor pulse is small.

An interesting phenomenon has been observed in PSR 1237 + 25 which may be related to the erratic amplitude behaviour of the Crab pulsar main pulse. The mean pulse profile for this pulsar is shown in the lower part of Figure 1. It is clear that there are at least five components making up the average profile. Occasionally however the central component becomes much more intense, generally dominating the pulse profile. This strong central pulse in PSR 1237 + 25 is narrow and only weakly polarised – similar to the Crab pulsar main pulse. The variation of position angle shown in the upper portion of Figure 1 suggests that we are looking directly down on a magnetic pole for this pulsar. If this interpretation is accepted, then the erratic central pulse is being emitted directly from the pole. These results suggest that for the Crab pulsar the line of sight may pass through a magnetic pole as for PSR 1237 + 25, with the main pulse being emitted from the pole.

Discussion

C. Heiles: Many pulsars have components which vary relative to each other by large factors. The Crab pulsar is qualitatively different, however; the main pulse occasionally becomes a factor of fully 2000 times the average pulse density.

R. N. Manchester: The erratic centre pulse from PSR 1237 can be up to about a 100 times as strong as the mean.

F. C. Michel: There appears to be a systematic tendency for the sweep in polarisation to occur when the total polarisation or intensity is small – is this a general feature?

R. N. Manchester: The polarised component is strong at the position of the P.A. swing in PSR 2045, and so the swing cannot be accounted for by a drop in the polarised power. For PSR 1237 the polarised component is low at the position of the swing in position angle. However, the P.A. of the two sides of the pulse is almost the same, so again the swing cannot be explained by this.

---

* Operated by Associated Universities, Inc., under contract with the National Science Foundation.

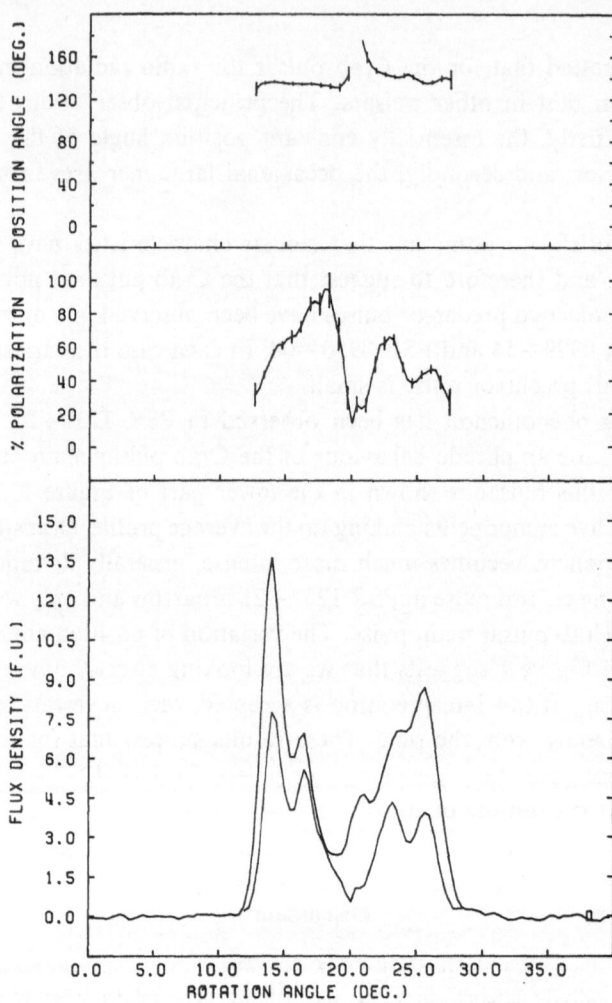

Fig. 1.   Mean pulse profile for PSR 1237 + 25 measured at 410 MHz. The lower portion of the figure gives the total intensity and linearly polarised component, the central portion, the percentage linear polarisation, and the upper portion, the position angle of the linear component.

3.7 TIMING OF THE PULSAR NP 0527

D. W. RICHARDS* and J. A. ROBERTS**

Cornell-Sydney University Astronomy Center, Arecibo Observatory, Arecibo, Puerto Rico

Abstract. Timing measurements have been made of NP 0527 in order to examine its physical relationship with NP 0532. 18 months of timing data have been used to obtain an accurate frequency, rate of change of frequency and position. The calculated 'slowdown age' of the NP 0527 is 3.0×10^6 years which rules out an origin in the supernova explosion of 1054 A.D. The upper limit of its transverse velocity is 0.046c which is significantly less than the value necessary if it had a common origin with NP 0532.

The question of a physical relationship between the pulsar NP 0532 and its neighbor, NP 0527, was first raised by Reifenstein *et al.* (1969). They suggested that NP 0527 was ejected from the supernova which produced NP 0532 and calculated that if NP 0527 has not been accelerated, it must be moving with a space velocity of ~ 0.15c. More recently, Gott *et al.* (1970) have proposed a model wherein both pulsars were initially components of a double star in the nearby association I Geminorum. According to this picture, NP 0527 became a supernova first and both stars 'ran away' from the association, NP 0532 being formed later. Both pulsars were calculated to have space velocities of ~ 150 km/sec.

In principle, timing observations of the pulses from NP 0527 can measure the proper motion, and distinguish between these two proposals. Timing observations also provide a measure of df/dt, the time rate of change of the pulsar repetition frequency, f, and a determination of the position of the pulsar.

We wish to report the preliminary reduction of 18 months of timing observations of NP 0527 made at the Arecibo Observatory. The rather grand scale of the pulses from NP 0527, in comparison with those from NP 0532, is illustrated by the typical pulse shape shown in Figure 1. One component of the double pulse is as wide as the full period of the Crab pulsar. This scale is reflected in the timing accuracies which are reduced by a factor of about 100, or roughly in the ratio of the width of the first component of NP 0527 (30 msec) to that of the main pulse in NP 0532 (300 μsec). The errors in the daily mean arrivals (derived as in Richards *et al.*, 1970) are about 1 msec as compared with about 10 μsec for NP 0532.

A set of pulse numbers, N, and barycentric arrival times, t, was derived for NP 0527 in much the same way as for NP 0532 (Richards *et al.*, 1970; Roberts and Richards, 1970), except that the topocentric arrival times were found by matching a transparent overlay to line-printer plots of the observations, and not by computerized cross correlation. The position and the dispersion constant used for the reductions to the barycenter and to infinite frequency were those given by Zeissig and Richards (1969).

* Present Address: Air Force Cambridge Research Laboratories, Bedford, Mass., U.S.A.
** Permanent Address: CSIRO Radiophysics Laboratory, Epping, N.S.W. 2121, Australia.

Davies and Smith (eds.), The Crab Nebula, 211–216. All Rights Reserved.

These barycentric arrivals were fitted to a polynomial of the form

$$N_{\text{fit}} = N_0 + f_0 t + \tfrac{1}{2}(df/dt)_0 t^2, \qquad (1)$$

where N_{fit} is a continuous variable ($1/2\pi$ times the pulse phase), N_0 is a constant, and the subscript 0 denotes values at $t=0$. The residuals from this fit had a quasi-sinusoidal form with a period of one year, indicating an error in the assumed position.

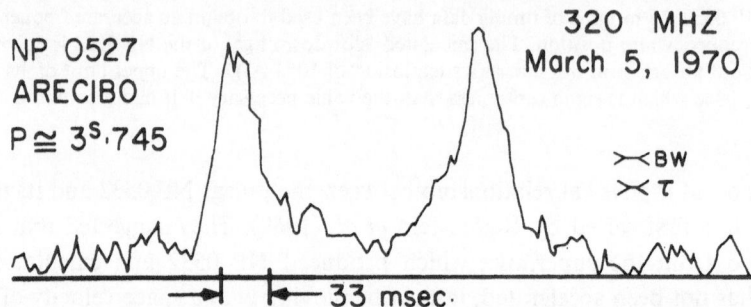

NP 0527
ARECIBO
$P \cong 3^{S}.745$

320 MHZ
March 5, 1970

BW
τ

33 msec·

Fig. 1.   Typical pulse shape for NP 0527 as obtained at Arecibo.

It turns out that the eccentricity of the earth's orbit is sufficiently great that a revised position could not be found by simply fitting a sinusoid to the residuals, so the following technique was used to determine corrections to the originally assumed right ascension, α, and declination, δ.

Let $\hat{\mathbf{s}}$ be the unit vector pointing towards the pulsar, and $\hat{\mathbf{s}}_0$ the corresponding unit vector determined by the first estimates of α and δ. The difference, $\Delta\mathbf{s}$, is defined by

$$\hat{\mathbf{s}} = \hat{\mathbf{s}}_0 + \Delta\mathbf{s}.$$

If $\mathbf{r}$ is the vector pointing from the solar system barycenter to terrestrial observing site at the time the pulse arrives, then the barycentric light travel time, b, is

$$b = \mathbf{r}\cdot\hat{\mathbf{s}} = \mathbf{r}\cdot\hat{\mathbf{s}}_0 + \mathbf{r}\cdot\Delta\mathbf{s},$$
$$= b_0 + \Delta b.$$

Here b_0 is the original estimate of b, and the correction Δb may be expressed in terms of the corrections $\Delta\alpha$, $\Delta\delta$ to be made to the right ascension and declination.

Thus $\Delta\mathbf{s}$ is given approximately by

$$\Delta\mathbf{s} = \frac{\partial}{\partial\alpha}(\hat{\mathbf{s}}_0)\,\Delta\alpha + \frac{\partial}{\partial\delta}(\hat{\mathbf{s}}_0)\,\Delta\delta,$$

or

$$\Delta\mathbf{s} = [-\hat{\mathbf{x}}\cos\delta\sin\alpha + \hat{\mathbf{y}}\cos\delta\cos\alpha]\,\Delta\alpha$$
$$+ [-\hat{\mathbf{x}}\sin\delta\cos\alpha - \hat{\mathbf{y}}\sin\delta\sin\alpha + \hat{\mathbf{z}}\cos\delta]\,\Delta\delta,$$

where $\hat{\mathbf{x}}, \hat{\mathbf{y}}, \hat{\mathbf{z}}$ are unit vectors in the Equatorial Cartesian set of coordinates (at epoch 1950.0). Writing

$$\mathbf{r} = \hat{\mathbf{x}}r_x + \hat{\mathbf{y}}r_y + \hat{\mathbf{z}}r_z,$$

we find

$$\Delta b = \Delta\alpha R_\alpha + \Delta\delta R_\delta, \tag{2}$$

where

$$R_\alpha = - r_x \cos\delta \sin\alpha + r_y \cos\delta \cos\alpha,$$
$$R_\delta = - r_x \sin\delta \cos\alpha - r_y \sin\delta \sin\alpha + r_z \cos\delta.$$

The functions R_α and R_δ can be computed when b_0 is calculated, and used later in a least squares solution for $\Delta\alpha$ and $\Delta\delta$. A fit to the function formed from the sum of the r.h.s. of equation (1) and f_0 times the r.h.s. of equation (2) yielded the residuals shown in Figure 2. The fitted frequency and rate of change, referred to epoch Julian Date 2 440 400.5 are:

$$f = 0.266\,987\,663\,61 \pm (87 \times 10^{-11})\,\text{Hz UTC},$$

$$\frac{df}{dt} = (-2.8481 \pm 0.0041) \times 10^{-15}\,\text{sec}^{-2}$$

The 'age' of the pulsar, $-f/(df/dt)$, is 3.0×10^6 yr. This appears to rule out an origin in the supernova explosion of 1054 AD as suggested by Reifenstein et al., independently of any consideration of the proper motion. The new right ascension and declination are:

$$\alpha\,(1950.0) = 05^h25^m52{.}^s08 \pm 0{.}^s07,$$
$$\delta\,(1950.0) = 21°56'32'' \pm 16''.$$

Because the pulsar lies almost in the ecliptic plane, and near the summer solstice, the declination is poorly determined. The formal one standard deviation errors from the least squares fit quoted above should be considered optimistic since there may be systematic errors. In this preliminary reduction no allowance was made for the different receiver time constants and bandwidths used on different days, and an old value of the dispersion constant, B, was used. To keep the effect of possible errors in B at a minimum, only observations at 318 MHz or higher have been used in this reduction. Possible time variation of B has been neglected.

Any proper motion of the source would produce quasi-sinusoidal residuals with a period of one year and an amplitude that increased linearly with the time. The fitting procedure determines the source position near the mid-point of the data span, so that proper motion residuals in Figure 2 would be zero near the mid-point and form a one-year sinusoid with amplitude increasing away from the mid-point, and a phase reversal at the mid-point. No such trend is evident, and to be within the errors the amplitude must be less than ± 8 msec after 9 months. For a distance of 2 kpc this gives an upper limit of 0.2c for any transverse velocity parallel to the ecliptic plane.

A much more stringent limit on the proper motion may be deduced from the measured value of df/dt, as Staelin (1970) has pointed out. Qualitatively Staelin's argument is the following. If the pulsar has a constant velocity with a component transverse to the line of sight, then the distance to the pulsar will be increasing in a non-linear fashion. This leads to a changing Doppler effect and hence a decrease of the apparent pulsar repetition frequency. Shklovsky (1969) and Detre (1969) have in fact suggested that this effect could account for all the observed frequency change for pulsars with periods ~ 1 sec. Here we place an upper limit on any transverse velocity by assuming all the measured df/dt is due to the motion of the pulsar.

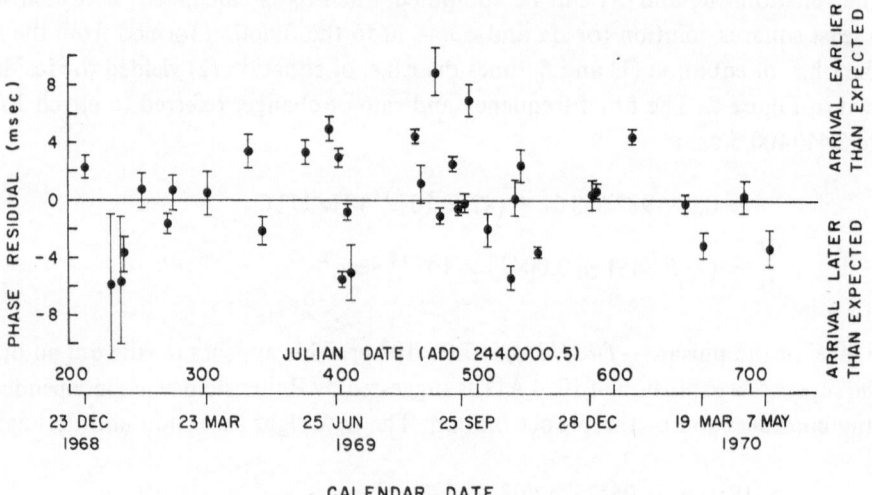

Fig. 2.   Phase residual for simultaneous quadratic and position correction fit. Some of the scatter is due to systematic effects (see text).

Resolve the (constant) velocity **v** of NP 0527 into components perpendicular $(v_p \hat{\mathbf{p}})$ and parallel $(v_s \hat{\mathbf{s}})$ to the line of sight. As before, $\hat{\mathbf{s}}$ points from the solar system barycenter to the pulsar. If f_0 is the (constant) pulsation frequency of the pulsar in its rest frame, then we observe a Doppler shifted frequency

$$f = f_0 (1 - \mathbf{v} \cdot \hat{\mathbf{s}}/c).$$

Thus

$$df/dt = - (f_0/c)\, \mathbf{v} \cdot d\hat{\mathbf{s}}/dt,$$

or

$$df/dt \approx - (f_0/c)\, \mathbf{v} \cdot \hat{\mathbf{p}}\, (v_p/r),$$
$$= - (f_0 v_p^2)/cr,$$

where r is the distance to the pulsar. Hence

$$v_p = [- (rc/f_0)\, df/dt]^{1/2}.$$

Using $r = 2$ kpc and f_0 and df/dt from the fit discussed above, we find

$$v_p \approx 13\,800 \text{ km/sec} = 0.046\,c.$$

The corresponding upper limit on proper motion is $1\rlap{.}''4$ arc/yr. This upper limit definitely eliminates the Reifenstein *et al.* model, but the limit is far above the values derived by Gott *et al.*

For their model Gott *et al.* calculated that df/dt due only to pulsar spin down would be $(0.64-1.43) \times 10^{-15}$ sec$^{-2} \pm 15\%$. The contribution to df/dt from their predicted transverse velocity is quite small and hence their estimate of df/dt falls short of the observed value by at least a factor of 2.

If NP 0527 is a neutron star of one solar mass, radius 10 km, and uniform density, the observed df/dt corresponds to an energy loss rate of 3.6×10^{31} erg/sec. The radio energy density of about 5×10^{-27} J m^{-2} Hz^{-1}/pulse (Zeissig and Richards, 1969) can account for only $\sim 6 \times 10^{29}$ erg/sec, assuming a bandwidth of 1000 MHz, a distance of 2 kpc, and a beam solid angle of $4\pi/50$ sterad. If we suppose that nearly all of the 3×10^{31} erg/sec is radiated in the optical, the absolute magnitude of the pulsar will be about $+10$. At 2 kpc, neglecting absorption, it will be fainter than magnitude $+21$, near the borderline of detectability in previous searches (Kristian, 1970; Willstrop, 1970).

Acknowledgements

The assistance of G. A. Zeissig in securing the observations is greatly appreciated. We also thank the MIT Haystack Observatory for provision of computer time; the staff of the Arecibo Observatory for assistance in all phases of the work; and M. E. Ash, I. I. Shapiro, and W. B. Smith of the MIT Lincoln Laboratory for ephemerides. Additional support under contract F-44620-69-C-0092 of the U.S. Air Force Office of Scientific Research is gratefully acknowledged. The Arecibo Observatory is operated by Cornell University under contract with the National Science Foundation, and with partial support from the Advanced Research Projects Agency.

References

Detre, L.: 1969, IAU Commission 27 Information Bulletin on Variable Stars No. 380.
Gott, J. R., Gunn, J. E., and Ostriker, J. P.: 1970, *Astrophys. J.* **160**, L91.
Kristian, J.: 1970, this symposium, Paper 2.3, p. 87.
Reifenstein III, E. C., Brundage, W. D., and Staelin, D. H.: 1969, *Phys. Rev. Letters* **22**, 311.
Richards, D. W., Pettengill, G. H., Counselman III, C. C., and Rankin, J. M.: 1970, *Astrophys. J.* **160**, L1.
Roberts, J. A. and Richards, D. W.: 1970, this symposium, Paper 2.11, p. 125.
Shklovsky, I. S.: 1969, *Astron. Zh.* **46**, 715, translated in *Soviet Astron.* **13**, 562 (1970).
Staelin, D. H.: 1970, personal communication.
Willstrop, R. V.: 1970, this symposium, Paper 3.9, p. 222.
Zeissig, G. A. and Richards, D. W.: 1969, *Nature* **222**, 150.

Discussion

J. G. Davies: To add to the confusion of this subject it is worth noting that the pulsar with the second longest known period (JP 2319) is $1\frac{1}{2}°$ from Cas A. If it were ejected from the supernova, it will have to be travelling near the speed of light and, further, its dispersion measure makes it rather unlikely that it is at the same distance as the radio source.

D. Staelin: Dr. Sutton and I wish to note that the suggestion of Reifenstein, Brundage, and Staelin (1969) that NP 0527 may have been ejected from the Crab Nebula can now be ruled out by the slow-down rate reported here by Richards. If NP 0527 had indeed been ejected from the Crab about 1000 years ago, then the transverse velocity of NP 0527 would be greater than 0.1 c. The resulting second-order Doppler effect would produce an apparent rate of change in period more than twice the measured value, and hence NP 0527 must have a different origin.

3.8 INTERSTELLAR SCATTERING AND
THE PULSE FROM THE VELA PULSAR

M. M. KOMESAROFF

Division of Radiophysics, CSIRO, Sydney, Australia

P. A. HAMILTON

Department of Physics, University of Tasmania, Hobart, Australia

and

J. G. ABLES

Division of Radiophysics, CSIRO, Sydney, Australia

Abstract. Measurements of PSR 0833−45 were made at wavelengths of 21, 48, 73 and 100 cm. Pulse broadening proportional to about the fourth power of the wavelength was found, consistent with multipath scattering in the interstellar medium. It is further concluded that the rotation measure was constant at 33.2 rad m^{-2} across this wavelength range, supporting the conclusion that the radiation originates in the vicinity of a magnetic pole.

1. Introduction

During February 1970 the pulsar PSR 0833−45 was observed simultaneously at a number of wavelengths between 20 and 120 cm using the ANRAO 210-ft telescope at Parkes. The results indicated a dramatic broadening of the pulse at wavelengths longer than about 75 cm (Ables *et al.*, 1970). If the duration Δt of the trailing edge of the pulse is expressed in the form $\Delta t \propto \lambda^{\alpha}$, where λ is wavelength, α is found to approach the value 4 asymptotically at the longer wavelengths. The effect was attributed to scattering from electron density inhomogeneities in the interstellar medium, most probably associated with the Gum Nebula.

Staelin and Sutton (1970) and Rankin *et al.* (unpublished) have found that for NP 0532 the pulse duration also increases rapidly at the longer wavelengths.

In May 1970 a second series of observations of PSR 0833−45 was undertaken at Parkes at a number of wavelengths between 20 and 100 cm. In addition to the instantaneous flux density, the linearly polarized component and its position angle were measured and averaged over several thousand pulses.

These later results provide further support for the scattering hypothesis. They also support the view that the intrinsic pulse shape and instantaneous position angle are substantially independent of wavelength when propagation effects have been eliminated. The latter result agrees with that of Radhakrishnan and Cooke (1969) and Radhakrishnan (1969) derived from observations at much shorter wavelengths; it therefore supports their conclusion that the radiation originates from the vicinity of a magnetic pole, and well within the velocity of light cylinder.

2. Equipment

The observing technique has been described by Ables *et al.* (1970). It involved several

Davies and Smith (eds.), The Crab Nebula, 217–221. All Rights Reserved.
Copyright © 1971 by the IAU.

receivers, each with its own linearly polarized feed. The receiver outputs were fed through a rapid multiplexer to a PDP-9 computer, which also received a pulse train at the Doppler-shifted pulsar period. The whole feed assembly could be rotated in position angle.

3. Results

The results for four of the observed wavelengths are shown in Figures 1–4. The observed values of flux density are indicated by filled circles and the position angle by open circles. Except for Figure 1 the full lines are not lines of best fit to the data

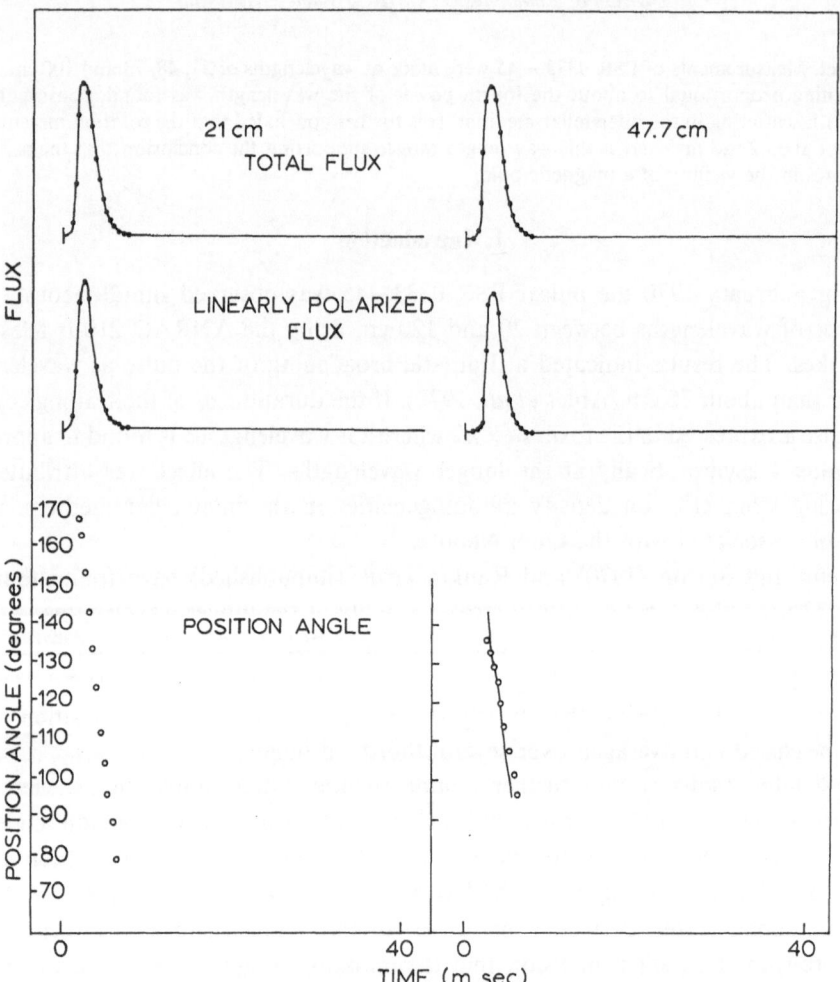

Fig. 1.   Observed pulse shape and polarization at 21 cm wavelength. The filled circles represent the observed values of total flux density and the polarized component. The continuous lines represent the curves of best fit to the data points. Instantaneous position angles are indicated by open circles. The period is 89.2 msec.

Fig. 2.   Observational results at 47.7 cm. The filled and open circles have the same significance as for Figure 1, but the continuous lines have been computed by the convolution process described in the text.

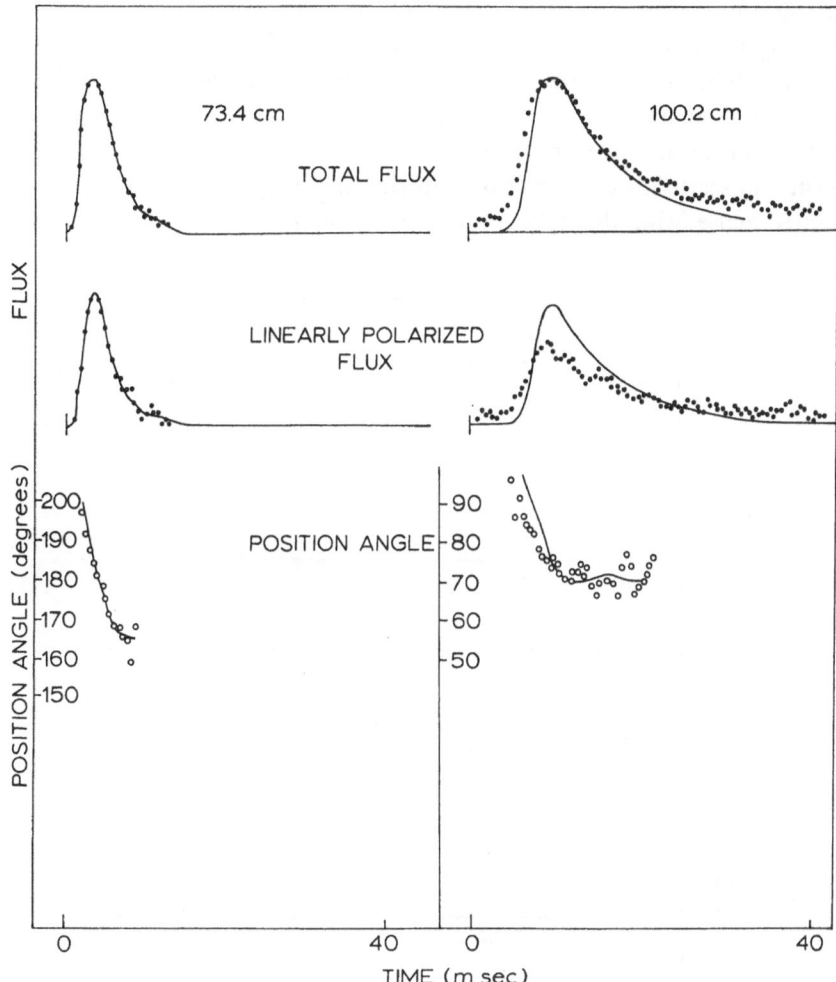

Fig. 3.   Observational results at 73.4 cm. The same convention has been used as in Figure 2.
Fig. 4.   Observational results at 100.2 cm. The same convention has been used as in Figures 2 and 3.

points, but are curves computed as indicated below. As pointed out previously, the pulse broadens with increasing wavelength. In addition, both the degree of linear polarization at the peak and the change of position angle within the pulse decrease as the wavelength increases.

4. Interpretation

The case of a plane wave traversing a 'thick' medium consisting of a larger number of randomly spaced irregularities along the line of sight has been considered by a number of authors. Fejer (1954), for example, has shown that the angular spectrum of the scattered component of the emergent radiation is approximately a two-dimensional Gaussian function. For a radio wave traversing a very tenuous irregular ionized medium, the angular spread of the emergent wave varies as λ^2.

According to arguments of the type adduced by Scheuer (1968), it then follows that a very short-duration pulse of radiation from a point source will show an approximately exponential decay after passing through such a scattering medium; this is the result of multipath transmission.

The effect of scattering on a pulse of finite duration may be computed by convolving the emitted pulse with the appropriate exponential function. If the scale of the

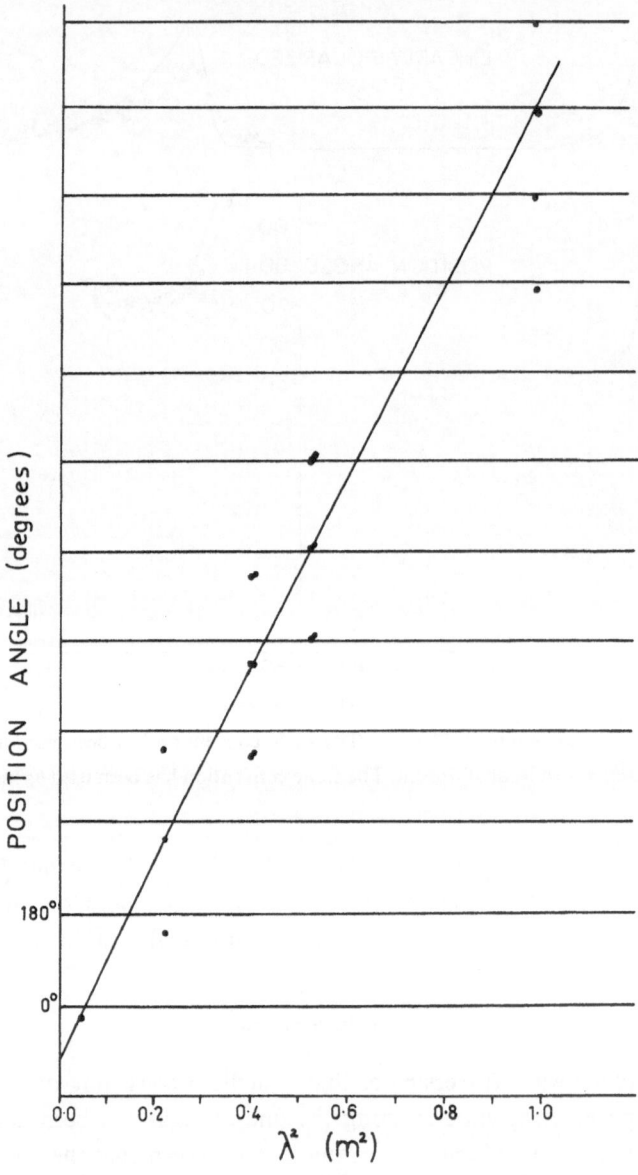

Fig. 5.   Position angle of polarization at pulse peak vs. λ^2. The slope of the line which passes through all points corresponds to a rotation measure of 33.2 rad m^{-2}.

irregularities is much larger than the observing wavelength, and if magnetic field effects may be neglected, the effect of scattering is independent of polarization. It follows that the characteristics of the scattered pulse, including its instantaneous polarization, may be derived by convolving each Stokes parameter of the emitted pulse with the same exponential function; the scale of this function will of course be proportional to λ^4.

From the results of Radhakrishnan and Cooke (1969) we conclude that scattering has negligible effect on the pulse from PSR 0833−45 at wavelengths of 20 cm and shorter. We have therefore derived the scale of the exponential convolving function at $\lambda = 100$ cm by trial, by comparing the total flux pulse at this wavelength with that at 20 cm. For the other wavelengths we have taken the scale width of the exponential to be proportional to λ^4.

5. Discussion

Following are the results of carrying out the convolutions of the assumed intrinsic pulse as described above:

(i) In all cases the calculated variation of polarization position angle agreed well with that observed (Figures 1(c) to 4(c)).

(ii) At 100 cm the predicted magnitude of the linearly polarized component of the pulse exceeded that observed by about 30%.

(iii) For wavelengths of 73 cm and shorter no significant difference was found between the predicted and observed total intensity at any instant during the pulse. The same was true of the polarized intensity (Figures 2b, 3b).

Finally, after making slight adjustments to the position angle at the pulse peak, which the convolution results indicated were necessary, it was found that the rotation measure was constant at 33.2 rad m^{-2} across the 5 to 1 wavelength range. The measured position angles are given in Figure 5. Taken in conjunction with the foregoing results this strongly indicates that the rotation of the position angle at the peak with wavelength is a propagation effect in the interstellar medium, and that the intrinsic position angle at any instant is independent of wavelength.

Acknowledgements

We would like to thank Mr. D. Smart for his assistance with the considerable amount of computer reduction involved, and Mrs. R. Raison for help with other aspects of the data reduction.

References

Ables, J. G., Komesaroff, M. M., and Hamilton, P. A.: 1970, *Astrophys. Letters* **6**, 147.
Fejer, J. A.: 1954, *Proc. Roy. Soc. London* **A220**, 455.
Radhakrishnan, V.: 1969, *Proc. Astron. Soc. Australia* **1**, 254.
Radhakrishnan, V. and Cooke, D. J.: 1969, *Astrophys. Letters* **3**, 225.
Scheuer, P. A. G.: 1968, *Nature* **218**, 920.
Staelin, D. H. and Sutton, J. M.: 1970, *Nature* **226**, 69.

3.9 SEARCHES FOR OPTICAL PULSARS

R. V. WILLSTROP

The Observatories, Madingley Road, Cambridge, U.K.

Abstract. The limitations of searches for pulsed optical radiation from pulsars and other objects are considered for the two cases (a) when the period is known from radio observations, and (b) when the period is unknown. Results of searches by the author and by others are summarised.

So far, radio searches for pulsars have been more successful than optical searches, by a factor of about 50 if we judge success by the number of positive identifications. However, in addition to measurements of NP 0532 many optical searches have established upper limits to the pulsed light flux reaching us from a number of other pulsars.

Let us consider the limitations of optical searches. The positions of some pulsars are already known within a few seconds of arc (Reichley *et al.*, 1970), and some searches have been made using photometer diaphragms as small as 10 arc sec in diameter. Photon counting equipment is normally used. Because the optical emission from NP 0532 is a broad band phenomenon, photomultipliers are usually used without any filter in searches for other optical pulsars, and the light of the night sky causes many more counts than the dark emission of the detector. It is therefore the statistical fluctuations in the night sky light transmitted by the diaphragm which limit the sensitivity of optical searches.

It is easy to show that, with a multichannel scaler giving time resolution equal to the undispersed radio pulse duration, and supposing that an excess of detected photons at one phase amounting to five standard deviations is certain to be recognised, the minimum detectable pulsed photon flux outside the Earth's atmosphere is

$$\mathscr{F}_{\min} = 5 \left(\frac{NSf}{AT\varepsilon} \right)^{1/2} \text{ photons cm}^{-2} \text{ sec}^{-1} \tag{1}$$

where N photons cm^{-2} sec^{-1} reach the Earth's atmosphere from each square second of arc of sky;

S square seconds of arc of the sky are searched;

f is the fraction of the pulsar period occupied by the undispersed radio pulse (the duty cycle);

A cm^2 is the collecting area of the telescope;

T sec is the observing time used;

and ε is the overall quantum efficiency, including the transmission of the atmosphere and telescope.

Clearly, searches to very faint limits depend on the determination of accurate positions by radio methods.

Davies and Smith (eds.), The Crab Nebula, 222–228. All Rights Reserved.
Copyright © 1971 by the IAU.

At a reasonably dark observing site the sky brightness is about $21^{m}.5$ per square second of arc, equivalent to $N = 6 \times 10^{-3}$ photons cm^{-2} sec^{-1} (arc sec)$^{-2}$ in the spectral range 3500 to 6000 A. Cocke and Disney (1970) searched recently for optical pulsars in the directions of CP 0328 and CP 0950 and claimed limiting sensitivities (apparent magnitudes) of $25^{m}.5$ and $24^{m}.5$ (later improved to $25^{m}.6$). Equation (1) gives limits of $25^{m}.5$ and $25^{m}.2$. It is reassuring to find such agreement between theory and practice.

Kristian reported (this symposium, Paper 2.3, p. 87) his observations of 11 pulsars, reaching limits between 20^{m} and 25^{m}, all with negative results. Lynds et al. (1968) very early set a limit of 26^{m} on CP 1919. Chiu (private communication) has searched the area around PSR 0833 − 46 to a limit of 24^{m}, using an image intensifier to provide spatial resolution as well as time resolution. I have yet another negative observation to report, of HP 1506. Mrs. Mitton, one of the students at the Cambridge Observatories, pointed out (private communication) that the radio position is 1.4 sec following and 7 arc sec south of a star of about 16^{m}. This coincidence seemed good enough to investigate further. A fainter star, about 19^{m}, 16 arc sec Nf the first, was also observed. The shading in Figure 1 represents the excess or deficiency in successive time bins, and the inset in each case shows the predicted effect of a 22^{m} star with all its light pulsed at the period of HP 1506. Evidently the light of these two objects is not modulated by more than about 1 per cent and 10 per cent respectively.

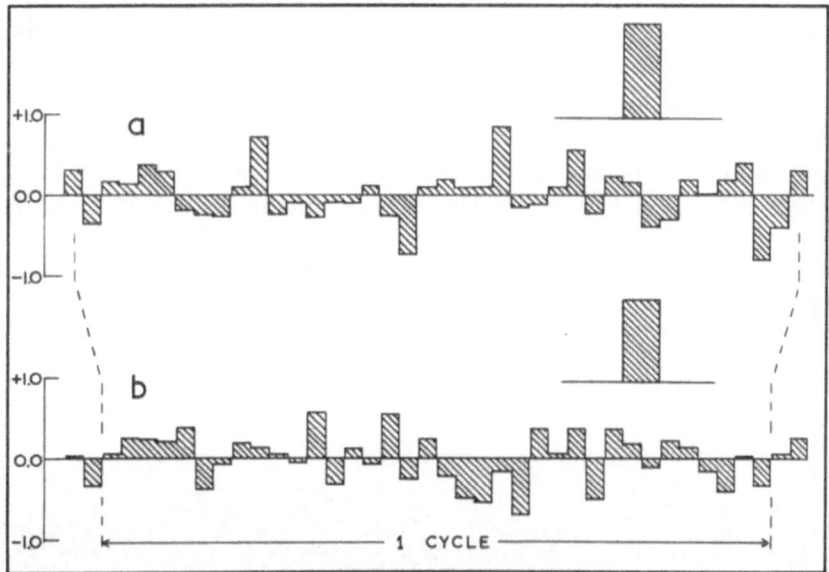

Fig. 1. Results of searches for optical variability in two stars near HP 1506 ($\alpha = 15^{h}08^{m}03^{s}.3$ $\delta = +55°42'50''$ (1950)). Ordinates: Variation in the rate of detection of photons from the star and sky, in per cent, as a function of phase. Abscissae: Phase, relative to an arbitrary origin. (a) 1970 June 3 $22^{h}53^{m}$ to $23^{h}30^{m}$: 52.072 readings/sec. Summation over 3001 cycles. 16^{m} star at $\alpha = 15^{h}08^{m}01^{s}.9$ $\delta = +55°42'57''$ (1950) (J. Mitton, private communication). (b) 1970 June 4 $22^{h}57^{m}$ to $23^{h}36^{m}$: 48.8176 readings/sec. Summation over 3154 cycles. 19^{m} star Nf the 16^{m} star. In both observations the photometer diaphragm was 25 arc sec in diameter.

I now wish to discuss the problem of detecting pulsed optical radiation in cases when no pulsed radio emission has been detected, and the period is therefore unknown. This problem arises because Cavaliere and Pacini (1970) have advocated a radio search for a pulsar associated with the supernova remnant Cas A, and Bahcall *et al.* (1970) have suggested optical and X-ray examination of newly reported supernovae nearer than 10 Mpc. The recent appearance of a supernova of 11^m in M 101 at 1.1 Mpc adds to the interest of their suggestion. Signal averaging methods must be used, to overcome the statistical fluctuations in the night sky light, but the multichannel scaler is not a suitable recording device because it depends on the period being known. Instead, the numbers of photons detected in equal short intervals must be recorded on magnetic tape, paper tape or cards, and subsequently analysed, by computer, for all independent periods, for example by the Cooley-Tukey fast Fourier transform. Recording equipment of this type is available at the Hale Observatories, Harvard, Cambridge (U.K.), and some other observatories.

Figure 2 shows the plotter output from a recent analysis of my earliest data on the Crab Nebula, obtained on 24 November 1968. The plot is based on observations covering 2^m48^s (16384 readings at 97.64 per sec) using a 36-inch telescope with a diaphragm 50 arc sec in diameter, in a comparatively bright sky. At the time of observation the period of this pulsar was known, but its position was still uncertain by ±10 min of arc, and it was fortunate that the diaphragm was placed accurately enough at the centre of the nebula to include the pulsar. Because of the strong interpulse the fundamental, at 30.22 Hz, is weaker than the second harmonic which is aliassed and appears at 37.20 Hz (97.64 − 60.44).

Using this equipment at the Cambridge Observatories and at the Royal Greenwich Observatory on the Isaac Newton Telescope I have examined a number of supernova remnants, white dwarfs and other objects. The results are summarised in Table I. With the exceptions of Nova (DQ) Her 1934 and the nucleus of M 31 all the fluc-

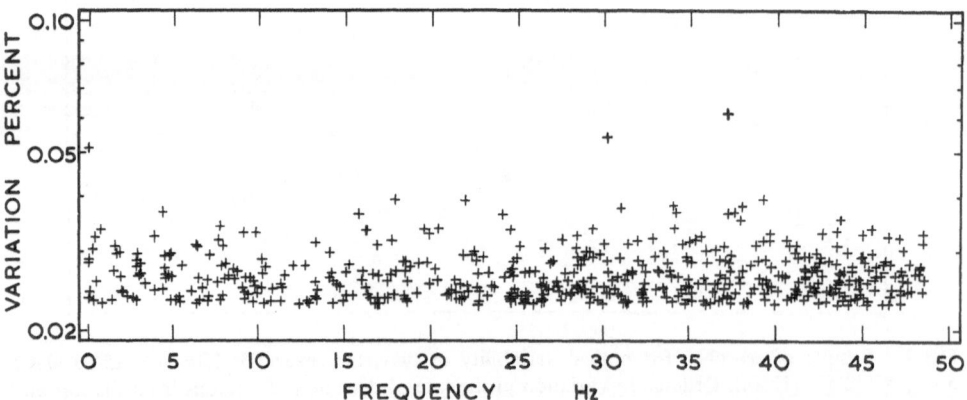

Fig. 2.   NP 0532 observed on 24 November 1968. Data analysed after the discovery of optical flashes. This fast Fourier transform is based on 16384 readings made at 97.64/sec using the Cambridge 92 cm reflecting telescope with a diaphragm 50 arc s in diameter. The fundamental, at 30.22 Hz, and the second harmonic, at 37.20 Hz, are both present.

tuations found were attributable to photon statistics or to irregular sky transparency. The observations of the nucleus of M 31, made on two different nights in 1969 August, showed a variation of the order of 0.15 per cent and with a period of 2.393 seconds. This was shown to originate in a periodic error in the drive of the telescope amounting to ±0.01 arc sec, which was caused by a slight eccentricity in one component of a small gearbox, rotating at 25 revolutions per sidereal minute. Such a small error is, of course, quite unimportant for other types of observation, and the fact that no variations were found at the frequency of rotation of the principal gears testifies to the work of the makers.

The limitations of this equipment are illustrated in Figure 3, which also indicates the observed limits of pulsed light from several pulsars. The upper limit to frequencies

TABLE I

Objects examined in search for regular fluctuations
(other than NP 0532)

Object	Date	Tel.	Diaph. arc sec.	Rate	Max. var. (per cent or mag.)	In frequency range Hz
White dwarf stars						
GRW + 70°5824	1969 May 16	98	13	97.6	0.44	0.0 –48.8
					0.21	0.025– 8.1
W 1346	1969 Aug 14	98	13	48.8	0.80	0.005–24.4
					0.48	0.005– 8.1
R 627	1970 Mar 8	36	25	24.4	0.64	0.0 – 0.006
					0.48	0.006–12.2
					0.21	0.009– 2.034
L 1244 − 26	1970 Mar 8	36	13	12.2	1.50	0.0 – 0.008
					0.33	0.008– 6.1
L 1409 − 4	1970 Mar 9	36	25	12.2	1.30	0.0 – 0.012
					0.30	0.012– 6.1
	1970 Mar 9	36	25	97.6	0.75	0.0 –48.8
Flare star						
AD Leo	1970 Mar 10	36	25	12.2	0.15	0.065– 6.1
Galaxies						
NGC 5548	1969 May 16	98	14	48.8	0.67	0.006–24.4
					0.28	0.025– 4.07
M 31	1969 Aug 14	98	7	48.8	0.24	1.3 –24.4
					0.14	0.26 – 6.1
M 31	1969 Aug 18	98	7	12.2	0.12	0.28 – 0.42
					0.10	0.42 – 1.525
NGC 4151	1970 Mar 5	36	25	97.6	0.60	0.0 –48.8
	1970 Mar 5	36	25	24.4	0.42	0.0 – 0.042
					0.26	0.042–12.2
					0.13	0.136– 2.034
M 82	1970 Mar 6	36	50	97.6	0.40	0.9 –48.8
	1970 Mar 6	36	50	48.8	0.27	0.0 –24.4
					0.10	0.023– 4.068
M 87	1970 Mar 7	36	25	24.4	0.27	0.062–12.2
					0.12	0.047– 2.034

Table I (Continued)

Object	Date	Tel.	Diaph. arc sec.	Rate	Max. var. (per cent or mag.)	In frequency range Hz
Planetary nebula						
M 57 Nucleus	1969 Aug 18	98	7	48.8	1.50	0.0 –24.4
					0.68	0.005– 4.068
Supernova remnants						
OA 184	1970 Feb	36	50	97.6	$16^m.4$	0.0 –48.8
(4 areas searched)						
OA 184	1970 Feb	36	240	97.6	$13^m.7$	0.0 –48.8
(7 areas searched)						
IC 443	1970 Feb	36	50	97.6	$16^m.7$	0.05 –48.8
(2 areas searched)						
IC 443	1970 Feb	36	240	97.6	$13^m.7$	0.0 –48.8
(7 areas searched)						
HB 9	1970 Feb	36	50	97.6	$16^m.5$	0.0 –48.8
(2 areas searched)						
Miscellaneous						
3C 273	1970 Feb 8	36	25	97.6	4.00	0.0 –48.8
BL Lac	1969 Aug 17	98	13	24.4	1.00	0.01 –12.2
					0.56	0.20 – 2.03
	1969 Aug 17	98	13	48.8	0.68	0.015–24.4
					0.36	0.100– 4.88
DQ Her	1969 Aug 18	98	13	48.8	0.88	0.01 –24.4
					0.51	0.006– 4.07
3C 386	1969 Aug 18	98	13	48.8	1.50	0.0 –24.4
					0.64	0.008– 4.07

that can be determined unambiguously is half the rate of the readings, the Nyquist limit. Higher frequencies are detectable, with reduced sensitivity, and by using different recording rates in turn it is possible to resolve almost all ambiguities. The extreme lower limit to frequencies that can be observed is one cycle in the time covered by 16384 readings (the current limit of the computer and reduction programme). In practice random changes in sky transparency introduce noise at very low frequencies and there is usually some reduction in sensitivity. Over the rest of the frequency range the sensitivity is fairly accurately estimated by Equation (1), setting the duty cycle, $f = 0.5$. For example, using the Isaac Newton Telescope with a diaphragm of 10 square seconds of arc and making 100 readings per second, the limit of detection should be about $22^m.7$. If a pulsar similar to that in the Crab Nebula, which is apparently about $16^m.5$, (Lynds *et al.*, 1969), were located in Cas A, at twice the distance of the Crab and with 6 or 7 magnitudes of absorption instead of 1.7, it would have an apparent (integrated) magnitude of $22^m.3$ or $23^m.3$. But Cas A is only one third of the age of the Crab, so it might be intrinsically much brighter. With such a small diaphragm it would take about 9 hours to search an area 20×20 arc sec. This search area covers only a small fraction of the area of the diffuse object, but so also did my original look at the Crab.

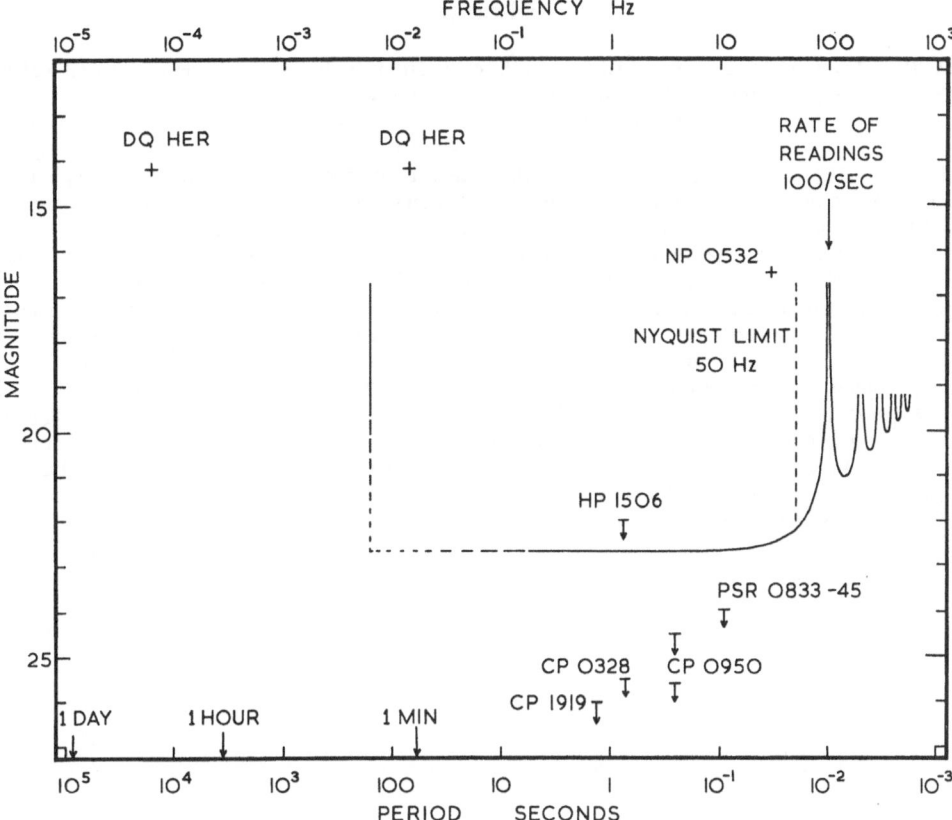

Fig. 3.   Ordinates: Apparent magnitude; Abscissae: Frequency in Hz (top scale) or period in seconds (bottom scale). Symbols: + indicates the observed periodicities and time-average brightness of DQ Her (Walker, 1954, 1956) and NP 0532 (Lynds *et al.*, 1969). ⊤ indicates the periodicities determined by radio observations and upper limits to the time average of any pulsed optical radiation from five other pulsars. See the text for details. The lines at 164 sec period, and at 22^m.7 then curving up to 100 Hz, indicate the limits of sensitivity of equipment built at Cambridge when used on the 249 cm (98-in.) Isaac Newton telescope with a photometer diaphragm 10 (arc sec)^2 in area, and recording 16 384 readings at 100/sec.

So I conclude that the future of optical searches for pulsars is not hopeless, though the past has been less fruitful than some of us hoped.

References

Bahcall, J. N., Rees, M. J., and Salpeter, E. E.: 1970, *Astrophys. J.* **162**, 737.
Cavaliere, A., Pacini, F.: 1970, *Astrophys. J.* **159**, L21.
Cocke, W. J., Disney, M. J.: 1970, *Bull. Am. Astron. Soc.* **2**, 190.
Lynds, R., Maran, S. P., Trumbo, D. E.: 1968, *Science* **161**, 42.
Lynds, R., Maran, S. P., Trumbo, D. E.: 1969, *Astrophys. J.* **155**, L121.
Reichley, P. E., Downs, G. S., and Morris, G. A.: 1970, *Astrophys. J.* **159**, L35.
Walker, M. F.: 1954, *Publ. Astron. Soc. Pacific* **66**, 230.
Walker, M. F.: 1956, *Astrophys. J.* **123**, 68.

Discussion

M. Rees: I would like to comment further on the calculations by Bahcall, Salpeter and myself which Dr. Willstrop has mentioned. We assume that the electromagnetic luminosity of a young pulsar can be obtained by scaling from the Crab pulsar according to the Ω^4 law predicted by the electromagnetic dipole theory (which is pessimistic insofar as it implies that only $\sim 10^{-4}$ of the available energy is channeled into the optical band). Pulses should be detectable from a newly formed pulsar, spinning with its break-up angular velocity (~ 50 times the rotation speed of the Crab pulsar), out to distances ~ 10 Mpc. The envelope would become transparent enough for optical pulses to shine through after 3 weeks to 3 months, depending on assumptions regarding the mass, velocity, and ionization of the expanding debris. The envelope would be more opaque to X-rays and to radio pulses, so these would probably remain undetectable until the pulsar has slowed down substantially from its initial rotation rate. Optical observations thus seem to stand the best chance of detecting extragalactic pulsars.

We believe that these results are sufficiently encouraging to justify a search for optical pulses, with millisecond periods, from the sites of all newly-reported supernovae at distances $\leqslant 10$ Mpc.

D. ter Haar: Calculations by Tsytovich, Buckee and ter Haar (*Phys. Letters* **32A**, 1970) suggest that, of those pulsars for which dP/dt is known apart from NP 0532, only PSR 0833 may pulse optically. Other pulsars may pulse in the infra-red. In these estimates we assume that the loss of rotational energy predominantly shows up at those frequencies where the brightness temperature is of the order of the effective temperature of the relativistic electrons.

3.10   RESULTS OF A SEARCH FOR VISIBLE PULSARS

P. HOROWITZ

Harvard University

C. PAPALIOLIOS and N. P. CARLETON

Smithsonian Astrophysical Observatory and Harvard University

Abstract. A number of interesting celestial objects, including X-ray sources, nova and supernova remnants, white dwarfs, infrared stars, planetary nebulae, radio sources, and some other peculiar nebulosities have been searched with Fourier and correlation techniques for the presence of optical pulsars. The sensitivity of these methods was established with synthetic data and by observations of the Crab Nebula pulsar, the minimum detectable signal (5 standard deviations above noise) being approximately 20th visual magnitude, time averaged. To this limit no new pulsars were found over the range of periods searched, from 5 msec to several seconds.

For the past two years we have been attempting to locate visible pulsars, both at known radiofrequency pulsar sites and elsewhere, using Harvard's 61-inch telescope at Agassiz Station. The search techniques are completely straightforward, consisting of (a) multi-channel scaling of the light intensity *modulo* the pulsar period, for known radio-frequency pulsars, and (b) performing a discrete Fourier transform on the sampled light intensity for other candidates where the period is not known *a priori*. There are many very interesting details connected with these techniques, but since they will appear shortly in a formal journal publication we will confine ourselves here to some of the motivation for the search, and the results.

Besides the known pulsars, we examined the objects listed in Table I, the most interesting of which are the supernova remnants. The Crab and Vela pulsars establish a connection between pulsars and supernova remnants, strengthened by the rough agreement of ages deduced by quite separate means. One therefore expects to find pulsars at the sites of galactic supernovae, and, furthermore, recent supernovae are likely to contain faster (and possibly visible) pulsars. These expectations are based largely upon the characteristics of the Crab pulsar – the youngest known pulsar and the only one visible at optical wavelengths, but simple theoretical arguments lead to the same conclusions. It is not difficult to estimate the peak energy of photons emitted through synchrotron radiation for several popular pulsar theories. In the 'near field' models one gets (Horowitz, 1970)

$$U_{peak}(\text{photons}) \approx 3 \times 10^7 \Omega^{3/2} B_{12}^{3/4} \text{ eV}$$

and in the 'light cylinder' theories

$$U_{peak}(\text{photons}) \approx 4 \times 10^6 \Omega^{7/4} B_{12}^{3/4} \text{ eV}$$

where Ω is the neutron-star's angular frequency, $(\Omega \gtrsim 10 \text{ rad/sec})$ and B_{12} is the surface field in units of 10^{12} gauss. These are surprisingly high energies, and seem to

Davies and Smith (eds.), The Crab Nebula, 229–233. All Rights Reserved.
Copyright © 1971 by the IAU.

TABLE I

Objects examined for visible pulses

Region	Object	Date	Aperture	Sample. msec.	No. of Samples	Lim. Magn. or % Mod.
Radiofrequency Pulsars						
CP 1919	Ryle's star (19^m)	4/68	7"	16	1.1×10^6	25^m (0.4%)
CP 0328	Calc. pos'n.	10/69	15"	2	5×10^5	~17^m
PSR 1929	Calc. pos'n.	7/69	15"	0.5	2.5×10^5	~17^m
CP 0834	Calc. pos'n.	11/68	15"	50	2×10^4	IR: Sky Lim.
CP 0950	Calc. pos'n.	11/68	15"	50	2×10^4	IR: Sky Lim.
CP 1133	Calc. pos'n.	11/68	15"	50	2×10^4	IR: Sky Lim.
HP 1507	Calc. pos'n.	7/68	4'	ROCKING CAMERA		~16^m
Planetary Nebulae						
38+12°1	Central star	6/69	15"	2.5	32K	0.3%
M27	Central star	5/69	8"	2.5	32K	0.9%
M57	Central star	5/69	8"	2.5	32K	2.8%
M97	Central star	5/69	8"	2.5	32K	1.4%
NGC 6833	Central star	6/69	8"	2.5	32K	0.4%*
NGC 6587	Central star	8/69	15"	2.5	32K	2.6%
NGC 7293	Central star	9/69	15"	2.6	128K	0.3%
NGC 7662	Central star	6/69	8"	2.5	32K	0.3%
NGC 7027	Central region	8/69	4.6"	2.5	128K	0.2%*
Nova Remnants						
Q Cyg (1876)	star	7/69	8"	2.9	64K	1.4%
Lac (1910)	star	7/69	8"	2.9	64K	1.4%
Sge (1913)	star	7/69	15"	2.9	64K	0.12%
Aql (1918)	star	7/69	15"	2.9	64K	0.7%
Per (1901)	star	8/69	15"	2.9	64K	0.7%
White Dwarfs						
G 126-27		7/69	15"	2.9	64K	0.5%
L 845-70		7/69	8"	2.9	64K	0.8%
F 108		7/69	8"	2.9	64K	0.3%
Ton 202		8/69	15"	2.5	32K	8.7%
X-Ray Sources						
Sco X-1	Sandage et al	7/69	8"	0.1	16K	6.0%
				2.5	128K	0.15%
"Cyg XR-2"	Giacconi et al	7/69	8"	0.1	16K	50%
				2.9	128K	1.1%
"GX3+1"	Blanco et al	8/69	15"	2.5	64K	0.6%
Supernova Remnants						
Kepler (1604)	many pos'ns.	6-7/69	15"	2.9	32K, 64K	19^{m}1
Tycho (1572)	many pos'ns.	8-9/69	8"	2.5	32K, 128K	20^{m}1
Cas A	many pos'ns.	7,8,9/69	8"	2.5	64K, 128K	19^{m}8-20^{m}1
NGC 6888	many pos'ns.	6-7/69	15"	2.5	32K	18^{m}4
3c386 (1230?)	Griffin	7/69	15"	2.9	128K	0.8%
Androm. (1885)	Engelmann	9/69	4.6"	2.6	128K	~20^{m}0
3c58	many pos'ns.	12/69	15"	1.25	64K	18^{m}4
Other Objects						
BL Lac		7/69	8"	2.9	128K	0.7%
YT 110	19^{h}09^m,+16°45'	7/69	15"	2.9	64K	0.14%*
YT 119	19^{h}44^m,+25°04'	7/69	15"	2.9	64K	0.16%*
YT 120	19^{h}44^m,+25°13'	7/69	15"	2.9	64K	0.3%*
YT 127	19^{h}53^m,+29°10'	7/69	15"	2.9	64K	0.5%*
NGC 3556	Supernova	2/17/69	8"	1.25	16K	13.0%
3c273	QSO	3/69	15"	1.4	16K	1.6%
R Cor Bor		8/69	30"	2.5	64K	0.04%
T Cor Bor		8/69	15"	2.5	64K	0.1%*
Tau IR Source	Neugebauer et al	9/69	8"	2.6	128K	20^{m}2
S 22	many pos'ns.	9/69	15"	2.6	32K	18^{m}4

indicate that optical synchrotron radiation is possible for most pulsars. We have not said anything about *intensity*, though, and that depends upon factors like the rate of injection of particles into the system, a process that could depend strongly upon Ω.

In order to learn how faint a pulsar could be detected with transform techniques, we made optical observations of the Crab pulsar. The Fourier transform of two minutes of data clearly shows the fundamental frequency and seven harmonics, with amplitudes about 30 standard deviations above the noise. By transforming longer strings of data, and summing the amplitudes of harmonically related frequencies. the pulsar shows itself as peaks more than 100 standard deviations above noise, implying that a similar pulsar $3^{m}.5$ fainter would be detected with this technique. From such observations, as well as from computer experiments, we have established the sensitivity of the search techniques, and the estimated limiting magnitudes for each object examined is listed in the table. The range of frequencies searched goes from the inverse of the observation time up to half the sampling frequency, though seeing fluctuations frequently reduced the sensitivity to signals with periods greater than a few seconds. In most cases periods from 5 msec to about 10 seconds were within the range of the search.

The remainder of Table I is mostly self-explanatory. The central stars of planetary nebulae, nova remnants, white dwarfs, and some of the "other objects" are stellar, and were easy to examine. They were chosen particularly for unusual characteristics, e.g., lack of any spectral lines (the white dwarfs) or suspected variability (NGC 7662). In the planetary nebula NGC 7027, no central star can be seen, but searches with a 4.6″ diameter aperture near the brightest region failed to disclose any pulsation. The X-ray sources SCO-X1 and CYG-XR2 were sampled at 0.1 msec intervals, in addition to the usual 2.5 msec rate, in order not to overlook a very short-period pulsation, that has been suggested (Tucker, 1969).

The supernova remnants, certainly the most interesting candidates, posed a special problem of large angular size and absence of striking visible features (as in the Crab). Faint nebulosity has been discovered near the position of Kepler's and Tycho's supernovae, and both Tycho's supernova and Cas A show a characteristic circular structure, a few minutes in diameter, at radiofrequencies. Cas A also has some nebulosity, is an X-ray source, and is widely believed to be a supernova remnant. In these cases, as well as the suspected remnants NGC 6888, S 22, and 3C58, many stars in the vicinity were searched. An extragalactic supernova was observed at a time close to its peak brightness (in NGC 3556), as well as the position of the widely observed supernova event in the Andromeda galaxy in 1885. The YT objects are peculiar nebulosities suggested by Dr. Y. Terzian of Cornell. The list even includes a quasar (3C273) and an infrared star, the 'Taurus Source' (Neugebauer *et al.*, 1965). Three radiofrequency pulsars were observed with Dr. Frank Low (University of Arizona) at 5μ–20μ, using his helium-cooled germanium bolometer attached to the 28″ reflector at Mt. Lemmon.

With the single exception of the Crab pulsar, NP 0531, no optical pulsar-like fluctuations were observed in any of the objects searched. Since observations of the Crab pulsar confirmed that the technique was adequate to detect similar pulsars at least 3 magnitudes fainter, a number of conclusions are possible:

(1) Not all supernova events produce neutron stars.

(2) Not all neutron stars are pulsars.

(3) Not all pulsars produce optical pulses.

(4) Not all optically pulsing pulsars are visible from an arbitrary direction.

(5) Other optical pulsars are outside the regions of our search (in period, brightness, region of sky, etc.).

It is difficult to choose among these speculative propositions. It is worthwhile noting, however, that the uniqueness of the Crab Nebula among recent supernova remnants, particularly in its X-ray and visible activity, suggests that (1) or (5) may be true. The failure to observe any visible objects at the sites of radiofrequency pulsars, albeit old ones gives support to (3). And the polarization properties of the one visible pulsar (Kristian *et al.*, 1970) strongly hint at (4).

We would like to make two final remarks. One is that the Crab Nebula, when viewed at radiofrequencies, is brightest at the *center*, whereas other supernova remnants appear as hollow shell sources. This property, as well as some of the other remarkable features of the Crab (X-rays, optical synchrotron radiation, compact radiofrequency source, etc.), is presumably due to the continued excitation by the remnant pulsar, and suggest that these other supernova remnants have no associated pulsars. Of the objects studied, only 3C58 and 3C386 have similar radio structure to the Crab, but 3C58 has no visible counterpart on Palomar Sky Survey plates, and neither source has the other peculiar characteristics of the Crab. Here, too, we found no pulsar.

The other remark concerns Prentice's (1970) suggestion that most pulsars may be runaway stars, having motion relative to the original gaseous remnant. He assigns a nearby nebula to each of a number of radiofrequency pulsars, and suggests that only young pulsars might still be found within their nebular remnants (Crab, Vela). Perhaps the remnant star (if any) has already moved appreciably from the center of the nebulosity seen in the supernova remnants examined and was thereby overlooked.

As for the other objects looked at (nova remnants, planetaries, etc.), if one believes that they are what they seem to be, and that current descriptions are accurate, then there is little reason to expect them to pulsate at these frequencies (though white-dwarf radial pulsations might well be expected, and have been looked for previously (Lawrence *et al.*, 1967)). However, the techniques developed in the course of this search are capable of detecting pulsations many orders of magnitude faster than conventional astronomical photometry, and that alone seems to justify some general looking around at familiar objects.

References

Horowitz, P.: 1970, 'Optical Studies of Pulsars', Thesis, Harvard University.

Kristian, J. and Visvanathan, N., Westphal, J. A., and Snellen, G. H.: 1970, *Astrophys. J.* **162**, 475.

Lawrence, G. M., Ostriker, J. P., and Hesser, J. E.: 1967, *Astrophys. J.* **148**, L162.

Neugebauer, G., Martz, D. E., and Leighton, R. B.: 1965, *Astrophys. J.* **142**, 399.
Prentice, A. J. R.: 1970, *Nature* **225**, 438.
Tucker, W.: 1969, *Nature* **223**, 1250.

Discussion

J. P. Ostriker: It seems that in some cases your limits are high enough to allow a substantial (10%–50%) periodic component in the objects investigated.

P. Horowitz: All values given are upper limits, since no pulsations were seen, except of course in the Crab (which we used as a calibrating source). The limits vary since they are a function of observation time, aperture, brightness of the object etc.

J. A. Roberts: Because the pulse shapes clearly have a very large harmonic content I would not expect a Fourier transform method to be very sensitive method of searching for pulsations. A technique, such as superposed epoch, which does not spread the signal into many places in the transform plane, would seem more suitable.

P. Horowitz: I agree that the Fourier Transform is not best suited to searching for pulsations of short duty cycle, but if you add the amplitudes of harmonically related components then the sensitivity is only the fourth root of the duty cycle poorer than the optimum method, i.e. 'stacking' *modulo* the correct period. The Staelin Fast Folding Algorithm could be used to accomplish the latter, but the Fast Fourier Transform is so much faster, when searching a wide range of frequencies, that we preferred to use it instead; in all our searches we summed harmonically related frequencies and examined the resulting array.

3.11 MAGNETISM IN WHITE DWARFS

J. D. LANDSTREET*

Department of Astronomy, Columbia University, New York, N.Y., U.S.A.

and

J. R. P. ANGEL**

Columbia Astrophysics Laboratory, Columbia University, New York, N.Y., U.S.A.

Abstract. Searches have been made for the normal and quadratic Zeeman effect and broad-band circular polarization in white dwarf stars. A positive effect has been found in Grw + 70°8247 whose continuum shows both linear and circular polarization.

It has been realized for some time that in gravitational collapse the magnetic flux passing through the star may be conserved, resulting in very high fields in the collapsed object (Woltjer, 1964). The discovery of pulsars, which are generally thought to be neutron stars with fields of order 10^{12} G, shows that magnetic flux must be largely conserved in collapse to a neutron star. If no flux is lost, the field strength is increased as $1/R^2$, requiring initial fields before collapse of order 100 G. Such a value is not inconsistent with the fields observed in normal stars, ranging from about 1 G in the sun to 10^4 G in magnetic A stars.

In collapse to a white dwarf, the reduction in radius is typically a factor of 100, giving the possibility of field amplification of up to 10000. Thus fields of between 10^4 and 10^8 G could be expected, depending on the initial field. Various techniques exist to detect fields in this range. In the range from 10^4 to 10^6 G, the normal Zeeman effect is strong enough to be detectable, even though the spectral lines are very broad in dwarfs. We have searched for the Zeeman effect in the Hγ absorption line of a dozen DA dwarfs, and have been able to set an upper limit of between 10^4 and 10^5 G on their fields (Angel and Landstreet, 1970a). Above a few times 10^5 G, the quadratic Zeeman effect would cause a detectable shift of spectral lines, as pointed out by Preston (1970). He concludes from the lack of such shifts that few if any white dwarfs with hydrogen absorption lines can have surface fields in excess of 5×10^5 G. Above a few times 10^6 G the quadratic Zeeman effect is so strong that spectral lines can be smeared out in a non-uniform field, or be so shifted as not to be identified. In this domain magnetic fields can be detected through the circular polarization of continuum radiation (Kemp *et al.*, 1970a). It is this effect which has led to the recent discovery of a very strong field ($\sim 10^7$ G) in one white dwarf, Grw + 70°8247. So far this dwarf appears to be unique. A survey of other dwarfs without spectral lines has revealed no continuum circular polarization down to a level about two orders of magnitude less than that seen for Grw + 70°8247 (Angel and Landstreet, 1970b).

* Present address: Department of Astronomy, University of Western Ontario, London, Ontario, Canada.
** Alfred P. Sloan Research Fellow.

The initial observations of circular polarization of the continuum light of this star were made by Kemp at Oregon and confirmed by us at Kitt Peak, and are reported in the paper by Kemp *et al.* (1970b). Immediately after the discovery, we made extensive observations at the 107-in. telescope of McDonald Observatory to obtain the wavelength and time dependence of the effect, and discovered linear polarization. (Angel and Landstreet, 1970b) In brief, we find that the polarization is quite constant in time scale between a few seconds and a few days. The wavelength dependence is very striking: the component of circular polarization, which changes only slightly over the range of visible light, drops off very sharply in the ultraviolet, being less than 1% at 3300 Å. The linear polarization peaks in violet light and drops to zero in the red. Our detailed spectral measurements of both the linear and circular polarization below 7000 Å agree to within a few percent with the later measurements in broader spectral bands by Gehrels. This agreement is very gratifying as the two techniques are quite different. The polarization measurements have been extended to the near infrared (9200 Å) by Gehrels, where the linear polarization reappears with position angle

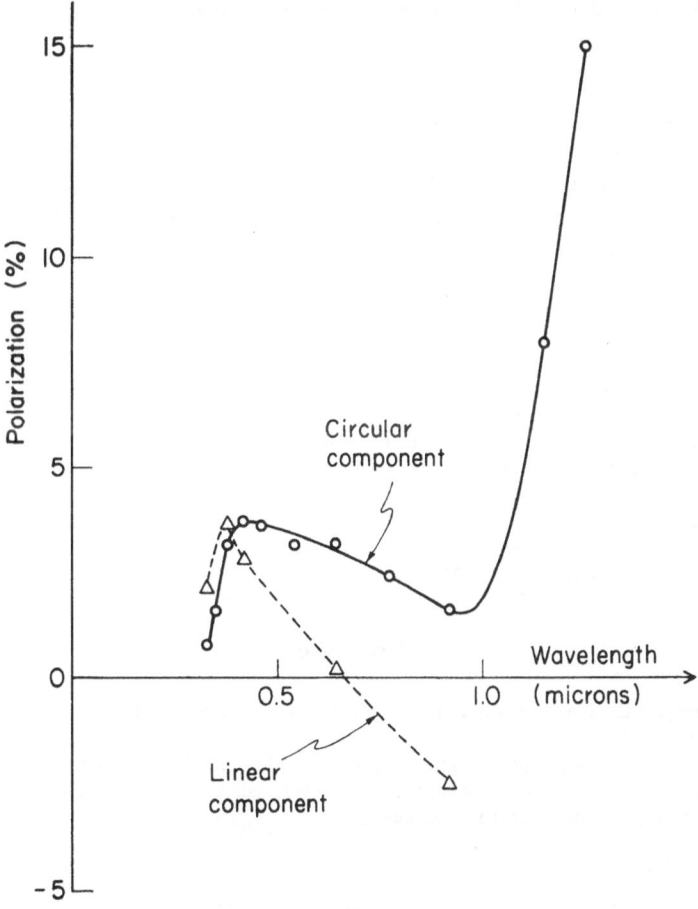

Fig. 1.   Circular and linear components of polarization of Grw + 70°8247.

rotated by about 90°, and to 1.2 μ by Kemp, where the circular polarization increases very steeply. All the data so far are compiled on Figure 1.

The complex wavelength dependence of the polarization is not yet understood, but there seems little doubt that a strong magnetic field is responsible for the effect. The phenomenon of elliptical polarization by preferential absorption in a magnetic field is well known, and has been investigated in laboratory measurements of solids (the Faraday and Kerr magneto-optic effects) and in a thin plasma (propagation through the ionosphere). Elliptical polarization of emitted thermal radiation is the inverse process required by detailed balance and has been demonstrated in the laboratory for emission from hot solids and flames (Kemp *et al.*, 1970a).

In the presence of a field, the opacity of the stellar atmosphere becomes polarization dependent, and in general will result in elliptical polarization. For the special case of propagation parallel to the magnetic field in an ionized medium, the absorption coefficients for the two senses of circular polarization are in the ratio $\{(\omega+\omega_B)/(\omega-\omega_B)\}^2$, where ω is the angular frequency of the propagated wave and ω_B is the cyclotron frequency eB/m (Ratcliffe, 1959). This difference in opacity will lead to a net circular polarization q of order ω_B/ω in the continuum emission.

Acknowledgements

This work was supported by the Research Corporation, the Air Force Office of Scientific Research under Grant AFOSR-70-1945 and Contract 49(638)1358, and the National Aeronautics and Space Administration under Grants NGR-33-008-102 and NGR-33-008-012. It is Columbia Astrophysics Laboratory Contribution No. 30.

References

Angel, J. R. P. and Landstreet, J.: 1970a, *Astrophys. J. Letters* **160**, L147.
Angel, J. R. P. and Landstreet, J.: 1970b, *Astrophys. J. Letters* **162**, L61.
Kemp, J. C., Swedlund, J. B., and Evans, B. D.: 1970a, *Phys. Rev. Letters* **24**, 1211.
Kemp, J. C., Swedlund, J. B., Landstreet, J., and Angel, J. R. P.: 1970b, *Astrophys. J. Letters* **161**, L77.
Preston, G. W.: 1970, *Astrophys. J. Letters* **160**, L143.
Ratcliffe, J. A.: 1959, *The Magneto-Ionic Theory*, Cambridge, England.
Woltjer, L.: 1964, *Astrophys. J.* **140**, 1309.

Discussion

J. Kristian: Doesn't the observed wavelength dependence of the polarisation drastically disagree with Kemp's theory?

R. Angel: Yes, but it should not be taken to give accurately the effect in a white dwarf, where there is radiative transfer. There are many arguments, however, which lead one to believe that V will be of order ω_c/ω, where ω_c is the cyclotron frequency.

3.12 OPTICAL PULSAR IN NGC 4254

P. A. FELDMAN

Queen's University, Ontario, Canada

Dr. Feldman described the optical variability which has been seen over the years in a region near the nucleus of NGC 4254 (M99) and he suggested that it might be a pulsar.

The following summary has been prepared by the Editors.

Lampland noted in 1921 that NGC 4254 showed signs of variability near its nucleus; these were secular changes in form and brightness. Walker made an investigation of Lampland's plate material for the period 1916 to 1948 and added his own observations in 1966–1967. He confirmed the existence of a variable object about 200 pc from the centre of NGC 4254. It had the characteristics of a blue stellar object with an absolute photographic magnitude of -14^m. Walker suggested that it might be a low luminosity quasar.

The present suggestion is that it may be a pulsar with about 10^7 times the luminosity of the Crab Nebula pulsar. If it is assumed that its magnetic field is about one tenth of Crab pulsar and that its efficiency in converting a dipole magnetic field into optical emission is 10 times more than for the Crab pulsar, then the pulsar in NGC 4254 would be easily detectable optically. However its radio flux would be 10^{-2} flux units and its X-ray emission would also be too weak to detect.

RELATION OF CRAB NEBULA TO OTHER SUPERNOVA REMNANTS

4.1 COMMENTS ON SUPERNOVA REMNANTS AND ANCIENT NOVAE

R. MINKOWSKI

Radio Astronomy Laboratory, University of California, Berkeley, Calif., U.S.A.

Abstract. The low expansion velocity of the Crab Nebula proves conclusively that the supernova of + 1054 was not type I. Only five supernova remnants found in radio surveys are connected with ancient novae, + 185, + 1006, + 1054, + 1572, + 1604. This corresponds to reasonable expectation. Most of the objects in historical records were comets, the remainder mostly ordinary novae. Most radio remnants are too old, some too heavily obscured to be found in historical records.

It seems unnecessary to recite here all the properties of the Crab Nebula which make it unique. It differs in every respect from all other supernova remnants. It is not surprising to find that the supernova of + 1054, whose remnant is the Crab Nebula, was not one of the two most frequent types of supernovae.

The supernova of + 1054 was considered for a long time as a supernova of type I. This assignment came about because it was not known that there is more than one type of supernova when Mayall and Oort (1942) showed that the available evidence was consistent with the interpretation that the nova of + 1054 was a supernova and that the Crab Nebula is its remnant. The information available at that time seemed to show that the supernova of + 1054 was similar to the supernova in IC 4182. This supernova later became the prototype of the supernova of type I, and this type was then assigned to the supernova of + 1054. This classification can no longer be maintained (Minkowski, 1966, 1968).

All we know about the lightcurve of the supernova of + 1054 is that it ceased to be visible in daytime at + 23 days, and was no longer visible at + 653 days. This indicates a decline of 8.5 mag., possibly less, between + 23 and + 653 days.

No other supernova has been observed for as long a period as that of + 1054. Kepler's nova was observed to + 356 days, Tycho's nova to + 457 days, the supernova in IC 4182 to + 635 days photographically, but visually only to + 100 days. Extrapolation leads to a decay between + 23 and + 653 days of 11.3 mag. for Tycho's nova, 11.0 mag. for Kepler's nova, and 10.6 mag. for the supernova in IC 4182 (on the assumption that the color did not change after + 100 days). The close agreement of these values demonstrates, of course, the well known similarity of the light curves which leads to the classification of Tycho's and Kepler's nova as supernovae of type I. There is a difference of 2.5 mag. between the decay of the supernova of + 1054 and of the supernovae of type I. In view of many uncertainties, this is not quite conclusive. It tends to contradict the classification of the supernova of + 1054 as type I, but does not rule it out entirely. A classification as type II can safely be ruled out; these supernovae show a much more rapid decline after the initial period.

The best argument against the classification of the supernova of + 1054 as type I is furnished by the low velocity of expansion of the Crab Nebula. Unfortunately the Crab Nebula is a very irregular object. This causes difficulties for precise determinations of the velocity of expansion and of the distance. Trimble (1968) estimates $+1450$ km sec^{-1} as the velocity of expansion in the centre of the nebula. If the roughly elliptical nebula is prolate, the velocity in the direction of the major axis is 2175 km sec^{-1} and the distance 2.0 kpc. These values are too large (Minkowski, 1970; Woltjer, 1970). A velocity of 1800 km sec^{-1} in the direction of the major axis and a distance of 1.7 ± 0.3 kpc are probably the best values. There is evidence that the expansion may be slightly accelerated, but no evidence that it might be decelerated. The absence of deceleration is not surprising because the mass of interstellar matter swept up by the nebula is not more than a few percent of the mass of the nebula. The present velocity of expansion cannot differ much from the initial value. This is quite different from the conditions in the remnant of a supernova of type I such as Tycho's nova where the average velocity – the ratio of diameter to age – is 13 000 km sec^{-1}. The accreted interstellar mass in this remnant may be larger than the original mass. There may have been strong deceleration. The initial velocity may have been much higher, perhaps of the order of 20 000 km sec^{-1} which would agree well with the appearance of the spectra of supernovae of type I near maximum. Supernovae of type II show velocities of the order 6000 km sec^{-1} by the widths of emission bands and in some cases by the presence of absorptions at the violet edge of the emission bands. The low velocity of expansion of the Crab Nebula shows conclusively that the supernova of $+1054$ was neither a supernova of type I nor an average supernova of type II. It must be one of the small fraction – about 10 per cent – of peculiar supernovae that Zwicky designates as type III, IV, and V. This conclusion agrees well with the unique properties of the Crab Nebula.

It cannot be assumed, however, that the Crab Nebula is so rare that it is the only object of its kind in the Galaxy. In particular, it seems unbelievable that there should be no observable objects of the same kind as the Crab Nebula but older. The search for such objects or sources is an interesting and important problem that can be approached in different ways.

A search for a nebula is not an efficient way to attack the problem. The probability is quite high that heavy obscuration may hide a nebula that is close to the galactic plane.

The search for a radio source is much more effective. A radio source identical to the Crab but at a distance of 10 kpc would be the 45th source in order of flux density in the 3CR catalogue, with an apparent diameter less than 1 arc min and the unusual spectral index -0.26. No such source is in the 3CR catalogue. At 20 kpc it would still be above the limit of the 3CR catalogue. The angular diameter would be less than 30 arc seconds. Such a source might be considered as extragalactic, but the spectral index would draw attention to it. No source closely similar to the Crab Nebula is in the 3CR catalogue.

If we search for older analogues to the Crab, we must ask how such objects are likely to look. The Crab Nebula is not a shell source like the 4 other sources that are remnants of known supernovae – +185; +1006; +1572; +1604 – and Cas A, a remnant of a relatively recent, but unobserved supernova. If the absence of the shell structure in the Crab is connected with the fact that the accreted mass and the deceleration are small, the Crab might transform itself into a shell source in the future, say during the next 10 000 years. It seems likely that the shape and structure of very old remnants depends more on the conditions in the ambient interstellar medium than on conditions of the original explosion. It might be very difficult to recognize very old analogues of the Crab. But one would expect that the transformation is slow and that there are sources older than the Crab which can still be recognized as similar. 3C58 is probably a source of this kind. It is not a shell source; its brightness distribution has a central peak. There is polarization, high in some areas (Weiler and Seielstad, 1970). The spectrum is not well determined; it might not follow a simple power law. Above 700 MHz the spectral index has a small negative value. All this shows great similarity to the Crab nebula. The angular diameters are similar, but around 1000 MHz the flux density of 3C58 is about 25 times lower. This is what one would expect for a source which is older and more distant by factors of the order two, but otherwise similar to the Crab nebula. Other such sources should be observable. MSH 15−56 might be an example. The investigation of such sources deserves more attention.

A third approach to the problem of finding analogues to the Crab is to identify remnants with ancient objects reported in the Annals of China, Japan, Korea and other historical records. This approach suffers from the same restriction as the optical search for nebulae: if the interstellar obscuration is heavy, neither the nebula nor its parent supernova can be seen. But many remnants are optically nebulosities of very low surface brightness and unobservable for this reason. In such cases the identification with an ancient supernova can confirm that a radio source is a supernova remnant and establishes its age. Five objects have been identified with ancient supernovae of +185, +1006, +1054, +1572 and +1604. They are all listed in the Annals. For the supernova of +1006 Arabic and European sources (Goldstein, 1965) add much evidence to that to be found in Oriental records (Goldstein and Ho, 1965). The supernova of +1572 is Tycho's, that of 1604 Kepler's. Korean sources give information on the light curve of Kepler's nova (Xi and Bo, 1965; Chu, 1968). The type of the supernova of +185 is doubtful. The supernova of +1006 was probably type I, Tycho's and Kepler's novae undoubtedly type I (Minkowski, 1968).

Compilations of the ancient data have been given by Hsi (1955), Ho (1962) and Xi and Bo (1965). Ho's listing is most complete and least influenced by suggested identifications of ancient objects with radio sources. Korean information has been collected recently by Chu (1968). These authors also give some information that is vital if misinterpretations are to be avoided. I will briefly summarize some important points and add some remarks on points that have been overlooked. To make a valid identification, it is not sufficient to pick out of one of these lists an object that might be in the proper position and that at a glance seems to be a supernova.

Chinese medieval astronomers recognized 21 different kinds of 'ominous' stars. The three classes which are most frequent and of most interest here are comets with tails ('hui'), comets without tails ('po'), and "guest stars" which include novae and super-novae. These designations, however, do not always define the true nature of the object. Sometimes different authors seem to quote different texts of the same record. Ho, for instance, quotes a record which designates Tycho's nova as 'hui'; other records quoted by Hsi, by Ho, and by Xi and Bo, call it a 'guest star'. A small comet without tail might easily be mistaken for a guest star. If a "guest star" is described as moving, it must have been a comet. Xi and Bo, however, give a record, which (at least in the NASA translation) states that Kepler's nova 'wandered about'. This might refer to no more than that the object was in the southwest in the beginning and in the south-east after the conjunction with the sun, as clearly stated in another record quoted by Hsi and by Xi and Bo. But if there would be no other evidence, this object might be rejected as a nova. If changes of position with time are clearly stated there can be no doubt that the object was a comet, no matter what its designation is. There are very rarely useful data on the brightness. Statements on color may sometimes refer to the true color, but colors have astrological significance (Needham, 1959). Positions are not very accurate, sometimes unreliable. Remember, for instance, that the position for the supernova of $+1054$ is given as southeast of ζ Tau, but the Crab Nebula is northwest of that star; one record puts it into the Pleiades (Duyvendak, 1942). One important point is that the Chinese astronomers used equatorial coordinates exclu-sively (Needham, 1959). Thus, an object described as having been below some star, was south of it, not at a lower altitude, a statement that would be useless unless the time of observation was stated.

Sometimes the period of visibility is stated. This is valuable information, but to use it properly we must know the brightness at which novae were discovered. This can be estimated roughly in the following way. Ho's catalog shows about 5 objects per century that were not clearly recognizable as comets. That is the same rate at which ordinary novae brighter than 0.5 mag. have occurred between 1900 and 1950. If the discovery brightness for ancient novae were much fainter than 0.5 mag., all ancient novae should be ordinary novae! If the discovery magnitude is about $+0.5$ mag., a supernova discovered at maximum must remain visible for at least about 200 days if it was type I, about 80 days or more if it was type II. Objects that were visible for much shorter periods cannot have been supernovae. Some may have been comets, but some must have been ordinary novae. It should not be forgotten that some ordinary novae have remained visible for more than 200 days – for instance N Aqu 1918 which remained visible for 260 days. An example of an ordinary nova might be the object of $+1431$, 4 January, which disappeared after 15 days and reappeared 100 days later. This is reminiscent of a slow nova like DQ Her.

One example of a suggested identification that seems to be invalid is that of the radio source CTA1 with the object of $+902$. If the statement on the position 'beneath' Hua Kai (Cassiopeia) is correctly interpreted as 'south', the position $\alpha = 1^h30$, $\delta + 65°$ (1950), does not at all agree well with that of CTA 1, $\alpha = 0^h02.6$, $\delta + 72°20'$ (1950).

Changes of position seem to be clearly indicated. The object of $+902$ was most likely a comet.

Another example is the suggested identification of the source 3C386 with the object of $+1230$ that has recently been discussed by Mackay (1970). The object of $+1230$ was undoubtedly a comet (Ho, 1962). The optical evidence leaves little doubt that the object is extragalactic. The observed red-shift $z = 0.0001$ is very small, but it must be corrected for the solar motion in the Galaxy to see how it fits into the red-shift-magnitude relation. The corrected value $z = 0.0008$ (Schmidt, 1965) is still too small to be a safe distance indicator. If we take it literally, it gives a distance of 2.4 Mpc (with $H_0 = 100$ km sec^{-1} Mpc^{-1}). If the interstellar absorption is 3 magnitudes, the absolute magnitude is $M_p = -14.9$. The total power between 10^7 and 10^{11} Hz becomes $10^{38.8}$ erg sec^{-1}. The object seems to be a dwarf galaxy with very low emitted power. We do not know much about such objects; an object of this kind in the Virgo cluster would have a flux density of 0.63 fu at 400 MHz.

We are thus left with only 5 ancient objects that are identified as supernovae. Katgert and Oort (1967) have shown that this number is consistent with the estimated frequency of one supernova per 25 years in the Galaxy. Virtually the same frequency has recently been found by Tammann (1970). It is thus not to be expected that many additional supernovae are to be found in the ancient records. A review of Ho's catalogue shows, however, about 90 ancient objects for which there is no indication that they might have been comets. What were these objects? Some may actually have been tail-less comets. But, as I have pointed out, a large fraction, possibly the great majority, may have been and probably were bright ordinary novae. Another question calls for an answer: there are now about 90 nonthermal sources known that are believed to be supernova remnants; why are only 5 of them identified as remnants of historical novae? The answer is simple. Many of these sources are in highly obscured regions where interstellar absorption would blot out any supernova, and many of these sources are older than about 3000 years so that the parent supernova cannot be in the ancient records.

The determination of the age of a remnant is a complex problem. Needed are: (1) a model for the expansion, (2) parameters which depend on the location of the remnant in the Galaxy, distance and ambient interstellar density, (3) parameters which depend on the type of the supernova, initial energy of the explosion or initial velocity of expansion and ejected mass. None of the parameters can be determined with great accuracy. Relevant observational data are angular size, and flux density.

Shklovsky (1962) has suggested as a model the similarity solution for a strong explosion in a gas of constant heat capacity, given numerically by Taylor (1950) and in analytical form by Sedov (1959). Radiation losses in the gas behind the shock front are assumed to be negligible. This assumption is probably fulfilled for all observed remnants. The discussion of supernova remnants by Milne (1970) shows that the distribution of linear diameters has the form to be expected on the basis of Sedov's solution whose use thus has now observational support.

For the age of a source we then have

$$t = 8.3 \cdot 10^{26} \, n_{\mathrm{H}}^{1/2} E^{-1/2} r^{5/2} \tag{1}$$

or

$$t = 3.9 \cdot 10^5 \, rv(t). \tag{2}$$

t is the age in years, r the radius in pc, n_{H} the number of interstellar H atoms per cm^{-3}, $n_{\mathrm{H}}/n_{\mathrm{He}} = 7$, and E the initial energy of the explosion in erg. $v(t)$ is the present velocity of expansion in km sec^{-1}.

Equation (2) looks deceptively simple, but it requires data which are not available except for one source, the Cygnus Loop (Minkowski, 1958). With the radius of 20 pc and the present velocity of expansion of 116 km sec^{-1} the age is 69000 years. Very similar to the Cygnus Loop is IC 443. For this object the available information is poorer; an assumed distance of 1.50 kpc leads to a radius of 10 pc. With the velocity of expansion of 65 km sec^{-1} (Lozinskaya, 1968), the age is 60000 years. Both objects are far too old to be remnants of supernovae to be found in any historical records.

The use of Equation (1) requires knowledge of E and n_{H}. The initial energy E must depend on the type of the supernova. Unless the distance is known, n_{H} cannot be found from the galactic distribution of H. There seem to be only three remnants with known ages and distances for which $E \cdot n_{\mathrm{H}}^{-1}$ can be determined with the aid of Equation (1). For Tycho's nova, a supernova of type I, En_{H}^{-1} is $1 \cdot 10^{51}$. For the Cygnus Loop En_{H}^{-1} is $5 \cdot 10^{50}$, for IC 443 it is $2 \cdot 10^{49}$. Both are believed to be remnants of supernovae of type II. For the radio remnants the type of supernova is not known. The distance can be found from the relation between surface brightness and linear diameter (Milne, 1970); such distances have uncertainties of about a factor 2. Individual ages from Equation (1) may be uncertain by a factor 10 and are not accurate enough to decide whether the age is less than 3000 years, the maximum age for objects in the historical records.

To obtain an estimate of the number of remnants in Milne's list that are younger than 3000 years, we assume $En_{\mathrm{H}}^{-1} = 5 \cdot 10^{50}$ as an average value that takes roughly into account that supernovae II are twice as frequent as supernovae I (Tammann, 1970). Equation (1) then permits us to compute the linear radius corresponding to an age of 3000 years. This gives a linear radius of 6 pc as the maximum radius for a source whose parent supernova might be found in the historical records. Since the historical records end about in +1600, sources must be excluded that are younger than 400 years and, according to Equation (1), smaller than 2.5 pc. Milne lists 25 sources with radii between 2.5 and 6 pc. Of these 10 are heavily obscured; their parent supernovae cannot have been seen. 12 are at declinations south of $-50°$ where the chance of the discovery of a supernova is small. This leaves three, plus a small number of objects with far southern declinations, to be compared with the five supernovae of +185, +1006, +1054, +1572, +1604 that are actually represented in Milne's list. It seems justified to conclude that very few, and perhaps no remnants are left that could be identified with old supernovae. Almost three quarters of the remnants are too old, the rest either too heavily obscured or too far south to be listed in the ancient records.

References

Chu, Sun-Il: 1968, *J. Korean. Astron. Soc.* **1**, 29.

Duyvendak, J. J. L.: 1942, *Publ. Astron. Soc. Pacific* **54**, 91.

Gardner, F. F. and Milne, D. K.: 1965, *Astron. J.* **70**, 754.

Goldstein, B. R.: 1965, *Astron. J.* **70**, 105.

Goldstein, B. R. and Ho Peng-Yoke: 1965, *Astron. J.* **70**, 748.

Ho Peng-Yoke: 1962, in *Vistas in Astronomy* **5** (ed. by A. Beer), Pergamon Press, London, Oxford, New York, Paris p. 127.

Hsi Tsê-Tsung: 1955, *Acta Astron. Sin.* **3**, 183 (1958, *Smithsonian Contr. Astrophys.*, **2**, 109).

Katgert, P. and Oort, J. H.: 1967, *Bull. Astron. Inst. Neth.* **19**, 239.

Lozinskaya, T. A.: 1968, *Astron. Zh.*, **46**, 245 (*Soviet Astron.* **13**, 192).

Mackay, C. D.: 1970, *Astrophys. Letters* **5**, 132.

Mayall, N. U. and Oort, J. H.: 1942, *Publ. Astron. Soc. Pacific* **54**, 95.

Milne, D. K.: 1970, *Australian J. Phys.* **23**, 425.

Minkowski, R.: 1966, *Astron. J.* **71**, 371.

Minkowski, R.: 1968, in *Nebulae and Interstellar Matter* (ed. by B. M. Middlehurst and L. H. Aller), University of Chicago Press, Chicago and London, p. 623.

Minkowski, R.: 1970, *Publ. Astron. Soc. Pacific* **82**, 470.

Needham, J.: 1959, *Science and Civilisation in China*, Vol. 3, The University Press, Cambridge, Section 20, f.1.

Schmidt, M.: 1965, *Astrophys. J.* **141**, 1.

Sedov, L. I.: 1959, *Similarity and Dimensional Methods in Mechanics* (transl. by M. Friedman), Academic Press, New York.

Shklovsky, I. S.: 1962, *Astron. Zh.* **39**, 206. (*Soviet Astron.* **6**, 162).

Tammann, G. A.: 1970, *Astron. Astrophys.* **8**, 458.

Taylor, G. I.: 1950, *Proc. Roy. Soc. London* **A101**, 159.

Trimble, V.: 1968, *Astron. J.* **73**, 535.

Weiler, K. W. and Seielstad, G. A.: 1970, *Bull. Am. Astron. Soc.* **2**, 224; 1971, *Astrophys. J.* **163**, 455.

Woltjer, L.: 1970, in 'Crab Nebula Symposium, June 18–21, 1969', *Publ. Astron. Soc. Pacific* **82**, 479.

Xi Ze-Zong and Bo Shu-ren: 1965, *Acta Astron. Sinica* **13**, 1 (NASA Tech. Trans. TT F-388).

Discussion

J. P. Ostriker: I would like to note that a recent analysis of pulsar observations gave a birth rate of about one per 35 years in the Galaxy, which within the errors agrees with the rate that you, Shklovsky, and Oort and Katgert derive for the occurrence of supernovae.

R. Minkowski: The frequency of pulsars that you suggest is indeed in complete agreement with the frequency of supernovae. One might wonder why there is not a pulsar in every supernova remnant. This may not be difficult to understand. First, beaming might make many pulsars unobservable. Second, the average distance of the known supernova remnants are larger, perhaps by a factor 5, than the distances of the known pulsars. Thus many pulsars in known supernova remnants might be too far away and too faint to be observable with the present means.

4.2 RADIO EMISSION FROM SUPERNOVA REMNANTS*

D. K. MILNE

Division of Radiophysics, CSIRO, Sydney, Australia

Abstract. Observations of the radio emission from supernova remnants are reviewed with emphasis on the dissimilarity between the Crab Nebula and the other remnants. From this we conclude that there may be several non-thermal sources in the Galaxy with the same centrally filled structure as the Crab. These are, however, more evolved, and clearly there is no other source of the same age and type as the Crab Nebula.

1. Introduction

The identification of the radio source Taurus A, the Crab Nebula, with the supernova of 1054 AD, is undoubtedly correct. However, as many have remarked, Tau A bears little resemblance to the other discrete non-thermal galactic radio sources also generally supposed to be the remnants of supernovae. The extremely high radio brightness of Tau A leaves little chance that another object of this type (and evolutionary stage) remains undiscovered in the Galaxy, but there is still the possibility that among the known supernova remnants (SNRs) there exist old, well-evolved, objects of the Tau A type. In this paper we review the results that have been obtained from radio observations of SNRs with this possibility in mind.

2. Basic Observable Quantities: Flux Density, Spectrum, Size

There are now more than 100 SNRs known in the Galaxy. A catalogue listing over 90 of the brightest objects was published recently (Milne, 1970a) and a dozen additional objects of low brightness have since been found in 408 and 5000 MHz surveys (Shaver and Goss, 1970). A complete catalogue of these objects with their radio parameters, angular size, 1 GHz flux density, spectral index and surface brightness is given in Table I. This is a revised version of Milne's catalogue with the Shaver and Goss objects added. A discussion of the possible evolutionary effects shown by the sources in the original catalogue has been given by Milne (1970a) and the general conclusions are not altered by the revisions or the inclusion of the additional SNRs. A brief account of this work is given here.

Firstly, the spectral index of these objects has an average value of -0.48 ± 0.1. There does not appear to be any relationship between spectral index and surface brightness or diameter (contrary to Harris' (1962) findings). There appears to be a relationship between surface brightness, Σ, and linear diameter, D. This relationship, derived from 15 SNRs with known distances and for an average type SNR, is

$$\Sigma = 9.52 \times 10^{-15} D_{(PC)}^{-4.54}\ \text{W m}^{-2}\ \text{Hz}^{-1}\ \text{sr}^{-1}, \qquad (1)$$

with the further possibility that for type I SNRs the surface brightness is lower than

---

* This paper was presented by Dr. V. Radhakrishnan.

Davies and Smith (eds.), The Crab Nebula, 248–262. All Rights Reserved.
Copyright © 1971 by the IAU.

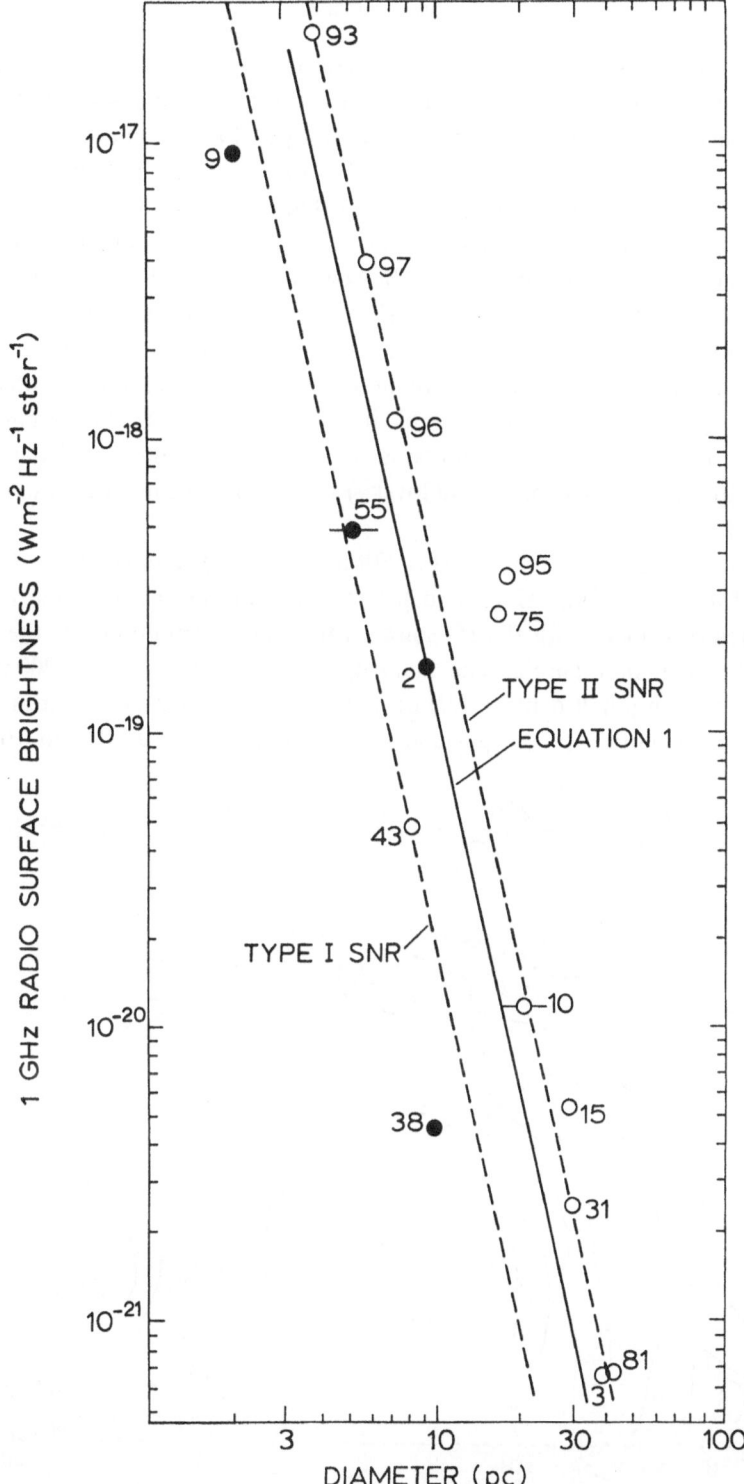

Fig. 1. The relationship between 1 GHz surface brightness and linear diameter for 15 SNRs whose distances are known. The filled circles represent those objects believed to be type I SNR. The numbers against each are the catalogue numbers of Table I. Possible evolutionary tracks for each type of SNR and the mean track (Equation (1)) are indicated in this figure.

for the type II objects of the same diameter. This relationship is displayed in Figure 1. The decrease in surface brightness with increase in diameter and with SNR type (and hence initial energy) is consistent with Shklovsky's (1960) evolutionary theory, although the power of D in Equation (1) (-4.54) is not as high as was predicted by Shklovsky (-6.0). Kesteven (1968) points out that if the emitting region were a shell expanding at constant thickness then this value of -4.5 would be correct. This assumption of constant shell thickness is however contrary to the observations quoted in Section 3, and an alternative model satisfying Equation (1) should be sought. Van der Laan's (1962a, b) shell models, whilst accounting for the structure and polarization in SNRs, are not able to explain the high-surface-brightness objects (e.g. Cassiopeia A) or the evolutionary track in the Σ–D plane. It does seem that the observed evolution of the radio emission supports a degradation of the magnetic field and particle energy density (Shklovsky) rather than an intensification (van der Laan).

Using the average Σ–D relation (Equation (1)) Milne computed the linear diameters and distances of the SNRs and showed that the galactic distribution has certain features coincident with the H I spiral arms. This distribution, for the SNRs in Table I, is shown in Figure 2. The majority of SNRs are within ± 200 pc of the galactic plane with a half-density thickness of 80 pc, a population I distribution. The total SNR contribution to the galactic radio power (from 10 MHz to 10 GHz) is

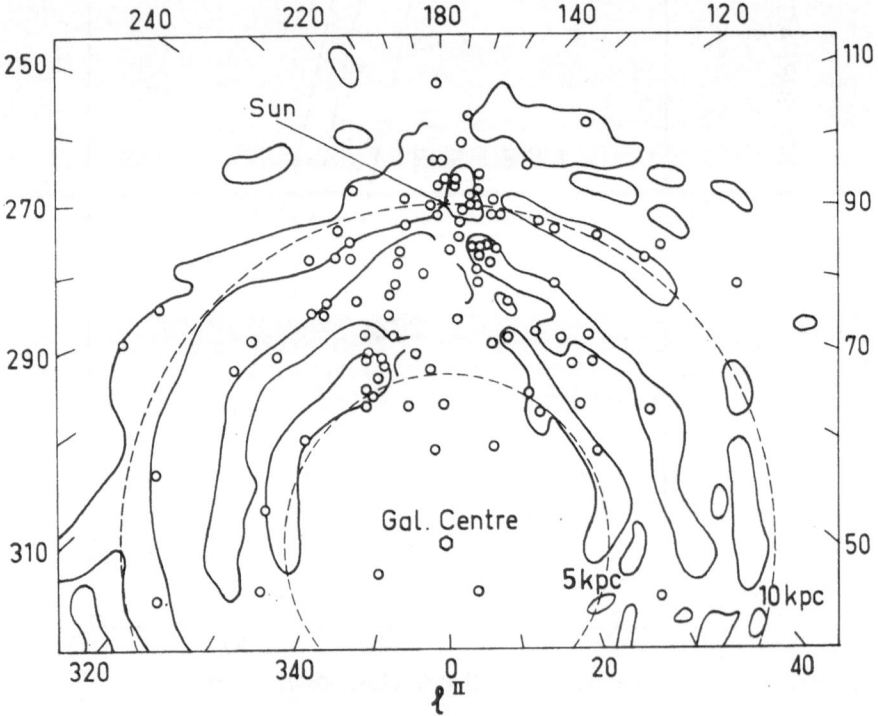

Fig. 2.   Galactic distribution of SNRs derived from the distances in Table I. The outlined regions indicate the distribution of neutral hydrogen.

TABLE I

Radio data for supernova remnants

(1) Catalogue No.	(2) Galactic source number	(3) Angular size $\phi_1 \times \phi_2$ (min arc)	(4) Flux density at 1 GHz S_0 (fu)[a]	(5) Spectral index at 1 GHz α	(6) Surface brightness at 1 GHz Σ unit[b]	(7) Diameter D (pc)	(8) Distance d (kpc)	(9) Ref.[c]	(10) Remarks
1	G119.5 + 10.0	125′ dia	38	−0.2	2.89E − 22	37.3	1.0		CTA 1
2	G120.1 + 1.4	6.0 × 7.0	58	−0.74	1.64E − 19	9.3	4.9	1	Tycho's Nova
3	G132.4 + 2.2	80′ dia	36	−0.7	6.70E − 22	31.1	1.4		HB 3
4	G156.4 − 1.2	300 × 120	225	−0.5	7.43E − 22	30.3	0.6		CTB 13
5	G160.5 + 2.8	120 × 140	150	−0.35	1.06E − 21	28.1	0.8		HB 9
6	G166.2 + 4.3	45′ dia	6.6	(−0.5)	3.88E − 22	35.0	2.7		VRO 42.05.01
7	G166.3 + 2.4	75′ dia	10.0	−0.5	2.12E − 22	40.0	1.9		OA 184
8	G180.0 − 1.7	180′ dia	120	−0.3?	4.41E − 22	34.0	0.7		S147
9	G184.6 − 5.8	3.0 × 4.2	1000	−0.25	9.45E − 18	3.8	1.7		Crab Neb., Tau A
10	G189.1 + 2.9	40′ dia	160	−0.45	1.19E − 20	16.5	1.4		IC 443
11	G193.3 − 1.5	80′ dia	27	−0.5	5.03E − 22	33.1	1.4		0607 + 17
12	G205.5 + 0.2	210′ dia	150	−0.5	4.05E − 22	34.7	0.6		Monoceros Neb.
13	G260.4 − 3.4	55′ dia	145	−0.5	5.70E − 21	19.4	1.2		Puppis A
14	G261.9 + 5.5	35′ dia	10.0	−0.38	9.70E − 22	28.6	2.9		0902 − 38
15	G263.4 − 3.0	220 × 180	1800	−0.30	5.40E − 21	19.6	0.4		Vela X, Y and Z
16	G284.2 − 1.8	23 × 12	25	−0.46	1.05E − 20	16.9	3.5		MSH 10 − 53
17	G289.1 − 0.4	3.0 × 3.2	20	−0.65	2.48E − 19	8.5	9.4		
19	G290.1 − 0.8	13 × 12	80	−0.5	6.10E − 20	11.5	3.2	2	MSH 11 − 61A
20	G291.0 − 0.1	10′ dia	25	−0.57	2.97E − 20	13.5	4.6	2	MSH 11 − 62
21	G292.0 + 1.8	2.8 × 2.7	15	−0.36	2.37E − 19	8.5	10.8		MSH 11 − 54
22	G295.1 − 0.6	13 × 14	29	−0.6	1.87E − 20	14.9	3.8		Kes 16
23	G296.3 + 10.0	86 × 75	49	−0.52	9.05E − 22	29.1	1.3		1209 − 51/52
23a	G298.5 − 0.2	27′ dia	10	−0.33	1.63E − 21	25.5	3.3	2	
24	G304.6 + 0.1	5.2 × 6.1	15	−0.50	5.62E − 20	11.7	7.2		Kes 17
26	G307.6 − 0.3	3′.7 dia	24	−0.45	2.09E − 19	8.8	8.2		
27	G309.6 + 1.7	6′.3 dia	47	−0.60	1.41E − 19	9.6	5.3		13S6A B }
28	G309.7 + 1.8	7.1 × 5.4	83	−0.60	2.58E − 19	8.4	4.7		13S6A A } (3.6)[d]

Table 1 (continued)

(1) Catalogue No.	(2) Galactic source number	(3) Angular size $\phi_1 \times \phi_2$ (min arc)[a]	(4) Flux density at 1 GHz S_0 (fu)[a]	(5) Spectral index at 1 GHz α	(6) Surface brightness at 1 GHz Σ unit[b]	(7) Diameter D (pc)	(8) Distance d (kpc)	(9) Ref.[c]	(10) Remarks
29	G310.6−0.3	7'.5 dia	9	−0.6	1.91E−20	14.9	6.9		Kes 20B ⎱ (3.3)[d]
30	G310.8−0.4	8 × 11	20	−0.35	2.71E−20	13.8	5.1		Kes 20A ⎰
30a	G311.5−0.3	3'.9 dia	3.7	−0.49	2.89D−20	13.6	12.0	2	
31	G315.4−2.3	40' dia	33	−0.5	2.46E−21	23.3	2.0		MSH 14−63
32	G316.3+0.0	13' dia	24.3	−0.47	1.71E−10	15.2	4.0	2	MSH 14−57
33	G320.4−1.2	26' dia	58.4	−0.53	1.03E−20	17.0	2.3	2	MSH 15−52
35	G322.3−1.2	2'.6. dia	4.2	−0.7	7.56E−20	11.0	14.6		Kes 24
36	G326.2−1.7	11 × 8.6	145	−0.24	1.84E−19	9.0	3.2		MSH 15−56
37	G327.4+0.4	10' dia	26	−0.78	3.10E−20	13.4	4.6		Kes 27
38	G327.6+14.5	30 × 22	25	−0.63	4.52E−21	20.4	2.8		1459−41
38a	G328.0+0.3	5'.9 dia	2.4	−0.55	8.20E−21	17.9	10.5	2	MSH 15−57
39	G328.4+0.2	4'.2 dia	17.5	−0.5	1.18E−19	10.0	8.2	2	Lupus Loop
40	G330.0+15.0	270' dia	340	−0.3	5.56E−22	32.4	0.4	2	
41	G332.0+0.1	12'.0 dia	9.1	−0.48	7.52E−21	18.2	5.2	2	MSH 16−51
42	G332.5+0.1	15' dia	26	−0.47	1.39E−20	15.9	3.7	2	1613−50
43	G332.4−0.4	9' dia	24	−0.43	3.53E−20	13.0	5.0	2	
43a	G333.0+0.3	2'.6 dia	8.3	−0.17	1.46E−19	9.5	12.6	2	
44	G336.7+0.5	11' dia	6.3	−0.49	6.20E−21	19.0	6.0	2	
45	G337.0−0.1	14' dia	17	−0.5	1.04E−20	17.0	4.2	2	Part of CTB 33
46	G337.3+1.0	11' dia	15	−0.2	1.48E−20	15.7	5.0	2	Kes 40
47	G337.8−0.1	6'.5 dia	17.7	−0.5	4.99E−20	12.0	6.4	2	Kes 41
47a	G338.1+0.4	12'.0 dia	4.5	−0.42	3.72E−21	21.3	6.1	2	
47b	G338.3−0.0	8'.9 dia	15.1	−0.65	2.27E−20	14.3	5.5	2	
47c	G338.5+0.1	8'.2 dia	28.3	−0.30	5.01E−20	12.0	5.1	2	
48	G342.1+0.1	30' dia	54	−0.5	7.15E−21	18.4	2.1	2	MSH 16−48
49	G348.5+0.1	8.1 × 5.3	84	−0.5	2.33E−19	18.6	4.5	2	CTB 37 A ⎱ (2.4)[d]
50	G348.7+0.3	3.8 × 6.0	47	−0.5	2.44E−19	8.5	6.1	2	CTB 37 B ⎰
51	G349.7+0.2	1.9 × 2.6	23	−0.6	5.54E−19	7.1	11.0		

Table 1 (continued)

(1) Catalogue No.	(2) Galactic source number	(3) Angular size $\phi_1 \times \phi_2$ (min arc)	(4) Flux density at 1 GHz S_0 (fu)[a]	(5) Spectral index at 1 GHz α	(6) Surface brightness at 1 GHz Σ unit[b]	(7) Diameter D (pc)	(8) Distance d (kpc)	(9) Ref.[a]	(10) Remarks
52	G355.3+0.1	7.5 × 9.5	30	−0.6	5.01E − 20	12.0	4.9		NGC 6383
53	G357.7−0.1	4.4 × 3.4	38	−0.6	3.02E − 19	8.1	7.2		MSH 17 − 39
54	G359.4−0.1	5.7 × 6.7	29	−0.4	9.05E − 20	10.6	5.9		
55	G4.5+6.8	2′.2 dia	20.0	−0.58	4.92E − 19	7.3	11.4		Kepler's Nova
56	G.53−1.1	12 × 18	38	−0.3	2.06E − 20	14.6	3.4		A4
57	G6.5−0.1	45′ dia	300	−0.40	3.97E − 20	15.1	1.3	4	A1, W28
58	G11.2−0.4	4′.6 dia	22	−0.52	1.27E − 19	9.8	7.3	2	
59	G18.9+0.3	7.0 × 14	35	−0.57	4.25E − 10	12.5	4.4		
60	G21.8−0.5	20′ dia	69	−0.52	2.05E − 20	14.6	2.5	2	MSM 18 − 18
61	G23.1+0.0	45 × 60	350	−0.22	1.54E − 20	15.6	1.0		MSH 18 − 113
61a	G23.6+0.3	9′.8 dia	7.6	−0.34	9.42E − 21	17.4	6.1	2	W41
61b	G24.5+0.2	7′.2 dia	10.0	−0.22	2.30E − 20	14.3	6.8	2	
61c	G24.7+0.6	14′.0 dia	13.4	−0.38	8.14E − 21	17.9	4.4	2	
62	G27.3+0.0	32′ dia	41	−0.4	4.77E − 21	20.2	2.2		
63	G29.7−0.2	2′.1 dia	11.1	−0.6	3.08E − 19	8.1	13.2	2	4C − 03.70
64	G31.9+0.0	3′.5 dia	22	−0.50	2.13E − 19	8.7	8.6		3C 391
65	G32.0−4.9	60′ dia	19	−0.45	6.28E − 22	31.5	1.8		3C 396.1
66	G33.0+0.1	31 × 10	25	−0.4?	9.16E − 21	17.3	3.4		
67	G33.7+0.0	7.3 × 5.6	12	−0.5	3.49E − 20	13.0	7.1		4C + 00.7
68	G34.6−0.5	28 × 35	230	−0.40	2.79E − 20	13.7	1.5		W44
69	G35.6−0.4	10′ dia	25	−0.3	2.98E − 20	13.5	4.7		
70	G37.6−0.1	30′ dia	48	−0.6	6.35E − 21	18.9	2.2		W47
71	G39.2−0.3	6′.7 dia	21	−0.44	5.54E − 20	11.8	6.1	2	NRAO 593
72	G39.7−2.0	50′ dia	48	−0.7	2.29E − 21	23.7	1.6		W50
73	G41.1−0.3	10′ × 8′	17	−0.3	2.53E − 20	14.0	5.3		3C 397
74	G41.9−4.1	165 × 140	150	−0.5?	7.74E − 21	30.0	0.7		CTB 72
75	G43.3−0.2	4′.8 dia	38.7	−0.47	2.00E − 19	8.9	6.4	2	W49B
76	G45.5+0.1	3.9 × 2.9	23	−0.4	2.42E − 19	8.5	8.7	2	Kuzmin 47

Table 1 (*continued*)

(1) Catalogue No.	(2) Galactic source number	(3) Angular size $\phi_1 \times \phi_2$ (min arc)	(4) Flux density at 1 GHz S_0 (fu)[a]	(5) Spectral index at 1 GHz α	(6) Surface brightness at 1 GHz Σ unit[b]	(7) Diameter D (pc)	(8) Distance d (kpc)	(9) Ref.[c]	(10) Remarks
77	G46.8 − 0.3	12′ × 5′	18	− 0.5	3.57E − 20	12.9	5.8		CTB 63
78	G47.6 + 6.1	60′ dia	12	(− 0.5)	3.97E − 22	34.8	2.0		
79	G49.2 − 0.5	26′ dia	160	− 0.25	2.82E − 20	13.6	1.8	1	Part of W51
80	G53.7 − 2.2	20′ dia	8.5	− 0.6	2.53E − 21	23.2	4.0		3C 400.2
81	G74.0 − 8.6	200 × 160	180	− 0.45	6.70E − 22	31.1	0.6		Cygnus Loop
82	G74.8 + 0.6	4.7 × 3.9	(10)	(− 0.15)	6.50E − 20	11.4	9.2		
83	G74.9 + 1.2	9.3 × 4.7	(18)	(− 0.5)	4.89E − 20	12.1	6.3		
84	G78.1 + 1.8	30 × 20	230	− 0.7	4.56E − 20	12.3	1.7		W66, CTB 91
85	G78.3 + 2.5	13.3 × 9.5	18	− 0.2	1.71E − 20	15.3	4.7		DR 3
86	G78.5 − 0.1	9.5 × 8.0	8	− 0.2	1.25E − 20	16.3	6.5		DR 12
87	G78.6 + 1.0	12 × 15	54	− 0.5	3.57E − 20	12.9	3.4		
88	G78.9 + 3.7	37 × 38	180	− 0.2	1.52E − 20	15.6	1.4		DR 1
89	G79.8 + 1.2	13 × 29	39	− 0.2	1.23E − 20	16.4	2.9		DR 11
90	G82.2 + 5.4	100 × 60	160	− 0.25	2.66E − 21	22.3	1.0	3	W63
91	G89.1 + 4.7	105′ dia	225	− 0.35	2.43E − 21	23.5	0.8	5	HB 21
92	G93.6 − 0.3	54′ dia	45	− 0.69	1.83E − 21	24.9	1.6		CTB 104
93	G111.7 − 2.1	4.0 × 3.8	3000	− 0.72	2.34E − 17	3.1	2.7		Cas A
94	G117.3 + 0.1	130′ dia	55	− 0.5	3.87E − 21	35.1	0.9		CTB 1
95	L.M.C.	1′.12 dia	3.5	− 1.01	3.33E − 19	7.9	24.5		N49 S
96	L.M.C.	0′.45 dia	1.9	− 0.50	1.12E − 18	6.1	46.5		N63 A
97	L.M.C.	0′.37 dia	4.6	− 0.50	4.00E − 18	4.6	42.8		N132 E

3SNRs in the L.M.C. (N49 S, N63 A, N132 E)

[a] 1 fu = 10^{-26} W m^{-2} Hz^{-1}

[b] W m^{-2} Hz^{-1} sr^{-1}

[c] References are: 1 Dickel (1969), 2 Shaver and Goss (1970), 3 Wendker (1968), 4 Milne and Wilson (1971), 5 Erkes and Dickel (1969)

[d] Possibly components of the one source, at the distances shown in parentheses.

5.5×10^{36} erg sec^{-1}, or about $\frac{1}{100}$ of the total radio emission from the Galaxy. The cumulative size distribution obtained for these objects, $D \propto [N(D' < D)]^{2/5} \propto t^{2/5}$ (where $N(D' < D)$ is the number of SNRs with diameters D' less than some given diameter D and t is the age of a SNR with diameter D), suggests that SNRs follow Sedov's (1959) treatment for an adiabatic explosion in a gas of constant heat capacity,

$$D_{(PC)} = 4.0 \times 10^{-11} (E_0/n_H)^{1/5} t_{(yr)}^{2/5}, \tag{2}$$

where E_0 (erg) is the initial energy of the explosion and n_H (cm^{-3}) is the ambient hydrogen number density in the medium.

3. Structural Characteristics of Supernova Remnants

Of approximately 55 objects listed in Table I for which observations of sufficient resolution have been made, 30 show a peripheral distribution of radio brightness indicating a shell structure, 6 show possible shell structure, a further 6 have a crescent structure which could indicate a rudimentary shell, and there are 3 well-resolved double sources. Thus a total of 45 SNRs exhibit a peripheral brightness distribution. There are possibly 9 objects which, although sufficiently well-resolved, do not appear to have any structure in their brightness distributions. The Crab Nebula is the brightest of these objects and has been observed with the highest resolution.

In Figure 3 we show characteristic contours (generally $\frac{1}{4}$, $\frac{1}{2}$ and $\frac{3}{4}$ power isotherms) of 43 of the resolved SNRs. The peripheral brightness distribution interpreted as shell structure can be seen in at least 36 of these. In this figure the SNR diagrams have been arranged in descending order of surface brightness and therefore, according to Figure 1, in order of increasing linear dimensions. The diagrams are (with the exception of SNR 93, Cas A) all drawn to the same linear scale corresponding to the distance given in Table I. The galactic plane is horizontal in these diagrams. There is nothing of an evolutionary nature immediately obvious in the structure of these objects nor does there seem to be any preferred orientation relative to the galactic plane. On this first point Shaver and Goss (1970) found for 19 well-resolved shells with diameters 3 to 40 pc that the relative shell thickness is fairly constant at near 15% of the diameter. However, these estimates are subjective and rather uncertain with the individual thickness/diameter ratios varying from 8% to 25%. It does seem though that the relative shell thickness is reasonably constant throughout Figure 3 and certainly they do not evolve at the constant shell thickness required by Kesteven (1968) in his interpretation of the Σ–D relationship (Figure 1, Equation (1)).

Another point raised by Shaver's work from 18 sources is that the brightest regions of a SNR lie each side of a diameter parallel to the galactic plane (Shaver, 1969). This result does not appear to be borne out in the examples shown in Figure 3. There are 30 SNRs in Figure 3 for which an axis of symmetry can be defined; the average angle made by these axes and the galactic plane is 45°. Nor does the situation improve much if we delete those SNRs which are more than 50 pc from the galactic plane; the average angle then is 52°, a slight but inconclusive shift towards Shaver's result.

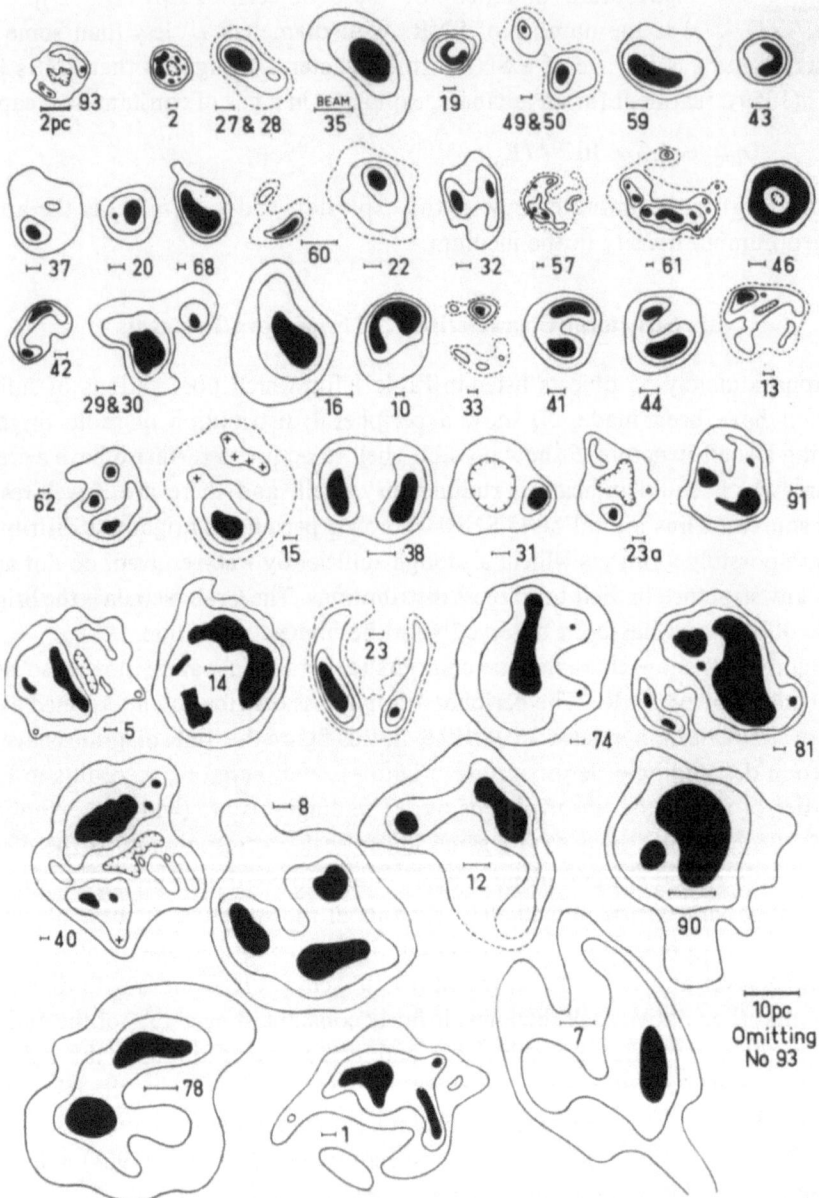

Fig. 3.   The structural characteristics of 43 well-resolved SNRs. The contour levels shown are quarter-power, half-power and three-quarter power (shaded): in a few cases one-eighth power contours are shown (broken lines) and for SNR 19 the 90 % contour level is shown shaded to indicate the shell. With the exception of SNR 93 (Cas A) the isotherms are all drawn to the same linear scale corresponding to the distances given in Table 1. The diagrams are arranged in order of surface brightness. The galactic plane is horizontal in each case. The observing beamwidth is indicated on each diagram.

Detailed analyses of the radial distribution of SNR brightness (e.g. Hill, 1967; Baldwin, 1967; Kesteven, 1968; Wynn-Williams, 1969; Rosenberg, 1970) show that for several SNRs the central part of the source is not as bright as expected from a uniform and isotropically emitting shell model fitted to the outer rim emission. Rosenberg has offered an explanation (for Cas A) in terms of a preferred direction of the synchrotron emission from the shell, this being due to a partial radial alignment of the magnetic field.

Lastly, one might expect that objects well off the galactic plane, where density variations of interstellar gas are less severe, would show the most uniform structure. However, in Figure 3, where $|z|$ ranges up to 700 pc, no obvious differences in structure are apparent. It is still possible that the expansion rate is greater for those objects away from the plane, but this should not affect their structural appearance, nor the Σ–D relationship.

4. Polarized Radio Emission from Supernova Remnants

Using a resolution sufficient to clearly resolve the shell structure, linearly polarized radio emission of the order of a few per cent is observed from most of the brighter SNRs. Seventeen SNRs are known to be polarized and detailed polarization maps at several frequencies have been constructed for at least 14 of these objects. The main feature is the low degree of polarization usually found, showing magnetic field disorder. In only a few cases is the degree high enough to suggest a simple model. A radial magnetic field is suggested for three of these sources: Cas A (Mayer and Hollinger, 1968), 1459−41 (SN 1006 AD) (Kundu, 1970) and IC 443 (Milne, 1971), although in this latter source there is a possibility that the magnetic field, initially parallel to the galactic plane, has been blown out by the expansion in the transverse directions (see Figure 4b). In other SNRs the magnetic field is directed predominantly along the shell (a tangential field). Examples of this are found in 1209−51/52 (Whiteoak and Gardner, 1968), W44 (Kundu and Velusamy, 1969), Vela X (Milne, 1968a) and W28 (Milne and Wilson, 1971 and Kundu, 1970).

With a particular source, in mind (1209−51/52) Whiteoak and Gardner (1968) interpret these two predominating field directions in terms of van der Laan's models, the radial magnetic fields being observed when the line of sight is along the ambient magnetic field and the tangential field when viewed transversely to the magnetic field. The SNRs in which the field is radial should show circular symmetry in their radio structure (a more complete shell) whilst tangential fields should be observed in SNRs exhibiting, ideally, a double crescent brightness distribution. There does not seem to be a great deal of verification of these principles in the examples we have. The situation is, however, not generally as simple as Whiteoak and Gardner suggest; local irregularities are common, and in many SNRs there are regions where the field is radial alongside other regions where a tangential field is suggested (e.g. W28 (Milne and Wilson, 1971), Puppis A and MSH 14−63 (Milne – unpublished data)).

It is only in those SNRs where a fairly uniformly directed field extends across the

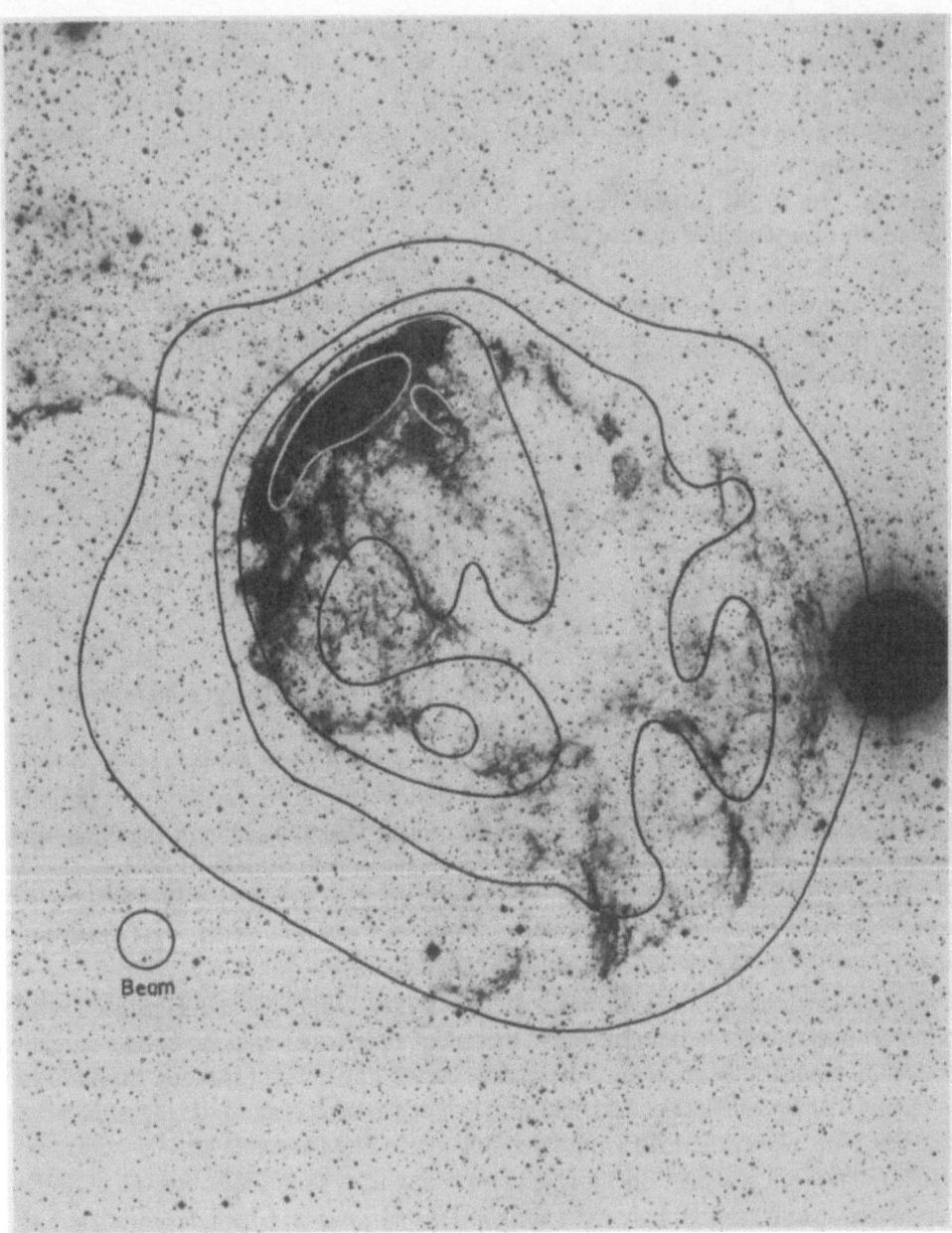

Fig. 4a. 5000 MHz isotherms superimposed on a red 48′ Schmidt photograph of IC 443 (Milne, 1971). The shell structure in this supernova remnant is well defined at both radio and optical wavelengths; the good radio-optical agreement shown here is the exception rather than the rule amongst SNRs.

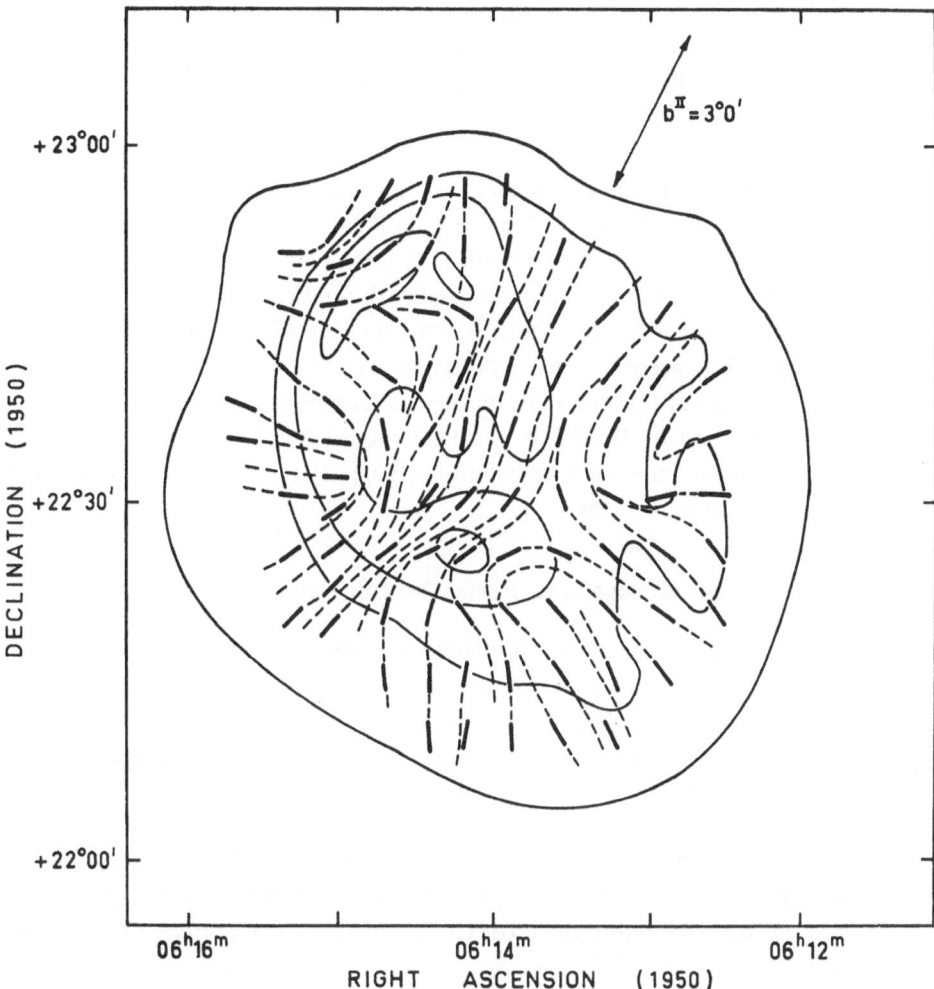

Fig. 4b.   The projected directions of magnetic field in IC 443; the contours are those of Figure 4a.
The magnetic field is predominantly radial (Milne, 1971).

source, or where the polarization is from a small region, that polarization is detectable
at low resolutions; examples of this are W44 and Tau A. One object which exhibits
relatively high polarization and has no shell structure is MSH 15−56 (SNR No. 36),
which consists of an $11' \times 9'$ central core imbedded in a halo $30'$ diam. (Milne, 1969);
this SNR has polarization up to 10% at 6 cm (Gardner *et al.*, 1969). It has a high
surface brightness and has been suggested as a possible X-ray source (Poveda and
Woltjer, 1968; Milne, 1970b).

5. Optically Identified Supernova Remnants

There are seventeen galactic SNRs well identified with visible nebulae. Most of these
can only be seen as a few faint, often sharp, filamentary wisps (e.g. Cas A, Pup A

and Kepler's Nova). In a few cases they are moderately bright and clearly exhibit a
shell structure (e.g. IC 443, S 147, the Cygnus Loop and Vela X). In most cases the
agreement between radio and optical brightness distribution is very poor in detail
but as a general rule one can say that the optical filaments outline the region containing the radio emission. It has been demonstrated that these filaments are 'sheets
seen edge on', possibly the best argument for this model being the observations of
temperature gradation within the filaments – difficult to justify physically with a
circular filament model (Parker, 1964, 1969; Milne, 1968b).

One exception to the rule of poor detailed radio-optical agreement is in IC 443
(Figure 4a), where the brightest radio emission coincides with the optical shell. Milne
(1971) finds that the spectral index is flatter (more thermal) around this shell and
most likely is a blend of thermal and nonthermal components. An appreciable free-
free radio contribution in IC 443 is in fact deduced by this author from the Hα
intensity; contrary to an earlier calculation (Hogg, 1964), this would explain the
detailed radio-optical agreement.

6. The Association of Supernova Remnants with Pulsars and X-Ray Sources

The discovery of a pulsar in the Crab Nebula (Staelin and Reifenstein, 1969) and in
Vela X (Large *et al.*, 1968) led to searches in other well-known SNRs but without
success. There is no acceptable positional agreement between the 41 pulsars listed by
Radhakrishnan (1969) and Large *et al.* (1969) and the SNRs in Table I except for the two
already noted. Large (1970) has in fact predicted that pulsars would be undetectable in
all but the closest SNRs, with the present limitations on sensitivity and dispersion.

A similar comparison with the X-ray sources has yielded far better but possibly
fortuitous results because of the large probable errors in the X-ray positions. Milne
(1970b) lists seven SNRs within the error circles for the X-ray sources. Of these three
have been identified with SNRs: Tycho's Nova, the Crab Nebula and Cas A. Of
special interest is the SNR close to Nor X − 2, MSH 15 − 56 (SNR 36), already singled
out in this review; this source has, like the Crab, high surface brightness (hence
comparatively young), fairly flat spectral index, high polarization and an absence of
radio structure. The other possibly significant suggested identification is that of
GX5 − 1 with A4 (SNR 56). The source GX5 − 1 has a well-established position (1·2′
error radius) and lies within SNR 56; it is suggested here that this is a definite identification.

7. Conclusions

Summarizing the radio observations we find that:
(1) There is an average evolutionary relationship

$$\Sigma = 9.52 \times 10^{-15} D^{-4.54}, \tag{1}$$

with variations from this probably dependent on the initial energy. There is no noticeable evolutionary effect on the spectral index.

(2) The expansion probably follows Sedov's equation

$$D = 4.0 \times 10^{-11} (E_0/n_H)^{1/5} t^{2/5}.$$ (2)

(3) Almost all of the resolved SNRs exhibit some form of peripheral brightness distribution indicating possible shell structure. The relative shell thickness is fairly constant.

(4) There seems to be no preferred galactic orientation and no latitude effects in the structure.

(5) Radio polarization of the order of a few per cent has been found in many SNRs, but generally the direction of polarization varies so much across the source that polarization is not observed until the shell is well resolved. The magnetic field distribution is mostly tangential; however, in many objects there are regions with tangential field adjacent to other regions in which the field is clearly radial.

(6) There is generally no detailed radio-optical brightness correlation. The source IC 443, one of the few objects that show a strong agreement, has a large thermal component.

(7) There are two known SNR-pulsar associations and possibly 7 SNRs which emit X-radiation.

Briefly summarizing the properties of the Crab Nebula in relation to the other SNRs, we have:

(1) It has a high surface brightness and its position (No. 9) on the Σ–D diagram (Figure 1) is well off to the low initial energy side of the average evolutionary track. It is still possible that it is an average type I SNR, if such a classification exists. From Equation (2) we obtain an initial energy/ambient hydrogen density ratio (E_0/n_H) of 10^{48} erg cm^3, considerably lower than any of the other SNRs with known ages.

(2) Taurus A has a spectral index of -0.25, flatter than most SNRs and possibly flatter than all of the SNRs with well-determined spectral indices.

(3) Even at high resolution Tau A exhibits a relatively amorphous brightness distribution (Hogg et al., 1969). In this respect it is unlike almost all of the other resolved SNRs. We have however pointed out that there may be eight other SNRs with no apparent structure and these may well form a Tau A type class.

(4) At low resolutions the percentage radio polarization from Tau A is greater than is found in most other unresolved SNRs, and the high resolution observations of Mayer and Hollinger (1968) show that the magnetic field is uniformly directed over most of the source.

(5) The Crab Nebula contains both a pulsar and a source of X-rays.

In conclusion, are there any other Tau A type objects within the Galaxy? Certainly there are no other known Tau A type objects at the same stage of evolution. Possibly MSH 15−56 (SNR 36) is a later stage in the evolution of these objects, and it is further possible that the flat spectrum SNRs 11−54 (SNR 21) and A4 (SNR 56) are also well-evolved members of this class. High-resolution searches locating more amorphous non-thermal galactic sources should yield the answer to this question.

References

Baldwin, J. E.: 1967, *IAU Symp.* **31**, 337.
Dickel, J. R.: 1969, *Astrophys. Letters* **4**, 109.
Erkes, J. W. and Dickel, J. R.: 1959, *Astron. J.* **74**, 840.
Gardner, F. F., Whiteoak, J. B., and Morris, D.: 1969, *Australian J. Phys.* **22**, 821.
Harris, D. E.: 1962, *Astrophys. J.* **135**, 661.
Hill, E. R.: 1967, *Australian J. Phys.* **20**, 297.
Hogg, D. E.: 1964, *Astrophys. J.* **140**, 992.
Hogg, D. E., MacDonald, G. H., Conway, R. G., and Wade, C. M.: 1969, *Astron. J.* **74**, 1206.
Kesteven, M. J.: 1968, *Australian J. Phys.* **21**, 739.
Kundu, M. R.: 1970, *Astrophys. J.* **162**, 17.
Kundu, M. R. and Velusamy, T.: 1969, *Astrophys. J.* **155**, 807.
Large, M. I.: 1970, *Astrophys. Letters* **5**, 11.
Large, M. I., Vaughan, A. E., and Mills, B. Y.: 1968, *Nature* **220**, 340.
Large, M. I., Vaughan, A. E., and Wielebinski, R.: 1969, *Nature* **223**, 1249.
Mayer, G. H. and Hollinger, J. P.: 1968, *Astrophys. J.* **151**, 53.
Milne, D. K.: 1968a, *Australian J. Phys.* **21**, 201.
Milne, D. K.: 1968b, *Australian J. Phys.* **21**, 501.
Milne, D. K.: 1969, *Australian J. Phys.* **22**, 613.
Milne, D. K.: 1970a, *Australian J. Phys.* **23**, 425.
Milne, D. K.: 1970b, *Proc. Astron. Soc. Austr.* **1**, 333.
Milne, D. K.: 1971, 'Radio Observations of the Supernova Remnants IC 443 and Puppis A',
    Australian J. Phys. (in press).
Milne, D. K. and Wilson, T. L.: 1971, *Astron. Astrophys.* **10**, 220.
Parker, R. A. R.: 1964, *Astrophys. J.* **139**, 493.
Parker, R. A. R.: 1969, *Astrophys. J.* **155**, 359.
Poveda, A. and Woltjer, L.: 1968, *Astron. J.* **73**, 65.
Radhakrishnan, V.: 1969, *Proc. Astron. Soc. Austr.* **1**, 254.
Rosenberg, I.: 1970, *Monthly Notices Roy. Astron. Soc.* **147**, 215.
Sedov, L. I.: 1969, *Similarity and Dimensional Methods in Mechanics*, Academic Press, New York.
Shaver, P. A.: 1969, Cornell-Sydney University Astronomy Centre, Preprint No. 137.
Shaver, P. A. and Goss, W. M.: 1970, *Australian J. Phys. Astrophys.*, Suppl. No. 14.
Shklovsky, I. S.: 1960, *Soviet Astron.* **4**, 243.
Staelin, D. H. and Reifenstein, E. C.: 1969, *Astrophys. J.* **156**, L121.
van der Laan, H.: 1962a, *Monthly Notices Roy. Astron. Soc.* **124**, 125.
van der Laan, H.: 1962b, *Monthly Notices Roy. Astron. Soc.* **124**, 179.
Wendker, H.: 1968, *Z. Astrophys.* **69**, 392.
Whiteoak, J. B. and Gardner, F. F.: 1968, *Astrophys. J.* **154**, 807.
Wynn,Williams, C. G.: 1969, *Monthly Notices Roy. Astron. Soc.* **142**, 453.

Discussion

L. Woltjer: How does one know that all these objects are really supernova remnants?

V. Radhakrishnan: One concludes that they must be from their spectrum, size, etc. There is no absolute proof as far as I know.

4.3 CASSIOPEIA A – THE YOUNGEST KNOWN
GALACTIC SUPERNOVA REMNANT

SIDNEY VAN DEN BERGH

David Dunlap Observatory, University of Toronto, Richmond Hill, Ontario, Canada

Abstract. Proper motion observations show that the explosion of the Cas A supernova took place in AD 1667 ± 8 (me). Individual moving knots have lifetimes ~ 10 yr. In these fast-moving knots oxygen and argon are overabundant by a factor of at least seventy with respect to hydrogen and nitrogen. Nitrogen is found to be overabundant in quasi-stationary flocculi. This suggests that these flocculi were formed from material that was ejected from the outer layers of the pre-supernova before its explosion in AD 1667.

Following up the first accurate position determination of Cas A by Graham Smith in 1951, Baade and Minkowski (1954) were able to identify this radio source with a remarkable new type of emission nebula. These early optical observations established that the remnant of Cas A consists of two entirely distinct components: (1) A rapidly expanding system of knots and filaments that radiate strongly in [O III] and in [S II] and (2) A number of quasi-stationary flocculi that emit [N II] and Hα.

Well-exposed plates of Cas A obtained in better than average seeing with the 200-in. telescope show approximately 130 nebulous knots. Of these about 100 are moving knots and approximately 30 are quasi-stationary flocculi. Intercomparison of plates taken some years apart (see Figure 1) shows that new knots are continually forming while old ones fade away. Available data on the decay of moving knots are adequately represented by the relation

$$n(t) = n(0) e^{-t/\tau} \tag{1}$$

with $n(0) \simeq 100$ and $\tau \simeq 10$ y. The main uncertainty in this result is due to variations in plate quality caused by differences in seeing. The survival lifetime of large filaments, such as No. 1 of Baade and Minkowski (1954), is undoubtedly much longer than that of the average knot. Intercomparison of plates obtained in very good seeing suggests that the decay timescale of the smallest structural details in the remnant of Cas A is probably considerably shorter than ten years. Such fast changes indicate that the recombination timescale $\tau \simeq (\alpha_{rec} \eta_e)^{-1}$ is short and hence that the density in moving knots is high. This conclusion is supported by observations of the intensity ratio of the [S II] lines $\lambda\lambda 6716, 6731$ which yields $\eta_e \simeq 1 \times 10^4$ cm^{-3}.

New moving knots appear predominantly in regions in which other moving knots and filaments are already present. As a result no significant changes in the *large-scale* distribution of luminous nebulosity has occurred in Cas A during the last nineteen years. This observation suggests that luminous moving knots are fragments of the former supernova that are ploughing through stationary interstellar cloudbanks at hypersonic velocities.

Davies and Smith (eds.), The Crab Nebula, 263–267. All Rights Reserved.
Copyright © 1971 by the IAU.

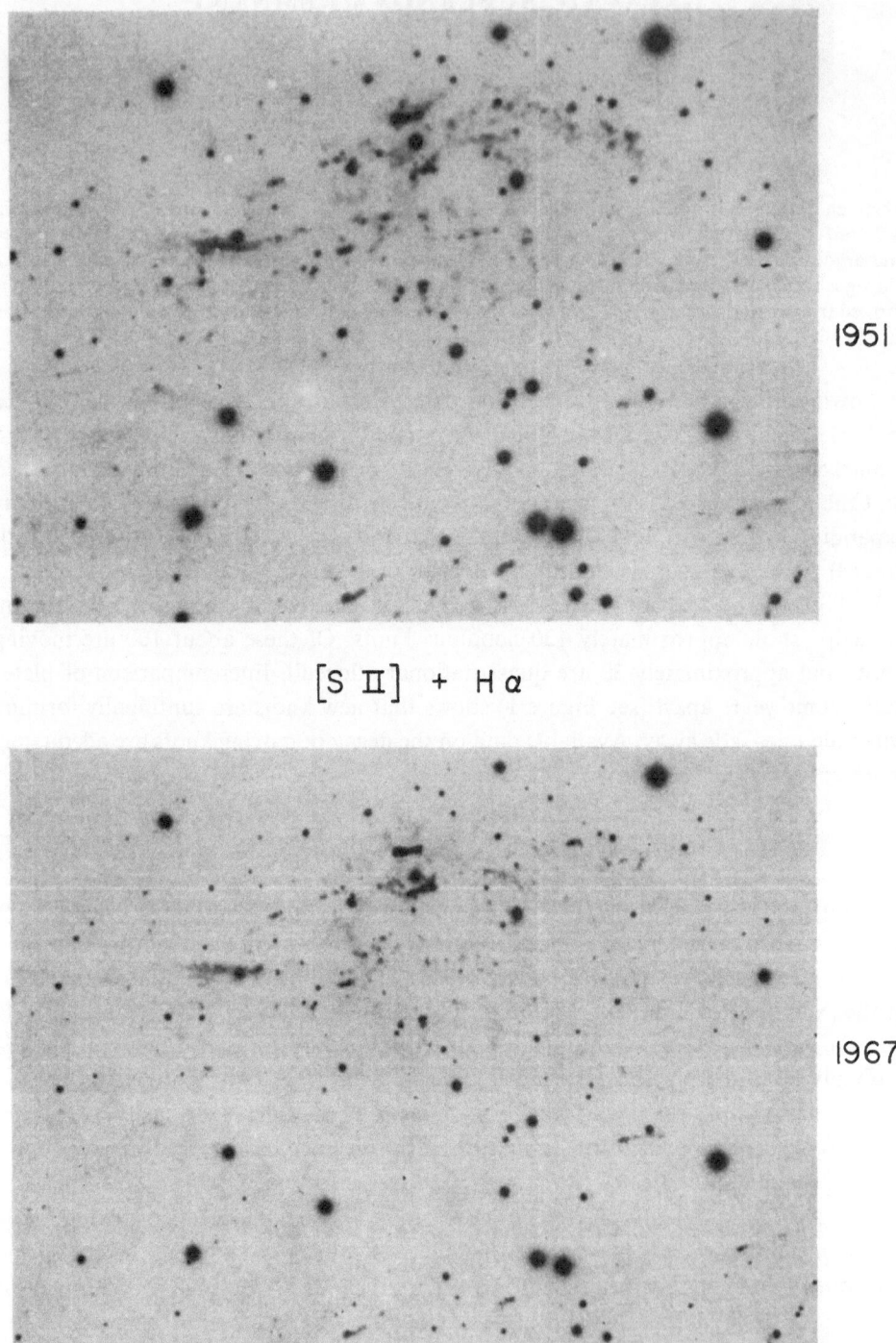

1951

[S II] + H α

1967

Fig. 1.   Intercomparison of these two plates, which were both taken in red light, shows the general expansion of the optical remnant of Cas A. Note the rapid changes that have taken place in the small-scale structural details of the moving knots.

If this conclusion is correct then the moving knots in Cas A should have a composition similar to that of the supernova from which they were formed. Furthermore the high densities that are observed in bright moving knots and the lack of observable deceleration (van den Bergh and Dodd, 1970) suggest that the material in moving knots cannot have been significantly contaminated by swept-up interstellar material.

Peimbert and van den Bergh (1971) have used observations of line intensity ratios to determine the chemical composition in moving knots and to estimate the absorption in the direction of Cas A. The absorption is found to lie in the range $5 < A_V < 7$ mag. In moving knots oxygen is overabundant with respect to nitrogen by a factor of at least seventy. The sulphur-to-oxygen ratio is about two times lower in Cas A than it is in the Orion Nebula. Furthermore oxygen and argon are found to be overabundant by a factor of at least seventy with respect to hydrogen.

Observations of stationary flocculi show that nitrogen is overabundant with respect to hydrogen and oxygen. This suggests that stationary flocculi were formed when a preexisting circumstellar envelope was engulfed by the expanding supernova shell.

The bright moving nebulosity in Cas A is strongly concentrated in *knots*. In this respect Cas A differs from the Crab Nebula in which the moving nebulosity is mainly

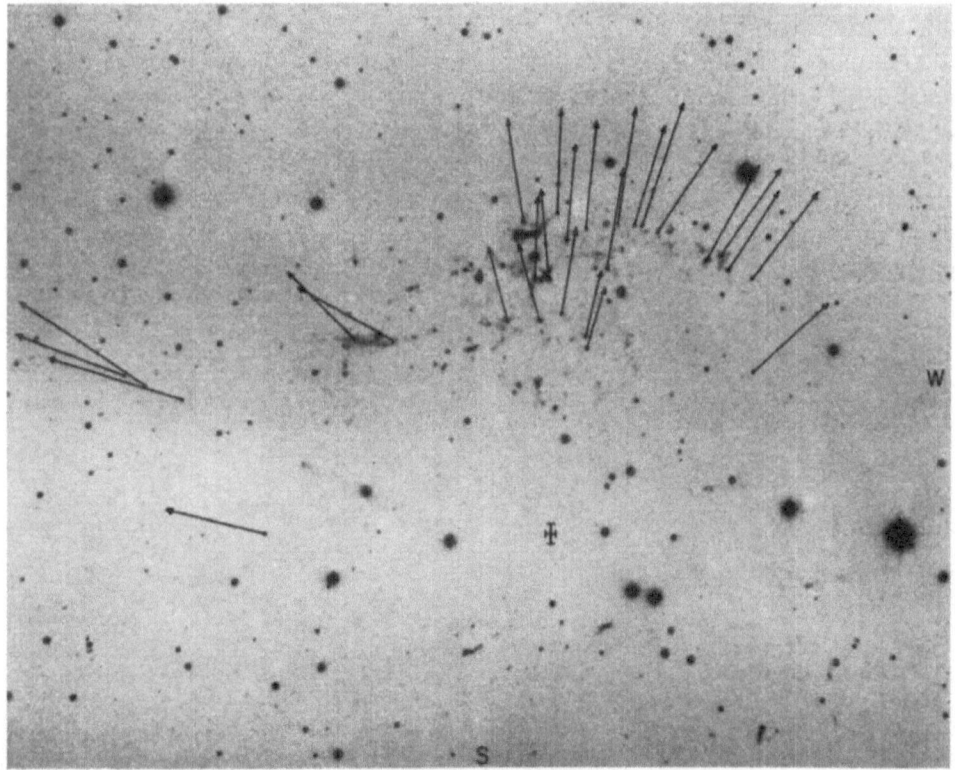

Fig. 2.   The figure shows 100 y proper motion vectors of individual knots of nebulosity. The centre of expansion of the optical remnant of Cas A is marked by a cross.
(Courtesy *The Astrophysical Journal*)

concentrated in *filaments*. This may indicate that magnetic fields play a larger role in the expanding shell of the Crab than they do in Cas A. Radial velocity observations confirm that the nebulosity in Cas A consists of discrete knots with differing velocities. Velocity differences of up to 3000 km sec^{-1} are observed for knots within a single filament. The origin of these large velocity differences remains a mystery. In most knots the half-width of the doppler profile of emission lines is about 200 km sec^{-1}. On direct plates typical moving knots are seen to have diameters of 2″ corresponding to linear dimensions of about $1 \times 10^{+17}$ cm.

Van den Bergh and Dodd (1970) have used sixteen 200-inch plates taken between 1951 and 1968 to study the expansion of the optical remnant of Cas A. Proper motions of individual knots are plotted in Figure 2. Assuming no deceleration the explosion of the supernova took place in AD 1667 ± 8 (me). In Table I the position of the centre of expansion that was obtained by van den Bergh and Dodd is compared with other determinations of the centre of Cas A. No star brighter than $V = 22.5$ is visible within eight standard deviations of the observed centre of expansion. For $A_V \leqslant 7$ mag this result implies that any stellar remnant of Cas A must be intrinsically fainter than the pulsar in the Crab Nebula. In making such a comparison one should, of course, remember that the apparent brightness of a pulsar depends on the inclination of its axis of rotation.

According to Woltjer (1957) $V \simeq 19$ mag (arc sec)$^{-2}$ for the brightest part of the amorphous nebulosity that surrounds the Crab pulsar. With $A_V \simeq 7$ mag the corresponding surface brightness in Cas A would be well below the level of detectability on currently available plates.

Photoelectric photometry of stars in the direction of Cas A is reported in van den Bergh (1971a). A detailed description of the spectra of moving knots and stationary flocculi in Cas A is given in van den Bergh (1971b).

TABLE I

Centre of expansion of Cassiopeia A

	Period	α(1950)	δ(1950)
van den Bergh and Dodd (1970)	1951–1968	23^{h}21^{m}11^s.4 ± 0^s.2	+ 58°32′18″.9 ± 3″.1
Minkowski (1959)	1951–1954	23 21 11 .8 ± 0 .4	+ 58 32 16 ± 8
Rosenberg* (1970)	1968	23 21 9 .5	+ 58 32 22

* Geometrical centre of radio source.

References

Baade, W. and Minkowski, R.: 1954, *Astrophys. J.* **119**, 206.
Bergh, S. van den: 1971a, *Astrophys. J.* (in press).
Bergh, S. van den: 1971b, *Astrophys. J.* (in press).
Bergh, S. van den and Dodd, W. W.: 1970, *Astrophys. J.* **162**, 485.
Minkowski, R.: 1959, *IAU Symp.* **9**, p. 315.
Minkowski, R.: 1970, private communication.
Peimbert, M. and Bergh, S. van den: 1971 (in preparation).
Rosenberg, I.: 1970, *Monthly Notices Roy. Astron. Soc.* **147**, 215.
Woltjer, L.: 1957, *Bull. Astron. Inst. Neth.* **13**, 293.

Discussion

R. C. Jennison: Several beautiful syntheses of Cassiopeia A have been performed in the range 1 GHz to 3 GHz in recent years but no structure has been mapped at lower frequencies since the original work by Jennison and Latham c. 1955. This low frequency (127 MHz) phase sensitive synthesis showed a spur which was shortly afterwards observed photographically by Minkowski as a flare in position angle 70°. Many years later it appeared on the Cambridge high frequency syntheses, not as a spur, but as a break in the shell wall at the same position angle. I have suggested (Nature **207** (1965) 740) that the spur has a spectrum strongly favouring low frequency radio emission. I note that the Crab has a break in the shell and a faint short spur at optical frequencies would suggest that it may also have a low frequency enhancement in this direction. We certainly cannot ignore the flares in these supernovae and I shall be glad if Dr. van den Bergh can comment on the optical flare in Cassiopeia A.

S. van den Bergh: It would certainly be quite important to repeat these old observations. Recent Cambridge observations at short wavelengths do not show radio radiation at the position angle of the optical jet. If the jet should show up at long wavelengths it would, of course have to have quite a steep radio spectrum. Comparison of the Cambridge one-mile telescope map of Cas A with optical observations does *not* show a close correlation between optical and radio 'hotspots'.

J. E. Felten: You contemplated values of the visual extinction ranging from three to seven in magnitude. Is there any way of getting a better idea what the visual extinction is in this direction?

S. van den Bergh: You could measure the relative intensities of the far infrared sulphur lines and the ultraviolet sulphur lines which originate in the same upper level and so you know the theoretical (unobscured) intensity ratio. These observations are difficult and have not been made.

A. T. Moffet: Is there any way to tell whether there might be large differences in the obscuration in front of different parts of Cas A? The optical filaments are all seen on the northern edge, while the radio source is quite symmetrical.

S. van den Bergh: No. Intercomparison of 200-in. plates taken in the ultraviolet and in the infrared (where the absorption is relatively small) show the same large-scale structure of the expanding system of knots. Another argument against a large difference in absorption in the northern and in the southern parts of the remnant is that the stationary Hα flocculi can be seen both to the north and to the south of the centre of the expansion. I might add that some moving knots have also developed in the southern part of Cas A recently.

L. H. Aller: Some years ago Minkowski found that the Cassiopeia A filaments exhibited a spectrum with strong [O III] lines but weak (or absent) H. We suggested that perhaps the filament was excited by collisions with the interstellar medium such that the ambient electron temperature was not high enough to collisionally excite H. The absence of H lines is difficult to understand. If doubly ionized oxygen is present, why do recombination lines of hydrogen not appear? Perhaps the ionization of oxygen is by radiation and the hydrogen recombination lines are too weak. Possibly the supernova material is deficient in H and what we observe is excited supernova material. But then why is the interstellar gas not excited as it is 'bulldozed' by the supernova shell? Admittedly the density of the bulldozed interstellar matter is low but we should see some H contribution. A very severe difficulty for theoreticians is imposed by the disappearance and re-appearance of filaments. Certainly the simple interpretation which Minkowski and I gave must now be discarded and strong timedependent effects must be included.

4.4 THE SECULAR BEHAVIOR OF X-RAY
AND RADIO EMISSION FROM SUPERNOVA REMNANTS*

WALLACE H. TUCKER

American Science and Engineering, Cambridge, Mass., U.S.A.

Abstract. Continuous injection models for the secular behavior of the radio and X-ray emission from supernova remnants are examined and compared with the observations. Among other things, it is concluded that (1) continuous injection probably occurs for at least 10 yr in every case and about 1000 yr in most supernova remnants, in which case the supernova remnants 3C392, W28, Pup A and IC443 should produce 1–10 keV X-ray fluxes $\approx 10^{-10}$ ergs/cm²-sec, and (2) the X-ray sources in the Crab Nebula, Cas A and Tycho can be explained in terms of a model wherein continuous injection occurs for 300 yr for the Crab Nebula, much less than 250 yr for Cas A and much longer than 400 yr for Tycho. Finally, it is shown that if Tycho and Cas A contain an X-ray star such as NP 0532, it is quite possible that the X-ray emission from those sources is predominantly due to the X-ray star.

1. Introduction

The purpose of this communication is to discuss the secular behavior of the radio and X-ray flux from supernova remnants which is to be expected on the basis of a model which assumes that relativistic electrons are continuously injected into the source for some time after the initial explosion. Ths results of this analysis, which are discussed in more detail elsewhere (Tucker, 1970), lead to the following conclusions: (1) a continuous injection model is capable of explaining the observed variation of the radio luminosity with the size of the source, if the injection time t_i is of the order of 2000 yr for most sources; (2) the intensity and spectral shape of the X-ray emission from the Crab Nebula, Cas A and Tycho are readily understood on the basis of this model if $t_i = 300$ yr for the Crab Nebula, $\ll 200$ yr for Cas A and $\gg 400$ yr for Tycho; (3) if Tycho is typical, then the supernova remnants 3C392, W28, Pup A and IC443 should produce X-ray fluxes of the order of $\frac{1}{10}$ that of Tycho; (4) if Cas A and Tycho contain a compact X-ray source similar to NP 0532, then it is quite possible that the X-ray emission from those objects is predominantly due to the compact source.

2. The X-Ray and Radio Emission from Relativistic Electrons in Supernova Remnants

The intensity of the synchrotron radiation from supernova envelopes can decrease with time for two reasons: (1) The energy losses suffered by the electrons as a result of adiabatic expansion and radiation and (2) the magnetic field strength can decrease with time. The relative importance of these effects depends on the energies of the electrons, the rate of production of relativistic electrons as a function of time and the evolution of the magnetic field strength with time.

* This paper was presented by Dr. E. M. Kellogg.

Assume that relativistic electrons are injected into the emitting region at a rate described by the equation

$$f(t) = 1/(1 + t/t_i)^2 \qquad (1)$$

This time dependence is consistent with the predictions of both the oblique rotator model (Gunn and Ostriker, 1969; Cavaliere and Pacini, 1970) and the low density limit of a relativistic stellar wind model (Michel, 1969; Goldreich and Julian, 1969), so that the results discussed below are not dependent on the validity of any particular pulsar model. In fact any law of the form $f(t) = 1/(1 + t/t_i)^x$ would give essentially the same results in the limits $t \ll t_i$ and $t \gg t_i$.

If it is assumed that the relativistic electrons are produced after the stellar envelope has been ejected, then the effective radius of the source is given by the position of the ejected shell for $t \leqslant t_s$, where t_s is the slowing down time, i.e., the time required for the interaction with the interstellar medium to have a noticeable effect on the motion. This happens in a time corresponding to a radius of the order of 4 parsec. For radii greater than this the ejecta are mixed in with the interstellar medium and the effective radius of the source is given by the position of the shock front which propagates ahead of the ejected shell*.

Since the dependence of r on t is different in the different phases of the expansion, the dependence of the luminosity on r will be different depending on the relative magnitude of the age of the source t, the injection time t_i, and the slowing down time t_s or equivalently r, r_i, and r_s. We know the value of r_s fairly well ($r_s \sim 4$ pc) and the value of r can be determined from the observations by a number of (usually indirect) methods, all of which give roughly the same values (see Minkowski, 1968; Shklovsky, 1968; Poveda and Woltjer, 1968; Westerlund, 1969a, b). The value of r_i or the corresponding time t_i is less certain, since it depends on the model for injecting energy into the nebula. In general if the energy input is due to a rotating neutron star which is slowing down, then $t_i \approx E_{r0}/L_0$ where E_{r0} is the initial rotational energy of the neutron star and L_0 is the initial rate of loss of rotational energy. This implies that in general the brighter sources should have shorter lifetimes, and vice-versa. Pulsar observations, when available, can provide a means for estimating t_i. However, pulsars are observed in only two supernova remnants, so in most cases t_i must be treated as a parameter to be determined indirectly with the help of observations. An estimate of t_i can be obtained by assuming that the luminosity at the maximum of the light curve is a good estimate of L_0, and that E_0 is of the same order as the gravitational binding energy; this implies values for t_i lying somewhere between a hundred and a thousand years. For example, the observations of NP 0532 imply that for the Crab Nebula, $t_i = 300$ yr. Using this value for t_i in Equation (1) yields a spectrum which is consistent with the radio and X-ray observations of the Crab, if we assume a magnetic field $B = 2 \times 10^{-4}$ G.

---

* This assumes that a shock front does in fact develop. In a direction transverse to the interstellar magnetic field this is surely a good assumption. In a direction parallel to the magnetic field it is not obvious what will happen. (For discussions of the dynamics of supernova envelopes see Spitzer and Tomasko, 1968; Kruskal et al., 1965, Colgate, 1967, and Shklovsky, 1968.)

For the cases of Cas A and Tycho, the radio and X-ray observations imply $t_i \ll 200$ yr and $t_i \gg 400$ yr, respectively (for more details, see Tucker, 1970).

The variation with t of the radio luminosity $L_{rad} (= \nu_{rad} L(\nu_{rad}))$ for the Crab Nebula, Tycho and Cas A is shown in Figure 1. Also shown are the positions of those supernova remnants for which the distance is fairly well known, and a broad shaded band having a slope of -0.67. Poveda and Woltjer (1968) have shown that it is probable that most of the other supernova remnants lie within the shaded area of the diagram. To get the curves going through the Crab Nebula, Tycho and Cas A, it was assumed that $r_s = 4$ pc in each case and $r_i = 0.3$ pc for the Crab Nebula and Cas A, and $r_i = 7$ pc

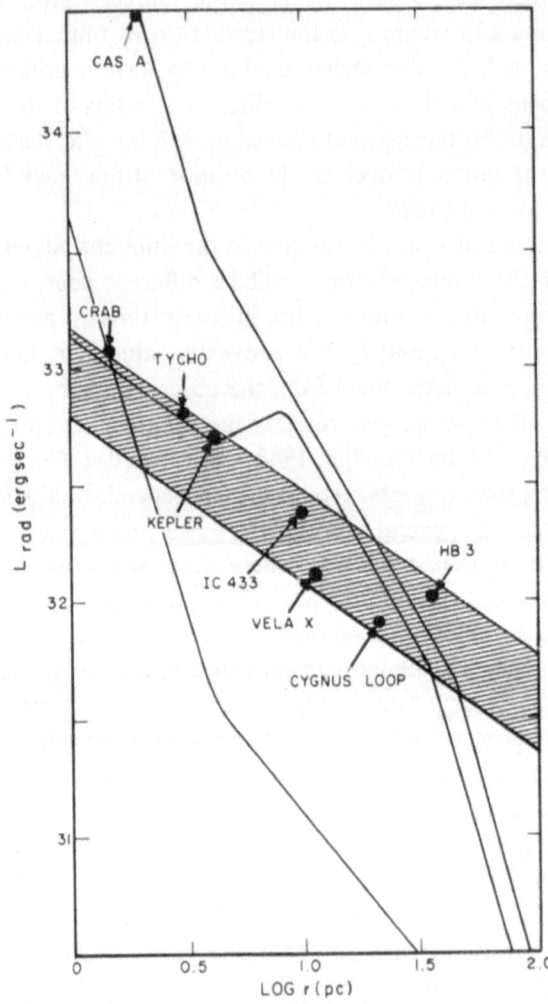

Fig. 1. The radio luminosity of supernova remnants as a function of their radius. Those remnants for which the distance is fairly well known are shown explicitly; the others most probably lie within the shaded area of the diagram. Data are taken from Poveda and Woltjer (1968). Also shown are curves illustrating continuous injection models for Cas A, the Crab Nebula and Tycho.

for Tycho, values suggested by a comparison of the observed spectra and the predictions of the continuous injection model.

Most sources appear to lie in the region traced out by the curve through Tycho, implying that $r_i \sim 7$ pc or $t_i \sim 2000$ yr for most sources. This would explain why the observed L_{rad} vs r curve does not exhibit any dependence on the radio spectral index α_r. If most sources are in the region $r_s < r \sim r_i$, the dependence of their radio luminosity on α_r is weak. Any spread in the values of r_s and r_i should wash out this dependence leading to a curve of the type observed. The observation of an increase of the radio intensity with time for sources having radii in the range 5–10 pc would lend support to this model.

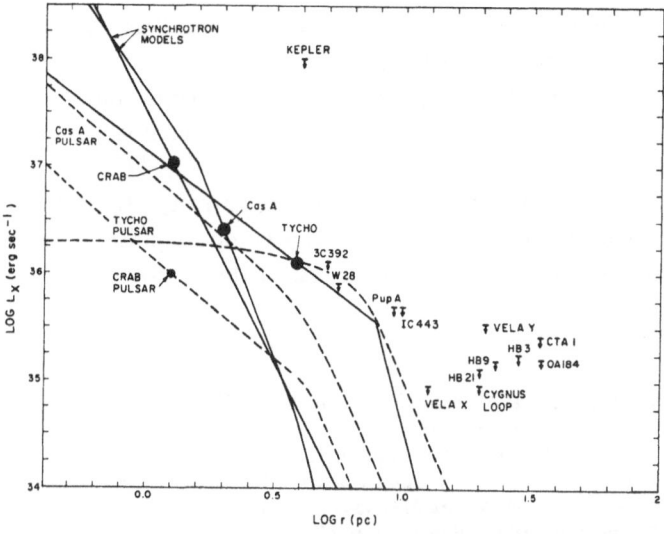

Fig. 2. The 1–10 keV X-ray luminosity as a function of the radius of the radio source for known supernova remnants (Gorenstein *et al.* 1970). Curves illustrating the models discussed in the text are also shown.

The variation with r of the 1–10 keV luminosity L_x for the Crab Nebula, Tycho, and Cas A are shown in Figure 2 together with observational data (Gorenstein *et al.*, 1970). Of course the curves for the Crab and Cas A must flatten for small r since at all times we must have $L_x < L_0$, where L_0 is the total power input of the central source.

The model that fits the observations from Tycho predicts a decrease with r according to $r^{-1.7}$, which suggests that at least one of the sources, 3C392, W28, Pup A or IC 443 should have an intensity at least $\frac{1}{10}$ that of Tycho.

It is well known that the Crab Nebula is not a uniform extended source. Rather, it is a complex source of the core-halo type, where in this case the core source is the pulsar NP 0532, which accounts for about 10% of the X-ray emission between 1 and 10 keV (Kellogg, 1970). Since this radiation is produced essentially at the site of the energy source of the nebula, we might expect the luminosity of the pulsar to follow rather closely the time dependence of the central energy source, which is given approximately by Equation (1). This implies that the luminosity of the pulsar X-ray source varies approximately as $1/r^2$ which is less steep than the $r^{-4.7}$ dependence for the extended source. The evolution of the pulsar X-ray source is shown by the dotted line in Figure 2, which shows that the pulsar X-ray source should become dominant for $r \gtrsim 3$ pc, or $t \gtrsim 2000$ yr. It is of interest to speculate whether a compact source (it may not be pulsing) dominates the emission from Cas A and Tycho. An estimate of the intensity of these hypothetical compact sources can be made as follows: If we extrapolate the luminosity of the Crab pulsar back in time according to Equation (1), an initial 1–10 keV luminosity $\sim 2 \times 10^{37}$ erg/sec is obtained. If we now assume that (i) the ratio of the initial 1–10 keV luminosity of the pulsar $L_{px}(0)$ to the initial total power output of the central source L_0 is a constant for all sources, and (ii) $L_0 t_i =$ constant for all sources with $t_i \sim 25$ yr for Cas A and 2500 yr for Tycho then the initial 1–10 keV compact source luminosities for Cas A and Tycho are of the order of 2×10^{38} erg/sec and 2×10^{36} erg/sec, respectively. An evolution in time according to equation (1) then leads to secular behavior of the compact sources as shown by the dotted curves in Figure 2.

These curves show that, if the assumptions are correct, a significant fraction of the emission from Cas A and Tycho is due to a compact pulsar-like source. Angular diameter measurements should provide a test of this hypothesis in the near future.

References

Cavaliere, A. and Pacini, F.: 1970, *Astrophys. J.* (Letters) **159**, L21.
Colgate, S.: 1967, *Astrophys. J.* **150**, 163.
Goldreich, P. and Julian, W.: 1969, *Astrophys. J.* **157**, 869.
Gorenstein, P., Gursky, H., Kellogg, E., and Giacconi, R.: 1970, *Astrophys. J.* **160**, 947.
Gunn, J. and Ostriker, J.: 1969, *Nature* **221**, 454.
Kellogg, E.: 1970, this symposium, Paper 1.5, p. 42.
Kruskal, M., Bernstein, I. and Kulsrud, R.: 1965, *Astrophys. J.* **142**, 369.
Michel, F.: 1969, *Astrophys. J.* **158**, 727.
Minkowski, R.: 1968, in *Nebulae and Interstellar Matter* (ed. by Aller and Middlehurst), Univ. Chicago Press, Chicago, Chapter 11.
Poveda, A. and Woltjer, L.: 1968, *Astron. J.* **73**, 65.
Shklovsky, I.: 1968, *Supernovae*, John Wiley and Sons, New York.
Spitzer, L. and Tomasko, M.: 1968, *Astrophys. J.* **152**, 971.
Tucker, W.: 1970, *Astrophys. Space Sci.* **9**, 315.
Westerlund, B.: 1969a, *Astron. J.* **74**, 879.
Westerlund, B.: 1969b, *Astron. J.* **74**, 882.

Discussion

J. E. Grindlay: Have you been able to estimate the current injection rate in Cas A and thus the present energy that may be put into high energy particles?

E. M. Kellogg: Cas A is assumed still to have electrons being injected. No absolute estimate of the injected energy or flux is made.

4.5 PULSARS AND CLOSE BINARY SYSTEMS

VIRGINIA TRIMBLE and MARTIN REES

Institute of Theoretical Astronomy, Cambridge, U.K.

Abstract. It is first considered what must happen if pulsars (i.e. neutron stars) are formed in close binary systems (CBS), and whether the resulting orbital motion and mass transfer should be observable. As this set of alternatives seems unlikely, there follow suggestions of how one might prevent the formation of neutron stars in close binaries. Finally, it is shown that 'runaway' pulsars with velocities larger than about 15 km/sec cannot be produced by isotropic supernova explosions within close binaries, and an alternative explanation is suggested for the observed correlation of periods of pulsars with their distances from the galactic plane.

1. Introduction

It is known (with decreasing levels of certainty) that: (1) an appreciable fraction of all stars occur in close binary systems (CBS). (2) some stars end their lives in supernova events, (3) some supernovae produce neutron stars, and (4) some neutron stars are observed as pulsars. At least a few pulsars might, therefore, be expected to occur in close binary systems.

2. Pulsars in Close Binary Systems

Two detectable effects should occur for pulsars which have close companions – orbital motion and accretion of material. Periodic changes in pulse arrival times with the amplitudes and periods expected from binary motion have not been observed for any pulsar. For the 15 of 50 known objects for which rates of change of period have been detected (Terzian, 1970), changes in orbital velocities of more than about 3×10^{-8} the velocity of light per 100 days can be excluded. This corresponds to separations greater than $40\sqrt{(M_1 + M_2)}$ AU, which is so large that the systems cannot be considered close, in either the intuitive sense or in the sense that mass exchange will occur as the component stars evolve. Corrected for projection effects and the statistics of small numbers, the available data show that not more than one pulsar in four or five can belong to a close binary system. Additional observations of period changes must either reveal binary motion or lower this limit considerably.

Consider now what must happen in a CBS where the less massive component is a neutron star. (Mass exchange in earlier phases of the history of the system will probably prevent the neutron star ever being the more massive component; Plavec, 1968a, b.) In due course, the ordinary star will evolve, increase its radius, and, if the system is sufficiently close, fill its Roche lobe. Matter is then free to flow toward the neutron star, which may or may not be a pulsar at the time. At least some of the matter will probably be captured and retained by the neutron star. This is the case for a white dwarf under similar circumstances (Giannone and Weigert, 1967).

There is, however, an upper limit to the rate at which matter can be accumulated on the neutron star. For any object radiation pressure more than balances gravity if $L/M \geqslant 6 \times 10^4$ in solar units. Since material falling onto a neutron star must get rid of about 10^{20} ergs/gram of gravitational potential energy, matter cannot be added faster than about 2×10^{18} g/sec or 3×10^{-8} $M_\odot$/yr. If the material attempts to fall straight down, anything in excess of this limit will be blown off by radiation pressure. In the case of the white dwarf, the primary energy source is nuclear reactions rather than gravitational potential energy, and the corresponding limit is about 3×10^{19} g/sec or 5×10^{-7} $M_\odot$/yr. The resulting system, alternately transferring and blowing off material, might look not unlike T Coronae Borealis, the only recurrent nova known to occur in a wide binary.

The real situation is probably not this simple. Material transferred in a CBS will carry considerable angular momentum (unlike, for instance, matter accreted by a neutron star from the interstellar medium, as discussed by Ostriker *et al.*, 1970). It will, therefore, not fall straight in, but form a disc around the neutron star, dropping down only as fast as viscosity transfers angular momentum from the inner to the outer parts of the disc. If viscosity were sufficiently small, the disc might persist much longer than mass transfer lasted (typically the dynamical or thermal time scale rather than the nuclear time scale) and all the matter eventually be captured. If only molecular viscosity counted, this would probably be the situation.

The relevant time scale, τ, is of the order 4×10^{-11} $R^2 N T^{-5/2}$ where R is the radius of the disc, N its average (particle) density, and T its temperature. For a solar mass or so of material filling the Roche lobe around the neutron star in a CBS at a temperature of 10^7 K, τ is about 3×10^9 y. Including turbulent and magnetic viscosity will decrease τ very considerably, and it is probable that material will fall in as fast as the L/M limits will let it. There is one further process to be considered. The 10^{20} erg/g available from gravitational potential energy would suffice to raise the temperature of the infalling material to nearly 10^{12} K. At such exalted temperatures, the cross-sections for processes producing neutrinos are no longer negligible. And, since any neutrinos formed will pass freely through the disc, energy radiated in this form is exempt from the L/M restriction. Thus a mechanism is perhaps available that will allow mass to be accepted by a neutron star as fast as its companion wants to get rid of that mass, even without storing it in a disc.

If sufficient mass deposition occurs to raise the neutron star above the largest stable mass for such a configuration, then either some violent event will take place (which may be presumed to disrupt the system) or the star will collapse to a black hole. Systems undergoing mass transfer and/or deposition should be short period (less than 10 yr or so) spectroscopic binaries showing lines of at most one star (and, perhaps, a disc), in which the invisible component is more massive than the Chandrasekhar limit. Tables of the 50 singleline spectroscopic binaries with well-determined orbits containing an unseen star larger than 1.4 $M_\odot$ are given by Trimble and Thorne (1969). None of these systems has a primary component which completely fills its Roche lobe, nor can any of them have matter falling from a disc to the neutron star at more than 10^{-3} of the limiting rate, unless neutrino energy loss is extremely efficient.

Matter falling on a neutron star should release its gravitational potential energy largely in the form of X-rays (Shklovsky, 1967, and others). None of the tabulated systems coincides with any X-ray source position in the catalogues of Dolan (1970), Matsuoka (1970), or Seward (1970). We can thus set limits on their X-ray luminosities. Consider system No. 27 of Trimble and Thorne's Table 1. This system is far enough from the galactic plane that X-ray source confusion is not a problem, and its distance (which is typical for these systems) can be determined fairly accurately. The primary is a Cepheid variable of period $8^d.38$ ($M_V \sim -3.7$) seen at apparent magnitude 5.87 to 7.02. The system is, therefore, about 1 kpc from us. Its X-ray flux at the earth is surely not greater than 5×10^{-9} erg cm^{-2} sec^{-1} in the 1.5 to 5 keV band. (Weak observed sources are typically around 2×10^{-9} erg cm^{-2} sec^{-1} in this energy interval.) Thus less than 10^{35} ergs/sec are being radiated as soft X-rays – or as any sort of X-rays unless the spectrum is very flat. This is less than 10^{-3} of the limiting L/M rate which we expect when radiation pressure rather than mass transfer rate or viscosity limits the mass deposition. Thus, in all of the tabulated systems, either no disc has been formed or all but 10^{-3} of the energy of accreting matter is being radiated in some form other than X-rays – presumably neutrinos. This degree of efficiency seems improbable.

The above estimate is a generous one; the secondaries having $M \gtrsim 2.0 M_\odot$ cannot be neutron stars, and matter falling onto a black hole may radiate more than 10^{20} erg/gram in some form or other.

Eclipsing systems have been excluded from consideration. One might imagine material lost by a primary forming a rapidly rotating disc around a black hole or neutron star which was sufficiently substantial to produce eclipses. Such a system would look rather like β Lyrae (which is probably still in a phase of rapid mass transfer) and the secondary component (in the absence of X-ray emission produced as the disc material falls in) could not be distinguished from an ordinary star surrounded by a similar disc. There are again no significant position coincidences of observed X-ray sources with the single-line eclipsing systems tabulated by Trimble and Thorne (1969).

More generally, there is a limit to the input rate of soft X-rays into the Galaxy from the temperatures of interstellar H I clouds. It is about 10^{40} erg/sec. The contribution from all neutron stars accreting matter (in or out of CBS's) must not exceed this. Neutron stars passing through the general interstellar medium may supply a significant fraction of this (Ostriker et al., 1970). In any case, if $L/M = 6 \times 10^4$ and M is about one solar mass, at most 40 objects can be fitted in under the upper limit.

Thus, if one pulsar is formed every 30 yr (Gunn and Ostriker, 1970), the time during which matter is accreted multiplied by the fraction of pulsars in close binaries is at most 1200 yr. If infalling matter is partially blown off by radiation pressure and/or neutrino losses are important, the accretion time will be no greater than the time during which the primary wants to lose mass. This is typically a thermal time scale and is in the range 10^4 to 10^6 yr. Thus at most 0.1 to 0.001 of pulsars formed in the recent past can have remained in CBS. If neutron stars which are not pulsars occur in close binaries these limits must be made more stringent. The formation of a disc in which viscosity is so low that matter continues to fall in at the limiting rate of 2×10^{18} g/sec until all the disc material has radiated away its potential energy (this takes 10^8 years or so) would lower the limit to one pulsar in 10^5 in a binary system.

One might get around the X-ray limit by postulating, ad hoc, a large (e.g. $R \sim 2 \times 10^{15}$ cm at $T = 600$ K) cloud around each such system which transforms it into an infra-red source (and incidentally accounts for the rarity of optical identifications of pulsars). Because the galactic infrared background is stronger than the X-ray background most pulsars formed might radiate at the limiting rate for 10^5 y or so. The binary systems discussed above could not, of course, have their X-rays hidden in this fashion because we see the objects optically.

The conclusion of Trimble and Thorne (1969) that we probably do not observe any binary systems containing neutron stars or black holes is thus reiterated. Nor can many of the pulsars formed in the past be members of CBS's without violating X-ray limits. The evidence of pulsar period constancy suggests that the vast majority of such objects now observed do not have close companions.

3. Pulsars not in Close Binary Systems

Evidently one of two things must happen – either most stars in CBS's do not undergo the sort of supernova event which produces pulsars or the explosion always disrupts the system – or both. It would not be surprising if binary components were to end their lives in some rather different fashion from single stars. The initial primary in a CBS, which always remains the more rapidly evolving star (Paczynski, 1967) although it becomes the less massive after mass exchange, will have lost much or all of its hydrogen-rich envelope before reaching the supernova stage. This may actually influence the course of interior evolution, for instance by preventing the explosive carbon ignition (Paczynski, private communication) which perhaps produces some supernovae, and it may well be critical in determining how much matter is expelled in the event, how much implodes, and how far. If Arnett's (1971) explanation of supernova light curves is correct, the loss of the hydrogen envelope will drastically change what we see of the event.

Alternatively, the system may be disrupted. Gunn and Ostriker (1970) have suggested that this usually happens, both the neutron star and the secondary flying off as runaways. This provides an explanation of the observed correlation of periods of pulsars with their distances from the galactic plane and of the supposedly anomalous proper motion of the Crab Nebula (but see elsewhere in this volume). The systems are presumed to start with a scale height of about 80 pc and the neutron stars to fly off with a Maxwellian distribution of velocities peaked at 100 km/sec.

Unfortunately, the phenomenon of mass exchange has not been taken into account in formulating this picture. In order for an isotropic expansion to unbind the system, it is necessary that the difference of the component masses, $M_1 - M_2$, be greater than $2M_{N1}$ (twice the mass of the neutron star formed). It is clear that, once mass transfer has transformed the initial primary into the less massive (but still more rapidly evolving) star, no possible explosion of this star, M_1, can fulfill the condition. It is, therefore, necessary that the star not fill its Roche lobe at any phase preceding supernova explosion.

This restriction allows upper limits to be placed on runaway velocities from systems of various masses, by requiring that the more massive star not be larger than the radius of its Roche lobe at any phase of stellar evolution which has been studied. Formulae given by Gott *et al.* (1970) can be combined to yield the velocity of the neutron star after the explosion,

$$V_{N1} = \frac{G^{1/2}M_2(M_1 - M_{N1})}{\mathbf{a}^{1/2}(M_1 + M_2)^{1/2}[M_{N1} + \{M_1^2 - 2M_{N1}(M_1 + M_2)\}^{1/2}]}$$

where **a** is the semi-major axis of the binary orbit, G the gravitational constant, and the M's as above. Radii of Roche lobes in terms of **a** are given by Kopal (1959), and maximum radii reached by stars of various masses can be taken from Paczynski (1970).

TABLE I

Upper limits to runaway velocities of neutron stars formed by supernovae in close binary systems

M_1 (solar masses)	M_2 (solar masses)	M_{N1} (solar masses)	$R_1{}^{max}$ (cm)	R^{lobe}/a	a^{min} (cm)	$V_{N1}{}^{max}$ (km/sec)
3	$\frac{1}{2}$	$\frac{1}{2}$	7.59×10^{13}	0.567	1.34×10^{14}	2.3
3	1	$\frac{1}{2}$	7.59×10^{13}	0.469	1.62×10^{14}	4.1
3	$\frac{1}{2}$	1	7.59×10^{13}	0.567	1.34×10^{14}	2.2
3	1	1	7.59×10^{13}	0.469	1.62×10^{14}	4.5
5	2	$\frac{1}{2}$	7.59×10^{13}	0.461	1.65×10^{14}	6.4
5	4	$\frac{1}{2}$	7.59×10^{13}	0.398	1.91×10^{14}	11.1
5	2	1	7.59×10^{13}	0.461	1.65×10^{14}	6.3
5	4	1	7.59×10^{13}	0.398	1.91×10^{14}	12.2
7	4	$\frac{1}{2}$	7.42×10^{13}	0.429	1.77×10^{14}	10.2
7	6	$\frac{1}{2}$	7.42×10^{13}	0.398	1.19×10^{14}	13.9
7	4	1	7.42×10^{13}	0.429	1.77×10^{14}	10.1
7	6	1	7.42×10^{13}	0.398	1.91×10^{14}	14.4

The resulting upper limits to V_{N1} are given in Table I. The primary masses have been chosen to be large enough that the stars will have a fairly small scale height in the Galaxy and small enough that the stars will be reasonably common objects. The secondary masses were chosen to give the largest possible runaway velocities consistent with the system becoming unbound.

None of these velocities is anywhere large enough to explain the observed period-scale height correlation, which must, therefore, be otherwise accounted for. Notice that ordinary runaway stars (once secondaries of systems with very massive – e.g. 50 $M_\odot$ – primaries; Blaauw, 1961) cannot be the ancestors of pulsars. In order to produce an age-scale height correlation, the high velocity must be acquired at or near the time the pulsar is formed. If the object had been given a high velocity at a time before the supernova event comparable with pulsar ages (10^6–10^7 y) then all objects would have the same scale height, independent of age (i.e., period).

A way of producing the required velocity is immediately suggested by the dynamics of the Crab Nebula – NP 0531 system. As reported elsewhere in this volume, the nebula and pulsar have proper motions differing by about $0.014 \pm 0.003''$/yr. This, at a distance of 1500 pc, amounts to 100 km/sec, and it has not been suggested anywhere that the star which became SN 1054 was other than single at the time of the explosion.

The period-scale height relation thus finds ready explanation in recoil of the neutron star from an asymmetric supernova explosion. Such asymmetry would also, of course, allow the neutron star to be expelled with larger velocity from a binary system, but the companion star is in no way necessary for the phenomenon to occur.

If on the other hand supernova explosions are usually asymmetrical enough to give the resulting neutron star a 'kick' of 100 km/sec then most of the objects starting out in binaries should be expelled.

The explanation suggested by Michel (1970) for high pulsar velocities also involves a close pair of stars – in this case both of them neutron stars, but as his binary systems have never passed through a main sequence phase (having been formed from a single star) mass transfer is irrelevant.

Acknowledgements

Many of the ideas in this discussion were contributed by Dr. B. Paczynski, who, however, assumes no responsibility for the authors' having made proper use of them. V. T. gratefully acknowledges a NATO Postdoctoral Research Fellowship.

References

Arnett, W. D.: 1971, *Astrophys. J.* **163**, 11.
Blaauw, A.: 1961, *Bull. Astron. Inst. Neth.* **15**, 265.
Dolan, J. F.: 1970, *Astron. J.* **75**, 223.
Gianone, P. and Wiegert, A.: 1967, *Z. Astrophys.* **67**, 41.
Gott, J. R., Gunn, J. E., and Ostriker, J. P.: 1970, *Astrophys. J.* **160**, L91.
Gunn, J. E. and Ostriker, J. P.: 1970, *Astrophys. J.* **160**, 979.
Kopal, Z.: 1959, *Close Binary Systems*, Chapman and Hill, London, p. 136.
Matsuoka, M.: 1970, Univ. of Tokyo, Space Aeronautics Sci. Report No. 445.
Michel, F. C.: 1970, *Nature* **228**, 1072 and this symposium, Paper 6.8, p. 378.
Ostriker, J. P., Rees, M. J., and Silk, J.: 1970, *Astrophys. Letters* **6**, 179.
Paczynski, B.: 1967, *Acta Astron.* **20**, 47.
Plavec, M.: 1968a, *Astrophys. Space Sci.* **1**, 239.
Plavec, M.: 1968b, *Advances Astron. Astrophys.* **6**, 201.
Seward, F. D.: 1970, Univ. Calif. Livermore Preprint, UCID 15622.
Shklovsky, I. S.: 1967, *Astrophys. J.* **148**, L1.
Terzian, Y.: 1970, privately circulated pulsar tables.
Trimble, V. L. and Thorne, K. S.: *Astrophys. J.* **156**, 1013.

PHYSICS OF THE CRAB NEBULA

5.1 PLASMA INTERACTIONS IN THE CRAB NEBULA

F. D. KAHN

Astronomy Department, University of Manchester, Manchester, U.K.

Abstract. Alfvén waves can be carried by the thermal plasma in the Crab Nebula. Each such wave perturbs the relativistic plasma present, in particular it strongly affects those particles which are in resonance with it. For a wave travelling parallel to the magnetic field the resonances are quite simple, but a wave travelling obliquely can give rise to multiple resonances, and can therefore couple together particles with quite different energies.

It is shown that the interaction with the relativistic plasma leads to an amplification of the Alfvén waves when the mean velocity of relativistic plasma relative to the thermal plasma exceeds the Alfvén speed. The rise time for the instability is quite short, and the waves, once excited, are highly effective in redistributing the directions of motion of the relativistic particles. The relativistic plasma therefore cannot stream freely through the thermal plasma. On the other hand the disturbances will never quite die out in the Crab Nebula as long as the Crab pulsar keeps injecting fresh plasma, and thus keeps producing new inhomogeneities in the relativistic plasma density.

1. Introduction

The Crab Nebula shows unmistakable signs
  (i)   of containing relativistic charged particles,
  (ii) of the presence of a magnetic field,
  (iii) of undergoing frequent injections of fresh relativistic particles produced in the Crab pulsar.

The nebula cannot therefore be in a state of equilibrium. Further, even though most of the internal energy of the system is due to relativistic particles and to an (electro) magnetic field, the overall expansion of the nebula takes place at a speed of around 10^8 cm sec^{-1}, much less than the speed of light. Since the gravitational potential within the nebula is negligibly small, it follows that the inertia of the thermal plasma present holds back the relativistic plasma and stops it from getting away much faster. It therefore becomes important to study interactions between relativistic and thermal plasmas.

I shall discuss one way in which such an interaction can arise, through a coupling via the Alfvén waves carried by the thermal plasma. The properties of Alfvén waves are easily derived. The perfect conductivity condition requires that a plasma always responds to low frequency fluctuations in such a way that there is no electric field with respect to the rest frame of the plasma, or that

$$\mathbf{E} + \frac{\mathbf{u}}{c} \wedge \mathbf{H}_0 = 0 \tag{1}$$

where $\mathbf{u}$ is plasma velocity, $\mathbf{H}_0 \equiv (0, 0, H_0)$ is the zero-order magnetic field, and $\mathbf{E}$ is the electric field.

Davies and Smith (eds.), The Crab Nebula, 281–291. All Rights Reserved.
Copyright © 1971 by the IAU.

Note that $\mathbf{E}$ is perpendicular to $\mathbf{H}_0$. The vector potential $\mathbf{A}$ of the Alfvén wave may be defined by

$$\mathbf{E} = -\frac{1}{c}\frac{\partial \mathbf{A}}{\partial t};$$ (2)

expressed in components

$$\mathbf{A} = (A_1, A_2, 0).$$ (3)

If the wave has a space-time dependence of the form $\exp i\,(k_1 x_1 + k_3 x_3 - \omega t)$, it follows from (1), (2) and (3) that

$$i\omega\,(A_1, A_2, 0) + H_0\,(u_2, -u_1, 0) = 0.$$ (4)

The linearized equation of motion for the plasma is

$$\varrho_0 \frac{\partial \mathbf{u}}{\partial t} = \frac{1}{c}\mathbf{j} \wedge \mathbf{H}_0,$$ (5)

where ϱ_0 is the density of the thermal plasma, and where the current density is denoted by

$$\mathbf{j} = \frac{c}{4\pi} \nabla \wedge (\nabla \wedge \mathbf{A}) = \frac{c}{4\pi}\{k_3^2 A_1, (k_1^2 + k_3^2)A_2, -k_1 k_3 A_1\}.$$ (6)

Equation (4) now becomes

$$-i\omega\varrho_0\,(u_1, u_2, u_3) = \frac{H_0}{4\pi}\{(k_1^2 + k_3^2)A_2, -k_3^2 A_1, 0\}.$$ (7)

Thus $u_3 \equiv 0$, and the wave motion splits into two modes. One mode is described by Mode I:

$$i\omega A_1 + u_2 H_0 = 0$$ (8)

and

$$i\omega\varrho_0 u_2 - \frac{k_3^2 H_0}{4\pi} A_1 = 0.$$ (9)

The waves of this mode obey the dispersion relation

$$\omega^2 = \frac{H_0^2}{4\pi\varrho_0} k_3^2 \equiv v_A^2 k_3^2,$$ (10)

and the components of the group velocity are

$$\partial\omega/\partial k_1 = 0, \qquad \partial\omega/\partial k_3 = \pm v_A.$$ (11)

Energy therefore propagates parallel to the three-axis with speed $\pm v_A$. The space-time dependence of the wave has the form

$$\exp i\{k_1 x_1 + k_3\,(x_3 - v_A t)\}.$$ (12)

The other mode of the wave is described by
Mode II:

$$i\omega A_2 - u_1 H_0 = 0 \tag{13}$$

and

$$i\omega \varrho_0 u_1 + \frac{1}{4\pi} (k_1^2 + k_3^2) H_0 A_2 = 0, \tag{14}$$

with the dispersion relation

$$\omega^2 = v_A^2 (k_1^2 + k_3^2). \tag{15}$$

The group velocity now has components

$$\partial \omega / \partial k_1 = v_A \sin \alpha, \qquad \partial \omega / \partial k_3 = v_A \cos \alpha,$$

where the wave-vector $\mathbf{k}$ makes an angle $\alpha \equiv \tan^{-1} k_1/k_3$ with the three-axis. The space-time dependence of the waves in this mode is

$$\exp i \{(k_1 x_1 + k_3 x_3) - (k_1^2 + k_3^2)^{1/2} v_A t\}. \tag{16}$$

The Alfvén waves are thus split into two linearly polarized modes, with different frequencies. Degeneracy occurs only if $k_1 (=k_2)=0$. In that case it is possible to describe the waves in terms of independent, circularly polarized modes.

2. Response of the Relativistic Plasma to Alfvén Waves

I shall now consider how the relativistic plasma is affected by the Alfvén waves, and later find a condition under which the resulting interaction leads to an instability. For the sake of simplicity the wave-vector $\mathbf{k}$ of the Alfvén wave is taken to be parallel to the three-axis. In that case one can assume that the wave is circularly polarized, and has a vector potential given by

$$\mathbf{A} = A (1, i, 0) \exp \{ik_3 (x_3 - v_A t)\}. \tag{17}$$

To illustrate a simple case, let the relativistic plasma present have a momentum distribution which is isotropic with respect to a frame of reference Σ_R, where Σ_R moves with a velocity $(0, 0, v_R)$ relative to the rest-frame Σ of the thermal plasma. With respect to Σ_R, the momentum distribution is

$$f_R = F_0 (p_{\|, R}^2 + p_\perp^2), \tag{18}$$

where $p_{\|, R}, p_\perp$ are the components of momentum parallel and perpendicular to the three-axis as seen from Σ_R. Relative to the frame of reference Σ, the distribution function becomes

$$f_0 = F_0 \left(p_\|^2 + p_\perp^2 - \frac{2 v_R \mathscr{E}}{c^2} p_\| \right)$$

$$= F_0 (p_\|^2 + p_\perp^2) - \frac{2 v_R \mathscr{E}}{c^2} p_\| F_0' (p_\|^2 + p_\perp^2). \tag{19}$$

This result is correct to order v_R/c, and therefore good enough, unless the relative mean velocity of the two plasmas is very large. In Equation (19), $p_{\parallel}$ is the three-component of momentum and $\mathscr{E}$ the energy of a particle, with respect to Σ. The linearized Vlasov equation for the relativistic plasma is

$$\frac{\partial f}{\partial t} + \dot{x}_3 \frac{\partial f}{\partial x_3} + \dot{\varphi} \frac{\partial f}{\partial \varphi} + \dot{p}_{\parallel} \frac{\partial f}{\partial p_{\parallel}} + \dot{p}_{\perp} \frac{\partial f}{\partial p_{\perp}} = 0. \tag{20}$$

I define

$$\varphi = \tan^{-1} p_2/p_1 \tag{21}$$

where p_1, p_2 are the one and two components of momentum.

Equation (20) is to be linearized and then solved. For this purpose note that $\partial f/\partial x_3$ and $\partial f/\partial \varphi$ both vanish in the undisturbed plasma, which is spatially homogeneous, and has axial symmetry in momentum space with respect to the three-axis. The coefficients of these two terms need only be calculated to zero order, since $\partial f/\partial x_3$ and $\partial f/\partial \varphi$ can only appear as first order quantities. The coefficients are $\dot{x}_3 = c^2 p_{\parallel}/\mathscr{E}$, and $\dot{\varphi} = -ecH_0/\mathscr{E}$; the first of these results is obvious from the relation of velocity and momentum, and the second describes the Larmor precession of a charged particle in the undisturbed magnetic field.

On the other hand, both $p_{\parallel}$ and $p_{\perp}$ stay constant in the undisturbed field, and therefore $\dot{p}_{\parallel}$ and $\dot{p}_{\perp}$ are first order quantities. It is then enough to insert for $\partial f/\partial p_{\parallel}$ and $\partial f/\partial p_{\perp}$ the zero-order quantities $\partial f_0/\partial p_{\parallel}$ and $\partial f_0/\partial p_{\perp}$.

The rates of change of $p_{\parallel}$ and $p_{\perp}$ can be found in terms of the vector potential **A**. A brief calculation yields that

$$\left. \begin{aligned} \dot{p}_{\parallel} &= \frac{ik_3 eAc}{\mathscr{E}} p_{\perp} e^{i\varphi} \\ \dot{p}_{\perp} &= \frac{ik_3 eA}{c} \left(v_A - \frac{c^2 p_{\parallel}}{\mathscr{E}} \right) e^{i\varphi}. \end{aligned} \right\} \tag{22}$$

When these quantities are substituted into the linearized equation, and use is made of the assumed space-time dependence (17) of the wave, it turns out that, correct to the first order, the change in the plasma distribution function is given by

$$f^{(1)} \left\{ \frac{k_3 c^2 p_{\parallel}}{\mathscr{E}} - kv_A - \frac{ceH_0}{\mathscr{E}} \right\} = \frac{k_3 eAc}{\mathscr{E}} \left\{ \left(p_{\parallel} - \frac{\mathscr{E}v_A}{c^2} \right) \frac{\partial f_0}{\partial p_{\perp}} - p_{\perp} \frac{\partial f_0}{\partial p_{\parallel}} \right\} \tag{23}$$

I now substitute for f_0 from equation (19) and find that

$$f^{(1)} = \frac{(2\mathscr{E}p_{\perp}/c^2)(eA/c)(v_R - v_A)F_0'}{p_{\parallel} - \mathscr{E}v_A/c^2 - eH_0/ck_3}. \tag{24}$$

$f^{(1)}$ becomes large for particles that resonate with the wave, that is for particles which satisfy

$$p_{\parallel} = \frac{\mathscr{E}v_A}{c^2} + \frac{eH_0}{ck_3}. \tag{25}$$

The resonance depends, essentially, on the $p_{\parallel}$, k_3 relation. It also depends on the sense of polarization of the wave. The present calculation is done in terms of a circularly polarized wave: to reverse the sense of polarization we need only change k_3 to $-k_3$. However, a plane polarized wave can be split into two circularly polarized waves with opposite senses of polarization, and it will therefore resonate with particles at two different values of $p_{\parallel}$.

The resonance condition (25) has a simple interpretation. The components of $\mathbf{p}_{\perp}$ for a particle in its zero order trajectory are

$$(p_1, p_2) = p_{\perp} \left\{ \cos\left(\frac{ceH_0}{\mathscr{E}} t + \varepsilon\right), \; -\sin\left(\frac{ceH_0}{\mathscr{E}} t + \varepsilon\right) \right\}, \qquad (26)$$

where ε is a phase factor. The one and two components of the vector potential are

$$(A_1, A_2) = A \left\{ \cos(k_3 x_3 - \omega t), \sin(k_3 x_3 - \omega t) \right\}$$

$$= A \left\{ \cos\left(\frac{k_3 c^2 p_{\parallel}}{\mathscr{E}} - \omega\right) t, \sin\left(\frac{k_3 c^2 p_{\parallel}}{\mathscr{E}} - \omega\right) t \right\} \qquad (27)$$

at the position of the particle in its zero order orbit. Resonance occurs when the scalar product $\mathbf{p}_{\perp}$. $\mathbf{A}$ has a phase factor which is independent of time, that is when

$$-\frac{ceH_0}{\mathscr{E}} = \frac{k_3 c^2 p_{\parallel}}{\mathscr{E}} - \omega. \qquad (28)$$

With ω set equal to $k_3 v_A$, conditions (25) and (28) are found to be equivalent.

3. Alfvén Waves which Propagate Obliquely

The interaction of the relativistic plasma with an Alfvén wave becomes somewhat more complex when the wave-vector $\mathbf{k}$ is not parallel to the magnetic field. As was mentioned before, the Alfvén wave now has two independent plane polarized modes, I and II. Each of these modes can be analysed into two circularly polarized waves, with opposite senses of polarization, but the component waves are now no longer independent, and must always be considered together.

Another important modification occurs in the resonance condition. The components of $\mathbf{p}_{\perp}$ are, once again

$$(p_1, p_2) = p_{\perp} \left\{ \cos\left(\frac{ceH_0}{\mathscr{E}} t + \varepsilon\right), \; -\sin\left(\frac{ceH_0}{\mathscr{E}} t + \varepsilon\right) \right\}, \qquad (29)$$

for the zero order particle trajectory; the components of the vector potential become

$$(A_1, A_2) = A \left\{ \cos(k_1 x_1 + k_3 x_3 - \omega t), \sin(k_1 x_1 + k_3 x_3 - \omega t) \right\}. \qquad (30)$$

The same substitution as before can be made for x_3, and for the other position variable it is found that

$$x_1 = \int v_1 \, dt = \frac{c^2}{\mathscr{E}} \int p_1 \, dt = \frac{cp_{\perp}}{eH_0} \sin\left(\frac{ceH_0}{\mathscr{E}} t + \varepsilon\right). \qquad (31)$$

In the complex notation the phase dependence of the vector potential at the position of the particle can now be expressed by

$$\mathscr{A} \equiv A_1 + iA_2 = A \exp i \left\{ \frac{k_1 c p_\perp}{eH_0} \sin\left(\frac{ceH_0}{\mathscr{E}} t + \varepsilon\right) + \left(\frac{k_3 c^2 p_\parallel}{\mathscr{E}} - \omega\right) t \right\};$$

(32)

the corresponding relation for $\mathbf{p}_\perp$ is

$$\mathscr{P} \equiv p_1 + ip_2 = p_\perp \exp\left\{ -i\left(\frac{ceH_0}{\mathscr{E}} t + \varepsilon\right) \right\}.$$

(33)

The presence of the sine term in the exponent in (32) somewhat changes the phase relation. On using the series expansion

$$e^{iz \sin \theta} = \sum_{-\infty}^{\infty} J_r(z) e^{ir\theta}$$

(34)

relation (32) may be re-written in the form

$$\mathscr{A} = A \sum_{-\infty}^{\infty} J_r\left(\frac{k_1 c p_\perp}{eH_0}\right) \exp ir\left(\frac{ceH_0}{\mathscr{E}} t + \varepsilon\right) \exp i\left(\frac{k_3 c^2 p_\parallel}{\mathscr{E}} - \omega\right) t$$

$$\equiv \sum_{-\infty}^{\infty} \mathscr{A}_r \exp ir\left(\frac{ceH_0}{\mathscr{E}} t + \varepsilon\right) \exp i\left(\frac{k_3 c^2 p_\parallel}{\mathscr{E}} - \omega\right) t.$$

(35)

The particle can now have a resonance with any one of the infinite number of components into which the expression for $\mathscr{A}$ has been analysed. The condition for resonance with the rth component is clearly

$$\frac{ceH_0}{\mathscr{E}} = r \frac{ceH_0}{\mathscr{E}} + k_3 \frac{c^2 p_\parallel}{\mathscr{E}} - \omega$$

or

$$p_\parallel = \frac{\mathscr{E}\omega}{c^2 k_3} - (r-1) \frac{eH_0}{ck_3}.$$

(36)

A given wave can now resonate with a much larger number of particles. The importance of the rth resonance is indicated by the value of the coupling coefficient $J_r(k_1 c p_\perp/eH_0)$. If $k_1 = 0$, then all these coefficients are zero, except that for $r = 0$. In that case condition (36) is identical with condition (25) (when ω/k is set equal to v_A).

However, interesting new effects can arise when there are high resonances, for which $r \gg 1$. The coupling coefficient $J_r(k_1 c p_\perp/eH_0)$ for such a resonance is appreciable only when

$$r \approx k_1 c p_\perp/eH_0,$$

(37)

that is when the argument of the Bessel function is approximately equal to its order. I shall now compare relation (37) with relation (36), but shall first simplify the latter by noting that the term $\mathscr{E}\omega/c^2k_3$ is small compared with the momentum of the particle, since the phase velocity of the Alfvén wave will certainly be small compared with the speed of light. Relation (36) is therefore, to a good enough approximation,

$$p_\| \doteq -(r-1)\frac{eH_0}{ck_3} \doteq -r\frac{eH_0}{ck_3}. \tag{38}$$

Two results now follow: first the wave-number k of the Alfvén wave is related to the momentum of the resonating particle by

$$k \doteq \frac{reH_0}{cp}. \tag{39}$$

In a high resonance a particle of given energy resonates with a wave having a much larger wave-number (and much smaller wavelength) than that of the wave with which the particle is in zero resonance. However, the resonant wave-number varies inversely as the momentum (or the energy) of the relativistic particle. It therefore is possible for a given wave to resonate at the same time with a particle of low energy, in a low resonance, and a particle of high energy, in a high resonance. By this means the high energy particles can be coupled to the low energy particles in the relativistic plasma, or, more generally, the whole range of energies can be coupled together.

It also follows from equations (37) and (38) that

$$p_\perp/p_\| \doteq -k_3/k_1 = -\cot\alpha. \tag{40}$$

High resonance therefore always occurs with particles which are travelling nearly at right angles to the wave-vector $\mathbf{k}$, or almost in planes of constant phase of the Alfvén wave.

4. The Growth Rate for Unstable Waves

The full treatment of the properties of oblique waves will clearly be rather complex. To illustrate the argument I therefore return, now, to Alfvén waves for which $k_1=0$, that is to waves whose wave-vector is parallel to the magnetic field. The disturbance caused by the wave sets up a current in the relativistic plasma given by

$$\mathbf{j} = N_R e \int_0^\infty \int_{-\infty}^\infty \int_0^{2\pi} \frac{c^2}{\mathscr{E}} (p_\perp\cos\varphi,\ p_\perp\sin\varphi,\ p_\|) f^{(1)} p_\perp\, dp_\perp\, dp_\|\, d\varphi. \tag{41}$$

In this formula N_R is the number of particles per cm³ in the relativistic plasma, and the form of $f^{(1)}$ is given by relation (24) in terms of the perturbing wave and the zero order particle distribution. The zero order distribution is assumed to be normalized to unity, and the phase factor $\exp i\{k_3(x_3 - v_A t)\}$ is assumed to be understood. The integration past the singularity of $f^{(1)}$, at $p_\| = (\mathscr{E}v_A/c^2) + (eH_0/ck_3)$, is carried out in the usual manner by deforming the contour in such a way that the calculation gives the value obtained when the frequency ω tends to the real axis from the upper half of the complex-ω plane. The result is that

$$\mathbf{j} = - i (\operatorname{sgn} k_3) \frac{2\pi^2 N_R e^2}{c} A (v_R - v_A)(1, i, 0) \int_0^\infty p_\perp F_0 (p_\parallel^{*2} + p_\perp^2) \, dp_\parallel \quad (42)$$

after some manipulation. $p_\parallel^*$ is the resonant value of $p_\parallel$. For comparison the electric field of the Alfvén wave is

$$\mathbf{E} = \frac{i k_3 v_A}{c} A (1, i, 0), \tag{43}$$

and $\mathbf{E} \cdot \mathbf{j}$ is negative when

$$|k_3| v_A (v_R - v_A) > 0. \tag{44}$$

In that case the plasma response is such that work is done by the plasma on the wave. In other words the wave draws energy from the plasma, and grows. The condition for instability can be written

$$|v_R| > |v_A|, \tag{45}$$

and this means that instability will occur whenever the diffusion speed v_R of the relativistic plasma, with respect to the thermal plasma, exceeds the Alfvén speed v_A. The growth rate σ of the instability is given by

$$\sigma = \frac{1}{2} \frac{\text{rate at which work is done on the wave/cm}^3}{\text{energy density of the wave}} \tag{46}$$

I make the estimate that a typical value of

$$\int_0^\infty p_\perp F_0 (p_\parallel^{*2} + p_\perp^2) \, dp_\perp$$

is of order

$$\frac{1}{2\pi p_\parallel^*} \sim \frac{c}{2\pi \bar{\mathscr{e}}},$$

where $\bar{\mathscr{e}}$ is a typical energy in the plasma. Further, from relation (25)

$$p_\parallel \sim \bar{\mathscr{e}}/c \sim e H_0 / c k_3,$$

and now substituting into equation (46) I find that, by order of magnitude,

$$\sigma \sim \frac{4\pi^2 N_R e v_A (v_R - v_A)}{c H_0} \sim \frac{\pi N_R e H_0}{c \varrho_0}, \tag{47}$$

if v_R is of the order of v_A $(\equiv \sqrt{H_0^2/4\pi\varrho_0})$.

Typical values for the Crab Nebula can be substituted into (47) by setting, for example,

N_R = number density of relativistic particles = 10^{-6} cm^{-3},
$H_0 = 10^{-4}$ G
ϱ_0 = thermal plasma density = 10^{-21} g/cm^3,

with the result that

$$\sigma \approx 5 \times 10^{-9} \; \text{sec}^{-1} \equiv 0.17 \; \text{yr}^{-1}.$$

The growth rate can therefore be quite large. It is even larger for larger assumed values of the relative diffusion speed v_R.

5. The Redistribution of Particle Momenta by the Wave

It remains to estimate how rapidly the particle momenta are redistributed by the ensemble of Alfvén waves which result from the instability. The major effect leading to the redistribution is the acceleration of the charged particles by the disturbance magnetic field $H^{(1)}$. There is also an acceleration due to the disturbance electric field $E^{(1)}$. But this field is of order (v_A/c) times $H^{(1)}$, and therefore small enough to be neglected here.

Consider now a typical particle whose energy is $\mathscr{E}$ and momentum $p = \mathscr{E}/c$. It resonates with the Alfvén wave whose wavenumber is $k_3 \sim eH_0/\mathscr{E} \equiv k^*$, say. A wave with a nearby wave-number, $k^* + \delta k$, will change its phase relative to the resonant wave at a rate such that the change in phase as observed at the moving particle is

$$\begin{aligned} \delta\varphi &= (v_\parallel - v_A) \, T \, \delta k \\ &\doteq v_\parallel T \, \delta k \end{aligned} \tag{48}$$

after time T. During the time interval T the waves which lie within a range

$$\delta k \approx 1/v_\parallel T$$

around k^* will therefore contribute coherently to the magnetic field deflecting the particle. If the Fourier decomposition of the magnetic field is given by

$$H = \int \mathscr{H}(k) \, e^{ik(x_3 - v_A t)} \, dk \tag{49}$$

then the effective magnetic field H_{eff} which acts on the particle during the interval T is

$$H_{\text{eff}} = \int_{\delta k} \mathscr{H}(k) \, e^{ik(x_3 - v_A t)} \, dk. \tag{50}$$

In a random field there are no correlations between contributions from different wavenumbers. Hence the expectation value of the effective magnetic field is

$$\langle H_{\text{eff}} \rangle = 0$$

and the mean square effective field is

$$\langle H_{\text{eff}}^2 \rangle = |\mathcal{H}^2(k^*)| \, \delta k . \tag{51}$$

The mean square deflection of a particle is therefore given by

$$\langle \delta\theta^2 \rangle = \frac{c^2 e^2 \langle H_{\text{eff}}^2 \rangle}{\mathcal{E}^2} T^2$$

$$= \frac{c^2 e^2 |\mathcal{H}^2(k^*)|}{\mathcal{E}^2 v_\parallel} T . \tag{52}$$

The characteristic time during which the particle momenta are redistributed is given by the setting $\langle \delta\theta^2 \rangle$ equal to unity, and becomes

$$T_{\text{redist.}} = \frac{\mathcal{E}^2 v_\parallel}{c^2 e^2 |\mathcal{H}^2(k^*)|} \approx \frac{\mathcal{E}^2}{ce^2 |\mathcal{H}^2(k^*)|} \tag{53}$$

with $v_\parallel \approx c$.

The mean free path for a relativistic particle is therefore of the order of

$$l = cT_{\text{redist.}} \approx \frac{\mathcal{E}^2}{e^2 |\mathcal{H}^2(k^*)|} . \tag{54}$$

If the spectrum of the Alfvén waves were known it would now be possible to calculate mean free paths for particles of various energies. However, it does not seem worth doing this in detail until the various possible interactions between the waves and the particles have been fully studied. I shall only make an order of magnitude estimate. It follows from the relation

$$k^* \approx eH_0/\mathcal{E}$$

that the width of the Alfvén wave spectrum Δk is related to the typical particle energy $\bar{\mathcal{E}}$ by

$$\Delta k \approx eH_0/\bar{\mathcal{E}} \tag{55}$$

If $\overline{H_w^2}$ is the mean square field due to the Alfvén waves, then

$$l \sim \frac{\mathcal{E}^2 \, \Delta k}{e^2 |\mathcal{H}^2(k^*)| \, \Delta k} \sim \frac{\bar{\mathcal{E}} e H_0}{e^2 \bar{H}_w^2} = \frac{\bar{\mathcal{E}} H_0}{e \bar{H}_w^2} . \tag{56}$$

If further the rms wave field $\sqrt{\overline{H_w^2}}$ is, say, β times the undisturbed field H_0, then

$$l \sim \bar{\mathcal{E}}/\beta e H_0 \sim 2 \times 10^{10} \, \beta^{-1} \text{ cm} \tag{57}$$

in the Crab Nebula. Quite moderate amplitudes of the Alfvén waves can thus result in very short mean free paths for the particles in the relativistic plasma.

6. Conclusions

It is very likely that Alfvén waves are readily excited and play an important role in the physics of the Crab Nebula. They lead to the following effects:

(i) The relativistic plasma can only diffuse slowly with respect to the thermal plasma. The pressure in the relativistic plasma therefore acts on the thermal plasma, and governs its motion.

(ii) The Alfvén waves redistribute the directions of motion of the particles in the relativistic plasma. The mean free path for a relativistic particle is quite short, and the wave particle interaction is therefore quite effective in making momentum distributions more isotropic.

(iii) However, since the mean free path for a plasma particle is finite, there can never be complete isotropy while there are inhomogeneities in the plasma density. In particular some diffusion of the relativistic plasma will occur, relative to the thermal plasma. As long as the Crab pulsar keeps injecting new plasma, inhomogeneities will continue to be produced and Alfvén waves will be generated.

(iv) Equation (54) suggests that particles with a large energy $\mathscr{E}$ will have a longer mean free path than particles with small energy. If so, the degree of anisotropy should be larger among the high energy particles than among the low energy particles of the relativistic plasma. This suggests that the particles with a large value of $\mathscr{E}$ in the plasma could still be unstable in places where the particles with low $\mathscr{E}$ are stable. As a result energy might possibly be transferred via an Alfvén wave, with which the high-$\mathscr{E}$ particles have a high resonance and the low-$\mathscr{E}$ particles a low resonance. In this way additional energy could be supplied to the electrons, and this may extend the time during which they can emit synchrotron radiation.

References

Details of the physics of the Crab Nebula are given in several other papers contributed to this symposium.

The theory of Alfvén waves and of plasma waves has been discussed by many authors, including

P. C. Clemmow and J. P. Dougherty, *Electrodynamics of Particles and Plasmas*, Addison Wesley, 1969.

The importance of the interaction of relativistic plasmas and Alfvén waves in the interstellar plasma has been emphasied by a number of authors. The effect is essentially the same as that in the Crab Nebula. References are

R. M. Kulsrud and W. P. Pearce: 1969, *Astrophys. J.* **156**, 445.

J. Skilling: 1970, *Monthly Notices Roy. Astron. Soc.* **147**, 1.

D. G. Wentzel: 1969, *Astrophys. J.* **156**, 303.

Skilling in particular gives an interesting discussion of the associated physical effects.

Discussion

J. P. Ostriker: Your mechanism is similar to the one proposed to stop the streaming of cosmic rays.

F. D. Kahn: Yes – it is a standard result.

5.2 RADIO POLARIZATION OF THE CRAB NEBULA

R. G. CONWAY

University of Manchester, Nuffield Radio Astronomy Laboratories, Jodrell Bank, U.K.

Abstract. Optical and radio measurements of polarization can be combined to find the configuration of the magnetic field throughout the Crab Nebula. The component in the line of sight is found from the radio rotation measure.

New measurements of polarisation at 11 cm with a resolution of $7'' \times 14''$ are combined with previous results to show that the rotation measure is fairly uniform and near -25 rad m^{-2} near the centre, but that it becomes irregular near the edges. It rises to 300 rad m^{-2} in the filaments, possibly indicating a concentration of electrons about 10–100 cm^{-3}.

The Crab Nebula is of interest historically on two counts: not only was it the first radio source to be identified with an optical object, it was also the first source in which the synchrotron mechanism was identified, by virtue of the linear polarization of the optical radiation. Figure 1 shows the map at $5''$ arc resolution obtained by Woltjer, giving the fractional polarization m and the position angle χ, quantities which are related to the Stokes parameters I, Q and U by the expression for the 'polarized intensity'.

$$mI \exp(2j\chi) = Q + jU.$$

The central portion of the nebula exhibits a more or less aligned polarization, with the E vectors running roughly parallel to the major axis of the nebula. The magnetic field thus appears to be predominantly along the minor axis direction. Surrounding the central area are a series of 'fans' in each of which the E vectors are arranged radially about a centre. Here the percentage polarization is high, and may exceed 50%. For the most part the fans are related to prominent filaments, which are seen by Woltjer as the seat of powerful currents, the magnetic field of which is arranged radially around the filament throughout a considerable volume. The polarization is not related to filaments however at the north end of the nebula.

No radio measurements yet exist at as high a resolution as the optical map. In principle the radio observations could complement the optical, since the optical polarization determines the direction of $B_\perp$, while the radio maps which are subject to modification by Faraday rotation, could give information on $B_\parallel$. In practice, however, the information on $B_\parallel$ is still incomplete and statistical rather than detailed. We shall discuss the information gained by studies of progressively finer resolution.

Burn (1966) has considered the integrated polarization of the nebula as a whole. The value of m decreases rapidly from 9% (optical) to a few percent (at $\lambda 3$ cm) thereafter decreasing slowly with wavelength. Burn assumes that this variation is caused by the Faraday effect, and is due to the fact that some parts of the nebula suffer more

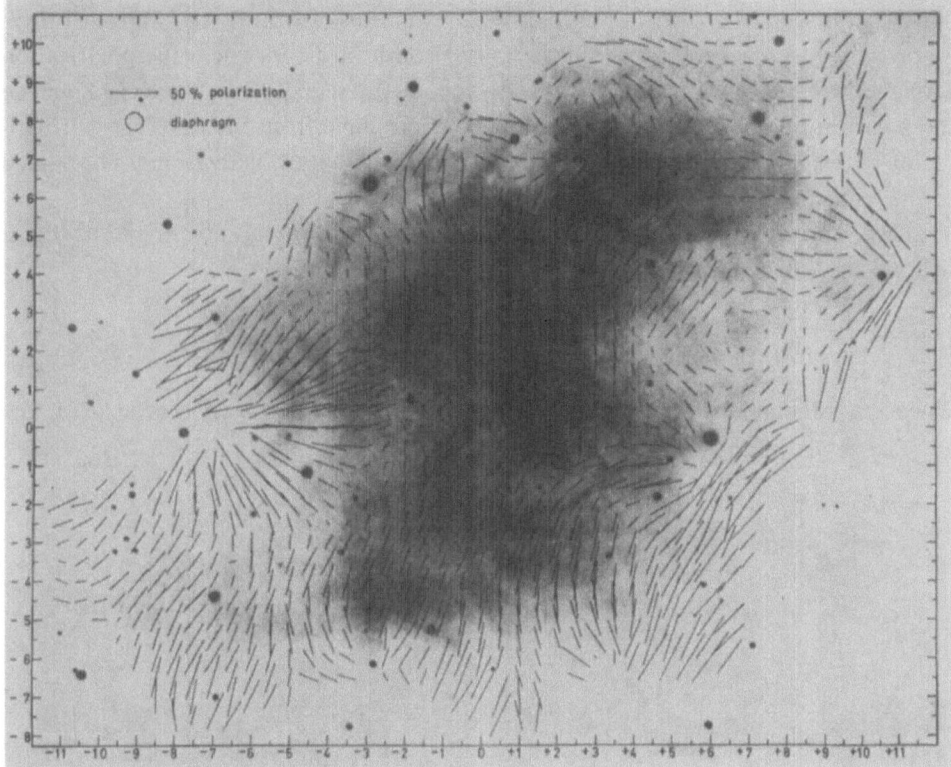

Fig. 1. Optical polarization of the Crab Nebula. The origin is at RA 05^{h}31^{m}31.5^s, Dec 21°58′55″, and the unit of distance is 11″ arc.

rotation than others. If the usual definition of rotation measure is adopted (Gardner and Whiteoak, 1966):

$$RM = 8 \times 10^5 \int N_e B \, dl \quad \text{radians}/m^2$$

then we may define a 'Faraday dispersion', Δ, as the range of rotation measures present from greatest to least. Burn shows that the variation of m with λ is determined principally by the dispersion, Δ, and deduces that the value of Δ for the Crab is ~ 1000 rad/m^2. Order of magnitude estimates for N_e and B in or around the filaments allow values of this order, though with considerable uncertainty.

Burn's assumption that the $m(\lambda)$ curve is entirely due to Faraday rotation may be questioned, since it is known that the electrons producing the radio radiation are distributed over a wider volume than those producing the optical radiation. Basically, as the radio data become available, one wishes to repeat Burn's analysis, point by point, across the face of the nebula, to overcome this objection.

Low-resolution maps of the nebula have been published by Downs and Thompson (1968), by Mayer and Hollinger (1967), by Allen (1967) and by Wright (1970) at wavelengths from $\lambda 1.55$ to $\lambda 21$ cm, all with resolution of from 1' to 2' arc. All these maps show a central region, from the pulsar southwards, in which the polarization is aligned, and another region, centred on the northern end of the major axis, where the polarization has a markedly different position angle from the rest. The northern region corresponds to the area on the optical map where the polarization is not related to the filaments.

The rotation measure may be deduced for the central area and for the northern region. The values are

$$\text{Centre} - 25 \text{ rad/m}^2$$
$$\text{North} - 40 \text{ rad/m}^2.$$

Since the integrated polarization is determined by the (stronger) central area, the

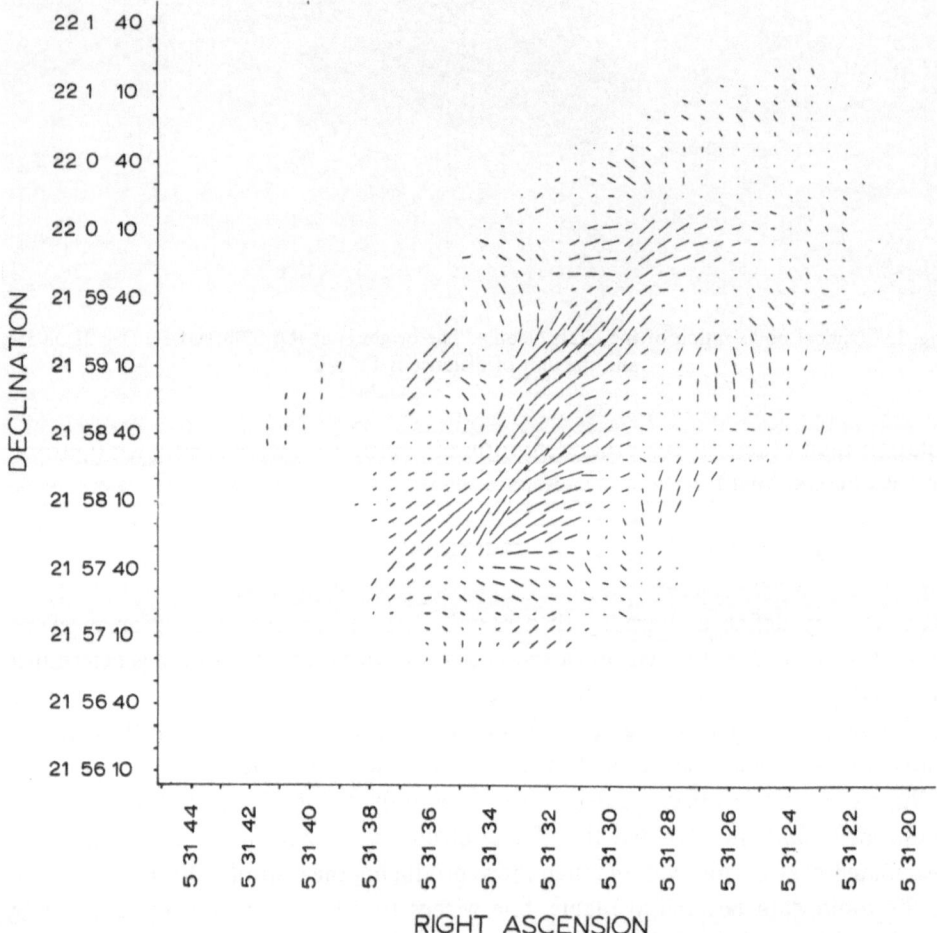

Fig. 2.   Radio polarization of the Crab Nebula at $\lambda 11$ cm. The bar lines are proportional in length to the polarized intensity, mI. The beam is $8'' \times 14''$, with the larger dimension roughly parallel to the major axis of the nebula.

value of *RM* for the centre is closely equal to that deduced by Gardner and Davies (1966) for the integrated polarization. Since this value is compatible with values for other nearby sources, Gardner and Davies ascribe the value $-25\,\text{rad/m}^2$ to the galactic Faraday rotation. The net rotation due to the nebula is unknown, but we may estimate it as $<50\,\text{rad/m}^2$.

Further information may be gained from a high resolution study at $\lambda 11$ cm. Figure 2 shows a map, with resolution $8'' \times 14''$, obtained with the NRAO 3-element interferometer at Greenbank, U.S.A., in 1967. As before, the central area has rather a uniform polarization. The northern region can also be seen, and for the first time, features corresponding to the great optical fans can be made out. These are areas of *low* polarization, notably two areas centred at $\alpha\,05^h31^m36^s$, $\delta\,21°58'40''$ and $\alpha 05^h31^m34^s$, $\delta 21°59'20''$, both of which lie in the great 'eastern fan' on Woltjer's map.

Detailed study of the p.a. on the two maps is rendered uncertain because of the different resolutions, but it is clear that rotation measures in excess of $300\,\text{rad/m}^2$ are to be found in such areas. As a result of the rapid variation, the smoothing by the $7''$ beam cuts down the apparent degree of polarization considerably. Thus the high rotation measures mentioned by Burn refer only to the edge regions, and those close to prominent filaments. In contrast, the central area appears to have rather uniform rotation measure, close to the galactic value of $-25\,\text{rad/m}^2$.

The degree of polarization at $\lambda 11$ cm is about 10–11% near the centre, which is less than the optical polarization for the same area (25–30%). Hence probably the Faraday mechanism is producing some depolarization on a small scale (less than the $7''$ beam). It would seem that this depolarization is caused by the 'fibres' or fine filaments at the limit of seeing. If a magnetic field of the order of 300 μG is assumed throughout the nebula, then these estimates of Faraday rotation suggest that the electron density is $\leqslant 3\,\text{cm}^{-3}$ in the body of the nebula as a whole, but rises to perhaps 10 or 100 in the fibres and filaments. We may compare these with the value of 1500 cm^{-3} given by Woltjer for 'a few strong filaments'.

References

Allen, R. J.: 1967, Ph.D. Thesis, Massachusetts Institute of Technology.
Burn, B. J.: 1966, *Monthly Notices Roy. Astron. Soc.* **133**, 67.
Downs, G. S. and Thompson, A. R.: 1968, *Astrophys. J.* **152**, L65.
Gardner, F. F. and Davies, R. D.: 1966, *Australian J. Phys.* **19**, 29.
Gardner, F. F. and Whiteoak, J. B.: 1966, *Ann. Rev. Astron. Astrophys.* **4**, 245.
Mayer, C. H. and Hollinger, J. P.: 1967, *Astrophys. J.* **151**, 53.
Wright, M.: 1970, *Monthly Notices Roy. Astron. Soc.* **150**, 271.

Discussion

N. Visvanathan: The optical polarization is uniform within 15 sec of arc of the pulsar. Is the same true of the radio polarization?

R. G. Conway: The optical polarization is not affected by Faraday rotation which causes small scale structure in the radio polarization.

5.3   TRANSFER OF ENERGY TO THE CRAB NEBULA FOLLOWING THE SPIN-UP OF THE PULSAR*

D. B. MELROSE

Department of Theoretical Physics, Faculty of Science,
Australian National University, Canberra, Australia

Abstract. Observed enhanced activity in the central region of the Crab Nebula following the spin-up of the pulsar is discussed from the point of view of the transfer of energy to relativistic electrons. It is argued that a rapid deposition of energy associated with the spin-up of the pulsar causes a radial energy flux which becomes a flux in hydromagnetic activity at about the regions where enhanced synchrotron emission is observed. It is shown that such hydromagnetic activity is rapidly damped by the relativistic electrons with energy being transferred to the relativistic electrons. This acceleration can account for the short synchrotron halflifetimes observed. The model predicts highly enhanced X-ray emission from the central region of the Nebula following a spin-up.

1. Introduction

Before the discovery of the pulsar in the Crab Nebula, two independent lines of reasoning led to the conclusion that an energy source was present.

Woltjer (1957) points out that the half-lifetime of the electrons radiating optical synchrotron is less than the age of the Nebula. The identification of the X-ray emission from the Nebula as synchrotron radiation indicates the presence of electrons with half-lifetimes as short as a year, e.g. less than the light travel time across the Nebula. This requires that these electrons either are being injected into the Nebula at the present time or are being accelerated in the Nebula at the present time. If the electrons are being accelerated rather than injected then simple arguments on the energy balance (Melrose, 1969) lead to the conclusion that the acceleration must be in a localized region. Furthermore, if one writes the rate of energy gain per electron in the form

$$\left\langle \frac{\mathrm{d}E}{\mathrm{d}t} \right\rangle = \frac{E}{T_A}, \tag{1}$$

where T_A may be a function of the energy E of the electron, then T_A must be less than the synchrotron half-lifetime of the electron for this acceleration to have any significant effect. The power required in injected or accelerated electrons is about 10^{38} erg sec^{-1}.

Scargle (1969) considers the so-called wisps in the central region of the Nebula and concludes that an energy source supplying a power comparable to that required to offset the synchrotron losses is required to account for the wisps. The wisps are interpreted as local enhancements in the synchrotron emission due to compressions of the magnetic field, see also Woltjer (1957), Shklovsky (1968). Although the three dimensional configuration of the compressions is unknown it appears that very strong com-

---

* This paper was presented by Dr. P. Stewart.

pressions (magnetic field strength increasing severalfold) are required to give the observed enhancement in synchrotron emission. The wisps vary over time scales of the order of weeks to months in the central region of the Nebula. They propagate with a velocity of about $0.2\,c$ which is interpreted as the velocity of propagation of the hydromagnetic compressional disturbance. Slower moving $(10^{-2}\,c)$ less intense enhancements in the synchrotron emission are observed in the outer regions of the Nebula. Scargle found that these enhancements vary quasi-periodically with a period of several years.

The discovery of the pulsar led to models for the electrodynamic acceleration of particles in the neighbourhood of the pulsar (Pacini, 1968; Ostriker and Gunn, 1968; Pacini and Rees, 1970). Such models may be able to account for the injection of relativistic electrons but do not obviously account for the excitation of the wisps.

Scargle and Harlan (1970) report increased activity in the wisps following the spin-up of the pulsar in September 1969. The increased activity at a given radial distance r occurred approximately at a time $t=r/c$ where t is the time elapsed from the spin-up. The energy associated with this enhanced activity was found to be between 10^{41} and 10^{45} erg, which is considerably greater than the change in the rotational energy of the pulsar resulting from the spin-up. Scargle and Harlan suggest that both the spin-up of the pulsar and the enhanced activity in the wisps result from an energetic event occurring in the neighbourhood of the pulsar. They suggest that the spinning of the pulsar builds up magnetic stresses in the surrounding magnetic field and that this magnetic energy is released by an instability. The quasi-periodic variations in the wisps suggest that such events occur quasi-periodically, consequently one would expect spin-ups of the pulsar to occur every several years.

The excitation of the wisps requires strong compressions of the magnetic field. Now, for example, an energy release of 10^{44} erg into a volume of 10^{52} cm^3 (about that inside the region where the wisps are observed) corresponds to an outward pressure which is greater than but comparable to the magnetic pressure in the field of 10^{-3} to 3×10^{-4} G in the region of the wisps. Thus the magnetic field in the region of the wisps would be strongly compressed by such an energy release.

These observations on the activity in the wisps indicate that the power requirements of the Nebula may be supplied through the quasi-periodic injection of energy into the region of the wisps in the form of strong localized compressions of the magnetic field. We assume this to be the case and consider the acceleration of electrons in the region of the wisps resulting from these compressions in the magnetic field. The basic requirements on the acceleration mechanism for it to be able to account for the observations are that the energy in the compressional disturbance be transferred to the relativistic electrons of a time scale of about a year and that the T_A in (1) be less than the synchrotron halflifetimes of the electrons. We find that these requirements can be satisfied using parameters deduced from the observations of the wisps. This enhances the plausibility of the interpretation of the spin-up and the activity in the wisps as resulting from a quasi-periodic release of energy in the immediate surroundings of the pulsar.

2. Scattering and Acceleration of Particles

In this section we give a semi-quantitative discussion of the acceleration of particles in the presence of periodic or quasi-periodic variations in the magnetic field strength. Acceleration results from the combination of two effects. If the magnetic field strength B changes on a time scale much longer than the gyroperiods of the particles then the quantity $p_\perp^2/B$ is conserved ($p_\perp$ = component of momentum perpendicular to $\mathbf{B}$), i.e. we have

$$\frac{dp_\perp^2}{dt} = \frac{p_\perp^2}{B}\frac{dB}{dt}. \tag{2}$$

Any initially isotropic distribution of particles becomes anisotropic. If the particles are scattered in any way, scattering time τ say, then this anisotropy is reduced. If the field varies between B and $(1+\varepsilon)B$ over a characteristic time T then the combination of these two effects leads to a mean rate of energy gain per particle given by

$$\left\langle\frac{dE}{dt}\right\rangle \sim \frac{\varepsilon^2\tau}{T^2}E \quad (\tau < T) \tag{3}$$

to within a factor of order unity. Such acceleration is most effective for $\tau \approx T$ and goes to zero in both the limits $\tau \ll T$ and $\tau \gg T$.

The simplest example of such an acceleration is magnetic pumping. Suppose that B varies according to

$$\frac{dB}{dt} = \frac{2\varepsilon}{T}CB, \qquad C = \begin{cases} +1 & \text{(compression)} \\ -1 & \text{(rarefaction)}, \end{cases} \tag{4}$$

i.e. that B increases linearly with time from B to $(1+\varepsilon)B$ in the first half-period $T/2$ and then decreases linearly with time in the second half-period. In the first half-period the average value $\langle \sin^2\alpha \rangle$ of $\langle p_\perp^2 \rangle / p^2$ is greater than $\frac{2}{3}$ and in the second half period $\langle \sin^2\alpha \rangle$ is less than $\frac{2}{3}$. We write

$$\langle \sin^2\alpha \rangle = \begin{cases} \frac{2}{3} + \delta & \text{(compression)} \\ \frac{2}{3} - \delta & \text{(rarefaction)} \end{cases}$$

and note that when the tendency to become anisotropic is balanced by the tendency of the scattering to isotropize the particles (which balance is possible for $\tau < T$) we have

$$\delta \sim \frac{2\varepsilon\tau}{T}.$$

However according to (1) the rate of change of $p_\perp^2$ is proportional to $p^2 \langle \sin^2\alpha \rangle$ and so the increase in $p_\perp^2$ in the first half-period exceeds the decrease in the second half-period. The net energy gain over one period is given by (3) to within a factor of order unity.

Magnetic pumping applies in particular to compressions propagating across the magnetic field. An analogous acceleration applies to compressions propagating along the magnetic field. Each such compression corresponds to a magnetic bottle which can trap some of the particles. Suppose that the compressions propagate with a velocity V and have a length $\lambda = VT$. A fraction ε of all particles approaching the neck of each bottle is reflected. There is an increase in the energy of the particle if the particle is reflected from an oncoming compression. There is a decrease in the energy of the particle if the particle overtakes the compression. However, as in Fermi acceleration, the particles are more likely to be reflected by meeting a compression head-on than on overtaking a compression. This leads to an average energy gain per particle of

$$\Delta E \sim \left(\varepsilon \frac{V}{c}\right)^2 E$$

each time the particle reaches a compression (the factor ε takes the probability of reflection into account). For $\tau \ll T$ relativistic particles diffuse the distance λ between compressions in a time

$$t \sim \frac{\lambda^2}{c^2\tau}, \quad \lambda = VT.$$

The mean rate of energy gain, $\Delta E/t$, again reduces to (3) within a factor of order unity.

This type of acceleration can also be regarded as a form of viscous damping of the compressions. The anisotropy induced by the compressional disturbance is out of phase with the disturbance due to the scattering of the particles. If τ in (3) is independent of the energy E then the rate of increase in the energy density in particles W_p is

$$\frac{dW_p}{dt} \sim \frac{\varepsilon^2\tau}{T^2} W_p.$$

Conservation of energy implies that the energy density W_c in the compressions decreases at the rate

$$\frac{dW_c}{dt} = -\frac{dW_p}{dt} \sim -\frac{\varepsilon^2\tau}{T^2} W_p,$$

or equivalently

$$\frac{1}{W_c}\frac{dW_c}{dt} \sim -\frac{\tau}{T^2}\left(\frac{W_p}{W_M}\right), \tag{5}$$

where $W_c \approx \varepsilon^2 W_M$ with $W_M = $ ambient magnetic energy density.

The scattering of relativistic particles is discussed by a number of authors with reference to the scattering of cosmic rays, see Lerche (1967), Wentzel (1968), Kulsrud and Pearce (1969), Tademaru (1969) for example. The basic scattering process is associated with the fact that an anisotropic distribution of relativistic particles is unstable to the coherent emission of (relatively high frequency) hydromagnetic waves. The coherent emission of such waves reduces the anisotropy at a rate proportional to the energy density in these hydromagnetic waves. We analyze a particular example of this scattering in Section 3. Here we give a somewhat heuristic discussion of the effects of this scattering in the spirit of the discussion of the acceleration mechanisms given above.

The coherent emission of hydromagnetic waves requires that the anisotropy be greater than v_A/c, i.e. the hydromagnetic maser turns on only when the anisotropy, e.g.

$$\delta = \frac{|f(p, \alpha = \pi/2) - f(p, \alpha = 0)|}{f(p, \alpha = \pi/2) + f(p, \alpha = 0)}$$

exceeds v_A/c where v_A = Alfvén velocity. This requires $\varepsilon > v_A/c$ for any compression to cause coherent emission of waves. Let us suppose the growth time for the waves is very much shorter than the period T of the compressions. Then the growth of the waves will proceed until the anisotropy is maintained near its minimum possible value, i.e. we then have an anisotropy

$$\delta \approx \frac{v_A}{c} \approx \frac{2\varepsilon\tau}{T}.$$

The implied effective scattering time is

$$\tau \sim \frac{v_A}{c} \frac{T}{\varepsilon}. \tag{6}$$

The acceleration (3) with this scattering leads to an acceleration described by

$$\left\langle \frac{dE}{dt} \right\rangle \sim \frac{v_A}{c} \frac{\varepsilon}{T} E \tag{7}$$

to within a factor of order unity. In Section 3 we derive the appropriate numerical factor for one particular situation.

In the region of the wisps the parameters v_A, ε, T, W_p, W_M can all be estimated from the observations. It is already clear from (7) that for $T \sim$ one month, $v_A \sim 0.2\,c$, $\varepsilon \sim 1$ the acceleration time T_A in (1) is of the order of a year as required. Furthermore we then infer from (5) and (6), with $W_p \sim W_M$ from observation, that the wisps would damp out over about a year, as observed, transferring their energy to the relativistic electrons. Our more detailed discussion in Section 3 is aimed at estimating the numerical factor in (7) and at deducing the conditions under which the growth of the waves is rapid enough to maintain the anisotropy at a value of order v_A/c.

3. Magnetic Pumping

We analyze the acceleration of relativistic particles due to magnetic pumping in the case where the field strength varies linearly with time according to Equation (4). We refer the interested reader to Melrose and Wentzel (1970) for a more detailed discussion of the details of the equations describing the growth of the waves and the scattering by the waves and to Melrose (1970) for the details of the expansion in spherical harmonics used below.

We assume that the relativistic electrons have a number density n and a distribution function

$$f(p) = \frac{(a-3)}{4\pi} p_0^{-3} \left(\frac{p}{p_0}\right)^{-a} \quad (a > 3, p > p_0). \tag{8}$$

Note that $f(p) \propto p^{-a}$ corresponds to an energy distribution $N(E) \propto E^{2-a}$.

With $\mu = \cos$ (pitch angle), the change in the distribution function due to the changes described by (4) is given by

$$\frac{\partial}{\partial t} f(p, \mu)\Big|_c = -\frac{\partial}{\partial p_\perp^2}\left[2\frac{\varepsilon C}{T} p_\perp^2 f(p, \mu)\right]$$

$$= -\frac{C\varepsilon}{T}\left(\frac{1}{p}\frac{\partial}{\partial p} - \frac{\mu}{p^2}\frac{\partial}{\partial \mu}\right)\{p^2(1-\mu^2)f(p, \mu)\}. \tag{9}$$

The anisotropy induced in this way is such that $f(p, \mu)$ is an even function of μ. Thus if we expand in spherical harmonics by writing

$$f(p, \mu) = \sum_{n=0}^{\infty} f_n(p) P_n(\mu), \tag{10}$$

where P_n is a Legendre polynomial, then f_n is non-zero only for even n.

Anisotropies which are even in μ lead to coherent emission of circularly polarized hydromagnetic waves rather than linearly polarized hydromagnetic waves, see Melrose and Wentzel (1970). To an excellent approximation these waves may be regarded as propagating exactly along the magnetic field lines. According to Melrose and Wentzel (1970) a distribution of electrons of the form (8) subjected to a compression $(C = +1)$ reduced by either emitting or absorbing waves of one handedness or the other. If $\Gamma = +1$ (-1) denotes absorption (emission) and $P = +1$ (-1) denotes right (left) hand circularly polarized waves then the condition for the anisotropy to be reduced is

$$C P \Gamma = -1. \tag{11}$$

If k denotes the component of the wave number along the field lines, then electrons with a given sign of μ interact with waves with a given sign of k only if the condition

$$\frac{\mu}{|\mu|}\frac{k}{|k|} = -P \tag{12}$$

is satisfied.

Consider the following sequence of processes. Suppose that at the start of a compression $(C = +1)$ waves of both handedness $P = +1$ are already present. According to (11) the anisotropy induced by the compression can be reduced by absorbing $(\Gamma = +1)$ the waves with $P = -1$. The anisotropy can also be reduced by the emission

$(\Gamma = -1)$ of waves with $P = -1$. Both processes occur. As the compression proceeds the ratio of the energy density in the waves with $P = -1$ to that in waves with $P = +1$ decreases. At the end of the compression phase most of the waves have polarization $P = +1$. When rarefaction $(C = -1)$ begins the anisotropy is reduced primarily by absorbing $(\Gamma = +1)$ these waves $(P = +1)$. As the rarefaction phase proceeds the ratio of the energy density in waves with $P = +1$ to that in waves with $P = -1$ decreases. At the end of the rarefaction phase most of the waves have handedness $P = -1$. The cycle then recommences.

This sequence of processes allows the anisotropy to remain small at every phase of the cycle. Any intrinsic damping of the waves has little effect provided that this damping is not great enough to prevent the growth of the waves. The only essential condition for this sequence of events to occur during cyclic variations of the field strength is that the waves grow and damp on a time scale much shorter than the period of compression.

Let $W(k)$ be the energy density in the waves in the range dk at k. We do not distinguish explicitly between waves of different handedness because the handedness is more easily discussed independently as above. The following equations describe the growth or damping of the waves and the scattering of the particles by the waves:

$$\frac{\partial}{\partial t} W(k) = -\gamma(k) W(k), \tag{13}$$

$$\gamma(k) = \frac{-2\pi^2 e^2 v_A n}{|k| c} \int_{-1}^{+1} d\mu \frac{(1-\mu^2)}{|\mu|}$$

$$\times \left[p^2 \left\{ \frac{k}{|k|} \frac{\partial}{\partial \mu} f(p, \mu) + \frac{v_A}{c} p \frac{\partial}{\partial p} f(p, \mu) \right\} \right]_{p = p_R}, \tag{14}$$

$$p_R = \frac{p_0 k_0}{|k \mu|}, \qquad p_0 k_0 = \frac{|e| B}{c}; \tag{15}$$

$$\frac{\partial}{\partial t} f(p, \mu)\big|_s = \frac{\partial}{\partial \mu} \left[D(p, \mu) \left\{ \frac{\partial}{\partial \mu} f(p, \mu) + \frac{k}{|k|} \frac{v_A}{c} p \frac{\partial}{\partial p} f(p, \mu) \right\} \right], \tag{16}$$

$$D(p, \mu) = \frac{\pi e^2 (1-\mu^2)}{p^2 c |\mu|} W(k_R), \tag{17}$$

$$k_R = \frac{p_0 k_0}{p |\mu|}. \tag{18}$$

The sign $k/|k|$ is determined by (11) and (12) with $\Gamma = \gamma(k)/|\gamma(k)|$.

It is impractical to attempt to follow the time development explicitly. However one can readily show by expanding (9) in spherical harmonics that the term $f_2(p)$ in (10) is the first to grow. Initially only this term need be considered in (14) and (16). As indicated by our discussion above, waves are always present once the cycle is initiated.

Keeping only the term $f_2(p)$ in the expansions of (9), (14) and (16) leads to a self-consistency problem in which the values of $f_2(p)$ and $W(k)$ are to be determined by the condition that the reduction of $f_2(p)$, due to scattering, balances the growth of $f_2(p)$, due to the change in the field strength. An important part of the solution of this self-consistency problem is the determination of the k-dependence of $W(k)$.

Expanding (14) and (16) in spherical harmonics and retaining only the lowest order non-trivial terms gives

$$\frac{\partial}{\partial t} f_0(p)\Big|_c = -C\frac{\varepsilon}{T}(a-3)\{\tfrac{2}{3}f_0(p) - \tfrac{2}{15}f_2(p)\},\tag{19}$$

$$\frac{\partial}{\partial t} f_2(p)\Big|_c = -C\frac{\varepsilon}{T}\frac{2a}{3}f_0(p),\tag{20}$$

$$\frac{\partial}{\partial t} f_2(p)\Big|_s = -\tau_1^{-1}f_2(p) - aC\Gamma\frac{v_A}{c}\tau_2^{-1}f_0(p),\tag{21}$$

where we use (8) and where

$$\tau_1^{-1}(p) = \frac{5}{2}\int_{-1}^{+1} d\mu\,(3\mu)^2\,D(p,\mu),$$

$$\tau_2^{-1}(p) = \frac{5}{2}\int_{-1}^{+1} d\mu\,\frac{\mu}{|\mu|}(3\mu)\,D(p,\mu).\tag{22}$$

By requiring

$$\frac{\partial}{\partial t} f_2(p)\Big|_c + \frac{\partial}{\partial t} f_2(p)\Big|_s = 0,$$

(20) and (21) allow one to find $f_2(p)$. On inserting this value of $f_2(p)$ in (13) with (14) one finds

$$\frac{\partial}{\partial t} W(k) = -\frac{1}{T_0} W(k) + CP\frac{A}{T_1},$$

$$T_0^{-1} \propto \left(\frac{k}{k_0}\right)^{a-3}, \qquad AT_1^{-1} \propto \left(\frac{k}{k_0}\right)^{a-5}.\tag{23}$$

For $T_0 \ll T$ the magnitude of $W(k)$ does not vary rapidly over a period of the compression; the variation of $W(k)$ with time is sketched in Figure 1. The mean value of $W(k)$ for $T_0 \ll T$ is

$$\langle W(k)\rangle \approx \frac{AT_0}{T_1} \propto \left(\frac{k}{k_0}\right)^{-2}.$$

This identification of the functional form of $W(k)$ allows us to determine the solution of the self-consistency problem. One can then proceed to consider the inclusion of

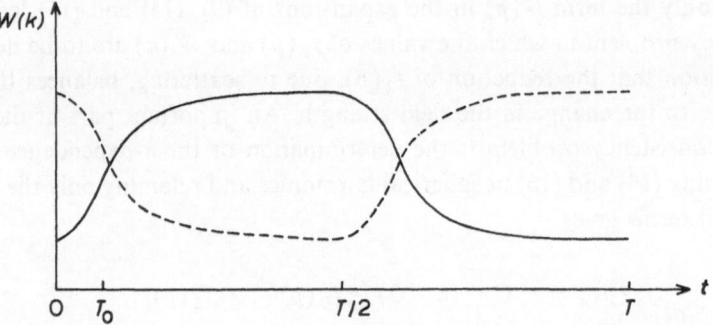

Fig. 1. $W(k)$ for waves of both handedness is sketched as a function of time over a period T of a compressional disturbance with time $t = 0$ corresponding to the onset of compression. The time $T_0 \ll T$ is indicated.

the term $f_4(p)$, resolve the problem and so on. However, as in the analogous problem discussed in Melrose (1970), the inclusion of more terms in the expansion in spherical harmonics leads to only minor numerical changes in the final result.

Retaining only the term $f_2(p)$ gives

$$W(k) = W_0 \left(\frac{k}{k_0}\right)^{-2} \quad (k < k_0),$$

$$W_0 = \frac{8}{15} \frac{p_0^2 C}{\pi^2 e^2} \frac{a(a-2)}{a^2 - 1} \frac{c}{v_A} \frac{\varepsilon}{T},$$

$$\tau_1^{-1} = \frac{15}{4} D_0, \qquad \tau_2^{-1} = 2 D_0,$$

$$D_0 = \frac{\pi^2 e^2}{p_0^2 c} W_0 = \frac{8}{15} \frac{a(a-2)}{a^2 - 1} \frac{c}{v_A} \frac{\varepsilon}{T}, \qquad (24)$$

$$\frac{f_2(p)}{f_0(p)} = - C \frac{a^2 - 1}{3(a-2)} \frac{v_A}{c},$$

$$\frac{1}{T_0} = \frac{2\pi^2 e^2 n}{p_0 k_0} \frac{(a-3)}{(a-2)} \left(\frac{v_A}{c}\right)^2 \left(\frac{k}{k_0}\right)^{a-3}.$$

Inserting the value of $f_2(p)$ so determined in (19) one finds that the rate of change of $f_0(p)$ includes the term

$$\left\langle \frac{\partial}{\partial t} f_0(p) \right\rangle = \frac{2}{45} \frac{(a-3)(a^2-1)}{a-2} \frac{v_A}{c} \frac{\varepsilon}{T} f_0(p) \qquad (25)$$

which has the same sign for both compression and rarefaction. This term describes magnetic pumping. In fact (25) leads to an acceleration of the form (1) with

$$T_A = \frac{45}{2} \frac{(a-2)}{(a^2-1)(a-3)} \left(\frac{c}{v_A}\right) \left(\frac{T}{\varepsilon}\right). \qquad (26)$$

The scattering times τ_1 and τ_2 in (24) are of the form (6) which we deduced from qualitative arguments. Similarly the acceleration (1) with (26) is of the form (7) but (26) includes an estimation of the undetermined factor of order unity in (7).

The condition for the above analysis to be applicable is that the time T_0, which is characteristic of the growth or damping of the waves due to the relativistic electrons, be much less than the time T. If the waves are intrinsically damped in the background plasma at a rate

$$\frac{\partial}{\partial t} W(k) = -\frac{1}{T_D} W(k) \tag{27}$$

then the above analysis applies only if

$$T_0 < \min[T, T_D]. \tag{28}$$

The condition $T_0 < T$ with $a < 3$ leads to an upper limit on the momentum of relativistic particles which can be accelerated by this process. This follows by noting the fact that particles with a given p interact with waves over a small range of k, and vice versa, around

$$pk \approx p_0 k_0. \tag{29}$$

The inequality (28) need not be extreme because for $T_0 \approx T$ the scattering time τ is also of the order of T. The acceleration (3) becomes more effective as the inequality $T_0 \ll T$ starts to break down. Only for $T_0 > T$ does the acceleration become ineffective.

4. Discussion

If the transfer of energy from the pulsar to the relativistic electrons occurs by the intermediate excitation of the wisps then a number of conditions need to be satisfied. Firstly, the time averaged power input into the wisps must balance the time averaged power lost by the Nebula (primarily synchrotron losses). Secondly, the acceleration must cause the energy per electron to increase faster than this energy decreases due to synchrotron losses, e.g. T_A in (1) must be of the order of a year. Thirdly, the electrons must remain in the region where acceleration occurs for a sufficiently long time, e.g. a year, to be significantly accelerated. Fourthly, the electrons must be accelerated to a sufficiently high energy ($>10^{14}$ eV) to account for the highest observed frequencies in the synchrotron emission. Fifthly, the wisps must lose their energy to relativistic electrons over a time of a year or so.

The energetic considerations remain uncertain because any estimation of the power input into the wisps depends on geometric assumptions. Only a surface brightness is observed; one must make an assumption on the configuration in three dimensions. All that can be said is that the mean power input into the wisps is consistent with this being the source of the mean power lost through synchrotron radiation.

The acceleration time T_A, see (26), with $a=4$, $\varepsilon \approx 1$ to 3, $T \approx 3 \times 10^6$ sec, $v_A \approx 0.2 \, c$ from observations of the wisps, is of the order of a year. With these numbers the scattering time (6) is of the order of 3×10^5 sec and so electrons diffuse over a distance of 10^{17} cm (to escape from the region where acceleration occurs) in a time

$$t \approx \frac{(10^{17})^2}{c^2 (3 \times 10^5)} \approx 3 \times 10^7 \text{ sec}.$$

Thus individual electrons remain in the region where acceleration occurs for about a year.

The maximum energy ($E_{max} = p_{max} c$) to which electrons can be accelerated follows from (28) and (29) with T_0 given by (24). If we set $p_0 \approx 10^{11}$ eV/c in (8) thereby restricting our attention to the electrons radiating at optical and higher frequencies (these require continued acceleration) then the observed power output requires $n \approx 10^{-7}$ cm^{-3} for these electrons. With $a = 4$ and $T = 3 \times 10^6$ sec the condition $T_0 < T$ requires

$$\frac{p}{p_0} < \frac{2\pi^2 e^2 ncT}{|e| B} \frac{(a-3)}{(a-2)} \left(\frac{v_A}{c}\right)^2 \approx 6 \times 10^3, \tag{30}$$

where we set $B = 3 \times 10^{-4}$ G. Thus $T_0 < T$ requires $E < E_{max} \approx 6 \times 10^{14}$ eV, i.e. electrons can be accelerated to energies in excess of 10^{14} eV as required.

It is believed that there is approximate equipartition of energy ($W \approx W_M$) in the Nebula. With equipartition equation (5) predicts that the wisps damp out over a time $T^2/\tau \approx 3 \times 10^7$ sec ≈ 1 yr. Thus it seems plausible that all the conditions listed above are satisfied by the processes under discussion.

The interpretation of the wisps as compressions and the above discussion of the acceleration processes associated with such compressions requires that X-ray emission be strongly enhanced following enhanced activity in the wisps. This results from two effects. Firstly the enhancement in surface brightness due to a compression is greater at X-ray frequencies than at optical frequencies. Secondly the shortness of the half-lifetimes of electrons radiating in the hard X-ray ($\gtrsim 100$ keV) region of the spectrum implies that this emission should be observed to die away in between periods of enhanced activity in the wisps. The acceleration envisaged here implies an upsurge in hard X-ray emission for about a year during enhanced activity in the wisps, then followed by a decrease in hard X-ray emission until the next period of enhanced activity. Furthermore the hard X-ray emission can arise only from the central region of the Nebula.

References

Kulsrud, R. M. and Pearce, W. D.: 1969, *Astrophys. J.* **156**, 445.

Lerche, I.: 1967, *Astrophys. J.* **147**, 689.

Melrose, D. B.: 1969, *Astrophys. Space Sci.* **4**, 165.

Melrose, D. B.: 1970, *Astrophys. Space Sci.* **6**, 321.

Melrose, D. B. and Wentzel, D. G.: 1970, *Astrophys. J.* **161**, 457.

Ostriker, J. and Gunn, J.: 1969, *Astrophys. J.* **157**, 1395.

Pacini, F.: 1968, *Nature* **219**, 145.

Pacini, F. and Rees, M. J.: 1970, *Nature* **226**, 622.

Scargle, J. D.: 1969, *Astrophys. J.* **156**, 401.

Scargle, J. D. and Harlan, E. A.: 1970, *Contributions from Lick Observatory* No. 313.
Shklovsky, I. S.: 1969, *Supernovae*, Wiley-Interscience Publ., New York.
Tademaru, E.: 1969, *Astrophys. J.* **158**, 959.
Wentzel, D. G.: 1968, *Astrophys. J.* **156**, 303.
Woltjer, L.: 1957, *Bull. Astron. Inst. Neth.* **14**, 39.

Discussion

I. Lerche: The result obtained must depend critically on the initial conditions which were assumed.

P. Stewart: Dr. Melrose assumed a wave distribution proportional to $(k/k_0)^{-2}$.

I. Lerche: There is no definite evidence for the distribution and it must be regarded as an assumption.

J. P. Ostriker: Electrons will tend to lose energy while protons will not. Hence there must be a build up of proton numbers.

P. Stewart: This makes the time scale for energy loss rather important.

F. C. Michel: How much compression does Melrose expect?

P. Stewart: A factor of 2 or 3.

F. C. Michel: This leads to acceleration by the same factor.

J. E. Felten: Would you expect the X-ray output to vary over a period of a year?

P. Stewart: Yes. Melrose makes the point that the X-rays will vary more than the light, where the particle lifetimes are longer. It even seems possible that the radio emission from the Crab Nebula changes in one month; if so, then the supersynthesis map will be incorrect.

A. S. Wilson: The beam size in the Cambridge maps corresponds to two months motion at the velocity of light. Any changes in the emission are unlikely to have any effect on the maps.

R. G. Conway: The observations at Green Bank took six months but there is no evidence for any changes during that time.

5.4 RADIATIVE IONIZATION OF THE FILAMENTS
IN THE CRAB NEBULA

KRIS DAVIDSON

Cornell-Sydney University Astronomy Center, Ithaca, N.Y., U.S.A.

and

WALLACE H. TUCKER

American Science and Engineering, Cambridge, Mass., U.S.A.

Abstract. We have attempted to explain the observed excitation conditions in the filamentary system of the Crab Nebula in terms of ionization and heating by high frequency radiation. It was found that it is possible to reasonably fit most of the observed lines by assuming either (1) an ultraviolet continuum which smoothly joins the optical and X-ray data and a gas composition in which the number densities of hydrogen and helium are equal, (2) a spectrum which drops off about as steeply as ν^{-2} and therefore does not smoothly fit the X-ray spectrum, or (3) an ultraviolet continuum which smoothly joins the optical and X-ray data plus strong emission-line features near 20 eV. A UV continuum which is much larger than in these models results in too much ionization in the filaments. The calculations suggest that most of the filaments consist of outer ionized regions with cores of neutral gas, so that the mass of the filamentary shell may be considerably larger than the 1.45 $M_\odot$ required to explain the emission line intensities.

1. Introduction

In this communication we summarize attempts to explain the observed excitation conditions in the filamentary system of the Crab Nebula in terms of ionization and heating by high frequency radiation. For a more detailed discussion of this work see Davidson and Tucker (1970).

2. The Calculations

The incident source spectra of four models are shown in Figure 1. Also shown are the observed optical (Minkowski, 1968 and references cited therein) and X-ray spectra Gorenstein *et al.*, 1970) renormalized by a factor of 2.95×10^6 (corresponding to a reduction in distance from 1700 pc to 1 pc; i.e., applicable to a filament which is apparently 2′ from the center of the Nebula). For the optical observations, a particular value for the interstellar extinction must be assumed. The visual extinction is probably about $A_v \approx 2$ magnitudes (Minkowski, 1968).

The geometry of each model is rather simplified – each case is plane-parallel with a boundary upon which a beam of ionizing radiation is normally incident. The gas pressure is uniform in each case (this should be a better approximation than would uniform density). The electron density of the region containing O II is about 1000 cm^{-3} in each case.

Model 1 is a simple power-law spectrum that smoothly connects the optical and X-ray data. It is rather similar to a model considered by Williams (1967). It encounters

Davies and Smith (eds.), The Crab Nebula, 308–313. All Rights Reserved.
Copyright © 1971 by the IAU.

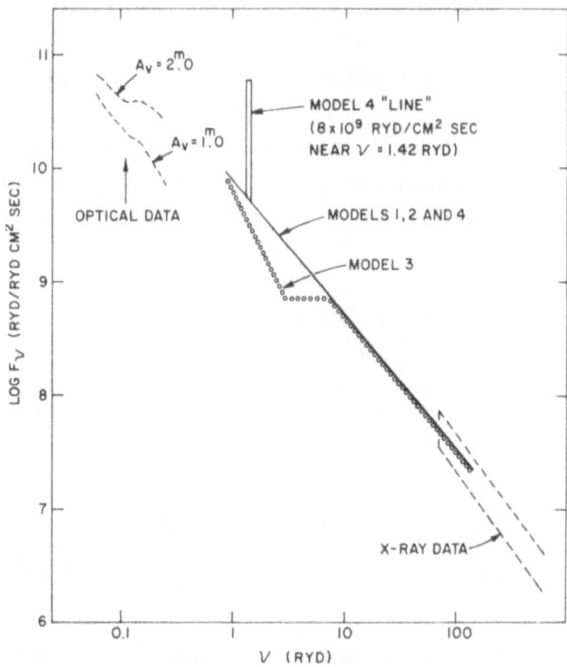

Fig. 1. Incident source spectra assumed for four calculated models. Observed optical and X-ray spectra are also shown, multiplied by 2.95×10^6 (i.e., appropriate to a filament which is $2'$ from the center of the Nebula). Optical data are due to O'Dell (1962); for X-ray data, see Gorenstein *et al.* (1970). The relative abundances by number for the various models are $n(Z)/n(H) = 1.0$ (H), 0.45 (He) 4×10^{-4} (C), 1.1×10^{-4} (N), 9×10^{-4} (O), and 5×10^{-4} (Ne) for models 1, 2 and 3 and $n(Z)/n(H) =$ 1.0 (H), 1.0 (He), 0.0002 (C), 0.0002 (N), 0.0006 (O), and 0.0005 (Ne) for Model 2.

the difficulty that the ionization ratio OII/OIII is several times too small to explain the ratio of the forbidden lines ([OII] $\lambda 3727$)/([OIII] $\lambda 5007$).

In Model 2 the radiation spectrum is the same as in Model 1, but the abundance of helium is taken to be equal to that of hydrogen. This will affect the line ratios, since most of the photons having energies above the ionization potential of helium will be quickly absorbed, so the zone where helium is ionized will be smaller than in Model 1. Since this is the region where OIII should occur, we would expect the OII/OIII ratio to be larger in Model 2 than in Model 1.

Models 3 and 4 have the same abundances as Model 1, but have different radiation spectra. In Model 3 the ionizing spectrum from 1 to 3 Rydbergs is a steep power law with spectral index 2, becomes level at 3 Rydbergs, and resembles the Model 1 spectrum above 7.6 Rydbergs. In this model there is no direct connection between the X-ray and optical spectra. The OII/OIII ratio should be larger in this model than in Model 1 since there are fewer photons capable of ionizing OII.

Model 4 has the same source spectrum as Model 1 plus a 'line' at 1.42 Rydbergs (19 eV). This line is below the ionization threshold of OII but above that of OI, so the amount of OII is increased without producing additional OIII. The peak in the spectrum could be produced by $2s-2p$ transitions in oxygen ions in a diffuse gas having an electron temperature of about 200000 K. The total luminosity in the peak is about 2×10^{37} erg/sec, so for a diffuse gas filling the nebula, this requires an electron density of about 20 cm^{-3} (Cox and Tucker, 1969). Such a gas would add about 20 pc cm^{-3} to the dispersion measure of the pulsar NP 0532. The expansion of the Crab Nebula should cause this 'internal' dispersion measure to decrease by about 0.2%/yr, giving a possible check on the presence of diffuse-ionized gas in the Nebula.

TABLE I

Calculated and observed emission line intensities for typical crab nebula filaments**

Model	1	2	3	4	Observed
Region producing [O II] $\lambda 3727$:					
n_H(cm^{-3})	810	1100	760	1040	–
n_e(cm^{-3})	1030	1020	900	1020	1000
T_e(K)	9960	9360	9900	8740	–
Ionized thickness (10^{16} cm)	2.9	2.3	2.1	5.0	–
Relative line intensities:					
[O II] $\lambda 3727$	7.2	10.5	10.0	10.0	9.7
[Ne III] $\lambda 3868$*	4.9	1.7	3.8	2.8	1.5
He I $\lambda 4471$	0.17	0.21	0.16	0.09	0.18
He II $\lambda 4686$	0.27	0.24	0.25	0.13	0.7
H I $\lambda 4861$	1.00	1.00	1.00	1.00	1.00
[O III] $\lambda 5007$*	30.5	16.2	22.3	17.9	14.6
[N II] $\lambda 6580$*	3.9	12.5	5.3	4.7	13.3
Ne II 12.8μ	1.7	1.0	1.8	2.1	–

* Two well-separated lines, e.g., [O III] $\lambda 5007$ means $\lambda 5007 + \lambda 4959$. Observed relative intensities are taken from Woltjer (1958).
** These models overestimate the H I-cooling rate; as a result the quoted values for the temperature and forbidden line intensities are too low by about 5% and 25%, respectively.

The filament parameters and the calculated relative emission line intensities of the models are listed in Table I. The observed relative intensities are also listed. The observed ratio ([O III] $\lambda 5007$)/([O II] $\lambda 3727$) is 1.5; note that this does not include a correction for the amount of extinction which is now thought to be present – such a correction might reduce the ratio to about 1.0.

The calculated value for the ratio of the oxygen lines is 4.2 in Model 1. This is roughly the same discrepancy found by Williams (1967).

In Model 2, in which the hydrogen and helium abundances are assumed to be equal, the ratio of the oxygen lines is in good agreement with the observations, as are the relative intensities of the Ne III and N II lines and the 4471 Å line of helium.

Models 3 and 4 represent attempts to obtain the correct line intensity ratios by adjusting the ionizing radiation spectrum. The relative line intensities in these models are also in good agreement with the observations.

A major difficulty with all the models involves the He II $\lambda4686$ recombination line; its observed intensity is several times larger than the predicted value. To drastically increase the intensity of this line, it would seem necessary to increase the incident flux above 4 Rydbergs (in order to increase the amount of He III). The observational measurement of He II $\lambda4686$ may be somewhat confused by a nearby [Fe III] line (Woltjer, 1958). Fe III should be found in the same region as O II; hence, relatively strong [Fe III] emission would not be surprising, given the prominence of the [O II] doublet. This may require a larger than average iron abundance, which is consistent with the popular theory that the large natural abundance of iron is due to the decay of ^{56}Ni following the expulsion of silicon-burning zones from supernovae (see Bodansky *et al.*, 1968, and references cited therein). In any case, further measurements of the intensity of the He II $\lambda4686$ line would be extremely useful in helping to determine the physical conditions in the filaments.

Note that the helium-line intensities produced by Model 2 are nearly the same as those in Model 1, despite the great difference between the helium abundances of the two models. This is because these are recombination lines. Therefore, if the helium abundance is sufficiently great, the intensity of He II $\lambda4471$ depends only on the total absorbed flux of ionizing photons involved, and not on the helium density.

These models and other similar calculations not discussed here show that a large increase in the ionizing flux is unacceptable, since too much ionization in the filaments would result. Assuming any plausible gas density and helium abundance, increasing the flux by a factor of more than about three beyond that which we have assumed would result in at least some filaments consisting entirely of ionized-helium zones producing very little [O II] and [N II] line emission.

Note that the thickness of the ionized zone in Models 2 and 3 is about half the total thickness of a typical filament. This may suggest that most of the bright filaments consist of ionized outer regions with cores of non-ionized gas. Such a model is supported by the fact that the intensity of [O I] $\lambda6360$ is equal to or somewhat greater than that of Hβ (Trimble, 1970). This observation is difficult to explain unless the filaments have neutral cores. In addition there seems to be no strong correlation between the spectra and the thickness of the filaments. If the radiation field were strong enough to ionize all the gas in a filament, then such a correlation should exist. The presence of neutral cores must raise the total mass of the filamentary system by an amount which is difficult to estimate because the calculations become very uncertain in this region.

The mass of the ionized portion of the filaments can be estimated from O'Dell's (1962) measurements of the Hβ flux from the Crab Nebula. He finds that the total Hβ flux is about 1.24×10^{-11} erg/cm^2-sec, which for a distance of 1720 pc and a visual extinction $A_v = 2$ magnitudes, would require about 1.5 $M_\odot$ in Models 2 and 3, and about 0.9 $M_\odot$ in Model 1. If we include the mass of the possible neutral filamentary cores of Models 2 and 3, then this estimate must be increased by more than a factor of two, raising the total mass of the filaments to more than 3 $M_\odot$.

3. Summary

In all the models the observed high intensity of the He II $\lambda4686$ line continues to pose problems. Additional observations should decide whether the theory or the observations are at fault on this point. In Models 2 and 3 the thickness of the ionized part of the filaments is less than the observed diameter of the filaments. This suggests that most of the filaments consist of ionized outer regions with cores of neutral gas.

The acceptance of any of the above models has interesting implications. Model 2 requires a helium abundance equal to that of hydrogen, a value which is about a factor 7 greater than the 'cosmic' abundances. A large abundance of helium is consistent with the idea that the filaments were formed by the shell ejected during the supernova explosion which produced the Crab Nebula. According to Model 2 the filaments should have neutral 'cores', so that the total mass of the shell may be considerably larger than previous estimates of about a solar mass.

In Model 3 there is no direct connection between the X-ray and optical spectrum, as is generally assumed. The most obvious explanation for the spectral shape of Model 3 is that it is due to synchrotron radiation from at least two ensembles of electrons having different energy distribution functions. In view of the difference in the radiative lifetime for the relativistic electrons producing the low frequency (radio-optical) and high frequency (X-ray) synchrotron radiation, this is a reasonable hypothesis. In this case the nice fit of the extrapolated X-ray spectrum to the optical data is purely coincidental.

Model 4 requires strong emission line features such as are produced by a 200 000 K plasma. The total luminosity required is $\sim 2 \times 10^{37}$ erg/sec and represents a major source of energy loss from the Nebula. This hypothesis may be subject to test by means of the dispersion measure of NP 0532.

References

Bodansky, D., Clayton, D., and Fowler, W.: 1968, *Astrophys. J. Suppl.* **16**, 299.
Cox, D. and Tucker, W.: 1969, *Astrophys. J.* **157**, 1157.
Davidson, K. and Tucker, W.: 1970, *Astrophys. J.* **161**, 437.
Gorenstein, P., Kellogg, E., and Gursky, H.: 1970, *Astrophys. J.* **160**, 199.
Minkowski, R.: 1968, *Nebulae and Interstellar Matter* (ed. by Aller and Middlehurst), Univ. Chicago Press, Chicago.
O'Dell, C.: 1962, *Astrophys. J.* **136**, 809.
Trimble, V.: 1970, preprint.
Williams, R.: 1967, *Astrophys. J.* **147**, 556.
Woltjer, L.: 1958, *Bull. Astron. Inst. Neth.* **14**, 39.

Discussion

L. H. Aller: The structure of the filaments is not uniform and there is a range of structure in each filament. Data from (S II) lines should be included in any future study.

V. Trimble: The filaments are necessarily ejected material as there is not enough material to be swept up. The abundances of N and O in the filaments are normal.

Model 4 would require a plasma and this would stop the filaments expanding at the observed rate. Model 2 seems promising.

J. A. Roberts: The change in dispersion measure predicted for Model 4 would be observable. It is in the opposite sense to the change which is observed.

L. Sartori: I am glad to see evidence for an abundance of He in supernovae as this supports our fluorescence theory.

E. M. Kellogg: Any increase in the helium abundance might destroy the good agreement with other line emissions.

5.5 SYNCHROTRON RADIATION IN HIGH MAGNETIC FIELDS

D. F. FALLA and A. EVANS

Department of Physics, The University College of Wales, Aberystwyth, U.K.

Abstract. We suggest that if there exist within the Crab Nebula localised condensations of material containing high magnetic fields, ($\gg 1$ G), the rapidly evolving synchrotron radiation power spectrum emitted by a single electron can give a radiation continuum with a spectral index having a unique value similar to that observed in the optical and low-energy X-ray regions. One implication of this result is that a simple comparison between the observed fluxes of optical and gamma radiation emitted by the Nebula is no longer meaningful, so that one cannot draw any immediate conclusion regarding the fundamental mechanism of electron and gamma ray production.

The radiation continuum for the Crab Nebula in the near infra-red, optical and low-energy X-ray regions may be represented approximately by the power-law formula $F(v) \propto v^{-n}$, with $n \approx 1$, a characteristic feature that it shares with other radiation sources associated with energetic cosmic events.

It is generally agreed that the continuum arises from synchrotron radiation by relativistic electrons. The usual explanation for the spectral index n is that it results from radiation, in a uniform magnetic field, by electrons that have an energy spectrum $N(E)dE \propto E^{-\alpha} dE$, where $\alpha = 2n + 1$. We show here that for regions of high magnetic field, n can be given a different interpretation, with the particular values $\frac{1}{2}$ and 1 having a special significance. We also consider the relevance of our arguments to the Crab Nebula.

The synchrotron radiation power spectrum, for an electron of energy γmc^2 in a magnetic field H, is described by the characteristic frequency

$$v_c = \left(\frac{3eH}{4\pi mc}\right)\gamma^2 . \tag{1}$$

For x defined as v/v_c, the power spectrum is given approximately by

$$P(v) \propto Hx^{1/3}, \quad \text{for} \quad x \leqslant \tfrac{1}{3}, \tag{2a}$$

and

$$P(v) \propto H \exp\left(-ax^{2/3}\right), \quad \text{for} \quad x > \tfrac{1}{3}, \tag{2b}$$

where a is a constant. The expression for the electron energy as a function of time,

$$\gamma = \gamma_0/(1 + \beta H^2 \gamma_0 t), \tag{3}$$

where γ_0 represents the initial electron energy and β is a constant, gives the radiation half-life for the electron as

$$t_{1/2} = (\beta H^2 \gamma_0)^{-1}. \tag{4}$$

Davies and Smith (eds.), The Crab Nebula, 314–320. All Rights Reserved.

The corresponding evolution of the synchrotron power spectrum can be represented by the variation of characteristic frequency with time,

$$v_c = (v_c)_0/(1 + t/t_{1/2})^2, \qquad (5)$$

where $(v_c)_0$ is the value of v_c for $\gamma = \gamma_0$.

It is normally assumed that v_c does not change appreciably over the relevant time interval, which is usually the period of observation, so that the radiation continuum can be derived from the synchrotron power spectrum by combining it with the electron energy spectrum. We examine here the situation for which the change in v_c during the relevant time interval is appreciable. In Figure 1, v_c has been plotted, from (5), for times that are comparable with the radiation half-life $t_{1/2}$.

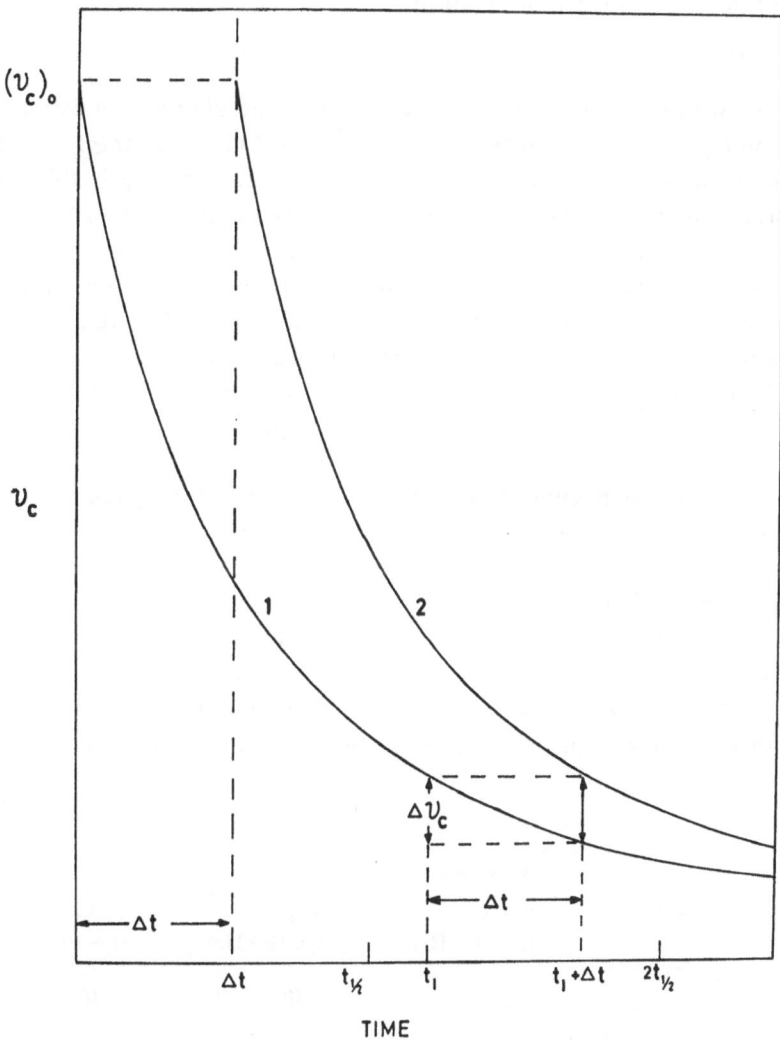

Fig. 1. Characteristic frequency as a function of time, for $t \sim t_{1/2}$.

Suppose that at time t_1 there commences an observation of the radiation produced by a single electron, and that the observation lasts for an interval of time Δt; during this interval, a change in characteristic frequency Δv_c occurs for the electron represented by curve 1 in the figure. For a system containing many electrons, a spread in characteristic frequency of the same magnitude Δv_c, for radiation observed at the instant of time $t_1 + \Delta t$, may have been produced either by monoenergetic electrons with $v_c = (v_c)_0$ continuously injected into the field during the interval from $t=0$ to $t=\Delta t$; or it may have been produced by electrons injected at a time $t=\Delta t$, with a whole range of initial energies, represented by the difference in ordinates of curves 1 and 2 at this t value.

For the case of the single electron observed over the interval from t_1 to $t_1 + \Delta t$, we suggest that for the evolutionary effect to be regarded as significant, $\Delta t \gtrsim t_{1/2}$. This condition can also be written in the form

$$H^{3/2} \Delta t \gtrsim \text{a constant}, \tag{6}$$

the value of which depends upon the frequency of the observed radiation. In the radio region ($v_c \approx 10^9$ Hz), a magnetic field $H \sim 10^{-4}$ G gives $\Delta t \gtrsim 10^6$ y: clearly, evolutionary effects are insignificant for a single electron radiating in this region of the spectrum. In the optical region however, $H \approx 100$ G gives $\Delta t \gtrsim 10^3$ sec, a time comparable with photographic plate exposure times. In the X-ray region ($v_c \sim 10^{18}$ Hz), $H \approx 3$ G gives $\Delta t \gtrsim 200$ sec, the detection time for X-radiation from the Crab Nebula in the lunar occultation experiment of Bowyer et al. (1964). If therefore there exist regions of the Nebula containing high magnetic fields $H \gg 1$ G, we may assume that, for a single electron, evolutionary effects are certainly significant for the X-ray region, and probably for the optical region as well, for the periods of observation normally employed.

In the case of a system containing many electrons, Δt can be given an alternative interpretation, as already indicated. In the Table I below, we show values of Δt calculated from (6) for two commonly assumed magnetic field values, for the X-ray, optical and radio regions.

If Δt in interpreted as the time over which monoenergetic electrons have to be continuously injected into the field for the equivalent spread Δv_c in characteristic frequency to be appreciable, then it is apparent that condition (6) could be fulfilled, even for these low magnetic fields, for electron injection times normally considered.

TABLE I

H (Gauss)	Δt (years)		
	X-ray ($v \approx 10^{18}$ Hz)	Optical ($v \approx 10^{15}$ Hz)	Radio ($v \approx 10^9$ Hz)
10^{-3}	> 1	> 50	$\gtrsim 10^4$
10^{-4}	> 30	$\gtrsim 10^3$	$\gtrsim 10^6$

While in our calculations of the theoretical radiation continuum now to be described, we consider only the case of a single electron observed over the time interval Δt, a similar mathematical method would be applicable to sources containing many electrons, with Δt given its alternative interpretation.

We consider a power spectrum of the form (2), in which v_c changes with time according to (4) and (5): the evolutionary effect is illustrated in Figure 2. The total

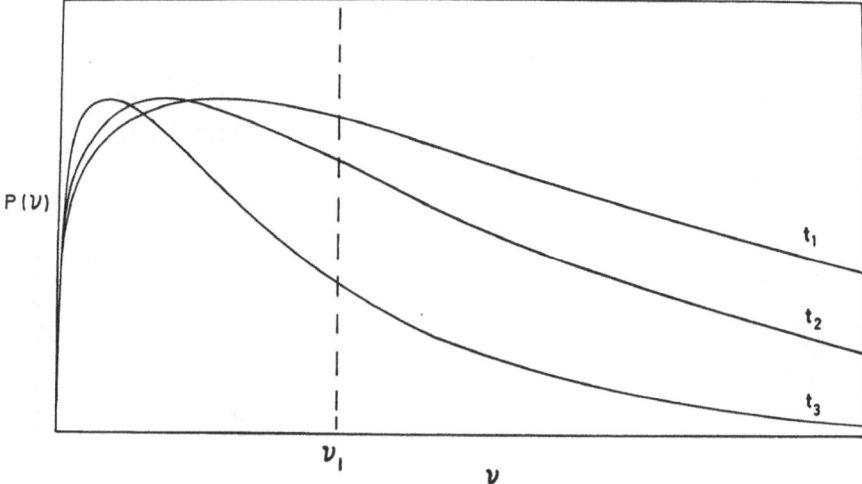

Fig. 2. Evolution of synchrotron radiation power spectrum, shown at successive moments of time t_1, t_2 and t_3.

radiation continuum produced by an electron of initial energy $\gamma_0 mc^2$, radiating in a uniform magnetic field H, has been derived by taking the frequency v as an independent parameter and, at each value of v, (for example, v_1 in the figure), computing the integral of the evolving power spectrum over the whole radiation lifetime of the electron. Figure 3a shows the radiation continuum for the optical and near infra-red regions, computed for one particular electron energy (given by $\gamma_0 = 75$), for several magnetic field values in the range from 2×10^4 G to 10^5 G; each curve gives the radiation continuum produced by a single electron.

If we perform the integration by an analytical method we find that, provided that the initial electron energy is sufficiently high, (a condition that can be expressed by the inequality $\log_{10} H\gamma_0^2 > 12, 9$ and 3 for the X-ray, optical and radio regions respectively), the integral can be written in the simple asymptotic form

$$F(v) = 2.63 \times 10^{-10} (vH)^{-1/2} \text{ erg Hz}^{-1}. \tag{7}$$

If the initial electron energy condition is not satisfied then this expression is only an approximation; in this case the more general formula used in deriving the curves of Figure 3a has to be taken.

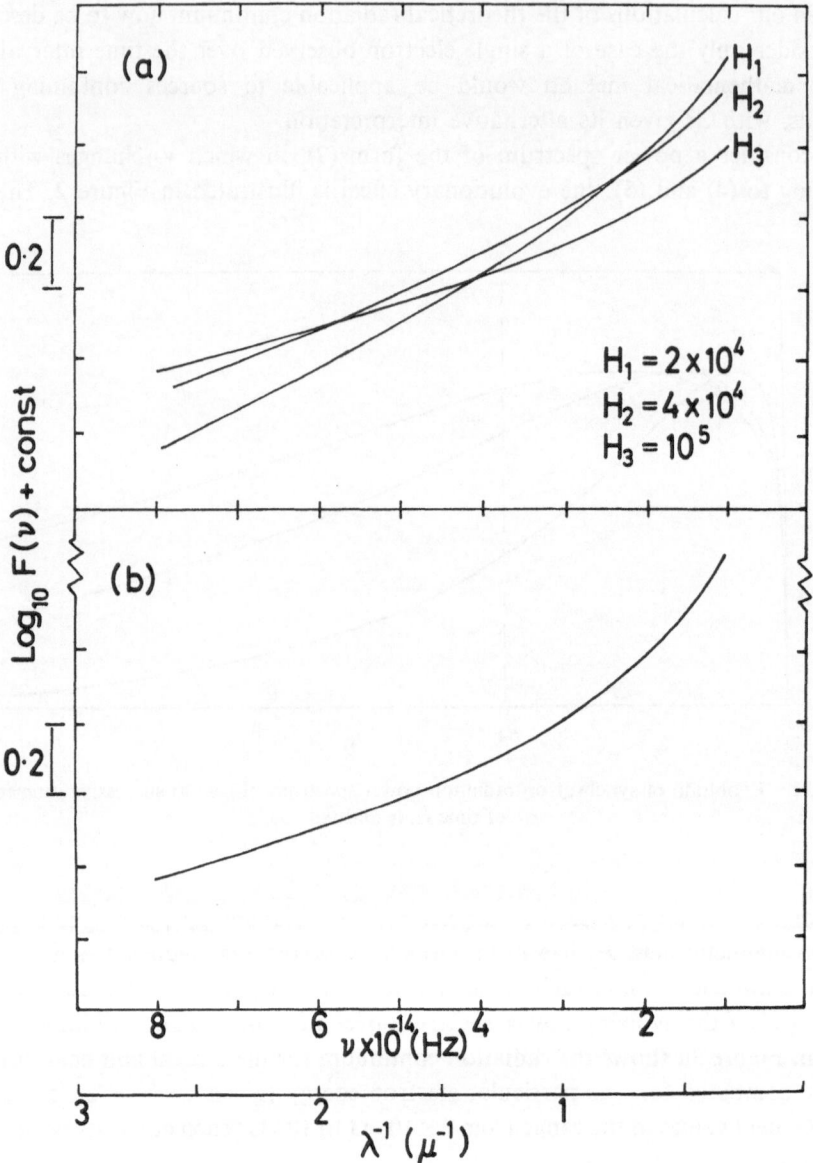

Fig. 3. $F(v)$ as a function of v, (a) for $\gamma_0 = 75$ and three different H values, (in gauss); (b) for $\gamma_0 = 75$ and a flat H distribution.

It seems that the only requirements for the validity of the theory described above are that the electron should be confined to a uniform high magnetic field region, and that it should be observed for the whole of its radiation lifetime; if these requirements are fulfilled then a spectral index $n \approx \frac{1}{2}$ is obtained. Furthermore, the theoretical radiation continuum flux $F(v)$ at the frequency v does not depend upon the initial energy of the electron: this has the important consequence that the spectral index for the radiation continuum is independent of the initial energy spectrum of the electrons injected into the magnetic field region.

For a complete radiation source it is necessary to integrate the computed radiation continua over an appropriate magnetic field distribution. We consider one particular case, – a flat distribution, in which all magnetic field values are equally probable. Inspection of the curves in Figure 3a reveals that for any frequency there exists an optimum H value at which the computed flux $F(v)$ is a maximum; the locus of these maxima is the upper envelope of the curves in Fig. 3a, and is shown in Fig. 3b. We take the envelope plotted in Fig. 3b as an approximate representation of the whole radiation continuum.

We find that for the optical and near infra-red regions, the envelope curve can be represented by the spectral index $n \approx 1$; we also find that an analytical method indicates that $n = 1$ identically. We suggest that this value of n has a special significance and arises in a manner quite different from that usually supposed: it could indicate that we are observing cosmic systems where different high magnetic field values occur with equal probability, and for which the radiating electrons have synchrotron power spectra that show a significant evolutionary effect.

High magnetic fields could be produced in localised condensations of material within the source. The radiating electrons either enter the magnetic field region from outside, or alternatively are generated *in situ* by the decay of pions produced in the nuclear interactions of protons incident upon the condensed material. (In low-energy pion decay, the electron energy distribution has a maximum at $\gamma_0 = 75$, the figure originally taken in our calculations). The rate of electron production *in situ* from a given proton flux is approximately proportional to the mass of the condensation, and does not depend upon the magnetic field contained by it; for the situation where the mass and magnetic field are completely uncorrelated, the suggested flat magnetic field distribution would be obtained.

We would expect this type of model to apply to all cosmic systems that contain high-energy protons together with condensations of material where there are high magnetic fields and where pion, and therefore electron, production can occur (Falla, 1970). A characteristic form for the radiation continua for sources associated with a wide variety of energetic cosmic events, including the Crab Nebula, has indeed been pointed out by Searle *et al.* (1965).

For the Crab Nebula in particular, our model has one clear implication: this is that the observed spectral index $n \approx 1$ is simply the characteristic value, the theoretical derivation of which we have described, so that it is not possible to obtain from the index any information concerning the energy spectrum of the interacting protons. Previously, comparisons between the synchrotron and gamma radiation fluxes had been made, with the object of testing the hypothesis that the relativistic electrons and gamma-ray photons arise from the decay of charged and neutral pions. In conventional synchrotron radiation models, where an electron energy spectrum with index α is involved, one can show that, for a given proton energy spectrum, the spectral index for the synchrotron radiation is exactly half that for the gamma radiation. For magnetic fields of magnitude $H \sim 10^{-4}$ G, calculations by Gould and Burbidge (1967) led to the conclusion that, in view of the low upper limit to the gamma-ray flux, observed by Fruin *et al.* (1964), the pion production process could not be responsible for the electrons that radiate in the optical to low-energy X-ray regions. On our model, however, where much larger fields are assumed to be present, the gamma radiation spectral index, which is determined by the spectral index for the interacting protons, is completely unrelated to the unique value $n = 1$ for the synchrotron radiation: a straightforward comparison between the fluxes of synchrotron and gamma radiation is therefore no longer meaningful.

We consider also the radiation emitted by the Crab pulsar, NP 0532. Electrons radiating in a region of uniform magnetic field of magnitude $\gtrsim 10^6$ G, even greater than the values we have assumed for the Nebula itself, would very probably, according to our arguments, be observed over the whole of their radiation lifetimes. For electrons with $\gamma \lesssim 10^4$ our model, which is based on standard synchrotron radiation theory, would still be applicable; for electrons with higher energies, however, it is probable that the synchrotron emission would be affected by strong radiation damping (Shen, 1970). An interpolation of the observed radiation continua for this pulsar between the optical and X-ray regions shows evidence (Trümper, 1970) for a spectral index $n = \frac{1}{2}$. We may note that the modification of the standard synchrotron radiation theory that we have discussed here would give this value of n for a single condensation of material with which a uniform high magnetic field is associated.

References

Bowyer, S., Byram, E. T., Chubb, T. A., and Friedman, H.: 1964, *Science* **146**, 912.
Falla, D. F.: 1970, *Astrophys. Letters* **6**, 77.
Fruin, J. H., Jelley, J. V., Long, C. D., Porter, N. A., and Weekes, T. C.: 1964, *Phys. Letters* **10**, 176.
Gould, R. J. and Burbidge, G. R.: 1967, *Handbuch der Physik* **46** (2), 265.
Searle, L., Rodgers, A. W., Sargent, W. L. W., and Oke, J. B.: 1965, *Nature* **208**, 1190.
Shen, C. S.: 1970, *Phys. Rev. Letters* **24**, 410.
Trümper, J.: 1970, *Astrophys. Letters* **5**, 271.

PHYSICS OF THE NEUTRON STAR

6.1 PHYSICS OF THE NEUTRON STAR

A. G. W. CAMERON

Belfer Graduate School of Science, Yeshiva University, New York, N.Y., U.S.A., and
Goddard Institute for Space Studies, NASA, New York, N.Y., U.S.A.

Abstract. Some characteristics of neutron star models are described. These include: the physics of the crystalline crust, the superfluid properties of the neutrons and protons, and the presence of rotational vortices in these superfluids. The possible relationship of some of these properties to the behavior of the Crab Nebula pulsar is described.

The discovery of pulsars has focused the attention of theoretical physicists upon neutron stars, with which the pulsars have been identified. Essentially every major field of theoretical physics is represented in the investigation of the neutron star (Cameron, 1970), and in most of these fields the problems posed lie at the very forefront of research, since the conditions in the neutron star are so extremely unlike those encountered in the laboratory. The Crab Nebula has long had the reputation of exhibiting a greater variety of physical phenomena than any other object in the sky, and this characteristic also seems to be present in the Crab Nebula pulsar, which exhibits a greater variety of phenomena than seen in any of the other pulsars.

In Figure 1 the pie-shaped diagram shows a section of a model of a neutron star. The star can be considered to have three major distinct regions. In the outermost region, the neutron-rich nuclei are present, which are expected to form a rigid crystalline lattice. Near the base of this crystal, neutrons become interspersed with the nuclei, and finally, just before the nuclei disappear with increasing density, some protons also become interspersed with the nuclei. The second region, lying toward the intermediate radial distances, consists of a neutron-proton-electron fluid. In the higher density regions of this part, negative mu mesons also become present. In the third and innermost part, hyperons also first put in their appearance, starting with the Σ^- hyperon (Langer *et al.*, 1969; Langer and Rosen, 1970).

The determination of the precise composition of the neutron star interior is intimately related to the determination of the appropriate equation of state. The calculation of the equation of state is based upon the usual and reasonable assumption that the interior of the star is at zero temperature, since the Fermi levels of all the degenerate particles are enormously high compared to the typical thermal energies of the particles in the interior. The determination of the equation of state requires the minimization of the total energy per baryon in each range of density in the interior. Among the energies which enter into the minimization calculation are the potential energies between the baryons. These potential energies are still very imperfectly understood, since densities both high and low compared to normal nuclear density are involved in the neutron star interior, and nuclear forces themselves are still relatively poorly understood. The models of neutron stars which are presented in this paper are based upon an equation of state for the full density range of the neutron star interior calcu-

Davies and Smith (eds.), The Crab Nebula, 323–333. All Rights Reserved.
Copyright © 1971 by the IAU.

COMPOSITION AND STRUCTURE OF
0.586 M$_\odot$ NEUTRON STAR
RADIUS = 13.7 KM

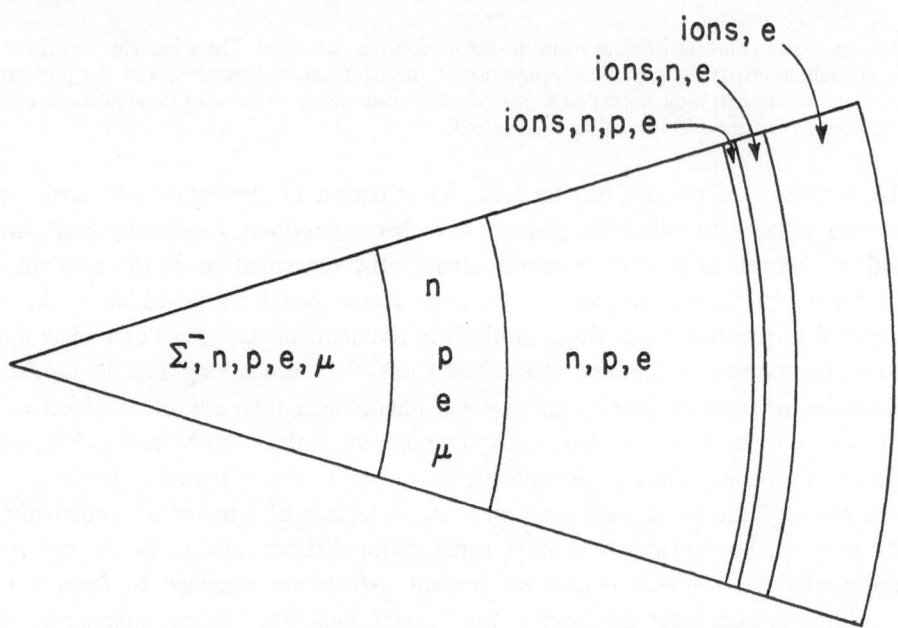

Fig. 1.   Section of a neutron star model showing the constituents. The distances are to scale.

lated by Langer, Rosen, Cohen, and myself (Langer *et al.*, 1969; Langer and Rosen, 1970). This equation of state is relatively stiff at greater than normal nuclear densities, and leads to rather high permitted masses for neutron stars. There have been other equations of state calculated in the last two years and applied to the determination of neutron star models, which differ quite markedly from our own. We are biased and prefer our own equation of state, because we know that it can reproduce the properties of ordinary nuclei rather well. Some of the other equations of state do not succeed so well in this respect.

The neutron star models which we have obtained possess the relation between central density and mass shown in Figure 2 (Cohen *et al.*, 1970; Cohen and Cameron, 1970). In the left portion of this figure there is a peak which corresponds to white dwarf stars. Those models on the left hand side of the peak are stable, whereas those on the right of the peak are unstable against collapse. The neutron star peak is the large one on the right; the intermediate structure which appears to have an additional peak does not correspond to stable models in hydrostatic equilibrium; all of the intermediate models between the white dwarf peak and the bottom of the neutron star peak are unstable against collapse.

The radii of the neutron star models are shown in Figure 3. Except for the lowest

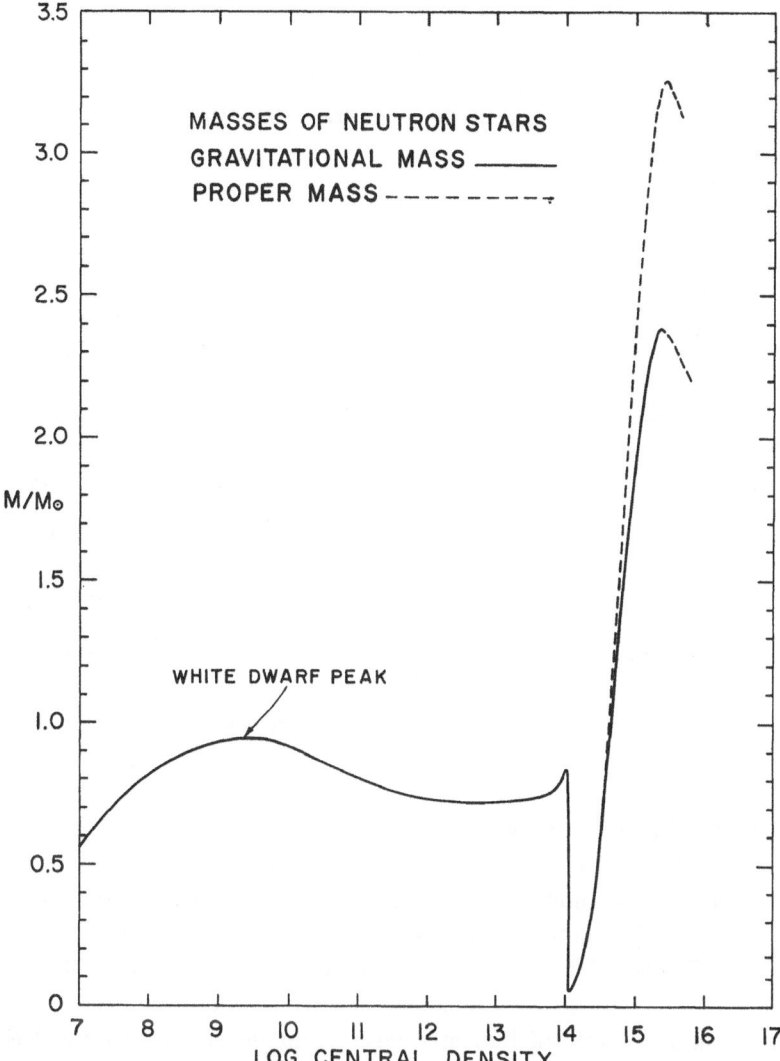

Fig. 2.   Masses of neutron star models, including the white dwarf range of central densities.

mass models, these radii are remarkably insensitive to the mass, being in the vicinity of 13 km.

The interior density distribution of a star with just over half a solar mass is shown in Figure 4. The density varies remarkably slowly over most of the interior, but there is a significant bulge at lower densities in the outermost part of the star. This bulge consists entirely of a mixture of ions and electrons, with no neutrons being present. This bulge would form the outer part of the crystalline crust.

From a model of a neutron star it is possible to calculate the moment of inertia, which is of interest for the determination of the rotational energy that the model may have for any given rotational period. It is necessary to make some small general

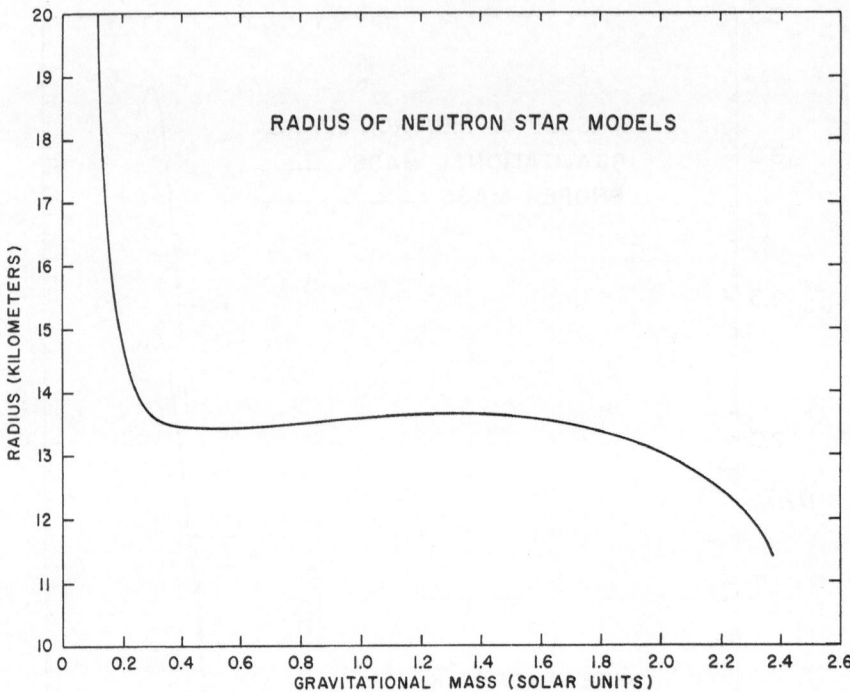

Fig. 3.   Radii of the neutron star models.

relativistic corrections to the calculated moments of inertia to take into account the dragging of the inertial frames which occurs at the high gravitational potentials present in neutron star models. Then for given values of the period and the rate of change of period of a neutron star model, the rate of loss of rotational energy can be calculated. For the Crab Nebula, we know that a minimum rate of energy input to the energetic electrons in the nebula is about 10^{38} erg per second. This must represent a minimum rate of loss of rotational energy, since additional rotational energy may be dissipated in other forms, such as the acceleration of energetic ions and the generation of internal heat. For the models shown in Figure 2, the minimum mass of neutron star which can be attributed to the Crab Nebula pulsar is about 0.4 solar masses. If the mass of the neutron star is much greater than this lower limit, close to the maximum allowable mass, then as much as a factor of 20 additional rate of energy loss would be present in the neutron star (Cohen and Cameron, 1969).

In the outer regions of the star, where ions are present, and except at the very outermost surface, the electrons are highly relativistic, and therefore their motions are only minutely perturbed by the localization of positive charges on the ions. Therefore two adjacent ions are not significantly shielded from each other by the electrons in which they are immersed, as would be the case in an ordinary plasma. Hence there is a large electrostatic repulsion between adjacent ions, which gives an interaction energy of about 1 MeV in a typical case. For temperatures at which the

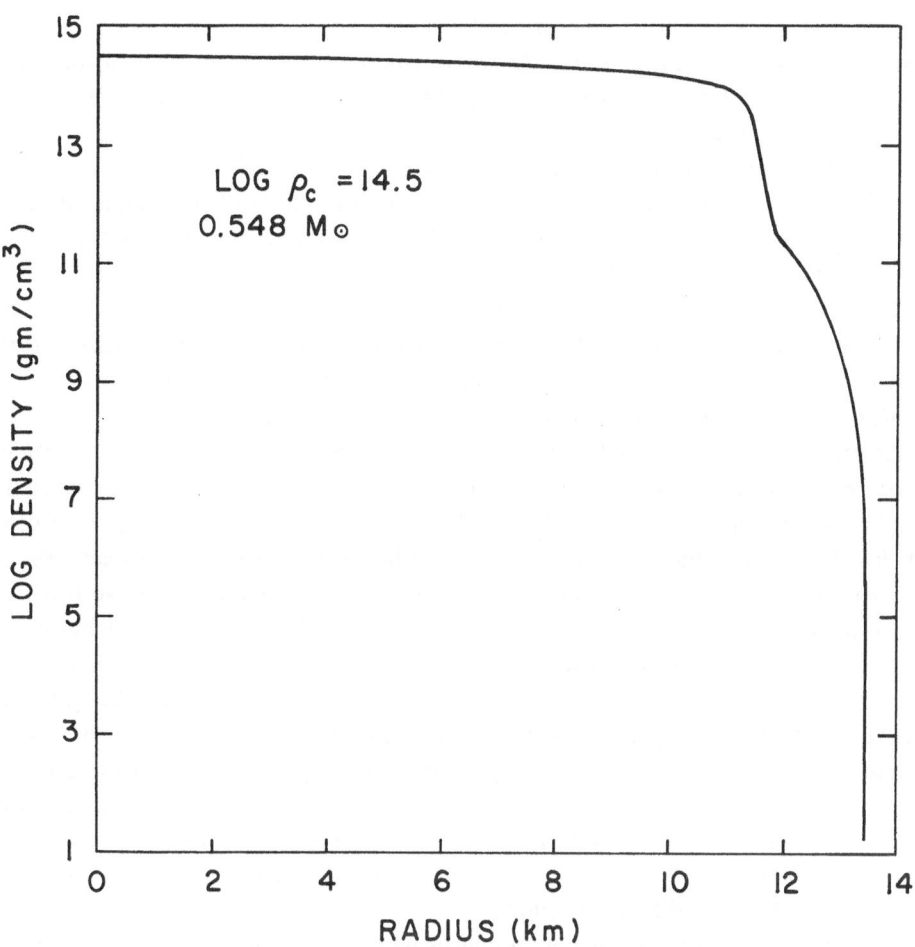

Fig. 4. Internal density distribution of a neutron star model.

typical kinetic energy of the ions is less than about one per cent of this value, the medium seeks a minimum energy configuration by freezing the ions into position in a lattice in which they are as far away from one another as possible, thus forming a crystal (Salpeter, 1961). This does not make much difference to the equation of state of the material, since most of the pressure is provided by the electrons, but it does provide the material with an enormous resistance to shear deformations (Ruderman, 1968). Toward the inner part of the ionic layer, the partial density in the form of neutrons greatly exceeds that in the form of ions, but nevertheless the crystalline lattice should still persist because of these coulomb effects.

Inside the ionic layer, the neutrons exist at less than the normal partial density of neutrons in nuclear matter, which is about 2×10^{14} g/cm^2. In this regime of lower density, nuclear forces are predominantly attractive, and this is expected to have a profound influence upon the properties of the neutron fluid. The attractive forces between the neutrons at the top of the Fermi sea are expected to make them act in pairs like bosons, which then gives the fluid superfluid properties. This introduces an energy gap at the Fermi surface, of 1 to 20 MeV in which no single particle states can exist. The magnitude of this energy gap is very large compared to the thermal energies anticipated in the neutron star interior, so the top of the Fermi sea of neutrons is not thermally rounded in the usual fashion. Only a small number of neutrons can be excited to a sufficiently high energy to populate the states lying above the energy gap (Ginzburg and Kirzhnits, 1965).

The protons are present in the medium just below the ionic layer with an abundance of about three per cent by number relative to the neutrons. The same attractive forces between protons act as between the neutrons, and it is therefore expected that the protons will also form a superfluid.

At greater than the normal nuclear density of neutrons, the nuclear forces become predominantly repulsive. Consequently, the normal superfluidity which is expected to exist at lower densities no longer exists. However, there is still one component of the nuclear force which has an attractive potential between neutrons or between protons, due to one of the p-state interactions, and it is likely that this attractive force will cause the neutrons to have an isotropic superfluidity, of a type which has never been observed in the laboratory (Ruderman, 1969a). In the deep core of the star, where many different types of hyperons may be present, the nature of the forces interacting between the various types of particles is not well enough known for any specific statements to be made about the expectations of superfluidity. However, it does appear that the repulsive forces are not strong enough to produce a nuclear ferro-magnetism which would result from an alignment of the spins of the particles. If such an alignment were to occur, magnetic fields of the order of 10^{15} G would be produced (Pearson and Saunier, 1969).

There are several immediate consequences which follow from the inability to have a normal thermal rounding of Fermi surfaces in the superfluid regions. One of these is the expectation that the nucleons will have a very small heat capacity; practically all of the heat capacity of the neutron star will result from a thermal rounding of the electron Fermi surface. The electrons do not form a superfluid. Another consequence is that many of the neutrino-antineutrino emission processes from the interior of the neutron star will be suppressed. These processes depend either upon the transformation of neutrons into protons and back again, or upon a scattering process in which a proton is induced to make a transformation from one cell of phase space to another. There is no available phase space for such processes except above the energy gap, so that the inhibition of such processes occurs by extremely large factors.

However, some modifications of these complete superfluid properties arise from the fact that the superfluids are rotating. A rotating superfluid is filled with vortices which have axes parallel to the axis of rotation. Each of the vortices has a very small core consisting of normal (not superfluid) material, and the particles in the vortices have a quantized rotation about each vortex axis. The whole assembly of vortices then rotates in a macroscopic fashion.

It is expected that the charged particle components of a neutron star will rotate rigidly together. Even if there were no initial magnetic field present in the interior which would assure this, any relative motion of the charged particle systems would generate currents that would in turn build up magnetic fields in the interior sufficient to assure the charged particle corotation. However, the neutrons are not affected by the presence of a magnetic field in this way, and hence the neutron superfluid is expected to be only very weakly coupled to the charged particle systems. This coupling occurs principally through the normal material in the vortex cores; one of the more effective mechanisms for the interaction is the scattering of electrons from the neutron magnetic moments (Baym *et al.*, 1969a). Additional friction between the neutrons and the charged particle systems arises from the interaction between the normal cores of the neutron and proton superfluids. For these reasons, as a rotating neutron star slows down, it is expected that the slowing of the neutron superfluid will lag behind the slowing of the charged particle system, so that the neutrons will be left rotating faster than the charged particle system and will be slowed down only as a result of the friction that does exist between the two systems (Baym *et al.*, 1969b).

The magnetic field which is likely to exist in a pre-supernova star will be greatly compressed during the formation of a neutron star, and field strengths of the order of 10^{12} G are expected to be produced. This is the magnitude of the neutron star magnetic field which is often invoked as being that required for the slowing down process in ordinary pulsars. The electrical conductivity in the interior of a neutron star is so enormously high, that the ohmic dissipation of the magnetic field would require a time long compared to the age of the Universe. Therefore the neutron star cannot exclude the magnetic field as a bulk effect. It is nevertheless expected that the superfluid protons will exhibit a form of the Meissner effect, in which superconductors exclude magnetic fields, by compressing the internal magnetic field into isolated bundles having a field strength somewhat greater than the average field strength, perhaps 10^{13} or 10^{14} G. The magnetic field will then interfere somewhat with the normal superfluid properties of the protons in the middle of the flux tubes, but the bulk of the protons between the flux tubes will be little affected by the presence of the field.

Furthermore, it is possible that the magnetic field may be permanently fixed in the interior of the neutron star by an electron effect which depends upon an initial field of the order of 10^{12} G being produced in the interior. This effect arises from the rotation of the electrons in perpendicular orbits about the field lines, which gives a magnetic moment arising from the circular motions of the electrons. These orbital magnetic moments will then be aligned in the interior of the star; they reinforce the magnetic field and can maintain the field against dissipative processes. This effect has been called Landau orbital ferromagnetism (Lee *et al.*, 1969).

There are two types of pulsar behavior which are generally attributed to the properties of the interior of a neutron star. These are the period discontinuities, resulting in a sudden decrease in the period of rotation, such as those which have been observed once each in the Vela and Crab pulsars, and also the sinusoidal phase variations of the pulse emission times which have been observed as residuals to fits to the pulsar rotational periods and their first two derivatives in the case of the Crab Nebula. These residuals have had a sinusoidal shape, and at first they were attributed to the possible presence of a planet near the Crab Nebula which would cause the center of mass of the pulsar to move backwards and forwards in the line of sight as the planet revolved around it with a period of about three months. However, the degeneration of this sinusoidal variation into a small amplitude stochastic variation about one cycle after the period discontinuity in the Crab Nebula pulsar is evidence against the influence of a planet in producing these variations. It is therefore very likely that the variation results from some internal behavior.

One suggestion, due to Ruderman (1969b), which has received a great deal of attention, is that the decrease of period which has occurred in the Vela and Crab pulsars has resulted from a large-scale fracture of the crystalline crust, resulting in a decrease of the moment of inertia of the star. In this picture, the neutron star was originally rotating much faster than it is at the present time, so that the crust would be formed with an appreciable equatorial bulge. As the neutron star slows down, the equatorial bulge will tend to flatten toward a sphere, but the high shear strength of the crystalline material will act to prevent this. Consequently, it would be expected that a large internal stress would be built up in the crust which would eventually exceed the breaking stress of the crystal, resulting in large-scale fracture of the crust and its sudden readjustment to a smaller equatorial radius.

One difficulty with this suggestion is that such a fracture would be expected to occur only a few times in the history of the slowing down of the pulsar. This makes it somewhat unlikely that two such events would have been seen in the two shortest period pulsars during the first two or three years of their observation. Furthermore, the coulomb crystal in the crust exists only as a result of the high pressure in the interior, which keeps the ions closely compacted into a lattice. In the absence of a magnetic field, it would then be possible for the non-crystalline material at the surface of the star to flow from the equator to the poles, thus relieving much of the internal stress by a redistribution of the mass. It remains to be shown that no anomalous resistive effects on the surface may occur which would allow this redistribution of mass to take place.

Another suggestion, put forward originally in an inapplicable form by Greenstein and myself (Greenstein and Cameron, 1969), is that an internal fluid instability could exist in the neutron star. In such a case, one would expect interior components of the neutron star to be rotating much faster than the surface; in this case the only available interior component that could be weakly enough coupled to the rest of the star would be the neutrons. If the angular velocity gradient in any region of the interior becomes sufficiently great, then the angular momentum per unit mass may decrease outwards, leading to a classical fluid instability which will result in interior convective motions and the outward transport of angular momentum, thus leading to an acceleration of the surface layers. Dr. Greenstein and I still think that a form of this instability is a possibility in the neutron star interior. If the friction between the superfluid neutron component of the star and the charged particle system has a sufficiently non-uniform distribution in radius, then the rotational vortices in the superfluid will become greatly stretched and highly deformed and tangled (Greenstein, 1970). This deformation may also occur if the ends of the vortices become anchored to irregularities in the coulomb crystal in the outer shell of the star. There is a tension exerted along the axes of the vortices, but this tension will be less in some portions of the neutron superfluid than it is at others, and therefore, it may be expected that variations will be produced in the angular velocity gradient in the interior. If the angular velocity gradient can become great enough so that a local outward decrease of angular momentum per unit mass is produced, then instabilities can set in which would mix outwards some of the superfluid neutron material possessing greater angular momentum per unit mass than that of the fluid which it displaces. In any case, since the neutrons are expected to be rotating faster than the charged particle systems, any sudden increased coupling between these systems will produce a sudden period decrease in the charged particle components of the star. Much too little is known about the detailed physics of the interior to allow a reasonable evaluation of this possibility at this time.

Although several suggestions have been made to account for the sinusoidal residuals of the pulsar phase, only one of the suggestions still appears to be possibly valid in view of the behavior of the residuals following the period decrease in the Crab Nebula pulsar. This is a suggestion due to Ruderman (1970) that an oscillation in radius of the rotational vortices of the neutron superfluid can occur. These vortices should exist in the form of a triangular lattice, providing that the tangling of the vortices is not too severe, and this lattice can undergo expansion and contraction, leading to increases and decreases of the rotation rate of the pulsar superimposed upon its bulk behavior. Indeed, Ruderman has estimated that the period of oscillation of the vortices should be a few months in the case of the Crab Nebula pulsar. This rotational variation becomes visible in the repetition rate of the pulsar emission only as a result of frictional coupling between the superfluid neutrons and the charged particle systems. If this is the correct explanation for the sinusoidal residuals, then we would have to guess that this oscillation mode of the interior is excited by some sort of internal nonlinear interaction, and the persistence of the variation for nearly a full cycle after the period decrease would then indicate that some considerable time is required before the internal effects of the period decrease to become propagated to the great bulk of the superfluid neutron interior. The tangled vortex superfluid instability suggested above as a possible explanation of the period decrease may possibly be reconciled with this situation, although the detailed physics is far from being worked out.

There is one other possible observation, which is very difficult, but which might be made to verify another consequence of the expected superfluid properties of a neutron star interior. The surface layers of a neutron star will have a greatly reduced opacity, compared to that of normal plasma, for the propagation of light along the direction of magnetic fields, and for the propagation of polarized light travelling perpendicular to magnetic fields with the electric vector perpendicular to the field direction. This results from the fact that the electron motions in the perpendicular direction in the strong magnetic fields under consideration are quantized with energy spacings large compared to thermal energies. Therefore the interaction of the electric vector of an electromagnetic wave with these electrons is strongly suppressed (Canuto, private communication). This should result in a greatly reduced temperature gradient in the outer layers of a neutron star, and hence an enhanced rate of cooling of the star. It is very difficult to make a realistic calculation of the opacity of the outer layers at very low temperatures, since the atoms will have their electronic orbits grossly distorted by the strong magnetic fields. Nevertheless, it would be expected that even the pulsar in the Crab Nebula should by now have cooled to a surface temperature of much less than 10^6 K.

However, the expected friction between the charged particle components in the star and the neutron superfluid must result in a portion of the lost rotational energy of the neutron star appearing as a heating of the neutron superfluid. The amount of heating will depend upon the rate at which the motion of the neutrons lags in slowing down behind that of the charged particle systems. If the rotation of the neutrons makes one lap with respect to the charged particles in a range of about 1 h to 1 sec, then the interior heating of the neutron superfluid should come into steady-state equilibrium with the surface radiation of the heat at a surface temperature in the range 1 to 10×10^6 K. Normally, one would not expect to see any visible light resulting from this radiation, since the plasma frequency at neutron star photospheric densities is greater than the frequency of visible light. However, with the expected situation in which the cyclotron frequency of the electrons in the magnetic field is itself much greater than the plasma frequency of the electrons, then radiation is allowed to come out at frequencies much less than the plasma frequency along the magnetic field directions. Hence it can be expected that a very faint object with a very blue color should be visible at the position of the Crab Nebula pulsar at times between the pulsar flashes. Very soft X-rays should similarly be emitted between the Crab Nebula X-ray pulses. However, the conditions of observation of these effects in the Crab Nebula are so adverse that it may be better to look for them in association with the Vela pulsar, even though the expected surface temperature of the Vela pulsar would be somewhat less for that neutron star.

Acknowledgements

This work has been supported in part by the National Science Foundation and by the National Aeronautics and Space Administration.

References

Baym, G., Pethick, C., and Pines, D.: 1969a, *Nature* **224**, 674.
Baym, G., Pethick, C., Pines, D., and Ruderman, M.: 1969b, *Nature* **224**, 872.
Cameron, A. G. W.: 1970, *Ann. Rev. Astron. Astrophys.* **8**, 179.
Cohen, J. M. and Cameron, A. G. W.: 1969, *Nature* **224**, 566.
Cohen, J. M., Langer, W. D., Rosen, L. C., and Cameron, A. G. W.: 1970, *Astrophys. Space Sci.* **6**, 228.
Cohen, J. M. and Cameron, A. G. W.: 1970, *Astrophys. Space Sci.* (in press).

Ginzburg, V. L. and Kirzhnits, D. A.: 1965, *Sov. Phys., JETP* **20**, 1346.
Greenstein, G. S. and Cameron, A. G. W.: 1969, *Nature* **222**, 862.
Greenstein, G. S.: 1970, *Nature* **227**, 791.
Langer, W. D., Rosen, L. C., Cohen, J. M., and Cameron, A. G. W.: 1969, *Astrophys. Space Sci.* **5**, 259.
Langer, W. D. and Rosen, L. C.: 1970, *Astrophys. Space Sci.* **6**, 217.
Lee, H. J., Canuto, V., Chiu, H. Y., and Chiuderi, C.: 1969, *Phys. Rev. Letters* **23**, 390.
Pearson, J. M. and Saunier, G.: 1970, *Phys. Rev. Letters* **24**, 325.
Ruderman, M.: 1968, *Nature* **218**, 1128.
Ruderman, M.: 1969a, New York University Phys. Dept. Tech. Rept. No. 6/69.
Ruderman, M.: 1969b, *Nature* **223**, 597.
Ruderman, M.: 1970, *Nature* **225**, 619.
Salpeter, E. E.: 1961, *Astrophys. J.* **134**, 669.

Discussion

R. Schwartz: The slide which Dr. Cameron showed on the mass-radius relation for neutron stars may be confusing. The increase in radius as M increases does not mean that the binding energy decreases. The binding energy is not $\sim GM^2/R$; R has to do with the ordinary (i.e. non-neutron) matter in the envelope, which is a tiny fraction of the mass. In this region, the central density increases as M goes up, despite the behaviour of R.

W. J. Cocke: One of the many arguments against a vibration model for pulsars was that vibrations would be quickly damped out by neutrino emission processes. Does the superfluidity phenomenon prevent this damping and thus allow vibrations to last for appreciable times?

A. G. W. Cameron: Superfluidity will certainly reduce neutrino emission associated with vibration. However if Σ^- particles are present, $N + N \rightleftharpoons P + \Sigma^-$ gives strong damping and may not be too much affected by superfluidity.

P. Horowitz: I'd like to make a comment on the demise of the 'starquake' theory, namely the frequency with which these jumps should be observed: If quakes are due to sudden crust readjustments in order to reduce the oblateness of the slowing star, then the mean time between quakes of the size observed in Vela is about 700 yr; for the Crab events of the magnitude of the September 1969, event should occur every 6 days. Thus it is remarkable that we were lucky enough to see the Vela jump, and even more remarkable that we see no further Crab jumps in a year of observing.

One further remark: If the Crab has a solid crust very much weaker than that necessary to produce jumps of the September event – for example, if it is full of cracks, or like gravel – then one gets very frequent small jumps, and this may be an explanation for the anomalous phase residuals reported by the radio and optical timing groups in the last two days.

6.2 GENERAL RELATIVISTIC THEORY OF
ROTATING NEUTRON STARS – A REVIEW

JEFFREY M. COHEN*

Institute for Advanced Study, Princeton, N.J., U.S.A.

Except in cosmology, astrophysicists are used to thinking of general relativistic effects as small (e.g., light bending, perihelion advance, red shift) and have generally left such problems to general relativists. However, the discovery of pulsars (Hewish *et al.*, 1968) may have changed this. Not only is general relativity necessary to treat rotating neutron stars, but relativity was also partly responsible for the elimination of pulsating white dwarfs as pulsar models.

Soon after the discovery of pulsars, pulsating white dwarfs were suggested as pulsar models. The idea was that, if the white dwarf were composed of a material with a high electron capture threshold (e.g. 13.5 MeV for C^{12}), the fundamental pulsation period might correspond to the pulsar periods. In Newtonian theory, the pulsation period decreases with increasing central density when there is no electron capture. However, when general relativistic effects were accounted for, (Skilling, 1968; Faulkner and Cribben, 1968; Cohen, 1968c) it was found that the pulsation reached a minimum and then increased again – the minimum period being larger than 1.33 sec. The reason for this general relativistic instability is that stars become unstable when the adiabatic index becomes $(\frac{4}{3}) + \varepsilon$, $\varepsilon > 0$. With the discovery of the Crab pulsar, all hope was abandoned of describing pulsars as pulsating white dwarfs.

It is presently believed that pulsars are rotating neutron stars (Gold, 1968). Because of their high density, rotating neutron stars must be treated via general relativity. Such a treatment is much more complicated than the corresponding Newtonian problem. In particular new effects arise such as the dragging along of inertial frames by rotating masses. The general relativistic equations describing a rotating neutron star are the following (Cohen and Brill, 1968)

$$
\begin{aligned}
-8\pi T^{00} &= B^{-1}C^{-1}[(C_r/B)_r + (B_\theta/C)_\theta + E^{-1}(CE_r/B)_r \\
&\quad + E^{-1}(BE_\theta/C)_\theta] + (E\Omega_r/2AB)^2 + (E\Omega_\theta/2AC)^2, \\
-8\pi(T^{11} - \tfrac{1}{2}T) &= B^{-1}C^{-1}[(C_r/B)_r + (B_\theta/C)_\theta] \\
&\quad + B^{-1}E^{-1}[(E_r/B)_r + (B_\theta E_\theta/C^2)] + A^{-1}B^{-1}[(A_r/B)_r \\
&\quad + (A_\theta B_\theta/C^2)] - \tfrac{1}{2}(E\Omega_r/AB)^2, \\
-8\pi(T^{22} - \tfrac{1}{2}T) &= B^{-1}C^{-1}[(C_r/B)_r + (B_\theta/C)_\theta] \\
&\quad + C^{-1}E^{-1}[(E_\theta/C)_\theta + (E_r C_r/B^2)] + A^{-1}C^{-1}[(A_\theta/C)_\theta \\
&\quad + (A_r C_r/B^2)] - \tfrac{1}{2}(E\Omega_\theta/AC)^2,
\end{aligned} \tag{1}
$$

* This work was supported in part by the US Air Force Office of Scientific Research, Office of Aerospace Research under AFOSR Grant 70-1866.

Davies and Smith (eds.), The Crab Nebula, 334–340. All Rights Reserved.

$$- 8\pi (T^{33} - \tfrac{1}{2}T) = B^{-1}E^{-1}[(E_r/B)_r + (B_\theta E_\theta/C^2)]$$
$$+ C^{-1}E^{-1}[(E_\theta/C)_\theta + (E_r C_r/B^2)] + A^{-1}E^{-1}[(A_r E_r/B^2)$$
$$+ (A_\theta E_\theta/C^2)] + \tfrac{1}{2}(E\Omega_\theta/AC)^2 + \tfrac{1}{2}(E\Omega_r/AB)^2 ,$$
$$- 8\pi T^{12} = C^{-1}E^{-1}[(E_r/B)_\theta - (E_\theta C_r/BC)]$$
$$+ A^{-1}C^{-1}[(A_r/B)_\theta - (A_\theta C_r/BC)] - (E^2\Omega_r\Omega_\theta/2A^2BC),$$
$$- 16\pi BCE^2 T^{03} = (CE^3\Omega_r/AB)_r + (BE^3\Omega_\theta/AC)_\theta .$$

where the quantities A, B, C, E, Ω are components of the metric

$$ds^2 = - A^2\,dt^2 + B^2\,dr^2 + C^2\,d\theta^2 + E^2\,(d\varphi - \Omega\,dt)^2 \qquad (2)$$

and where the components of the stress-energy tensor are given by

$$T^{\mu\nu} = (\varrho + pc^{-2})\,U^\mu U^\nu + \eta^{\mu\nu}\,pc^{-2}. \qquad (3)$$

In the metric (2), ds represents the distance between two nearby points. Although this set of equations represents a reduction from 10 non-linear equations to 6 it is still rather complicated. But they are valid even for strong gravitational fields and rapid rotation.

Fortunately, a simpler method is available for treating rotating neutron stars. This method (Brill and Cohen, 1966; Cohen and Brill, 1968; Cohen, 1965, 1967, 1968a), which assumes only slow rotation, is valid even for strong gravitational fields. It allows a fully relativistic treatment of rotating stellar models and has been applied to such problems by various authors, e.g. Cohen and Brill (1968), Hartle and Thorne (1968), Cohen and Cameron (1969).

At first sight one might think that a star, such as the Crab pulsar, rotating thirty times a second, is not slowly rotating. However, for a typical neutron star of radius 13 km, mass 1.3×10^{33} gm, and rotational period 33 msec or more, the conditions for slow rotation are fulfilled. These conditions are that the 'centrifugal' force acting on any element of the star be small compared to the gravitational force and that the velocity of any element be small relative to the light velocity.

Once the non-rotating models are known, only one equation needs to be integrated in order to describe a slowly rotating neutron star. The non-rotating models are obtained by integrating the equations

$$m_r = 4\pi r^2 \varrho , \qquad (4a)$$

$$c^2\varphi_r = Gmr^{-2}\left[\frac{1 + 4\pi r^3 pm^{-1}c^{-2}}{1 - 2Gmr^{-1}c^{-2}}\right], \qquad (4b)$$

$$p_r = - \varrho c^2\varphi_r[1 + p\varrho^{-1}c^{-2}]. \qquad (4c)$$

Here m is the mass, ϱ the density, p the pressure, r the radius, c the light speed, G the gravitational constant, and φ the 'gravitational potential'. These equations are very similar to the corresponding Newtonian equations which can be obtained by setting the expression in brackets equal to 1. Using an improved equation of state, these equations have been applied to obtain neutron star models (Cohen *et al.*, 1969a; Cohen and Cameron, 1970). I will not discuss these models here since this is being done at this symposium by Cameron (1971). For a discussion of the stability of general relativistic models, see, e.g. Chandrasekhar (1964), Taub (1962), Cohen *et al.* (1969a), Bardeen *et al.* (1966), Cocke (1965).

Given the non-rotating equilibrium model, one can describe a slowly rotating neutron star by integrating the single equation (Brill and Cohen, 1966; Cohen and Brill, 1968; Cohen, 1968a).

$$[A^{-1}B^{-1}r^4\Omega_r]_r = -16\pi BA^{-1}(\varrho + pc^{-2})(\omega - \Omega)Gc^{-2}. \tag{5}$$

Here A and B are given

$$A = e^\varphi, \qquad B = 1 - 2Gmr^{-1}c^{-2},$$

$\omega(r)$ is the angular velocity of the star, and Ω is the angular velocity of inertial frames along the rotation axis. In general relativity, as in Newtonian mechanics, an inertial frame has the property that 'Coriolis' and 'centrifugal' forces vanish in this frame. However, unlike in Newtonian mechanics, inertial frames in the vicinity of a rotating star are dragged along by the star and rotate relative to those far from the star. The angular velocity of inertial frames can be determined, e.g., by measuring the angular velocity of the axis of a gyroscope (Brill and Cohen, 1966, Cohen and Cohen, 1969).

Using the above results, it is easy to compute quantities of astrophysical interest such as the angular momentum J and rotational energy E_{rot} of the star. The latter quantity is quite useful for determining, e.g., a lower limit on the mass of a pulsar as can be done for the Crab pulsar (Cohen and Cameron, 1969).

As is well known in quantum mechanics and Newtonian mechanics, symmetries give rise to conserved quantities. For example, momentum and energy conservation are consequences of space and time translation invariance while angular momentum conservation is a consequence of rotational invariance. For a slowly rotating star, the angular momentum is given by

$$J = \int\limits_0^R dr\, d\theta\, d\varphi\, \sin^3\theta \varrho r^4\omega\,[(1 + p\varrho^{-1}c^{-2})BA^{-1}(1 - \Omega\omega^{-1})] \tag{6}$$

where R is the radius of the star. This expression (6) differs from the corresponding Newtonian expression by the quantity in brackets. Note that the pressure p as well as the density ϱ contributes to the angular momentum. Also, the motion of inertial frames, and the red shift ($z = e^{-\phi} - 1 = A^{-1} - 1$) enter into the general relativistic expression for the angular momentum.

In general relativity, the total energy of a finite size star is equal to its gravitational (Schwarzschild) mass (see, e.g., Landau and Lifshitz, 1962; Møller, 1962). Thus, it is reasonable to define the rotational energy of a star to be the total mass of the rotating star minus the mass of the same star when it is not rotating. The rotating and non-rotating stars have the same number of baryons. When we take into account baryon conservation, gravitational red shift, the contribution of space curvature to the volume element, and the motion of inertial frames, we obtain the following positive definite expression for the rotational kinetic energy of a slowly rotating star

$$E_{rot} = (4\pi/3) \int_0^R dr \, \varrho r^4 \omega^2 \left[(1 + p\varrho^{-1}c^{-2}) \, BA^{-1} (1 - \Omega\omega^{-1})^2 \right]$$

$$+ \int_0^R dr \, r^4 \Omega r c^2 (12ABG)^{-1} + GJ^2 R^{-3} c^{-2}. \qquad (7)$$

If the bracketed factors in the first integral are set equal to 1 and second two terms set equal to zero, the familiar Newtonian expression results. From inspection of Equation (7), it may be interesting to note that the expression for the rotational energy contains contributions from the pressure, red shift, and motion of inertial frames.

Various expressions for the rotational energy can be obtained by, e.g., multiplying Equation (5) by Ω, integrating over all space, and combining the result with Equation (7) yielding

$$E_{rot} = (4\pi/3) \int_0^R dr \, \varrho r^4 \omega^2 \left[(1 + p\varrho^{-1}c^{-2}) \, BA^{-1} (1 - \Omega\omega^{-1}) \right]. \qquad (7a)$$

For uniform rotation, the expression for the rotational energy takes the simple form

$$E_{rot} = J\omega/2. \qquad (7b)$$

This result (7b) can be obtained via physical arguments (Zel'dovich, 1970; Thorne, 1970) and via variational principles (Hartle, 1970).

The expression (7b) has exactly the same form as it does in Newtonian mechanics even when E_{rot} and J contain large general relativistic contributions. That this is reasonable can be seen from the following physical argument: surround the star with a concentric shell of negligible mass and with radius sufficiently large that the shell resides in flat space. If the shell is coupled to the star in such a way that the two rotate together, then we can work with the shell without worrying about what is inside. Since the gravitational field in the vicinity of the shell is weak, a Newtonian treatment can be used to obtain the familiar Newtonian relation (7b).

If the energy source of the Crab nebula is the Crab pulsar (Wheeler, 1966; Finzi and Wolf, 1969), then a lower limit on the mass of the Crab pulsar can be obtained by equating the rate of loss of rotational energy with the observed electromagnetic energy emitted by the Crab. When this is done, with general relativistic effects taken into account, a lower limit of about 0.4 solar masses is obtained (Cohen and Cameron, 1969).

For completeness, a table of rotational properties of a typical neutron star model is included here. From Table I, one can see that general relativistic effects can be quite large. The ratio of radius to gravitational radius can be as small as about 1.65 and the angular velocity of inertial frames at the center of the star can be as high as about 79 per cent of the angular velocity of the star (Figure 1).

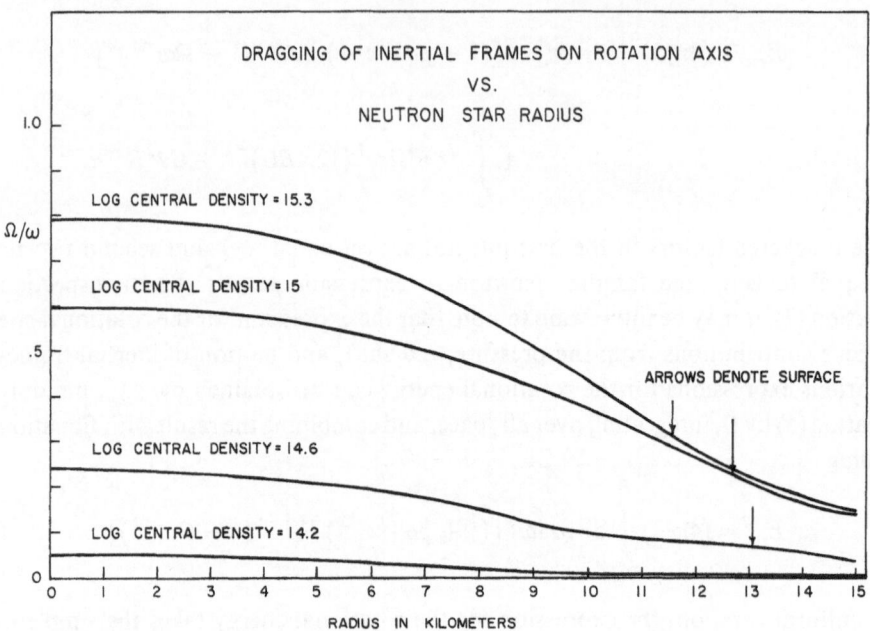

Fig. 1. Dragging along of inertial frames on rotation axis as a function of radius. The angular velocity of inertial frames along the rotation axis is equal to Ω while ω is the angular velocity of the neutron star. Both of these angular velocities are measured by an observer in an inertial frame far from the star. Outside the star Ω is given analytically by $\Omega = 2GJr^{-3}c^{-2}$.

Besides influencing the structure and stability of neutron stars, and giving rise to an induced rotation of inertial frames, general relativity also alters the magnetic field exterior to the neutron star. A frozen-in magnetic field has been postulated by various authors to explain the slowing down and the emission from pulsars (Goldreich and Julian, 1969; Gunn and Ostriker, 1969; Canuto and Chiu, 1968; Michel, 1969). Instead of falling off as $r^{-(2+l)}$ as in flat space, each magnetic multipole varies as a hypergeometric function of radius. In Table II is given the ratio of the magnetic field components in flat space to those which take general relativity into account. The general relativistic contributions become more pronounced as the multipolarity increases. Further details can be found elsewhere (Anderson and Cohen, 1970), see also Ginzburg and Ozernoi (1965) and Cruz et al. (1969) for a discussion of the dipole case.

From the above, it seems that general relativistic effects can be quite important in neutron star theory and should not be neglected.

TABLE I

Rotational properties of selected neutron star models, V_γ potential

Log central density gm/cm³	Radius/ Gravitational radius	J/ω gm/cm²	Central dragging of inertial frames $\Omega_{c/\omega}$	Surface dragging of inertial frames $\Omega_{s/\omega}$
14	498	1.93×10^{48}	0.024	1.6×10^{-4}
14.1	538	3.29×10^{43}	0.028	4.9×10^{-6}
14.2	58.7	5.48×10^{43}	0.046	1.1×10^{-3}
14.3	24.2	1.27×10^{44}	0.073	6.1×10^{-3}
14.4	12.9	2.78×10^{44}	0.11	0.017
14.5	8.34	5.16×10^{44}	0.16	0.032
14.6	5.59	8.95×10^{44}	0.23	0.054
14.7	4.13	1.37×10^{45}	0.21	0.080
14.8	3.30	1.84×10^{45}	0.39	0.11
14.9	2.71	2.30×10^{45}	0.47	0.14
15	2.27	2.69×10^{45}	0.56	0.18
15.1	1.99	2.94×10^{45}	0.65	0.22
15.2	1.79	3.02×10^{45}	0.72	0.25
15.3	1.65	2.94×10^{45}	0.79	0.29
15.4	1.57	2.77×10^{45}	0.84	0.32

TABLE II

Ratio of curved space magnetic field components to flat space components. Radial ratio is R_1; transverse ratio is R_2

$(r/2m) - 1$	$l=1$		$l=2$		$l=3$		$l=4$	
	R_1	R_2	R_1	R_2	R_1	R_2	R_1	R_2
1.0×10^3	1.0	1.0	1.0	1.0	1.0	1.0	1.0	1.0
1.0×10^2	1.0	1.0	1.0	1.0	1.0	1.0	1.0	1.0
1.0×10	1.1	1.1	1.0	1.0	1.2	1.2	1.3	1.3
5.0	1.1	1.2	1.3	1.4	1.4	1.4	1.5	1.6
4.0	1.2	1.2	1.3	1.4	1.5	1.5	1.7	1.7
3.0	1.2	1.3	1.4	1.5	1.7	1.7	1.9	2.0
2.0	1.3	1.5	1.7	1.8	2.1	2.2	2.5	2.6
1.0	1.6	1.9	2.3	2.6	3.3	3.5	4.5	4.8
6.0×10^{-1}	2.0	2.5	3.3	3.8	5.2	6.3	8.2	8.9
1.0×10^{-1}	4.0	6.7	12	17	28	36	69	85
1.0×10^{-5}	30	9.3×10^2	1.7×10^2	3.1×10^3	8.3×10^2	1.1×10^4	2.6×10^3	1.2×10^4
1.0×10^{-10}	64	3.0×10^5	4.0×10^2	9.7×10^5	2.0×10^3	3.3×10^6	9.5×10^3	1.2×10^7

References

Anderson, J. L. and Cohen, J. M.: 1970, *Astrophys. Space Sci.* **9**, 146.
Bardeen, J., Thorne, K. S., and Meltzer, D.: 1966, *Astrophys. J.* **145**, 505.
Brill, D. R. and Cohen, J. M.: 1966, *Phys. Rev.* **143**, 1011.
Cameron, A. G. W.: 1970, this symposium, Paper 6.1, p. 323.
Canuto, V. and Chiu, H. Y.: 1968, *Phys. Rev.* **173**, 1210.
Chandrasekhar, S.: 1964, *Astrophys. J.* **140**, 417.

Cocke, W. J.: 1965, Ann. Inst. Henri Poincaré, **2**, No. 4, 283.

Cohen, J. M.: 1965, Summer Institute on Relativity Theory and Astrophysics (Cornell); notes appear in *Lectures in Applied Math.* **8**: *Relativity Theory and Astrophysics* (ed. by J. Ehlers), A.M.S., Providence, 1967.

Cohen, J. M.: 1967, *J. Math. Phys.* **8**, 1477.

Cohen, J. M.: 1968a, *Phys. Rev.* **173**, 1258.

Cohen, J. M.: 1968b, *J. Math. Phys.* **9**, 905.

Cohen, J. M.: 1968c, N. Y. Pulsar Conference, 20–21 May.

Cohen, J. M. and Brill, D. R.: 1968, *Nuovo Cimento* **56B**, 209.

Cohen, J. M. and Cameron, A. G. W.: 1969, *Nature* **224**, No. 5219, 566.

Cohen, J. M. and Cameron, A. G. W.: 1970, *Astrophys. Space Sci.* (in press).

Cohen, J. M., Langer, W., Rosen, L., and Cameron, A. G. W.: 1969a, *Astrophys. Space Sci.* **6**, 228.

Cohen, J. M., Lapidus, A., and Cameron, A. G. W.: 1969b, *Astrophys. Space Sci.* **5**, 113.

Cohen, J. M. and Cohen, M. D.: 1969, *Phys. Teacher* **7**, No. 4, 241.

de la Cruz, V., Chase, J. E., and Israel, W.: 1970, *Phys. Rev. Letters* **24**, 423.

Faulkner, J. and Gribbin, J.: 1968, *Nature* **217**, 734.

Finzi, A. and Wolf, R. A.: 1969, *Astrophys. J. Letters* **155**, L107.

Ginzburg, V. L. and Ozernoi, L. M.: 1965, *Soviet Phys. – JETP* **20**, 689.

Gold, T.: 1968, *Nature* **218**, 731.

Goldreich, P. and Julian, W. H.: 1969, *Astrophys. J.* **157**, 869.

Gunn, J. and Ostriker, J.: 1969, *Nature* **221**, 454.

Hartle, J. B.: 1970, to be published.

Hartle, J. B. and Thorne, K. S.: 1968, *Astrophys. J.* **153**, 807.

Hewish, A., Bell, S. J., Pilkington, J. D. H., Scott, P. F., Collins, R. A.: 1968, *Nature* **217**, 709.

Landau, L. and Lifshitz, E.: 1962, *Classical Theory of Fields*, Addison-Wesley, Reading, Mass.

Michel, C.: 1969, *Astrophys. J.* **157**, 1183.

Møller, C.: 1962, *Theory of Relativity*, Oxford Press, New York.

Skilling, J.: 1968, *Nature* **218**, 923.

Taub, A. H.: 1962, *Colloq. Intern. Centre Nat. Rech. Sci.* **91**, 173.

Thorne, K. S.: 1970, to be published.

Wheeler, J. A.: 1966, *Annual Rev. Astron. Astrophys.* **4**.

Zel'dovich, Ya.: 1970, to be published.

Discussion of this paper was deferred until after the paper by Bonazzola (6.4).

6.3 DYNAMICAL STABILITY OF NEUTRON STARS

G. CHANMUGAM and M. GABRIEL

Institut d'Astrophysique, Université de Liège, B4200 Cointe-Ougree, Belgium

Abstract. The Nemeth-Sprung equation of state is modified and used to obtain neutron star models. Contrary to the results of some authors it is found that neutron stars with central densities $\lesssim 10^{14}$ g cm^{-3} are dynamically stable. It is suggested that some pulsars may belong to this category of stars.

1. Introduction

Neutron star models determined as accurately as possible are most valuable for understanding the behaviour of pulsars. The equation of state of neutron star matter at densities greater than about 5×10^{14} g cm^{-3} is poorly known. Fortunately however, there has been much progress recently in the theory of nuclear matter so that our understanding of neutron star matter at densities less than about 5×10^{14} g cm^{-3} should be better than when some of the earlier models on neutron stars were constructed (Tsuruta, 1964, Meltzer and Thorne, 1966). In this work we use the Nemeth-Sprung (1968) equation of state, which we modify, in order to discuss neutron star models and their stability.

2. Equation of State

Nemeth and Sprung have used Brueckner theory to derive the equation of state of neutron star matter consisting of electrons, protons, neutrons and muons. In their work they use the soft core Reid (1967) potential which is more recent than the Levinger and Simmons potentials (1961) used in constructing neutron star models (Tsuruta, 1964, Cohen *et al.*, 1970). Nemeth and Sprung (1968) have calculated four equations of state (their results contain some errors which we have corrected) of which they believe that their equation of state (2b) is the best one. It is this equation of state that we use here. Furthermore, since their equation of state does not include the effects of nuclear clustering we have included these effects, using the binding energy formula of Green (1954) and essentially following the methods of Langer *et al.* (1969).

We find for increasing values of $\varrho \gtrsim 3 \times 10^{11}$ g cm^{-3} that the system consists of electrons, neutrons and nuclei. The nuclei becoming increasingly neutron rich as ϱ increases. At $\varrho \approx 4.6 \times 10^{13}$ g cm^{-3}, protons appear while the nuclei disappear at $\varrho \approx 4.7 \times 10^{13}$ g cm^{-3}. Muons appear at $\varrho \approx 1.8 \times 10^{14}$ g cm^{-3}. These thresholds are in essential agreement with the results of Langer *et al.* except that we find that the phase with electrons, protons, neutrons and nuclei exists over a narrower density range. At low densities the effects of nuclear forces are small and hence the difference between our equation of state and that of Langer *et al.* becomes negligible (Figure 1). It was also found that when nuclear clustering is included, the Nemeth-Sprung equation of state gives higher pressures than when clustering is not included. This is because the additional protons in the nuclei, when nuclear clustering is included,

Davies and Smith (eds.), The Crab Nebula, 341–345. All Rights Reserved.

have to be neutralised by additional electrons which contribute significantly to the pressure, particularly at low densities. The adiabatic index is however smaller when nuclear clustering is included. It is important to realize that the treatment of the phase with nuclei leaves much to desired. Coulomb interactions, lattice effects, shell model effects and effects of the neutron sea on the nuclei have been neglected in this calculation. Electromagnetic many body forces, which have been suggested to be of possible

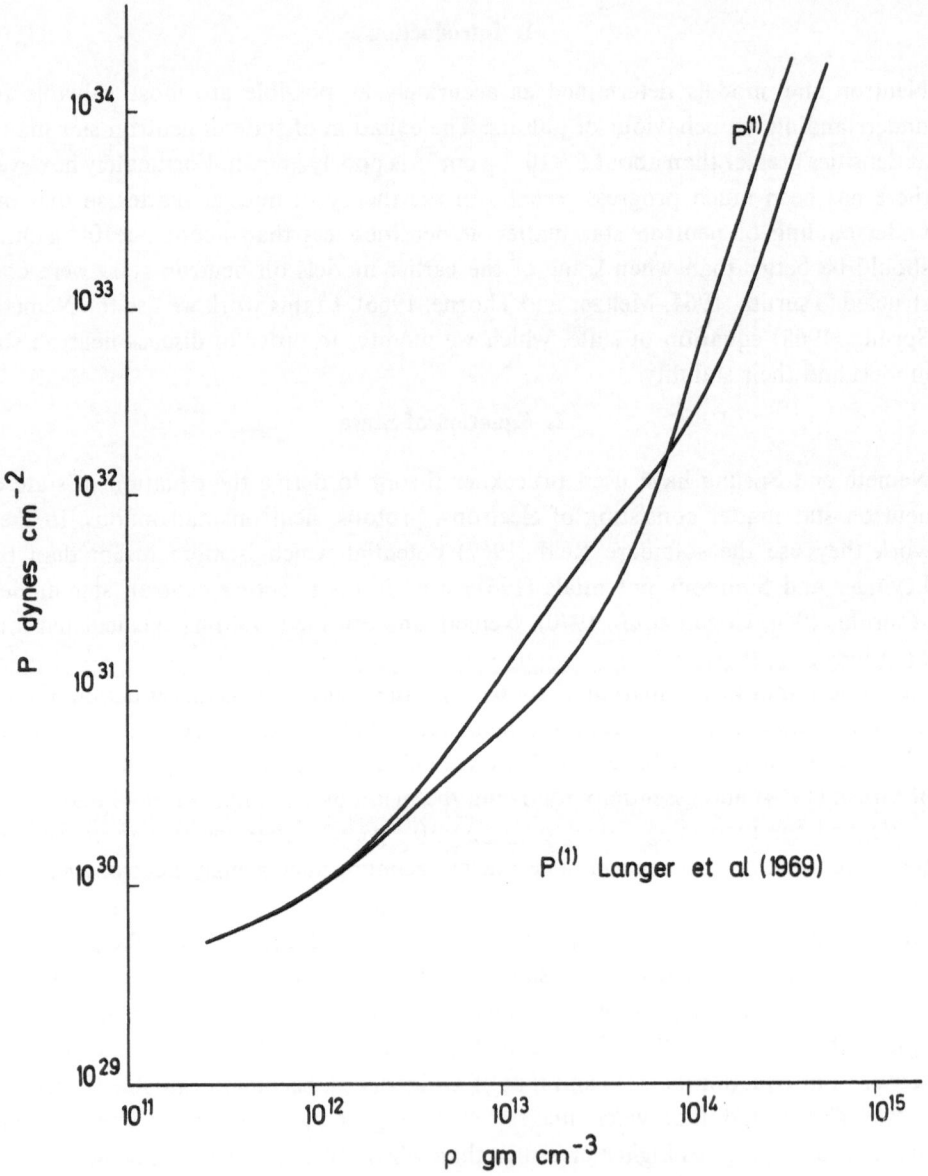

Fig. 1. Pressure as a function of density, from the modified Nemeth-Sprung equation of state, compared with that from the results of Langer *et al.* (1969).

importance in dense electron gases as found in neutron stars (Chanmugam and Schweber, 1970), have also been neglected.

3. Neutron Star Models

We have used the modified Nemeth-Sprung equation of state to integrate the general relativistic stellar structure equations to obtain neutron star models (Table I) with central densities ϱ_c up to $4.8 \times 10^{14} \text{ g cm}^{-3}$. The models have a minimum mass of $\approx 0.07 \, M_\odot$ at $\varrho_c \approx 2.4 \times 10^{14} \text{ g cm}^{-3}$. For $\varrho_c \gtrsim 2.4 \times 10^{14} \text{ g cm}^{-3}$ we find that the masses of neutron stars are smaller than generally believed earlier. In particular our models have masses about one fifth of those given by Cohen et al. (1970). The recent models of Wang et al. (1970) also have small masses. We find that our models have radii in rough agreement with the results of Cohen et al. but are much smaller than those of Wang et al. This is probably because the masses of these neutron stars are sensitive to the equation of state at high densities whereas the radii are sensitive to the equation of state at low densities. The periods of radial oscillation and the thickness of the crystalline layer are in reasonable agreement with results of Cohen et al.

TABLE I

Neutron star models

Central density in g cm^{-3}	Mass in solar units	Radius in km
4.77×10^{14}	0.23	10.6
3.92×10^{14}	0.16	12.0
3.17×10^{14}	0.11	14.9
2.52×10^{14}	0.08	53.2
1.98×10^{14}	0.83	3580.0
1.51×10^{14}	0.80	1415.8
$5.8 \ \times 10^{13}$	0.74	1057.3
4.35×10^{13}	0.73	1030.7
2.71×10^{13}	0.71	930.6
8.24×10^{12}	0.70	711.1

Cohen et al. and Wang et al. find that their models have a minimum mass at $\varrho_c \approx 10^{14} \text{ g cm}^{-3}$. They also find that their models with central densities $\varrho_c \lesssim 10^{14} \text{ g cm}^{-3}$ are unstable. In their calculations they use the adiabatic index along the equation of state γ_E to determine the criteria for dynamical stability. However, the period of oscillations of neutron stars are in general much smaller than the nuclear reaction rates (Meltzer and Thorne, 1966) so that one should use the adiabatic index at constant composition γ_c in determining the criterion for stability. We find that γ_c is sufficiently large (Figure 2) so that the models with $10^{13} \lesssim \varrho_c \lesssim 10^{14} \text{ g cm}^{-3}$ are dynamically stable. These models have masses $\approx 0.75 \, M_\odot$ and radii ≈ 1000 km (Table I). For $\varrho_c \lesssim 10^{13} \text{ g cm}^{-3}$ it is possible to have long periods of the order of the nuclear reaction rates so that the γ to be used is amplitude and period dependent and lies between γ_E and γ_c, and the problem becomes more complicated.

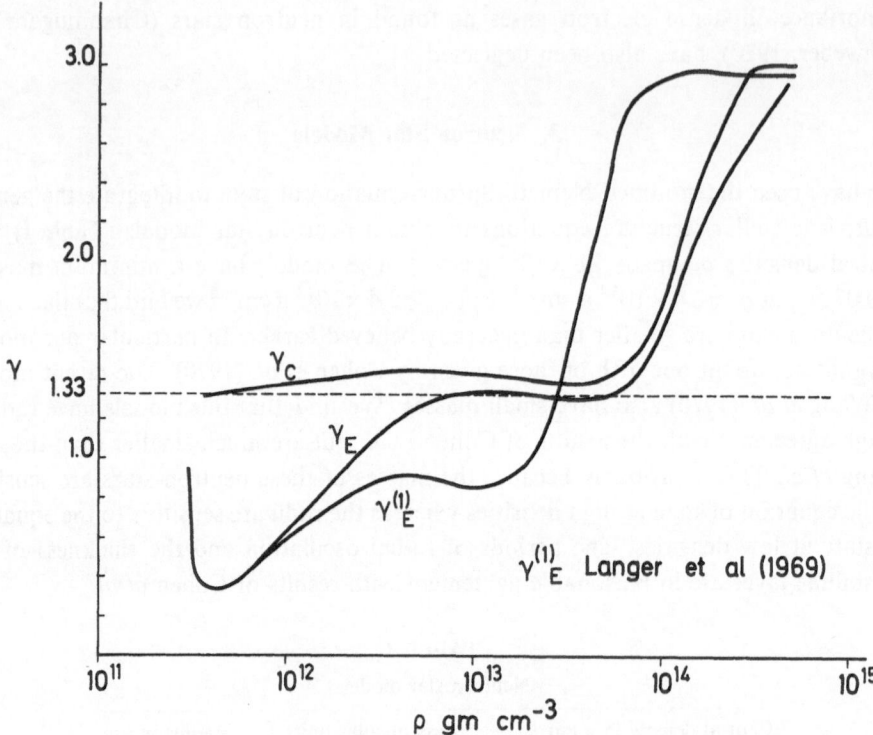

Fig. 2.   The adiabatic indices γ_E (along the equation of state) and γ_c (at constant composition) from the modified Nemeth-Sprung equation of state, compared with γ_E from the results of Langer *et al.* (1969).

4. Application to Pulsars

If we assume that the Crab pulsar is a rotating neutron star we find for our models with $\varrho_c \gtrsim 2.4 \times 10^{14}$ g cm^{-3} that the rotational energy loss is insufficient to account for the 10^{38} erg sec^{-1} necessary to drive the Crab nebula. Despite the small masses and because of the large radii in their models Wang *et al.* do find sufficient rotational energy loss to drive the Crab nebula. However, it should be noted that Wang *et al.* use the equation of state of Langer *et al.* at low densities and do not explicitly calculate effects due to nuclear clustering. Since the potentials used by Wang *et al.* are different from that used by Langer *et al.* their equation of state at low densities is not self consistant. Hence similar criticisms apply to their values for radii. They also do not include effects due to the presence of protons, muons and electrons.

Ginzburg *et al.* (1969) have suggested that some pulsars may be rotating and vibrating, the period ≈ 1 sec corresponding to rotation and the period $\gtrsim 10^{-2}$ sec corresponding to vibration. This theory has been criticized on the grounds that neutron stars with periods of vibrations $\gtrsim 10^{-2}$ sec occur only over a very narrow range for models near the mass minimum. However this criticism applies only to models where γ_E is used in discussing the stability. But if, as in our work, γ_c is used, stars with periods $\gtrsim 10^{-2}$ sec occur over a wider range of density, adding support to the theory of Ginzburg *et al.*

In conclusion we would like to suggest that there exist stable neutron stars with central densities $\varrho_c \lesssim 10^{14}$ g cm^{-3}, and it is likely that at least some pulsars belong to this category.

Acknowledgements

We would like to thank Prof. P. Ledoux for valuable discussions. One of us (G.C.) thanks the Université de Liège for a research fellowship.

References

Chanmugam, G. and Schweber, S. S.: 1970, *Phys. Rev.* **A1**, 1369.

Cohen, J. M., Langer, W. D., Rosen, L. C., and Cameron, A. G. W.: 1970, *Astrophys. Space Sci.* **6**, 228.

Ginzburg, V. L., Zheleznyakov, V. V., and Zaitsev, V. V.: 1969, *Astrophys. Space Sci.* **4**, 464.

Green, A. E. S.: 1954, *Phys. Rev.* **95**, 1006.

Langer, W. D., Rosen, L. C., Cohen, J. M., and Cameron, A. G. W.: 1969, *Astrophys. Space Sci.* **5**, 259.

Levinger, J. S. and Simmons, L. M.: 1961, *Phys. Rev.* **124**, 916.

Meltzer, D. W. and Thorne, K. S.: 1966, *Astrophys. J.* **145**, 514.

Nemeth, J., Sprung, D. W. L.: 1968, *Phys. Rev.* **176**, 1496.

Reid, R. V., Jr.: 1967, Thesis, Cornell University.

Tsuruta, S.: 1964, Thesis, Columbia University.

Wang, C. G., Rose, W. K., and Schlenker, S. L.: 1970, *Astrophys. J.* **160**, L17.

Discussion of this paper was deferred until after the following paper by Bonazzola.

6.4 MODELS OF ROTATING NEUTRON STARS
IN GENERAL RELATIVITY

S. BONAZZOLA

Observatoire de Paris, 92-Meudon, France

and

G. MASCHIO

Instituto di Matematica Applicata, Universita di Roma, Rome, Italy

Abstract. We describe a numerical method of integration of Einstein's equations coupled with the hydrodynamic equations for a perfect fluid. We give, also, some preliminary results.

We point out that the analogous problem was considered by Ostriker and Mark (1968) in the Newtonian case; by Thorne (1969) in General Relativity, in the weak angular velocity approximation; by Barden and Wagoner (1969) for disk configurations in the pressure vanishing case.

1. Introduction

We shall describe a numerical method of integration of Einstein's equations coupled with the hydrodynamic equations for a perfect fluid. We shall give, also, some preliminary results.

We want to point out that the analogous problem was considered by Ostriker and Mark (1968) in the Newtonian case; by K. Thorne (1969) in General Relativity, in the weak angular velocity approximation; by Bardeen and Wagoner (1969) for disk configurations in the pressure vanishing case.

2. Assumptions

We begin by listing some properties that we should expect to characterize the geometry of the space and the perfect fluid.

(I) Space-time should have Euclidean topology.

(II) Space-time should be stationary and with axial symmetry. It should be flat at spatial infinity. Hence the space-time should admit a global 2-parameter abelian group of motion.

(III) The fluid should be considered as the source of the metric.

(IV) The fluid is assumed to be perfect so that the energy momentum tensor has the form

$$T^{\alpha\beta} = (\mu + P) U^{\alpha}U^{\beta} + Pg^{\alpha\beta}$$
$$U_{\alpha}U^{\alpha} = 1$$

and we suppose the existence of such a state equation (in parametric form $\mu(p)$, $P(p)$,

Davies and Smith (eds.), The Crab Nebula, 346–351. All Rights Reserved.
Copyright © 1971 by the IAU.

$N(p)$) that the quantity

$$N = \int\limits_0^p \frac{\mathrm{d}P}{\mathrm{d}p'} \frac{\mathrm{d}p'}{\mu(p') + P(p')}$$

is a regular function of p when ϱ, μ, P tend to zero.

3. Properties of a Rotating Gas

Let

$$x^1 = \varrho, \qquad x^2 = \phi, \qquad x^3 = z, \qquad x^0 = ct$$

be the cylindrical coordinates adapted to the axial symmetry.

The form of the metric is

$$\mathrm{d}s^2 = g_{00}c^2\,\mathrm{d}t^2 + 2g_{02}c\,\mathrm{d}t\,\mathrm{d}\phi + \varrho^2 g_{22}\,\mathrm{d}\phi^2 + g_{11}\,\mathrm{d}\varrho^2 + g_{33}\,\mathrm{d}z^2$$

where the $g_{\alpha\beta}$ depend only on ϱ and z.

We have from the motion equations of the fluid:

$$\nabla_\alpha T^\alpha_\beta = 0 \Leftrightarrow U^\alpha \nabla_\alpha U_\beta + \partial_\beta N = 0 \tag{1}$$

so that the vector $U^\alpha \nabla_\alpha U_\beta$ has to be a free curl vector.

We shall try to satisfy the Equation (1) by choosing the vector U^α in the following form

$$U^\alpha = (0, \Omega, 0, 1)/\Gamma$$
$$= (g_{00} + 2g_{02}\Omega + \Omega^2 g_{22})$$

where Γ^2 is obtained by the condition $U_\alpha U^\alpha = -1$.

From the Equation (1) we obtain

$$\frac{1}{2} \frac{\Omega^2 \partial_\alpha g_{22} + 2\Omega \partial_\alpha g_{02} + \partial_\alpha g_{00}}{\Omega^2 g_{22} + 2g_{02}\Omega + g_{00}} + \partial_\alpha N = 0. \tag{2}$$

We can satisfy the equation (2) in the following cases:

A. RIGID MOTION

This case is characterized by $\Omega = $ constant. We have the General Relativistic Bernouilli's theorem (Boyer, 1965; Lichnerowicz, 1955).

$$N + \tfrac{1}{2} \log(g_{00} + 2\Omega g_{02} + \Omega^2 g_{22}) + K = 0 \tag{2a}$$

where K is an integration constant.

As an important consequence of this theorem, we can obtain without any integration, the energy momentum tensor as a function of two parameters, Ω and $\mu(0, 0)$ knowing the $g_{\alpha\beta}$ (K can be related to the density of the fluid at the center, $\mu(0, 0)$).

B. DIFFERENTIAL MOTION

From the Equation (2) we have

$$\partial_\alpha \left(N + \tfrac{1}{2} \log \left(\Omega^2 g_{22} + 2\Omega g_{02} + g_{00} \right) \right) - \frac{(\Omega g_{22} + g_{02}) \partial_\alpha \Omega}{\Omega^2 g_{22} + 2\Omega g_{02} + g_{00}} = 0 .$$

If we put

$$\frac{\Omega g_{22} + g_{02}}{\Omega^2 g_{22} + 2\Omega g_{02} + g_{00}} = f(\Omega)/\Omega \tag{3}$$

where $f(\Omega)$ is an arbitrary function of Ω, we obtain the General Relativistic Bernouilli's theorem for a fluid in differential rotation

$$0 = \partial_\alpha \left(N + \tfrac{1}{2} \log \left(\Omega^2 g_{22} + 2\Omega g_{02} + g_{00} \right) + \int \frac{f(\Omega) \, d\Omega}{\Omega} \right) \tag{2b}$$

The solution of the Equation (3) gives us

$$\Omega = \Omega \left(g_{\alpha\beta} (\varrho, z) \right) .$$

Let us give two examples of differential motion
(I) $f(\Omega) = \Omega_0 = \text{constant}$
We have

$$\Omega = \frac{(2 - \Omega_0) \, g_{02} \mp \sqrt{g_{02}^2 (\Omega_0 - 2)^2 + 4 g_{00} g_{22} (\Omega_0 - 1) \Omega_0}}{2 g_{22} (\Omega_0 - 1)} .$$

By taking in account that, when ϱ tends to zero, g_{02} and g_{22} tend to zero as ϱ^2, we have

$$\lim_{\varrho \to 0} \Omega \propto \frac{1}{\varrho}$$

i.e. a vortex.

(II) Except in a few simple cases, it is not easy to solve the Equation (3). For this reason, we consider a system composed of shells of fluid in rigid rotation. The shells have different angular velocities.

For that, we have to solve the following problem: find in a fluid in rigid motion, the two surfaces S across which the angular velocity Ω can jump. To have a stationary motion, it is easy to show that the pressure P of the fluid must be continuous across S. Let Δ be the jump of Ω across S. Let us call V_1 the region in which the angular velocity is equal to Ω and V_2 the region in which the angular velocity is equal to $\Omega + \Delta$.

From the equation $P_1(S) = P_2(S)$ we obtain the equation of the surface S

$$\frac{g_{22} (\Delta^2 + 2\Omega_1 \Delta) + 2 g_{20} \Delta}{g_{00} + g_{22} \Omega^2 + 2 g_{20} \Omega} = \text{const.} \tag{4}$$

Of course, in the Newtonian limit, we have the Poincaré's theorem: the surfaces across which the angular velocity can change are cylinders. This result is quite useful for numerical applications.

4. Integral form of Einstein's Equations

The Einstein's equations can be written in integral form

$$\xi = -\frac{1}{4\pi} \int G_1(x, x') [\chi A^2 (T_{\overline{0}\overline{0}} - \tfrac{1}{2} T_\alpha^\alpha) - Q_{00}] \varrho' \, d\varrho' \, dz' \, d\phi'$$

$$\Phi = -\frac{1}{4\pi\varrho} \int G_3(x, x') \left[\frac{2\chi A^2}{B\varrho^2} T_{\overline{0}\overline{2}} + Q_{02} \right] \varrho'^2 \, d\varrho' \, dz' \, d\phi'$$

$$CB = 1 + \frac{1}{2\pi\varrho} \int \chi (T_1^1 + T_3^3) A^2 BC\varrho' G_2(x, x') \, d\varrho' \, dz'$$

$$\log(AB) = \frac{1}{4\pi} \int [A^2\chi (T_{\overline{2}}^2 + T_{\overline{2}}^2 + T_0^0 - T_{\overline{0}}^{\overline{0}}) + Q_{11}] G_0(xx') \, d\varrho' \, dz$$

(5)

where

$$\xi = \tfrac{1}{2} \log(g_{00}), \qquad \Phi = g_{02}/(g_{00}\varrho^2)$$

$$B = \sqrt{\overline{g_{00}}}; \qquad A = \sqrt{-g_{11}}, \qquad C = \frac{1}{\varrho}\sqrt{-g_{22} + (g_{02})^2/g_{00}}$$

$Gn(x; x')$ are the Green's functions of the elliptic operators

$$\partial^2\varrho + \partial^2 z + \frac{n}{\varrho} \frac{\partial}{\partial\varrho} \quad (n = 0, 1, 2, 3)$$

Let $\gamma_\alpha, \gamma_{\alpha\beta}$ be

$$\gamma_\alpha = g_\alpha/\sqrt{\overline{g_{00}}}; \qquad \gamma_{\alpha\beta} = g_{\alpha\beta} - \gamma_\alpha\gamma_\beta$$

The quantities $T_{\overline{00}}, T_{\overline{0\alpha}},$ and $T_{\overline{\alpha\beta}}$ are defined by the relations:

$$T_{\overline{00}} = \gamma_\alpha\gamma_\beta T^{\alpha\beta}; \qquad T_{\overline{\beta 0}} = \gamma_\alpha\gamma_{\delta\beta} T^{\alpha\delta}$$

$$T_{\overline{\alpha\beta}} = T^{\gamma\delta}\gamma_{\gamma\alpha}\gamma_{\beta\delta}$$

The terms $Q_{\alpha\beta}$ are the quadratic terms of the derivatives of $g_{\alpha\beta}$.

We can integrate the System (5) in a selfconsistent way: we first obtain the functions ξ, Φ, A, C by using a trial distribution of matter ($g_{\alpha\beta}$ being the flat metric); then by using the hydrodynamical Equation (2a) or (2b) we compute a new distribution of matter, we get the terms $Q_{\alpha\beta}$ by using the quantities ξ, Φ, A, C; we compute again the functions ξ, Φ, A, C, the new distribution of matter and so on.

We do not discuss here the convergence of the method.

We have got ready a computer program in order to obtain solutions of the Einstein's equations.

The calculations were performed on an Univac 1108 Computer. The method converges very well, and the time spent to get one solution is about five minutes. The number of the cycles is about four. The precision is 4 figures for g_{00}, g_{11}, g_{22} and 3 figures for g_{02}.

The precision decreases a little when the eccentricity of the fluid increases.

Until now, only a few solutions were obtained. We can give some results in the case of rigid motion and for a free Fermi gas equation of state: for a star of 0.6 solar mass the maximum of the eccentricity (the ratio between maximum radius and minimum radius) is 1.5, and the period 3 msec (only 10 times less then the period of the Crab pulsar). The ratio $K = QM/J^2$ is $\simeq 10^4$ (Q is the quadrupole momentum, M is the mass of the star, J is the angular momentum).

We remember that K is equal to 1 in the Kerr's metric. Because of the large value of K in our solution, it seems that a perfect fluid in rotation cannot be a source of the Kerr's metric. Much work has still to be done: we need more solutions to get relationships of the kind: $N(\mu_0, \Omega)$, $M(\mu_0, \Omega)$, $W(\mu_0, \Omega)$, $J(\mu_0, \Omega)$ where $N, M, W, J, \mu_0, \Omega$, are respectively the barionic number of the star, the global mass, the global kinetic energy, the global angular momentum. All these global quantities must be obtained as functions of the central density n_0 and the angular velocity Ω.

Another relationship would be quite interesting: the critical mass of the star as a function of the angular velocity, for rigid motion, as for differential motion.

5. Applications to Different Problems

The solutions obtained for the equilibrium configurations make possible the approach to a new class of problems.

(a) Stability of a rotating object in general relativity, by using the equilibrium solution as a background.

(b) Problem of gravitational collapse: we remember that for the problem of the evolution of the Einstein's equations, the Cauchy's data cannot be given arbitrarily: they have to be a solution of the equation

$$R^{0i} = \chi T^{0i}.$$

We can show that the $g_{\alpha\beta}$ for the equilibrium solutions are good Cauchy-data for the collapse problem, relative to a very well defined physical situation. This is the first step to solve the more difficult problem of the collapse.

References

Bardeen, J. M. and Wagoner, R. V.: 1969, *Astrophys. J.* **158**, L65.
Boyer, R. H.: 1965, *Proc. Cambridge Phil. Soc.* **61**, 527.
Lichnerowicz, A.: 1955, *Théories de la gravitation et de l'electromagnétisme*, Masson, Paris.
Ostriker, J. P. and Mark, J. W. K.: 1968, *Astrophys. J.* **151**, 1075.
Thorne, K.: 1969, *Astrophys. J.* **153**, 807.

Discussion

A. G. W. Cameron (to Dr. Chanmugam): The vibrational periods calculated for a model which are imaginary using a composition-changing equation of state can certainly be found real for a composition-preserving equation of state. This simply means that a model can vibrate in a condition in which it is secularly unstable and is on the way to collapse.

G. Chanmugam: The question of long term effects and secular stability is as in the case of ordinary stars is a very difficult one as non-linear effects are important and needs detailed separate discussion.

D. W. Richards: Can the various theorists come to any agreement on the upper and lower limits of the mass and radius for the star NP 0532? If so, what are these limits?
Unanimous reply: No.

J. P. Ostriker (to Dr. Cohen): If I remember an earlier version of this work correctly, you found a different moment of inertia appropriate for angular momentum (J) and kinetic energy (T). That result seems inconsistent with the relation $T = \frac{1}{2} \omega J$ which you gave today (and which is I believe right). Has your opinion changed?

J. M. Cohen: These results are the same. My expressions for the angular momentum and rotational energy are the *same* as those in *J. Math. Phys.* **8** (1967) 1477, and those in *Astrophys. Space Sci.*

V. Canuto (to Dr. Chanmugam): Did you compute the value of the nuclear compressibility? It seems to me that it is more sensitive to compare second derivatives of the energy than first, since they all look alike and yet they give quite different stable neutron star masses.

G. Chanmugam: We use Nemeth and Sprung's results. They do match the binding energy with nuclear matter but I don't know whether they consider the compressibility.

6.5 THE EFFECTS OF NUCLEAR FORCES ON THE MAXIMUM MASS LIMIT OF NEUTRON STARS

SACHIKO TSURUTA*

Goddard Space Flight Center, Greenbelt, Md., U.S.A.

Abstract. The original models of neutron stars must be improved by including effects of nuclear interaction. This paper compares the models reached by various groups, and presents an improved model by the Kyoto group. The maximum mass varies between $0.2\ M_\odot$ and $3\ M_\odot$ in the various models. The V_γ model is recommended for use in the absence of further information on the equation of state at high densities.

Recently the problem of the maximum mass limit of a stable neutron star has drawn the attention of many people, because it may seriously affect the models of pulsars and other phenomena which are most likely caused by the presence of neutron stars. In this brief report I wish to compare the neutron star models by various groups, including the most recent results I am aware of, those by the Cornell group (Boozer-Salpeter) and those by the Kyoto group (S. Ikeuchi, T. Mizutani, S. Nagata, R. Tamagaki, and C. Hayashi).

In Figure 1, the curve marked IDEAL represents the models with no nuclear interactions, originally constructed by Oppenheimer and Volkoff (1939). The models by Harrison *et al.* (1965) approximately lie on the same curve in the neutron star region. The dotted curves marked V_β and V_γ are the models constructed by Tsuruta and Cameron (1966) and subsequently used by Thorne and others at CIT [see, e.g. Hartle and Thorne, 1968] in their calculations of the moments of inertia of neutron stars, etc. I decided to show these curves also, because these results (especially the V_γ models) have been used by Ostriker and others [see, e.g. Ostriker and Gunn, 1969] in their pulsar studies. The solid curves show neutron star models constructed after the pulsar discovery. The Curve (1) is by Cohen *et al.* (1970), (2) is by the Cornell group (Boozer-Salpeter), (3) is by Wang *et al.* (1970), and the Curves (4) through (7) are obtained by the Kyoto group.

In all models shown here except the first (that marked IDEAL), nuclear interactions are taken into account. In the models marked as V_β and V_γ, the nuclear potentials of V_β and V_γ type, respectively, by Levinger and Simmons (1961) are used. In the Models (1), the modified Levinger-Simmons neutron gas models are used. Boozer and Salpeter (the Curve (2)) made use of the neutron gas calculations (2b) by Nemeth and Sprung (1968) with the soft-core Reid potentials. Wang *et al.* (the Curve (3)) took the average of the soft-core Reid potentials and several other potentials, but they all give the similar equations of state. The results by the Kyoto group are obtained in the following way. For densities ϱ less than $\varrho_0 \simeq 8 \times 10^{14}$ g/cm^3 where the neutron matter calculations from the concept of 'nuclear potential' becomes unreliable, the one-boson exchange hard-core potentials constructed by Kishi were used in the

* NAS-NRC Research Associate presently at Goddard Space Flight Center.

Davies and Smith (eds.), The Crab Nebula, 352–355. All Rights Reserved.

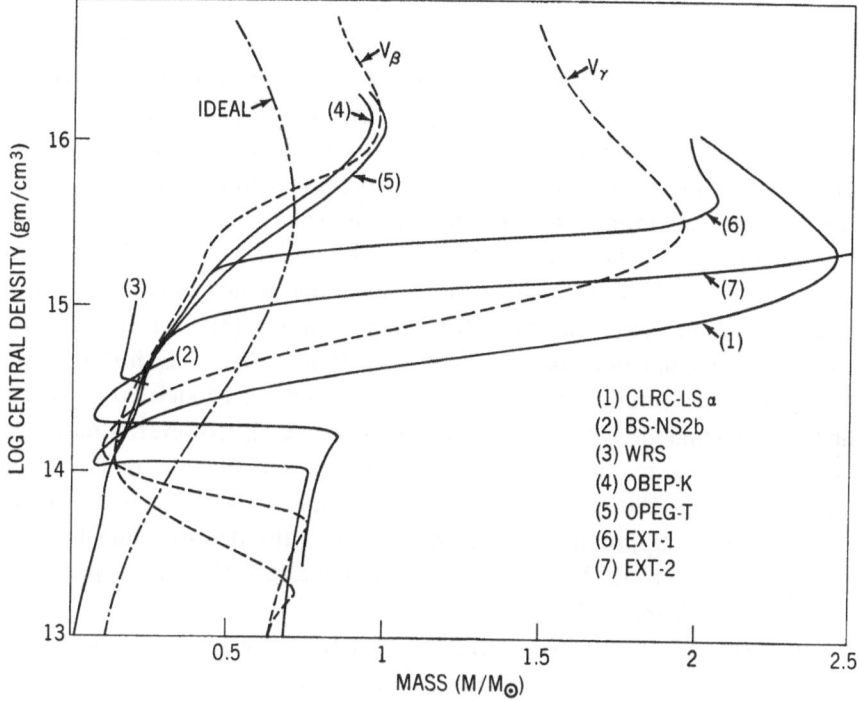

Fig. 1.   Neutron star models. See text for description of various models.

Models (4), called OBEP-K, and the one-pion exchange potentials with a Gaussian type soft core constructed by Tamagaki were used in the Models (5) called OPEG-T. For $\varrho > \varrho_0$ in these models, the equation of state obtained in this manner was parabolically extrapolated in the logarithmic scale. In the models (6) and (7), the equation of state $P = \varepsilon/3$ and $P = \varepsilon$, respectively, (where ε is the energy density) were used for $\varrho > \varrho_1 \simeq 5 \times 10^{15}$ g/cm^3 where the Zel'dovich type scalar or vector interactions are assumed to become applicable, the Method (4) was used for $\varrho \lesssim \varrho_0$, and the intermediate regions have been interpolated. Similar results are obtained if the OPEG-T or the Hamada-Johnston potentials are used for $\varrho \lesssim \varrho_0$.

We see, first of all, that drastically different masses are obtained depending on how the inter-particle interaction problem is treated. In the Models (2) and (3) the calculations are terminated where the nuclear potential approach is thought to become unreliable. If the calculations are extended to higher densities, the Curve (2) reaches the mass peak of about 1 $M_\odot$, similar to the Curves (4) and (5) by the Kyoto group. Near the breakdown point, ϱ_0, the Models (3), (4), (5), (6) and V_β have small mass values of only around 0.2–0.4 $M_\odot$, while the other models shown in the figure have larger masses. At the mass peak, the Curve (7) reaches almost 3 $M_\odot$. It may be pointed out that these models obtained by different methods nevertheless possess a few common points as mentioned below. As the mass increases from the minimum value to approximately 0.2 $M_\odot$, the corresponding radius decreases quickly from about

300 km to about 10 km or so. This occurs a little below or around the nuclear density, depending on the models. For densities of around 10^{15} g/cm³ the radius is around 10 km, and for densities of around 10^{16} g/cm³ the radius is around 5 km. Our studies also show that the effects of the presence of hyperons and protons on the mass limit are much smaller than the effects of nuclear forces, if the interactions among all baryons are basically similar. It will be hard to draw definite conclusions from the above results, but the following comments may be valid.

In the regions where the concept of 'nuclear potential' is still valid (for approximately $\varrho \lesssim 10^{15}$ g/cm³), using realistic potentials alone is not enough. For instance, quite different results are obtained in the Models (2) and (3), even though the both groups used the similar potentials (the soft-core Reid potentials). In this respect (of applying realistic potentials in a realistic way), I feel that the Models (5), the OPEG-T, are the best recommended (among the models shown here). However, it may be noted that many-body interactions are generally neglected and that the nuclear potential term is treated non-relativistically in the work done so far. The net effect of the corrections to these approximations seems to lower the densities and increase the masses near the mass peak. Thus the V_γ models used already in various pulsar studies seem to be not far from reality.

The exact behavior in the regions above about 10^{15} g/cm³ seems to be beyond the knowledge of present-day physics. Since the maximum mass of some of the models lie in these high density regions, it is hard to answer the question of how high masses stable neutron stars can have. It will be fortunate if such a question can be answered rather from the observational side. Will it be an impossible dream, if one contemplates that more detailed pulsar observations might help the break-through of the difficulties facing us today in particle physics and some other fields?

In the pre-pulsar discovery periods, I did not see much sense in going beyond our V_β and V_γ models of neutron stars. But, today, with more observations of pulsars becoming available, better theoretical work with the cooperation of experts in various fields, including the effort to construct more realistic neutron star models, seems very desirable.

Acknowledgements

I wish to thank Professors Hayashi, Nagata, Salpeter and Tamagaki, and Messrs. Boozer, Ikeuchi and Mizutani, for giving me the results of their calculations before publication (see also Mizutani, 1970).

References

Cohen, J. M., Langer, W. D., Rosen, L. C., and Cameron, A. G. W.: 1970, *Astrophys. Space Sci.* 6, 228.
Harrison, B. K., Thorne, K. S., Wakano, M., and Wheeler, J. A.: 1965, *Gravitation Theory and Gravitational Collapse*, University of Chicago Press, Chicago, Ill.
Hartle, J. B. and Thorne, K. S.: 1968, *Astrophys. J.* **153**, 807.
Levinger, J. S. and Simmons, L. M.: 1961, *Phys. Rev.* **124**, 916.
Mizutani, T.: 1970, M.A. Thesis, Tokyo University.
Nemeth, J. and Sprung, D. W. L.: 1968, *Phys. Rev.* **176**, 1496.

Oppenheimer, J. R. and Volkoff, G. M.: 1939, *Phys. Rev.* **55**, 374.
Ostriker, J. P. and Gunn, J. E.: 1969, *Astrophys. J.* **157**, 1395.
Tsuruta, S. and Cameron, A. G. W.: 1966, *Can. J. Phys.* **44**, 1895.
Wang, C. G., Rose, W. K., and Schlenker, S. L.: 1970, *Astrophys. J. Letters* **160**, L17.

Discussion

G. Börner: I would like to report some recent results obtained by Bethe *et al.* in Copenhagen. They obtain stable neutron star models in the range:

density	mass	radius
$\varrho = 1.2 \times 10^{14}$ g/ccm	$0.09 \, m_{\odot}$	228.5 km
$\varrho = 3.8 \times 10^{15}$ g/ccm	$1.67 \, m_{\odot}$	8.8 km

Nuclei exist up to densities of 1.7×10^{14} g cm^{-3}, so the crystal lattice formed by nuclei covers the region where you would expect superfluid neutrons.

W. A. Fowler: I wish to address a remark to all of the neutron star theorists. Neutron stars cannot be completely 'rigid' and deviations from the $-\,d\omega/dt \sim \omega^3$ law may indicate that they are indeed not rigid. It is my belief that nuclear physics data will not yield a correct equation of state for many years, if ever. Thus the mean value throughout the star of $\gamma = -\,d\ln p/d\ln V$ should be calculated for all models. There is some chance that the pulsar observations may yield small but accurately measurable effects which might determine $\langle \gamma \rangle$. In this way it might be possible to reduce the welter of equations of state now under theoretical consideration.

What I am trying to say backwards is that there is no hope of determining the proper equation of state from nuclear physics. Some of the observations, on the other hand, may depend on small effects in the integral structure, and may tell us a great deal about the equation of state.

A. G. W. Cameron: Nuclear properties are very useful for putting constraints on the equations of state used for neutron stars. The nuclear potential should reproduce the volume binding energy, the equilibrium density, the symmetry energy coefficient, and the compressibility of nuclei. It need not reproduce scattering data, except at low densities, since the nuclear-adjusted potential includes effects due to three-body forces.

6.6 NUCLEAR FORCES IN HIGH DENSITY MATTER

M. R. McNAUGHTON

*Laboratory for Astrophysics and Space Research, Enrico Fermi Institute,
Chicago, Ill., U.S.A.*

Abstract. The conditions for superfluidity or ferromagnetism in neutron stars are presented and discussed (but not derived). It is suggested that present estimates relating to these are in error and that the predictions made contradict at least one of three sets of nuclear physics data cited in the text. This is due to neglecting the action of the exclusion principle.

A comparatively simple method for calculating the strength of nuclear forces in the presence of many-body effects is outlined. Some preliminary results are presented together with projected future developments.

1. Introduction

Figure 1 shows the composition of neutron star matter at nuclear densities and zero temperature. This is computed assuming the particles obey the exclusion principle, are in chemical equilibrium, and form a medium that has over-all charge neutrality. Otherwise the interparticle forces have been neglected. The most significant feature of the diagram is the behaviour of the proton kinetic energy. This is very low compared to the neutron kinetic energy until Σ^- particles are formed, then it rises to a similar value.

When strong interactions between the particles are introduced, the physical effects can be roughly divided according to whether the interaction between particles of the same type is attractive or repulsive. If the forces are sufficiently attractive then the corresponding particles will form a superfluid; on the other hand, repulsive forces may cause ferromagnetism. The equations regulating these two effects are found to be

$$\delta_{\bar{k}} = -\frac{1}{2} \sum_{\bar{k}'} \frac{\delta_{\bar{k}'} \langle \bar{k}, -\bar{k} | T^{(S)} | \bar{k}', -\bar{k}' \rangle}{[\delta_{\bar{k}'}^2 + (E_{\bar{k}'} - E_{\bar{k}_F})^2]^{1/2}} \qquad (1)$$

$$1 \leqslant N(0) \{ T^{(S)} - T^{(t)} \}. \qquad (2)$$

$\delta_{\bar{k}}$ is the superfluid energy gap experienced by a particle in momentum state $\bar{k}$ which has an energy $E_{\bar{k}}$. $N(0)$ is the density of states per unit energy range close to the Fermi surface. The first equation is the well known first order B.C.S. equation. The matrix element it contains is derived from the scattering of particles with momentum $(\bar{k}, -\bar{k}'')$ into states with momentum $(\bar{k}, -\bar{k})$. The second equation is the criterion for saturated ferromagnetism. The matrix elements represent the forces acting between typical particles in the singlet or triplet configurations. The criterion is that the forces be less repulsive if all the particles have the same spin direction, this state will then be energetically favoured and a net magnetic dipole will be associated with the preferred spin direction.

The difficulty in utilizing Equations (1) and (2) results almost entirely from correctly evaluating the matrix elements. Present estimates are usually based on neglecting the exclusion principle (superfluidity; Hoffberg et al., 1970), or treating the particles as repulsive spheres (ferromagnetism; Brownell and Collaway, 1969), or taking an over-simplified model interaction (superfluidity; McNaughton, 1969). It is suggested below that all of these approximations are in error.

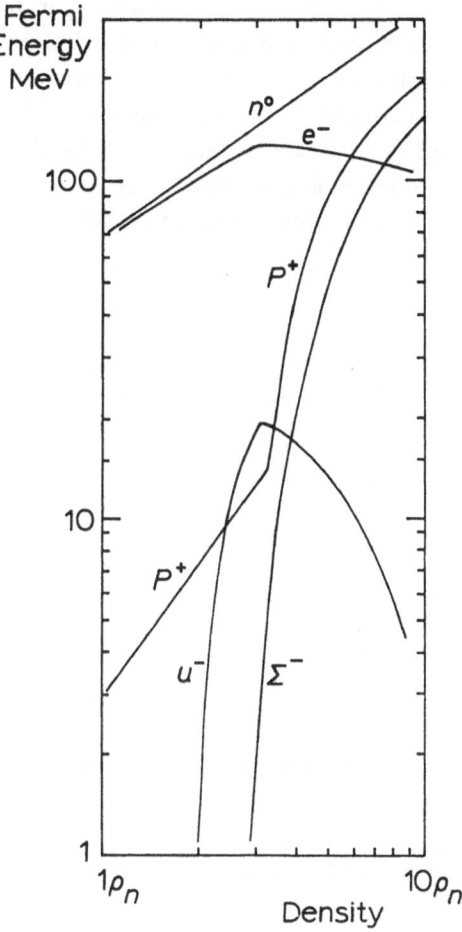

Fig. 1.   The Fermi energy of the particles present in neutron star material as a function of the total density. The density is plotted in multiples of the equilibrium density of ordinary nuclear matter, ϱ_n, $(\varrho_n = 3.7 \times 10^{14} \text{ g cm}^{-3})$.

Since a theory of nuclear forces cannot be tested directly in the astrophysical context it is advisable to look for laboratory situations to which it can be applied. There are three types of experimental data for which such a theory might reasonably be expected to account. These are:

(i) phase shifts determined from nuclear scattering experiments.

(ii) the nuclear matter binding energy

(iii) even-even nuclei pairing strengths

Item (i) is a measure of the strength of two body forces without the interference of other particles or the exclusion principle. It may be used to adjust free parameters in an expression for the bare two-nucleon interaction. Item (ii), the nuclear matter binding energy, measures the strength of the attraction seen by typical particles in a fully degenerate nucleon gas at one particular density – the density of ordinary nuclei near their center. Numerically it is about -16 MeV per particle at the density $\varrho_n \equiv 3.7 \times 10^{14}$ g cm^{-3}. The third item is determined from nuclear structure experiments and is basically a measure of the superfluid energy gap found at densities corresponding to the surface layers of nuclei. That the pairing (energy gap) occurs *only* at the nuclear surface ($\varrho \ll \varrho_n$) can be seen by noting that the observed pairing energy falls off rapidly with increasing nuclear size (Figure 2). Heavier nuclei have a smaller surface to volume ratio so that low density phenomena are less apparent.

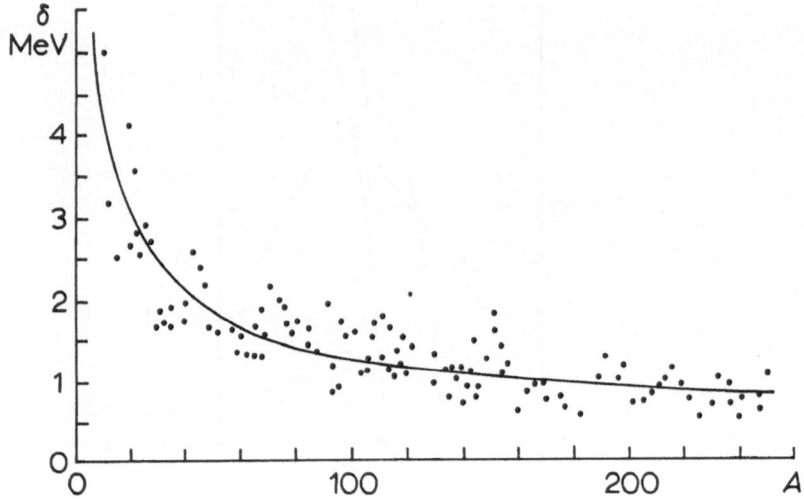

Fig. 2.   Pairing strengths measured in nuclei as a function of atomic number, A. After averaging over effects caused by partially closed shells, the results fit $\delta = 16.2 \times A^{-0.551}$.

2. Perturbation Diagrams, the V matrix to T matrix transformation

The simplest nuclear interaction is drawn in Figure 3a. It represents particles of momentum $(\bar{k}, -\bar{k})$ scattering into states $(\bar{k}', -\bar{k}')$. If the nuclear potential, $V(r)$, contains an infinitely repulsive core, $V(r) \to \infty$ for $r < r_c$, then the corresponding matrix element, $\langle \bar{k}', -\bar{k}' | V | \bar{k}, -\bar{k} \rangle$, will be infinite. To obtain a finite result, an infinite set of diagrams (Figure 3b, part (2), (3), (4) etc.) must be added together. Qualitatively, this corresponds to allowing for the correlation of the plane wave function about the repulsive core of the other particle. Alternatively, it means summing a perturbation series to infinite order.

If the sum of this series of diagrams is designated $\langle \bar{k}', -\bar{k}'| T(\omega) |\bar{k}, -\bar{k}\rangle$ then it obeys the integral equation

$$\langle \bar{k}', -\bar{k}'| T(\omega) |\bar{k}, -\bar{k}\rangle = \langle \bar{k}', -\bar{k}'| V |\bar{k}, -\bar{k}\rangle$$

$$- \sum_{\bar{k}''} \frac{u^2_{\bar{k}''} u^2_{-\bar{k}''} \langle \bar{k}' - \bar{k}'| V |\bar{k}'' - \bar{k}''\rangle \langle \bar{k}'' - \bar{k}''| T(\varpi) |\bar{k} - \bar{k}\rangle}{[E_{\bar{k}''} + E_{-\bar{k}''} + \omega]}$$

$$- \varpi = E_{\bar{k}} + E_{-\bar{k}} \tag{3}$$

$u^2_{\bar{k}''}$ and $E_{\bar{k}''}$ measure the availability and total energy of the momentum state $\bar{k}''$. In a laboratory scattering situation they are replaced by 1 and $(\hbar k)^2/2m$ respectively. It is convenient to represent the sum of these interactions by a single wavy line diagram (Figure 3b, part (1)).

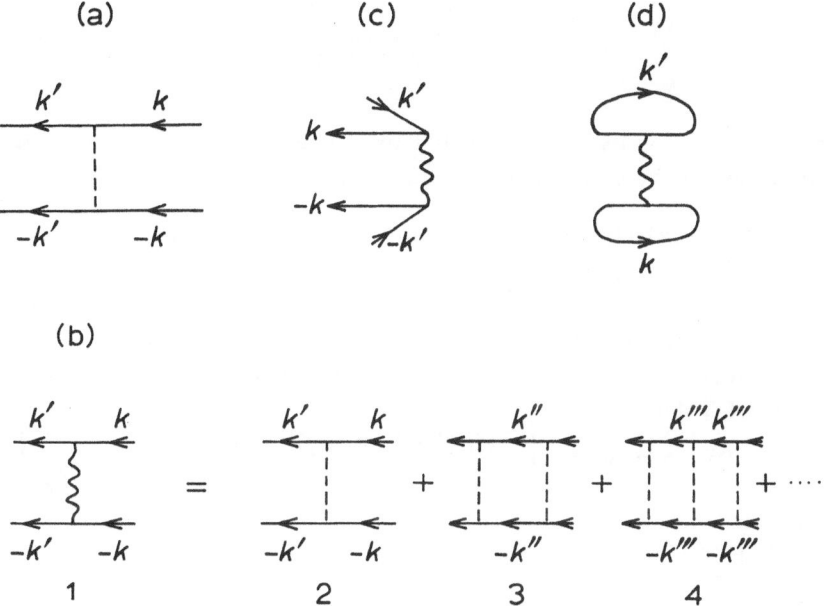

Fig. 3. (a) shows a simple scattering diagram, the corresponding matrix element is $\langle -\bar{k}', \bar{k}'| V | -\bar{k}, \bar{k}\rangle$. (b) illustrates how the T matrix is defined. (c) and (d) are the diagrams that define the matrix elements in equations (1) and (2) respectively.

The matrix elements occurring in Equations (1) and (2) obey very similar equations. The corresponding diagrams are drawn in Figures 3c and 3d. The backward pointing lines refer to holes in the Fermi sea and give a negative contribution to the energy denominators in the equations corresponding to Equation (3). The present work is directed to solving equations of this type having proper regard to the exclusion principle, ($u^2_{\bar{k}''}=0$ for $k''<k_F$), and to the influence of the other particles on the self consistent energies, ($E_{k''} \neq (\hbar k)^2/2m$).

The starting point in the calculation is a separable potential for the bare two body interaction:

$$V_{k'k} \equiv \langle \bar{k}', - \bar{k}' | V | \bar{k}, - \bar{k} \rangle = \sum_\alpha \lambda_\alpha W_\alpha (k') W_\alpha (k). \qquad (4)$$

The functions $W_\alpha (k)$ are chosen to be easy to use and the constants λ_α are then adjusted so that the phase shifts predicted agree with the experimental values. The solution to Equation (3) can then be written in the form

$$T_{k'k} (\omega) \equiv \langle \bar{k}', - \bar{k}' | T (\omega) | \bar{k}, - \bar{k} \rangle = \sum_{\alpha\alpha'} A_{\alpha\alpha'} W_\alpha (k') W_{\alpha'} (k). \qquad (5)$$

The matrix A is found to obey

$$A^{-1} = (D + M) \qquad (6.1)$$

with

$$D_{\alpha\alpha'} = \lambda_\alpha^{-1} \delta_{\alpha\alpha'} \qquad (6.2)$$

$$M_{\alpha\alpha'} = \sum_{\bar{k}''} \frac{u_{\bar{k}''}^2 u_{-\bar{k}''}^2}{E_{\bar{k}''} + E_{-\bar{k}''} + \omega} W_\alpha (k'') W_{\alpha'} (k''). \qquad (6.3)$$

To evaluate the integrals in (6.3) we borrow an idea from the reference spectrum method of nuclear theory (Day, 1967) and put

$$E_{\bar{k}} = (\hbar k)^2 / 2m_1^* + u_1, \quad k < k_F$$
$$E_{\bar{k}} = (\hbar k)^2 / 2m_2^* + u_2, \quad k > k_F$$

where u and m^* are constants.

Only one form of $V_{k'k}$ has been investigated to date, this is a Puff type potential for the singlet S state:

$$V_{k'k}^{(S)} = 2 (2\pi)^3 \underset{\lambda_c \to \infty}{\text{Lt}} \left[\lambda_c \frac{\sin (k'r_c)}{k'} \frac{\sin (kr_c)}{k'} - \frac{\lambda}{(k'^2 + \beta^2) (k^2 + \beta^2)} \right]$$

$\hbar = 2m = 1$
$\lambda = 0.886 \text{ F}^{-3}$
$\beta = 1.62 \text{ F}^{-1}$
$r_c = 0.257 \text{ F}$

The details of matching this to the experimental phase shifts are not reproduced here but it is important to obtain a good fit, not only at energies corresponding to k and k' ($\simeq 10$ MeV), but also at much higher energies ($\simeq 400$ MeV) if the integral in equation (6.3) is to be correct. Using potentials of this form the reaction matrix, $T_{k'k}$, can be found analytically. The constants u and m^* may be determined self consistently by computing the value of the diagrams that give rise to the particle binding energies. Because of the integrals involved in (6.3), the expressions are lengthy and are best displayed in numerical form as graphs. In practice the excited particle states may be treated as free particles ($u_2 = 0$, $m_2^* = 1$). The corresponding quantities for states below the Fermi surface are typically $u_1 \simeq -100$ MeV and $m_1^* \simeq 0.6$.

3. Numerical Results

Figure 4 shows the behaviour of $T_{kk}^{(S)}$ against k for a neutron-proton gas ('nuclear-matter') at a density $\varrho = \varrho_n$. The result obtained by ignoring the exclusion principle and many-body forces,

$$T_{kk}^* = -\left(\frac{4\pi}{mk}\right)\delta(k) \qquad \delta(k) = \text{phase shift}. \tag{8}$$

is also shown. The agreement is very poor.

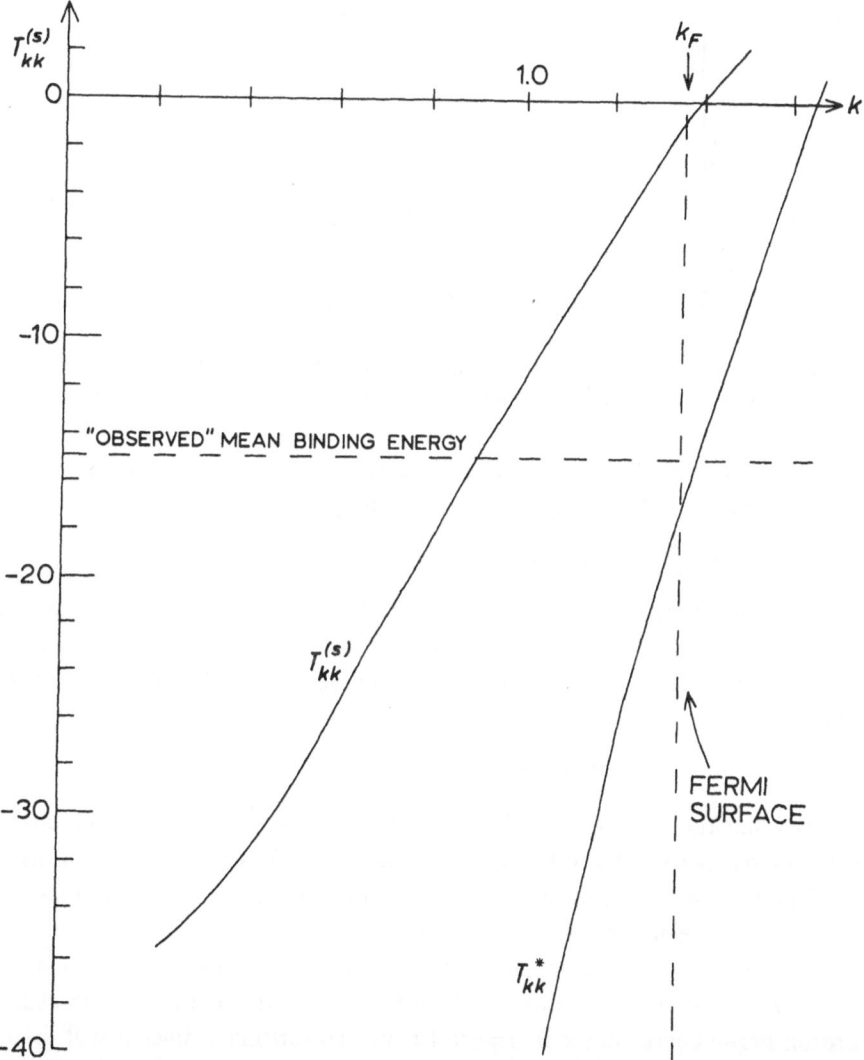

Fig. 4. T_{kk} as a function of momentum, k. The units are derived with $2m = h = 1$; T_{kk} is in F and k is in F^{-1} (1 F = 10^{-13} cm). T_{kk}^* gives a poor approximation to the observed mean binding energy, it also predicts a large negative value of the matrix element at the Fermi surface.

The scattering approximation gives much more negative values for $T_{kk}^{(S)}$ and, as a result, a very large net binding energy ($\simeq 130$ MeV). This may be understood by inspecting the phase shift against momentum diagram, Figure 5. The lowest momentum states see the most attractive potential, but these are just the states that are excluded from the summation over intermediate states by the exclusion principle.

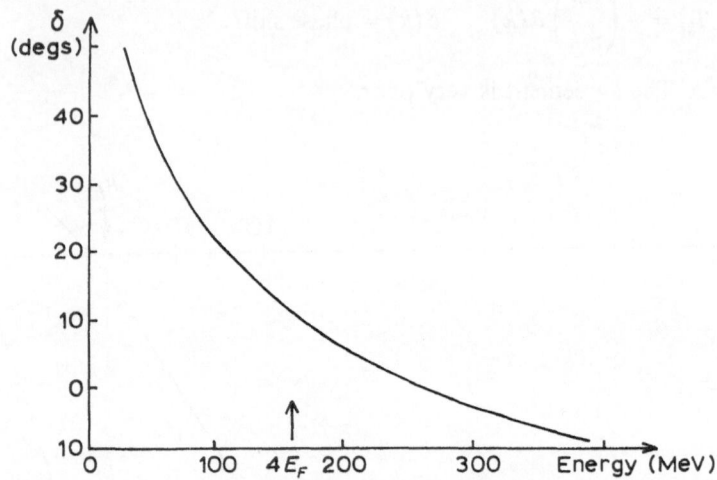

Fig. 5.   Phase shift, δ, observed in the $^1S_0$ configuration plotted against the incident particle energy. The point marked $4E_F$ corresponds to the Fermi energy used in Figure 4.

The size of the superfluid energy gap calculated from Equation (1) depends on the value of $T_{kk}^{(S)}$ near the Fermi surface. For a finite gap, we require,

$$N(0)\, T_{k_F k_F}^{(S)} \gtrsim -1. \tag{9}$$

The energy gap predicted by including the exclusion principle is vanishingly small. Equation (8), on the other hand would predict a large effect at nuclear density and consequently an even–even nuclei pairing energy of several MeV independent of the nuclear size.

5. Conclusions, Future Development

We believe that the techniques of Section 2 give results demonstrably in agreement with the three experimental tests referred to in Section 1. The step 'backwards' from the well known phase shifts before re-calculating the T matrix is an unfortunate necessity if useful results are to be obtained.

Further work is expected to show that the critical Fermi energy above which equation (9) fails is in the region $30 < E_F < 40$ MeV. If this is the case, the shape of the proton Fermi energy curve in Figure 1 is indeed fortunate, since it will single out the protons at $\varrho < 2 \times 10^{15}$ g cm^{-3} for an energy gap of several MeV. The exact details of the $\lambda = \lambda(E_F)$ relation will have no observationally significant results. That the neutrons may satisfy the Inequality (9) at sufficiently low densities is not important since a neutron star with an average density less than ϱ_n is unlikely to be stable (Wang *et al.*, 1970).

Hoffberg *et al.* have suggested that superfluidity may result from the coupling of particles with their spins parallel. In this case the interaction proceeds via the P state potential which remains attractive at higher values of E_F after the S state interaction has turned repulsive. While the S state interaction cannot enter into the first order matrix element in Equation (1), it may enter through higher order diagrams (Figure 6).

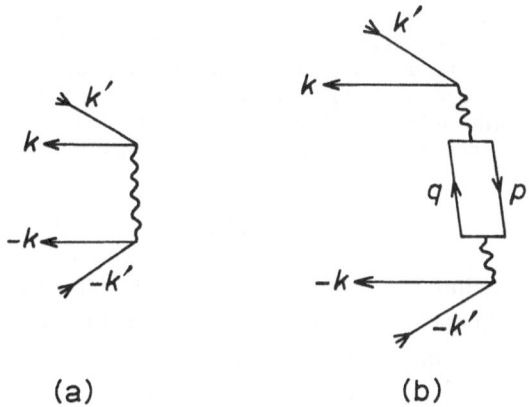

(a)   (b)

Fig. 6.   (a) shows the first order matrix element of the B.C.S. equation. (b) shows the second order matrix elements that enter into higher order approximations.

Present indications are that the large repulsive contribution of the higher order diagram may overwhelm the intrinsically smaller attractive contribution of the first order diagram. However, more work is needed on this point.

The parameters of the two-body triplet interaction $V_{k'k}^{(T)}$ have been found but the inequality for ferromagnetism has not yet been tested.

References

Brownell, D. H. and Callaway, J. A.: 1969, *Nuovo Cimento* **60B**, 169.
Day, B. D.: 1967, *Rev. Mod. Phys.* **39**, 719.
Hoffberg, M., Glassgold, A. E., Richardson, R. W., and Ruderman, M.: 1970, *Phys. Rev. Letters* **24**, 775.
McNaughton, M. R.: 1969, Thesis, Manchester University.
Wang, C. G., Rose, W. K., and Schlenker, S. L.: 1970, *Astrophys. J.* **160**, L17.

Discussion

V. Canuto: Are phase shifts a good method for determining the potential to be used?

M. McNaughton: They do not yield a unique solution for the potential but any potential used *must* satisfy these results. The point of citing nuclear matter binding energy etc., is to further distinguish between the potentials and select the correct one.

V. Canuto: Is the result determined for S waves only?

M. McNaughton: Yes. To consider the P and higher states suggested by Hoffberg *et al.*, one must go beyond the simple B.C.S. equation. It is possible for 3rd order diagrams to give a large (and repulsive) contribution since the 3rd particle is not confined to any special configuration with the two main particles, i.e. the repulsive S state interaction can enter and overcome the P state interaction.

6.7 PERIODICITY AND LUMINOSITY OF THE 'PULSAR' MODEL OF QUASARS*

WILLIAM A. FOWLER

Institute of Theoretical Astronomy Cambridge, U.K., and California Institute of Technology Pasadena, Calif., U.S.A.

Abstract. The diagnosis of pulsar-like symptoms in the quasar 3C 345 has renewed interest in differentially rotating supermassive stars (SMS). Quasi-periodic behavior follows from differential rotation. In reviewing early work on this quasar model it is found that SMS models with rotational periods of the order of a few hundred days and with pulsation periods approximately one-quarter as long can supply thermal and magnetic dipole radiation of the order of 10^{48} erg sec^{-1} for somewhat more than 10^6 yr. The surface field required for the magnetic dipole radiation is approximately 10^3 G. The SMS mass taken as typical in the calculations is $3 \times 10^9\ M_\odot$.

1. Introduction

This paper will not attempt to explain the Crab pulsar. Under the circumstances it is therefore somewhat unique. On the other hand, it might very well be said that the Crab pulsar explains this paper. Now that others (Morrison, 1969; Cavaliere *et al.*, 1969; Woltjer, 1971) have diagnosed pulsar-like symptoms in quasars there is renewed interest in supermassive stars (SMS) which Hoyle and Fowler (1963a, b) suggested some seven years ago as energy sources for radio galaxies and, after their discovery, for quasars. In later papers Fowler (1964, 1965, 1966a, b, c, d) treated nuclear energy generation in SMS in some detail. In addition, in connection with the binding energy equivalent of the rotational energy, it was especially noted in Fowler (1966c), p. 355 that "Since this energy must be lost during contraction it is another source of the observed energy emissions in quasars." This present paper will attempt to amplify this point in light of current ideas concerning the transformation of gravitational-rotational energy into the radio, infrared, optical, X-ray and cosmic-ray emissions from quasars. The discussion will be limited to the Kelvin-Helmholtz contraction stage of SMS prior to the onset of nuclear burning since it is during this stage that the theoretical periods of rotation and pulsation seem to match those of the observed quasi-periodicities in 3C 345 and other quasars.

In their first discussions of rotating SMS, Hoyle and Fowler noted that angular momentum problems must arise during contraction and that angular momentum and rotational energy would necessarily be transferred to surrounding material. At the same time, in the presence of magnetic fields, electromagnetic processes would generate radiation and relativistic particles. The suggested mechanism for this involved the toroidal winding of the magnetic field between the star and the surrounding material. Following Hoyle *et al.* (1964), Pacini (1967, 1968), and Gold (1968), this idea has

---

* Supported in part by the Office of Naval Research [Nonr-220(47)].

been replaced, in current fashion, by radiation and acceleration involving the rotation of a field as simple as that of a dipole but with retardation effects at the velocity of light circle.

Through the work of Feynman (1963), Iben (1963), Chandrasekhar (1964), and Fowler (1964) it soon came to be realized that non-rotating SMS suffered from a general relativistic instability which led to hydrodynamic collapse early in their evolution. In order to achieve lifetimes for hydrostatic stability of the order of 10^6 yr Durney and Roxburgh (1967) and Fowler (1966b, c, d) investigated rotational effects and found that uniform rotation would stabilize masses of the order of $10^6\ M_\odot$ up to and through hydrogen burning. In addition Fowler found that differential rotation would stabilize masses of the order of $10^9\ M_\odot$ up to and through hydrogen burning. These studies led to estimates for the rotation and pulsation periods of SMS of the order of several tens to a few hundred days. It may prove significant that these periods are of the order of those attributed to the light variations in 3C 345 and other quasars by Kinman *et al.* (1968).

2. The Period of Rotation of SMS

It is usually assumed that angular momentum loss commences when

$$\omega_R^2 R^3 = \alpha^2 GM \tag{1}$$

and maintains this relation thereafter. The notation is standard except that ω_R is the peripheral, equatorial angular velocity in the case of differential rotation and α is of the order of unity or somewhat less and is model dependent. For differential rotation with cylindrical symmetry in an object with polytropic index $n=3$ as discussed by Stoeckly (1965), $\alpha^2 = 0.456$. For this value Jeans' criterion for instability holds at the center of the SMS. The rotational period can be written as

$$P_R = \frac{2\pi}{\omega_R} = \left(\frac{32\pi^2}{x^3}\right)^{1/2} \frac{GM}{\alpha c^3}$$

$$= 1.3 \times 10^{-4} x^{-3/2} \frac{M}{M_\odot} \text{ sec} = 1.5 \times 10^{-9} x^{-3/2} \frac{M}{M_\odot} \text{ day} \tag{2}$$

where

$$x = \frac{R_s}{R} = \frac{2GM}{Rc^2} \tag{3}$$

is the ratio of the Schwarzschild radius, R_s, to the coordinate radius, R. The variation of $P_R(x)$ is shown in Figure 1. Note that for a given x and α, the period is linear in M. This explains the great difference possible in the rotation rates of neutron stars and SMS even though α is considerably smaller for neutron stars.

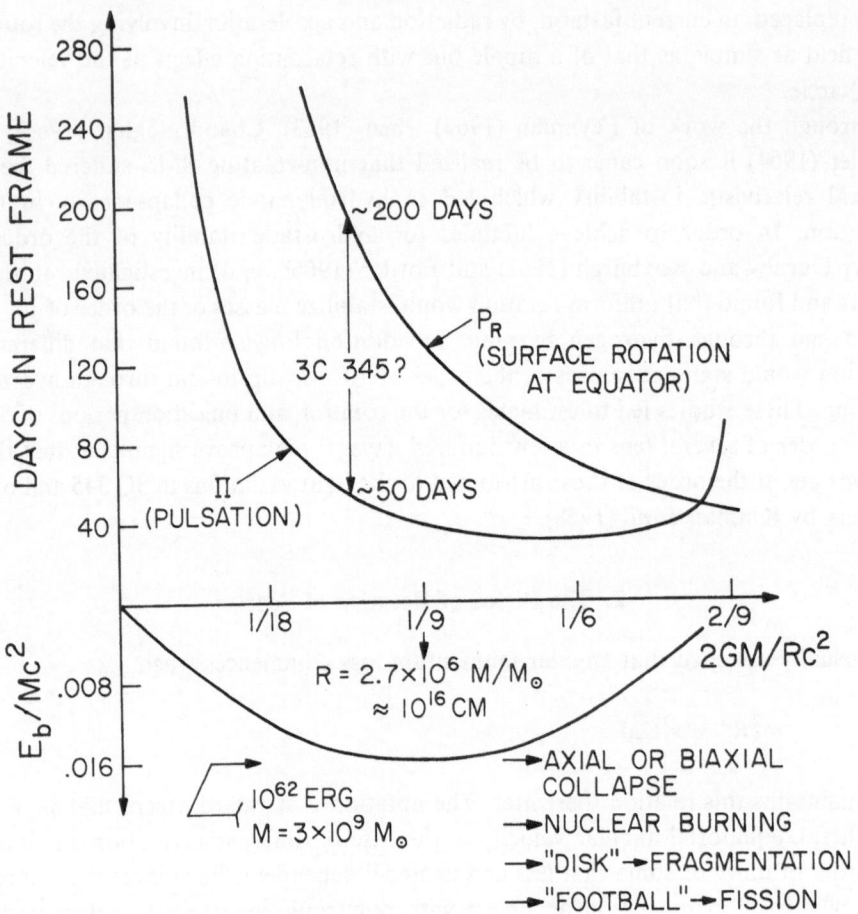

Fig. 1.   Rotation period (P_R), pulsation period (Π), and binding energy/rest mass (E_b/Mc^2) as a function of $2GM/Rc^2$ for an SMS with $M = 3 \times 10^9\,M_\odot$. Observational data for 3C 345 is fitted at $2GM/Rc^2 = {}^1\!/_{12}$ for which $P_R \sim 200$ days, $\Pi \sim 50$ days and $E_b \sim 10^6$ erg.

3. The Binding Energy of SMS

The binding energy of a differentially rotating SMS in hydrostatic equilibrium governed by Equation (1) is given by Fowler (1966c) as

$$\frac{E_b}{Mc^2} = \tfrac{1}{4}K^2\alpha^2 x - \zeta x^2 \qquad\qquad (4)$$

where ζ is a measure of post-Newtonian general relativistic effects and where KR is an 'effective' radius of gyration for differential rotation. Both ζ and K depend on the polytropic index n. In addition, ζ depends upon the volume average over the star of Γ_4 which is defined by $u/p \equiv (\Gamma_4 - 1)^{-1}$ with u the internal energy per cubic centimeter and p the pressure. I have found that the numerical calculations of Tooper (1966) and of Ipser (1969) can be expressed approximately by

$$\zeta \approx \frac{27}{16(5-n)^2} \left\langle \frac{\Gamma_4 - \frac{1}{3}}{\Gamma_4 - 1} \right\rangle \Rightarrow \left(\frac{9}{8}\right)^2 \quad \text{for} \quad \Gamma_4 = \tfrac{4}{3}, n = 3 \tag{5}$$

The dimensionless parameter K appears in the expressions for the rotational energy, Ψ, as follows

$$\Psi = \tfrac{1}{2}K^2 M R^2 \omega_R^2 = \tfrac{1}{2}K^2 \alpha^2 GM^2 R^{-1} \tag{6}$$

and

$$\frac{\Psi}{Mc^2} = \tfrac{1}{4}K^2 \alpha^2 x. \tag{7}$$

It is assumed that radiation pressure greatly exceeds gas pressure in SMS so that the effective ratio of specific heats and all adiabatic coefficients are approximately equal to $\tfrac{4}{3}$ and the internal structure is thus that of a polytrope of index $n=3$. In this case the Stoeckly model (1965) gives $K^2 = 2.47$ so that $K^2 \alpha^2 = \tfrac{9}{8}$. Note that K^2 is considerably larger than $k^2 = 0.075$ where kR is the true radius of gyration of a polytrope with index $n=3$. The moment of inertia, I_z, of the SMS about the axis rotation is given by

$$I_z = k^2 M R^2. \tag{8}$$

Equation (8) contains k^2 and not K^2 since I_z depends on the mass distribution and not on the angular velocity distribution.

If the mean square angular velocity throughout the SMS is defined by

$$\Psi \equiv \tfrac{1}{2}k^2 M R^2 \langle \omega^2 \rangle \tag{9}$$

then

$$\frac{\langle \omega^2 \rangle^{1/2}}{\omega_R} = \frac{K}{k} = 5.74 \tag{10}$$

whereas the ratio of the central angular velocity to that at the equatorial periphery is given by Fowler (1966c) as

$$\frac{\omega_c}{\omega_R} = 10.9. \tag{11}$$

Here and in what follows all numerical values hold for $n=3$.

In the Stoeckly model (1965) it is assumed that a gas cloud with uniform density $(n=0)$ and uniform angular velocity equal to ω_R condenses internally under conservation of angular momentum to a polytrope of index n which in the case of interest is equal to 3. The square of the radius of gyration for uniform density is $\tfrac{2}{5} R^2$ so that the conserved angular momentum, Φ, is given by

$$\Phi = \tfrac{2}{5}M R^2 \omega_R \tag{12}$$

and the angular momentum per unit mass by

$$\frac{\Phi}{M} = \frac{2}{5} R^2 \omega_R. \tag{13}$$

In the case of angular momentum loss it is assumed that the SMS evolves through a sequence of Stoeckly models governed by Equation (1) so that

$$\Phi = \tfrac{2}{3}\alpha (GM^3 R)^{1/2} \tag{14}$$

and

$$\frac{\Phi}{GM^2/c} = \frac{2\sqrt{2}}{5} \alpha x^{-1/2} . \tag{15}$$

In addition, Equations (6) to (12) still hold so that

$$\Psi = \tfrac{5}{4} K^2 \Phi \omega_R = 3.09 \ \Phi \omega_R \tag{16}$$

$$= \frac{25}{8} \frac{K^2 \Phi}{MR^2} = 7.72 \frac{\Phi^2}{MR^2} . \tag{17}$$

If the mean angular velocity throughout the SMS is defined by

$$\Phi \equiv k^2 MR^2 \langle \omega \rangle \tag{18}$$

then

$$\frac{\langle \omega \rangle}{\omega_R} = \frac{2}{5k^2} = 5.33 . \tag{19}$$

The numerical results for k, K, and ζ given above were calculated assuming sphericity, i.e., no deformation due to the rotation, although Stoeckly (1965) has considered deformation in detail for $n=\tfrac{3}{2}$. In the Stoeckly models with cylindrical symmetry this procedure is exact for k and K. In all equations R should be set equal to the equatorial radius. Changes in ζ will occur with deformation but it is my belief that these will not be large.

4. The Maximum Binding Energy of SMS

Differentiation of Equation (4) indicates that E_b/Mc^2 reaches a maximum value equal to

$$\frac{E_b^{\max}}{Mc^2} = \frac{K^4 \alpha^4}{64\zeta} = \frac{1}{64} \tag{20}$$

at

$$x(E_b^{\max}) = \frac{K^2 \alpha^2}{8\zeta} = \frac{1}{9} \tag{21}$$

or $R=9 \ R_s$. This is illustrated in Figure 1. Since E_b is the negative total energy of the SMS, it is taken positive along the negative abscissa of the figure. Numerically

$$E_b^{\max} = 3 \times 10^{52} (M/M_\odot) \ \text{erg} \tag{22}$$

Beyond the maximum in E_b or the minimum in total energy hydrostatic equilibrium requires that energy be supplied to the SMS from some source. If nuclear burning has not started or has terminated, the normal losses of energy from the rotating SMS will lead to axial or biaxial collapse at this stage. The eventual onset of nuclear burning may halt the collapse for a period but the ultimate result will be a disk-shaped object unstable to fragmentation or a football-shaped object unstable to fission. Hydrogen burning through the CNO bi-cycle does not occur before the maximum in E_b in SMS with $M > 10^9\ M_\odot$ so that quasi-hydrostatic equilibrium of the Kelvin-Helmholtz contraction stage holds over the interval $0 \leqslant x \leqslant K^2 \alpha^2 / 8\zeta$. At the end of this interval P_R approaches

$$P_R(E_b^{\max}) = \frac{128\pi\zeta^{3/2}}{K^3\alpha^4} \frac{GM}{c^3} = 4.1 \times 10^{-8} \frac{M}{M_\odot}\quad \text{day} \tag{23}$$

which is the order of 100 days for M somewhat in excess of $10^9\ M_\odot$.

5. The Period of Pulsation of SMS

The fundamental mode of radial pulsation of a star has a period which depends critically on its physical properties, in particular, on Chandrasekhar's adiabatic coefficient, $\Gamma_1 \equiv -d \ln p / d \ln V$. As noted previously the main pressure support in a SMS is due to radiation and $\Gamma_1 \approx \frac{4}{3}$. In this case Fowler (1966c) has shown that

$$\sigma^2 = \frac{4E_b}{3I_z} = \frac{Mc^2}{3I_z} [K^2\alpha^2 x - 4\zeta x^2] \tag{24}$$

Thus

$$\left(\frac{\sigma}{\omega_R}\right)^2 = \left(\frac{P_R}{\Pi}\right)^2 = \frac{2}{3}\left(\frac{K}{k}\right)^2 \left[1 - 4\left(\frac{\zeta}{K^2\alpha^2}\right)x\right] \tag{25}$$

where σ is the pulsational angular frequency and Π is the period. Equation (25) leads to the numerical results

$$\left(\frac{P_R}{\Pi}\right)^2 = \frac{2}{3}\left(\frac{K}{k}\right)^2 [1 - 4.5x]$$

$$= \left(\frac{2}{3} \to \frac{1}{3}\right)\left(\frac{K}{k}\right)^2 = 22 \to 11 \quad \text{for} \quad 0 \leqslant x \leqslant \frac{1}{9} \tag{26}$$

The variation of Π with x is illustrated in Figure 1. Note that the radial pulsations are stable until σ^2 becomes negative at $x = K^2\alpha^2/4\zeta = 2x(E_b^{\max}) = \frac{2}{9}$. However, as noted previously, energy must be supplied if collapse is to be avoided when $E_b^{\max}$ is reached. The ratio $(P_R/\Pi) = (\frac{2}{3})^{1/2}\ (K/\kappa)$ for $x = 0$ varies considerably with polytropic index. For $n = 2, 3, 3.5$ and 4, $P_R/\Pi = 2.2, 4.7, 8.0$ and 15 respectively at $x = 0$.

6. Discussion of the Periods and Binding Energy

Over the Kelvin-Helmholtz contraction stage Equation (26) indicates that $P_R/\Pi \approx 4$. This is illustrated in Figure 2. This ratio is of some interest in connection with the observations on 3C 345 where Kinman et al. (1968) attribute a rest frame period of

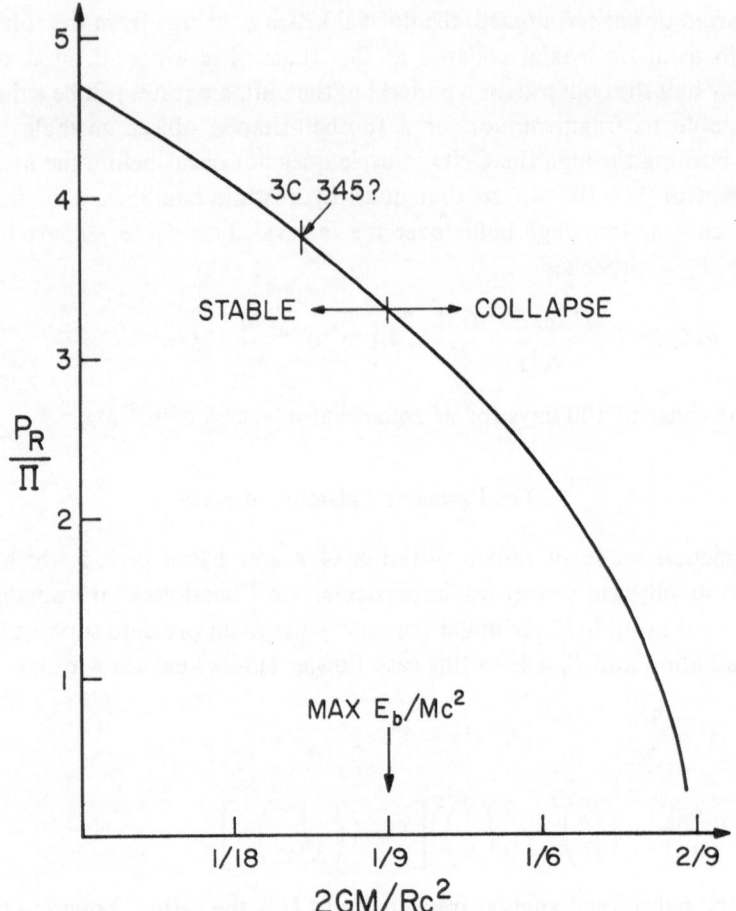

Fig. 2. The ratio of rotation period to pulsation period for SMS as a function of $2GM/Rc^2$.

200 days to rotation and find some additional evidence for a second period of $\sim$ 50 days. As indicated in Figures 1 and 2 a fair fit to these observations is obtained for $x = \frac{1}{12}$ and $M = 3 \times 10^9 \, M_\odot$.

It is important to note that the ratio P_R/Π is relatively independent of the uncertain parameter α. Furthermore the result $P_R/\Pi \approx 4$ for differential rotation is quite different than the customary results for uniform rotation $(K = k)$ without the general relativistic correction, namely $P_R/\Pi = (\frac{2}{3})^{1/2} = 0.82$. If the pulsar search-light model is applied literally to quasars then it must be recalled that, for differential rotation, P_R holds only for 'hot spots' on or near the equator. The rotational periods at the poles are of the order $P_R/10$ on the Stoeckly model. Differential rotation implies shear and thus centers of activity are not expected to survive more than a few complete rotations. This may be relevant to the 'quasi-periodic' behavior of 3C 345 and other quasars.

In the models under discussion $\alpha^2 = 0.456$ on the basis of the Jeans' criterion at the center of the star, namely $\omega_c^2 = \frac{4}{3}\pi G\varrho_c$, and the equipotential and equidensity contours at the center of the star are flat. Using the Roche approximation I have been able to show that the equatorial radius is slightly less than six times the polar radius. Thus the SMS is considerably deformed and it is of interest to consider lower values for α^2 for which the spherical approximation is more accurate. If, for example, α^2 is decreased by $\sqrt{10}$ to 0.144 then the ratio of the equatorial to polar radii for the contours at the center is 1.2 and at the surface 2.5. However, it will be noted from Equation (23) that M must be decreased by a factor of 10 to maintain the same period as before the change in α^2. Moreover, under these circumstances E_b^{max}/Mc^2 from Equation (20) is decreased by 10 and E_b^{max} by 100. This means, for example, a reduction in mean luminosity by a factor of 10 over a period shortened by a factor of 10. In this connection, however, it is possible to invoke magneto-turbulence to compensate for the reduction in α^2 and to restore the original values for P_R and E_b^{max}. This possibility has been treated by Bardeen and Anand (1966).

Uniform rotation leads to a considerable reduction in E_b^{max} since Equations (20) and (21) are replaced by

$$\frac{E_b^{max}}{Mc^2} = \frac{k^4\alpha^4}{64\zeta} = \frac{1}{14\,400} \qquad (27)$$

and   (uniform rotation)

$$x(E_b^{max}) = \frac{k^2\alpha^2}{8\zeta} = \frac{1}{135} \qquad (28)$$

In the numerical evaluation, $\alpha^2 = 1$, which corresponds to the Jean's criterion at the surface and to an equatorial radius approximately $\frac{3}{2}$ times the polar radius. Thus E_b^{max} is reduced by more than a factor of 200. Even so, $E_b^{max} = 1.3 \times 10^{50}$ $(M/M_\odot)$ which is not to be sneezed at.

In any case it is of crucial importance to monitor quasars continuously to ascertain whether regular or quasi-regular periods are characteristic of these objects. The differentially rotating SMS model has readily calculable periods as indicated here. If future observations ultimately establish such periods then considerable support will be given to the SMS model for quasars. It must also be remembered that SMS may exhibit relaxation oscillations (Fowler, 1966a) and sporadic flare phenomena.

7. The Luminosity of SMS

Hoyle and Fowler (1963a) showed that the thermal luminosity of a stable SMS is proportional to the mass according to the approximate relation

$$l_{th} \equiv \frac{L_{th}}{Mc^2} = 10^{-16} \text{ sec}^{-1} \qquad (29)$$

or

$$L_{th} = 2 \times 10^{38} \left(\frac{M}{M_\odot}\right) \text{ erg sec}^{-1} \qquad (30)$$

Equation (30) leads to $L_{th} = 6 \times 10^{47}$ erg sec^{-1} for $M = 3 \times 10^9 M_\odot$ as suggested for 3C 345 above. When combined with equation (22) this leads to a Kelvin-Helmholtz quasi-static contraction time of 1.5×10^{14} sec or 5×10^6 yr if no other type of radiation occurs. This interval is increased by 10^6 yr if hydrogen burning occurs during the stage of stable pulsations before $2x(E_b^{max})$ is reached.

In accordance with current fashion it is now appropriate to add to L_{th} the radiative magnetic dipole luminosity on the assumption that the SMS has a surface magnetic field with component, B, normal to the axis of rotation. According to Deutsch (1955) this luminosity is given by

$$L_{md} = \frac{2\varepsilon}{3c^3} B^2 R^6 \omega_R^4 \tag{31}$$

where the factor $\varepsilon \equiv \langle \omega^4 \rangle / \omega_R^4 \approx K^4/k^4 \approx 10^3$ has been introduced in recognition of the differential rotation which has cylindrical symmetry in the Stoeckly model (Stoeckly, 1965) and thus extends to the surface. There is considerable ambiguity in this procedure and even more in the nature of the magnetic field which can cut across the differentially rotating cylindrical shells. There is the difficult problem, too, associated with the survival of differential rotation in company with the magnetic field as well as with internal convection and turbulence. Notwithstanding these difficulties considerable insight can be gained into the behavior of the SMS under the energy and angular momentum loss implied by Equation (31). It will be found in what follows that the field B, necessary to yield $L_{md} \sim L_{th}$, is only a few thousand gauss and that the magnetic field energy is thus quite small compared to the rotational energy.

It might have been better to have used an expression due to Goldreich and Julian (1969) for L_{md} in Equation (31). In their expression the $\frac{2}{3}$ in Equation (31) becomes $\frac{1}{8}$ but, more important, B is the strength at the poles of a magnetic field parallel to the axis of rotation. When there is no winding of the field in differential rotation in Equation (35) to come, change the numerical coefficient from 24 to 128.

On the assumption of the conservation of magnetic flux it is possible to write

$$F_0 = B_0 R_0^2 = BR^2 \tag{32}$$

where F_0 is a constant measure of the flux and the subscript designates some fiducial time. Then (1), (3), (31) and (32) yield

$$L_{md} = \frac{2\varepsilon \alpha^4 F_0^2 G^2 M^2}{3c^3 R^4} \tag{33}$$

and

$$l_{md} \equiv \frac{L_{md}}{Mc^2} = \frac{x^4}{\tau} \tag{34}$$

where τ, which has the dimensions of time, is given by

$$\tau = \frac{24 G^2 M^3}{\varepsilon c^3 \alpha^4 F_0^2} = \frac{24 k^4 G^2 M^3}{c^3 K^4 \alpha^4 F_0^2} \tag{35}$$

The total luminosity must be equated to the time rate of decrease of the total energy of the system and thus to the rate of increase of the binding energy according to the equation

$$\frac{dE_b}{dt} = L_{tot} = L_{th} + L_{md}. \tag{36}$$

Although E_b must contain the post-Newtonian term which sets a limit on the energy which can be extracted from the SMS, namely E_b^{max} from Equation (20), yet the gravitational and rotational red-shift corrections to L_{tot} can be neglected. The correction factor is $\sim(1-x)$ which varies from unity down to $\frac{8}{9}$.

Substitution of Equations (4), (29) and (34) into Equation (36) leads to the following convenient equation

$$\frac{dt}{dx} = -\frac{R}{x}\frac{dt}{dR} = \frac{K^2\alpha^2\left(1 - 8\zeta x/K^2\alpha^2\right)}{4\left(l_{th} + x^4/\tau\right)} \tag{37}$$

where it will be recalled that l_{th} is a constant $= 10^{-16}$ sec^{-1}. For arbitrary reasons x rather than R has been used in the right-hand side of Equation (37). This equation can be integrated to yield

$$t = \frac{K^2\alpha^2 X}{4l_{th}}\left[\frac{1}{4\sqrt{2}}\ln\frac{x^2 + xX\sqrt{2} + X^2}{x^2 - xX\sqrt{2} + X^2}\right.$$
$$\left. + \frac{1}{2\sqrt{2}}\tan^{-1}\frac{xX\sqrt{2}}{X^2 - x^2} - \frac{4\zeta X}{K^2\alpha^2}\tan^{-1}\frac{x^2}{X^2}\right] \tag{38}$$

where X is the value of x at which $l_{md} = l_{th}$ so that from equation (34)

$$X^4 = \tau l_{th}. \tag{39}$$

In Equation (38) it has been taken that $t=0$ at $x=0$ or $R=\infty$. Simple expressions are obtained for t in two interesting cases as follows:

Case I, $l_{th} \gg l_{md}$

$$t = \frac{K^2\alpha^2 x}{4l_{th}}\left(1 - \frac{4\zeta}{K^2\alpha^2}x\right) = 3 \times 10^{16}x\left(1 - 4.5x\right)\ \text{sec} \tag{40}$$

where it is taken that $t=0$ at $x=0$ or $R=\infty$. The time finally reached at $x=x\left(E_b^{max}\right) = K^2\alpha^2/8\zeta$ is then

$$t_{th} = \frac{K^4\alpha^4}{64\zeta l_{th}} = \frac{E_b^{max}}{L_{th}} = 1.5 \times 10^{14}\ \text{sec} = 5 \times 10^6\ \text{years} \tag{41}$$

as found previously and as expected since L_{th} has been taken constant.

Case II, $l_{md} \gg l_{th}$

$$1 - \frac{t}{T} = \left(\frac{x_0}{x}\right)^3\left(1 - \frac{12\zeta x}{K^2\alpha^2}\right) = \left(\frac{x_0}{x}\right)^3\left(1 - 13.5\ x\right) \tag{42}$$

$$= \left(\frac{R}{R_0}\right)^3\left(1 - \frac{27GM}{Rc^2}\right) = \left(\frac{\omega_{R0}}{\omega_R}\right)^2\left(1 - 13.5x\right) = \left(\frac{B_0}{B}\right)^{3/2}\left(1 - 13.5x\right) \tag{43}$$

where, in this case, $x = x_0$ at $t=0$ neglecting the relativistic term and

$$T = \frac{K^2\alpha^2\tau}{12x_0^3} = \frac{3}{32}\frac{\tau}{x_0^3} \tag{44}$$

The time elapsed between $x = K^2\alpha^2/12\zeta$ and $x(E_b^{max}) = K^2\alpha^2/8\zeta$ or between $l_{md} = (\frac{2}{3})^4 l_{md}^{max} \approx 0.2 l_{md}^{max}$ and l_{md}^{max} is

$$\Delta t_{md} = \frac{64\zeta^3}{3K^4\alpha^4}\tau = \frac{512\zeta^3 k^4 G^2 M^3}{c^3 K^8 \alpha^8 F_0^2} = 3.65 \frac{G^2 M^3}{c^3 F_0^2} \tag{45}$$

Equations (38), (40) and (43) indicate that R decreases with time while x, ω_R, and B increase with time. The rotational energy then increases with time according to Equations (6) or (7) while the angular momentum decreases according to Equations (14) or (15). This is in striking contrast to 'rigid' neutron stars for which R remains constant as ω_R, Ψ and Φ decrease. Neutron stars spin down under angular momentum loss while SMS contract and spin up. In the latter case the rotational energy increases but the binding energy also increases at the expense of gravitational energy. In the Newtonian approximation for an SMS in units of GM^2/R the various energies can be expressed for $K^2\alpha^2 = \frac{9}{8}$ as

$$\Omega : E_b : \Psi : H = \frac{3}{2} : \frac{9}{16} : \frac{9}{16} : \frac{1}{4} \tag{46}$$

where Ω is the gravitational binding energy and H is the internal heat energy. From energy conservation $E_b = \Omega - H - \Psi$ and from the virial theorem $\Omega = H + 2\Psi$ so that $E_b = \Psi$ in the Newtonian approximation. In the 3C 345 example $R \approx 10^{16}$ cm and $GM^2/R \approx 2 \times 10^{62}$ erg so $\Omega = 3 \times 10^{62}$ erg, $E_b = \Psi \approx 10^{62}$ erg, and $H \approx 5 \times 10^{61}$ erg. Note that $\Psi/\Omega = \frac{3}{8}$ but when deformation is taken into account Ψ remains unchanged while Ω increases by $\sim 6^{1/3}$ so $\Psi/\Omega \Rightarrow 0.2$. General relativistic effects reduce the binding energy at maximum $(x = \frac{1}{4})$ by a factor of two to $E_b^{max} = \frac{1}{2}\Psi$.

In order to indicate as simply as possible the behavior of SMS with both thermal and magnetic dipole luminosity, Figure 3 has been drawn for the case where $L_{md} = L_{th}$ at $X = \frac{1}{12}$ with $M = 3 \times 10^9 M_\odot$ as in the calculations for 3C 345 above. From Equation (39) this is equivalent to setting

$$\tau = \frac{(3K^2\alpha^2/32\zeta)^4}{l_{th}} = \frac{1}{12^4 l_{th}} = 5 \times 10^{11} \text{ sec} = 2 \times 10^4 \text{ years} \tag{47}$$

Thus

$$\Delta t_{md} = 2\left(\frac{3}{32}\right)^3 \frac{K^4\alpha^4}{\zeta l_{th}} = 5 \times 10^5 \text{ years} \tag{48}$$

It then follows that

$$L_{tot} = 6 \times 10^{47}\left[1 + (12x)^4\right] \text{ erg sec}^{-1}$$
$$\Rightarrow 2.5 \times 10^{48} \text{ erg sec}^{-1} \quad \text{at} \quad x = \frac{1}{9} \tag{49}$$

and

$$t = 1.32 \times 10^6 \left[\ln\frac{(12x)^2 + 12\sqrt{2x} + 1}{(12x)^2 - 12\sqrt{2x} + 1} + 2\tan^{-1}\frac{12\sqrt{2x}}{1 - (12x)^2}\right.$$
$$\left. - \frac{3\sqrt{2}}{2}\tan^{-1}(12x^2)\right] \text{ years} \Rightarrow 4.4 \times 10^6 \text{ years} \quad \text{at} \quad x = \frac{1}{9} \tag{50}$$

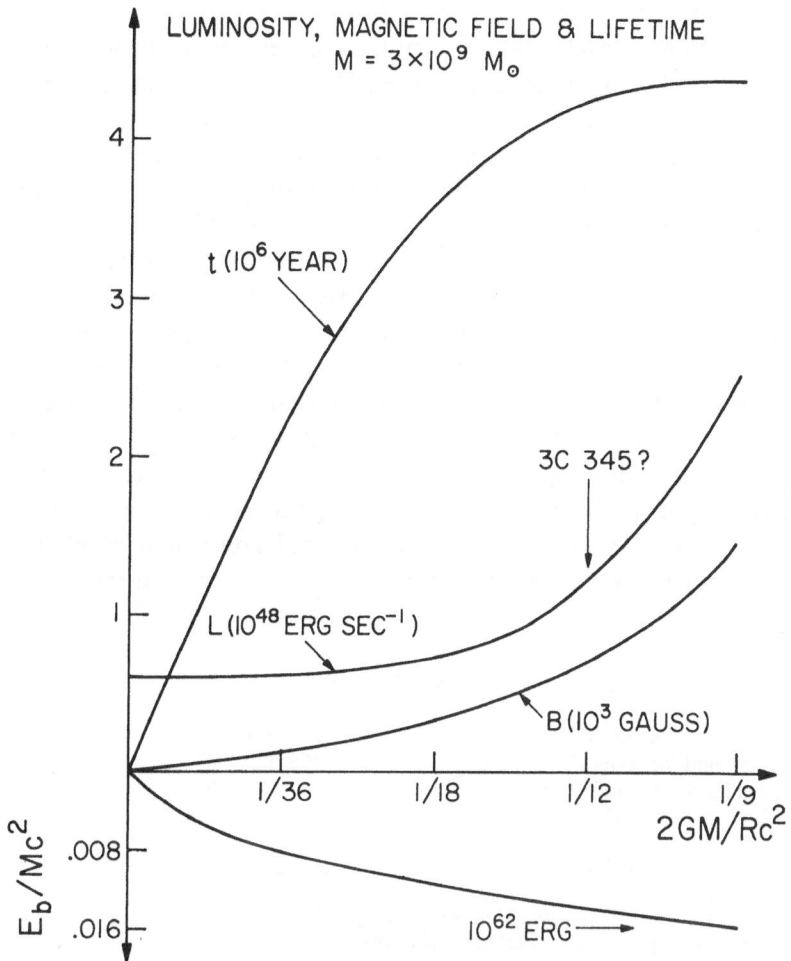

Fig. 3.   Luminosity, magnetic field and lifetime of a SMS with $M = 3 \times 10^9 \, M_\odot$ and $L_{md}/Mc^2 = L_{th}/Mc^2 = 10^{-16} \, \mathrm{sec}^{-1}$ at $2GM/Rc^2 = 1/12$.

as illustrated in Figure 3. The necessary magnetic field can be calculated from Equations (35), (32) and (47) with the result

$$B^2 = \frac{3}{2}\left(\frac{32}{3}\right)^4 \frac{\zeta^4 k^4 c^5}{K^{12}\alpha^{12}G^2 M} x^4 l_{th} \tag{51}$$

and

$$B = 6.2 \times 10^9 \left(\frac{M_\odot}{M}\right)^{1/2} x^2$$

$$\Rightarrow 1400 \, \mathrm{G \ at} \ x = \tfrac{1}{9} \quad \text{for} \quad M = 3 \times 10^9 \, M_\odot \tag{52}$$

as illustrated in Figure 3. A simple calculation shows that the magnetic field energy, $B^2 R^3/6$, if B extends throughout the SMS, is equal to $\sim 3 \times 10^{53}$ erg which is small compared to the rotational energy $\Psi \sim 10^{62}$ erg. This justifies the neglect of the magnetic field in all considerations except the magnetic dipole luminosity.

In summary it has been found that the SMS model for quasars with rotational periods of the order of a few hundred days and with radial pulsation periods approximately one-quarter as long can supply thermal and magnetic dipole radiation of the order of 10^{48} erg sec^{-1} for somewhat more than 10^6 yr. The surface field required for the magnetic dipole radiation is approximately 10^3 G. The mass of the SMS taken as typical in these calculations is $3 \times 10^9 \, M_\odot$. Approximately $\frac{1}{64}$ of the rest mass energy or $\sim 10^{62}$ ergs is released during the Kelvin-Helmholtz contraction stage which is stabilized by differential rotation. Bardeen and Wagoner (1969) have shown that considerably more of the rest mass energy (up to 40%) can be released in the collapse to a disk shaped structure but this happy circumstance has not been treated in this paper. The periods during this collapse stage will be considerably shorter than discussed in this paper. However, the results of Bardeen and Wagoner do justify in some measure the neglect in this paper of higher order rotational and gravitational terms. Salpeter and Wagoner (1970) have recently circulated a preprint in which the relationship between supermassive disks and stars is discussed in some detail.

References

Bardeen, J. M. and Anand, S. P. S.: 1966, *Astrophys. J.* **144**, 953.
Bardeen, J. M. and Wagoner, R. V.: 1969, *Astrophys. J.* **158**, L65.
Cavaliere, A., Pacini, F., and Setti, G.: 1969, *Astrophys. Letters* **4**, 103.
Chandrasekhar, S.: 1964, *Phys. Rev. Letters* **12**, 114, 437E.
Deutsch, A. J.: 1955, *Ann. Astrophys.* **18**, 1.
Durney, B. R. and Roxburgh, I. W.: 1967, *Proc. Roy. Soc.* **A296**, 189.
Feynman, R. P.: 1963, private communication.
Fowler, W. A.: 1964, *Rev. Mod. Phys.* **36**, 545, 1104E.
Fowler, W. A.: 1965, *Proc. Amer. Phil. Soc.* **109**, 181.
Fowler, W. A.: 1966a, *Some Recent Advances in the Basic Sciences*, Proceedings of the Third Annual Science Conference Belfer Graduate School of Science, Academic Press, New York p. 11.
Fowler, W. A.: 1966b, *Astrophys. J.* **144**, 180.
Fowler, W. A.: 1966c, *High Energy Astrophysics*, Proceedings of the International School of Physics 'Enrico Fermi', Varenna, 1965, Academic Press, New York, p. 313.
Fowler, W. A.: 1966d, *Perspectives in Modern Physics* (ed. by R. E. Marshak), John Wiley & Sons, New York, p. 413.
Gold, T.: 1968, *Nature* **218**, 731.
Goldreich, P. and Julian, W. H.: 1969, *Astrophys. J.* **157**, 869.
Hoyle, F. and Fowler, W. A.: 1953a, *Monthly Notices Roy. Astron. Soc.* **125**, 169.
Hoyle, F. and Fowler, W. A.: 1963b, *Nature* **197**, 533.
Hoyle, F., Narlikar, J. V., and Wheeler, J. A.: 1964, *Nature* **203**, 914.
Iben, I., Jr.: 1963, *Astrophys. J.* **138**, 1090.
Ipser, J. R.: 1969, *Astrophys. J.* **158**, 17, and private communications.
Kinman, T. D., Lamla, E., Ciurala, T., Harlan, E., and Wirtanen, C. A.: 1968, *Astrophys. J.* **152**, 357.
Morrison, P.: 1969, *Astrophys. J.* **157**, L73.
Pacini, F.: 1967, *Nature* **216**, 567.
Pacini, F.: 1968, *Nature* **219**, 145.
Salpeter, E. E. and Wagoner, R. V.: 1970, Preprint CRSR 388, Cornell University.
Stoeckly, R.: 1965, *Astrophys. J.* **142**, 208.
Tooper, R. F.: 1966, *Astrophys. J.* **143**, 465, and private communications.

Woltjer, L.: 1971, Semaine d'Etude, *Les Noyaux des Galaxies*, Pontifical Academy of Sciences, The Vatican, to be published.

Discussion

J. P. Ostriker: What is the ratio of kinetic to gravitational energy? If this ratio is more than $\frac{1}{4}$ and previous calculations may be used as a guide, then the model will be unstable to a non-axisymmetric ($m = 2$) perturbations – that is fission will occur.

W. A. Fowler: The ratio is approximately $\frac{3}{8}$ from calculations based on spherical symmetry. Deformation does not change the rotational energy but the gravitational energy roughly doubles so a better estimate for the ratio is 0.2.

J. A. Roberts: Would the field of 1000 G exist in the region producing the radio synchrotron emission, since this should produce easily detectable circular polarization.

W. A. Fowler: That might be the case but Dr. Cavaliere can give the answer.

A. G. Cavaliere: May I add a comment on this. The radio-emission phenomena in QSS (van der Laan, Kellermann and Pauliny-Toth 'clouds') probably originate in lumped masses ejected beyond the 'critical surface' with radius $r \approx c/\Omega$ existing around these massive rotating cores as well as around rotating neutron stars. There the general magnetic field is much lower than the surface value ($B \sim 1/r^2$ or $1/r^3$ up to the critical surface, and $\sim 1/r$ from there on). The internal field of the clouds, from the cited model, is also smaller (< 1 G) when they become transparent.

6.8 MULTIPLE PULSAR EJECTION IN SUPERNOVA EVENTS*

F. CURTIS MICHEL**

Institute of Theoretical Astronomy, Cambridge, U.K.

Abstract. Fragmentation in the collapse of a supernova core, followed by energy loss in neutron star formation, is shown to lead to disruption of the resulting system. The elements of the system, some of which should be pulsars, can attain velocities of the order of 10^3 km/sec if currently quoted parameters are correct.

Although pulsar formation is generally attributed to supernova events, only two pulsars are convincingly associated with known supernova remnants (the Crab Nebula and Vela). Other pulsars are close enough to remnants to suggest association (Prentice, 1970), but only if they are moving away from the remnant at velocities of the order of 10^3 km/sec. Another problem seems to be that a supernova core may well be too massive to form a gravitationally stable object (Arnett, 1967), (e.g. 'Neutron star'), and gravitationally collapsing objects ('black holes') do not presently seem promising candidates to be pulsars.

Fowler and Hoyle (1963) suggested some time ago that the symmetric ejection of radio-luminous material from galaxies could be caused by asymmetrical processes occuring in the collapse of very massive objects. Indeed, it is known that such an imploding system is Rayleigh-Taylor unstable. Thus the more massive core could fission into several less massive objects. The same idea can be applied to a supernova event, wherein the core fragments into some distribution of neutron stars, black holes, and general debris (planetesimals, dust, etc.). The number of fragments should be few, since the disparity between core mass and neutron star masses is not large. Such a multiple system would be readily detectible either directly as superimposed pulsars or indirectly from the orbital perturbations to the observed pulsar's period. At present, there is only a suggestion (Michel, 1970; Richards *et al.*, 1970) that some pulsars have companions and then only of planetary mass.

Our point here is to show that the resultant system can be expected to become unbound, with the component objects (not all of which need be pulsars) ejected with velocities of the order of 10^3 km/sec. Consider the binary fission of a supernova core into equal mass neutron stars. The collapse time to form two stars is expected to be of the same order as the time for the two stars to fall together. Furthermore, the conservation of the angular momentum of the original core would instead place them in highly eccentric orbits as shown in Figure 1. The semimajor axis would be comparable to the initial core radius, while the closest approach distance would be determined by the initial angular momentum of the core. Since final condensation occurs at the highest system density, this will also be when the fragments are closest together.

* This research was supported in part by NASA grant NGR 44-006-012.
** Permanent Address: Space Science Department, Rice University, Houston, Texas, U.S.A.

At closest approach, the kinetic energy relative to the barycenter can be as high as a tenth of the rest mass (for 'contact' trajectories), yet the system is only slightly bound. If now the rest mass of the system is rapidly reduced by a fraction f, the system would become unbound (Michel, 1970; Richards *et al.*, 1970; Blaauw, 1961; Michel, 1963) by f times the gravitational binding energy less the total binding energy. The velocity at infinity would then be given from

$$V^2/c^2 = GM(2f - 1 + e)/2a, \tag{1}$$

where a is the closest approach distance to the barycenter, f is the mass loss fraction,

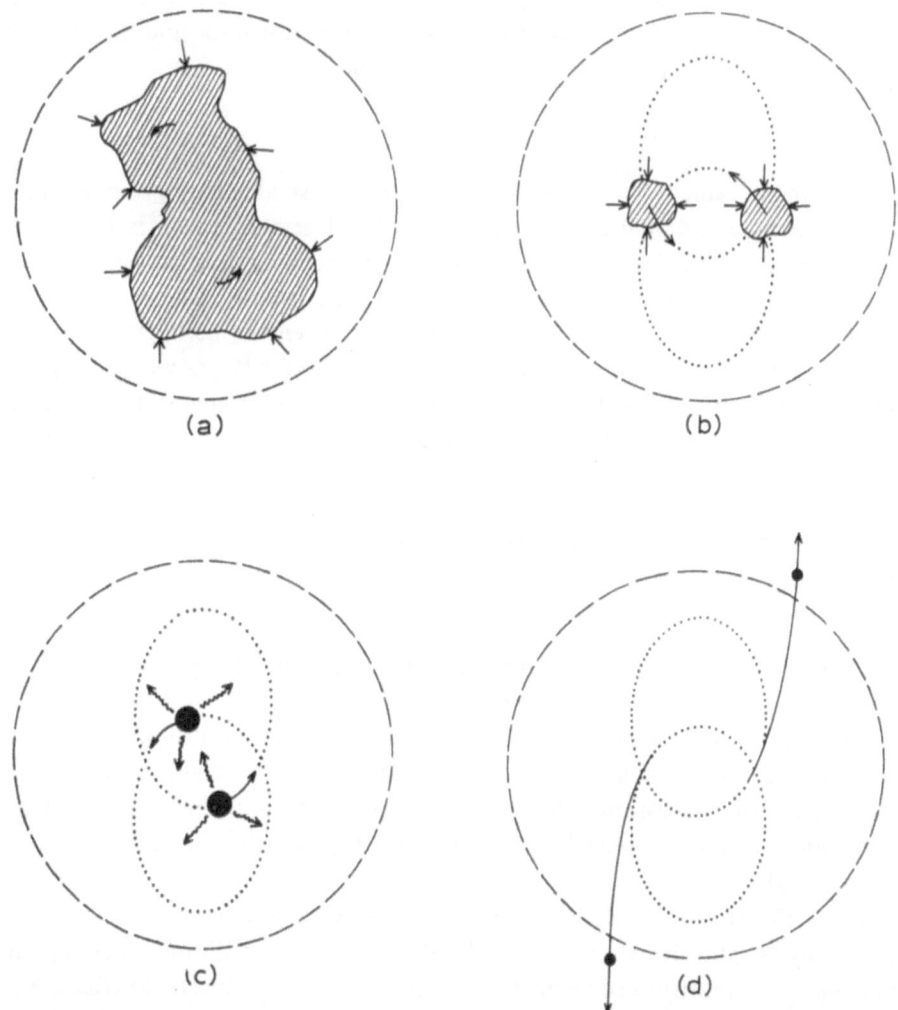

(a)   (b)

(c)   (d)

Fig. 1. Schematic development of run-away pulsars at the phases (a) Collapse and fragmentation, (b) Condensation, (c) Radiation and (d) Disruption. Dashed line shows boundary between ejected shell and collapsing core. Dotted lines show bound orbits after separation but before radiation. A complete orbit is not executed.

and e is the orbital eccentricity (nearly unity for the model described above). The eccentricity can be estimated roughly from

$$(1 - e)/(1 + e) \approx a/R_{\text{core}} \geqslant R_{\text{pulsar}}/R_{\text{core}} \tag{2}$$

which gives values for $1 - e$ of about 5×10^{-3} for present supernova models. Thus escape velocities are obtained if $f \geqslant (1-e)/2$ or about 2.5×10^{-3}. If non gravitational forces between the two fragments could be neglected, we would have

$$1 - e = 2w_0^2 r_0^3/GM \tag{3}$$

where w_0 is the initial rotation frequency of the core (radians/second) and r_0 is the initial distance between the center of mass of each fragment-to-be and the total center of mass $(r_0 \approx \frac{1}{2} R_{\text{core}})$. Thus for one solar-mass objects

$$1 - e = 1.8 \times 10^{-12} w_0^2 R \ (\text{km})^3 \tag{4}$$

and for nominal values $(w_0 = 0.1, R = 10^3)$, $1 - e$ is even smaller than given in Equation (2). Since nongravitational forces must act if the closest approach is less than the neutron star diameter, 'contact' trajectories should be favoured. Note that relatively little energy is required to adjust the closest approach distance. The rest-mass energy fraction released in the neutron star formation event is perhaps of the order (Wheeler, 1966; Tsuruta and Cameron, 1966) of $f = 3 \times 10^{-2}$, which would give a recession velocity of 2×10^4 km/sec for $a = 10$ km. The mechanism is less efficient if the energy is lost slowly, and a correction may be roughly estimated using the replacement

$$f \to f^* \approx f (t_a/t_{\text{decay}})^{2/3} \ (t_{\text{decay}} > t_a) \tag{5}$$

where

$$t_a^2 = a^3/GM \ (1 + e) \tag{6}$$

is the time for one radian of orbital motion at periapsis, and t decay is the characteristic time to release energy in the formation event.

Figure 2 plots the values of f, e, and a required to produce runaway pulsars at a velocity in excess of 10^3 km/sec. We see that nominal energy release values and stellar dimensions permit recession velocities of the order of or greater than 10^3 km/sec. Similar ideas could be applied also to massive objects (Fowler and Hoyle, 1963; Michel, 1963).

It is not clear from the available data whether such runaway motions are required for the pulsars. The absence of observed pulsars associated with the other supernova remnants could be attributed to the formation of non pulsar objects (M. Rees, private communication) or of pulsars that are not beamed towards us. On the other hand, the mechanism proposed here seems capable of producing runaway pulsars whether they are yet required by the data or not. The detection of such objects would thereby be suggestive of fragmentation in the supernova event.

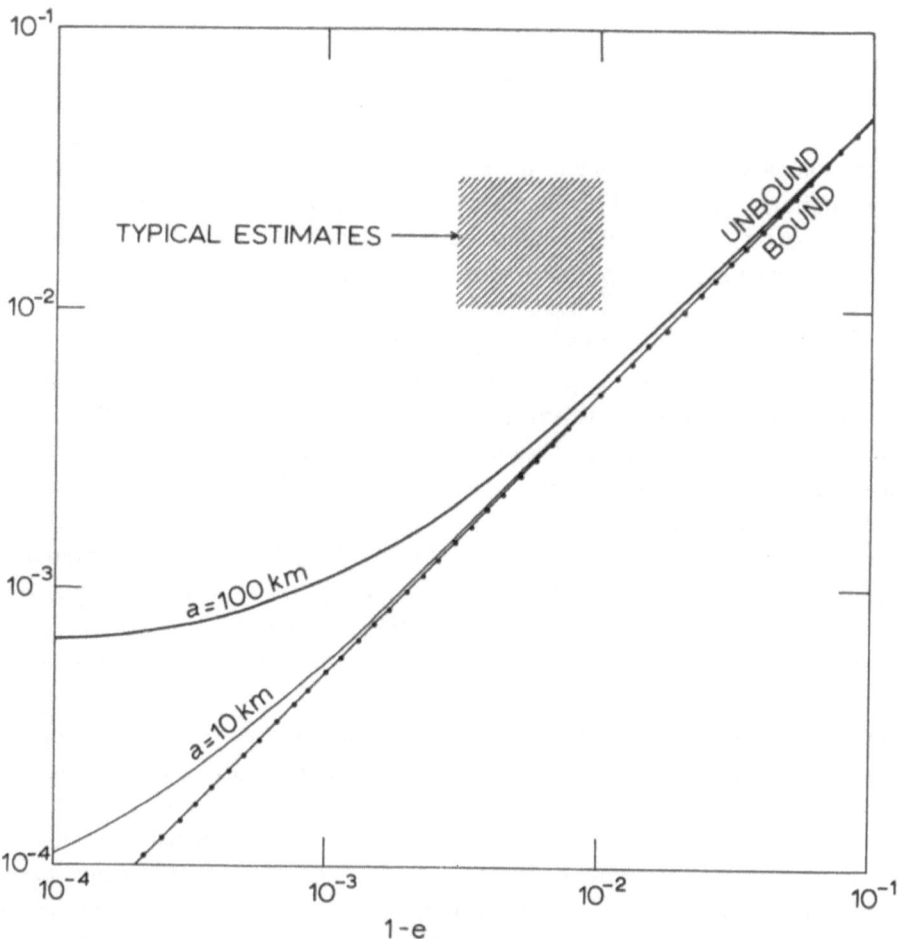

Fig. 2.   Regions of f and e that permit disruption for various values of a assuming one solar mass pulsars. The mass-loss fraction f should be replaced by f^* if the emission is slow, see Equation (5). Above the solid lines for the appropriate value of a, the velocity exceeds 10^3 km/sec, while below the dotted line, the system remains bound.

References

Arnett, D.: 1967, *Can. J. Phys.* **45**, 1621.

Blaauw, A.: 1961, *Bull. Astron. Inst. Neth.* **15**, 265.

Fowler, W. A. and Hoyle, F.: 1963, *Nature* **197**, 533.

Michel, F. C.: 1963, *Astrophys. J.* **138**, 1097.

Michel, F. C.: 1970, *Astrophys. J. Letters* **159**, L25.

Prentice, A. J. R.: 1970, *Nature* **225**, 438.

Richards, D. W., Pettengill, G. H., Counselman, C. C. III, and Rankin, J. M.: 1970, *Astrophys. J. Letters* **160**, L1.

Tsuruta, S. and Cameron, A. G. W.: 1966, *Can. J. Phys.* **44**, 1895.

Wheeler, J. A.: 1966, *Ann. Rev. Astron. Astrophys.* **4**, 393.

6.9 EMISSION OF GRAVITATIONAL WAVES FROM THE PULSAR

R. RUFFINI*

Institute for Advance Study, School of Natural Sciences, Princeton, N.J., U.S.A.

Dr. Ruffini gave a brief survey of the present experimental evidence for the existence of gravitational waves, and discussed the possibility that the shapes of neutron stars might depart from rotational symmetry, leading to the emission of gravitational waves.

The following summary has been prepared by the Editors.

The concept of gravitational waves, based on Einstein's theoretical work on General Relativity, is now widely accepted. The experimental evidence for their existence is less certain, although the experiments carried out by Weber since 1950 have given promising results. The first of these used the Earth as a detector, and gave only an upper limit for the intensity of waves with period of 54 min. The second, using an aluminium bar resonating at 1600 Hz, gave promising results; furthermore there are now suggested to be coincidences between times of detection at widely separated sites. It turns out that the frequency of this detector is close to the frequency of the quadrupole mode of vibration of a neutron star $\omega \sim (\pi G \varrho)^{1/2}$. Furthermore, the detector is directional, and there seems to be an anisotropy suggesting that the gravitational waves originate in the plane of the Galaxy.

The shape taken up by a body in very fast rotation has been studied by many people, including MacLaurin, Jacobi, Lord Kelvin, Poincare, Darwin, and Jeans. For an incompressible fluid with constant density the sequence of shapes taken up as the rotation speed increases is described by MacLaurin (flattened ellipsoids), then to a Jacobi sequence of ellipsoids in which the lengths of the three principal axes are all different, followed by pear-shaped configurations. We must investigate the possibility that a neutron star can have a Jacobi configuration, which can radiate gravitational waves. Jeans continued the analysis for a fluid with equation of state $P = k\varrho^\gamma$. He showed that if $\gamma > 2.2$ the star can follow the Jacobi sequence, but it does not if $\gamma < 2.2$.

In the Harrison-Wheeler equation of state $\gamma > 2.2$ only in the region of low density, and gravitational radiation seems unlikely. In the Cameron equation of state, on the other hand, $\gamma > 2.2$ for densities greater than 10^{13} g cm^{-3}, which applies to about 90% of the star.

To give a quantitative estimate of the gravitational radiation emitted by a neutron star initially in a Jacobi ellipsoid the usual formula (linearized theory) relating the energy flux to the third derivative square of the quadrupole moment have here been used. This treatment shows the following results:

(a) A neutron star, initially in the Jacobi sequence, would *decrease* its angular

* Partly supported by NSF Grant GP 7669.

momentum due to the emission of gravitational radiation and *increase* its angular velocity up to the encounter with the Maclaurin sequence.

(b) Typical amount of emission of gravitational radiation $\simeq 10^{52}$ ergs/sec should be expected in the first seconds of formation of a neutron star.

(c) An observation of this emission of gravitational radiation could give interesting indirect information on the effective γ of the neutron star ($\gamma \geqslant 2.2$).

6.10   YOUNG PULSARS – PULSED NEUTRON SOURCES

J. TRUEMPER

Max-Planck-Institut für Physik und Astrophysik, Institut für Extraterrestrische Physik,
Garching b. München, Germany

Abstract. The prospects of observing a pulsed neutron flux at very high energies from pulsars are discussed. Most likely candidates are the Vela pulsar at 2×10^{16} eV and the Crab Pulsar at 10^{17} eV.

It has been pointed out by several authors (Gold, 1969; Goldreich and Julian, 1969; Ostriker and Gunn, 1969) that pulsars are likely cosmic ray accelerators and that the total power as well as the maximum energies of the produced particles are large enough to explain the cosmic ray spectrum up to the highest energies which are of the order 10^{20} eV. Since the maximum particle energies will depend on the rotational period it is expected that at energies larger than 10^{16} eV the pulsar activity is limited to a rather short time after pulsar birth.

Shen and Pollack (1969) have shown that nuclei accelerated by pulsars will be effectively disintegrated at energies of 10^{15} to 10^{20} eV due to photonuclear reactions with the blue-shifted pulsed photon flux.

Now, during these photonuclear processes, a large number of high energy photo-neutrons might be produced and we want to discuss in the following the prospects of observing them.

The decay mean free path of a neutron having a Lorentz factor γ is $\lambda = \gamma c \tau$ where τ is the rest lifetime of the neutron (1000 sec). Neutrons will reach us without appreciable decay losses from the Crab pulsar at 10^{17} eV, from the Vela pulsar at 2×10^{16} eV and from the galactic center at 10^{18} eV.

The neutron flux would be pulsed at emission and it is an interesting matter of fact that the neutrons would preserve their pulse structure during space travel. The delay at earth of a neutron with respect to a photon starting simultaneously is given by $\Delta t = d/2c\gamma^2$ where d is the distance of the pulsar. Δt turns out to be less than a few tens of microseconds for the sources and energies given above.

Moreover, it can be easily seen that deflection of neutrons by the interaction of their magnetic moment with the galactic magnetic field inhomogenities can be completely neglected.

Hence, young pulsars might show up as pulsed neutron point sources at very high energies where they might be detected by air shower techniques. The expected neutron flux from distant pulsars would not be large enough to reveal the pulse structure by shower observations alone. However, as has been shown above the neutrons should arrive in coincidence or with a constant phase with respect to the photon pulses.

In view of their ages and distances it seems to be doubtful whether neutrons could be detected from the known pulsars. The most likely candidates are the Crab pulsar at 10^{17} eV and the Vela pulsar at 2×10^{16} eV.

Davies and Smith (eds.), The Crab Nebula, 384–385. All Rights Reserved.
Copyright © 1971 by the IAU.

References

Gold, T.: 1969, *Nature* **221**, 25.
Goldreich, P. and Julian, W.: 1969, *Astrophys. J.* **157**, 869.
Ostriker, J. P. and Gunn, J.: 1969, *Astrophys. J.* **157**, 1395.
Pollack, J. B. and Shen, B. S. P.: 1969, *Phys. Rev. Letters* **23**, 1358.

ENERGY CONSIDERATIONS AND THE ELECTRODYNAMIC LINK BETWEEN THE PULSAR AND THE NEBULA

7.1 RELATIONSHIP BETWEEN PULSAR AND NEBULA

L. WOLTJER

Department of Astronomy, Columbia University,
New York, N.Y., U.S.A.

Abstract. The magnetic field and the relativistic electrons in the Crab Nebula cannot have originated at the time of the supernova explosion. The energy density in the magnetic field is so large that it must have been generated using the energy supply in the pulsar. The energies of the electrons are so high, and their lifetimes correspondingly are so short, that they must have been accelerated, again using the pulsar energy. The efficiency of these processes must be high, but there is an adequate energy supply.

1. Introduction

For a long time the origin of the nebular magnetic field and the acceleration process for relativistic electrons have posed the most difficult problems in attempts towards an understanding of the Crab Nebula. Though much remains to be done, it seems that the discovery of the pulsar in the Nebula has brought us close to a solution of these problems.

2. The Nebular Magnetic Field

Several modes of origin can be excluded. The magnetic field cannot have resulted from a simple expansion of the field in the presupernova; during a spherical expansion the ratio of gravitational to magnetic energy remains constant and this ratio now being 10^{-6} the field never could have been contained in a stable object. A similar problem is encountered if one considers an inverse square field drawn out from (but still anchored in) the central star. Turbulent amplification of a weak field is in principle possible, but appears incompatible with the filamentary kinematics; the expansion velocities far exceed those of non radial motions.

Recently Goldreich and Julian (1969) have concluded that the electric fields associated with the rotation of a neutron star are sufficiently strong to lift electrons and positive ions from its surface. Because the positive and negative particles follow different trajectories, large scale poloidal currents may arise (if there is no background medium with high conductivity) which would result in mainly toroidal magnetic fields. Crude estimates suggest that fields of the order of 10^{-4} G may be obtained.

An alternative suggestion has been made by Gunn and Ostriker (1969) namely that no static magnetic field is present and that the relativistic electrons radiate in the fields of electromagnetic waves emitted by the rotating magnetic neutron star. This latter model is in a sense a limiting case, strictly valid if the pulsar is surrounded by a vacuum. Only a fully self consistent solution – which will not be easily obtained – can tell whether the real situation is closer to the wave solution or to

Davies and Smith (eds.), The Crab Nebula, 389–391. All Rights Reserved.

the static case, that is whether displacement currents or charged particle currents dominate.

3. Particle Acceleration

The relativistic electrons in the Crab Nebula lose energy by radiation and by expansion of the nebular volume. Because the electrons which radiate most of the energy have life times less than the age of the Nebula, about the same amount of relativistic particle energy as is lost has to be resupplied. It is a most striking result indeed that the loss of the rotational energy of the pulsar is of the same order as this energy requirement. Because no other comparable energy source appears to be available it seems that the rotational energy is transformed rather directly and highly efficiently into relativistic electron energy.

If we place the Nebula at a distance of 2 kpc the radiative loss amounts to $1.6 \times \times 10^{38}$ erg/sec (mainly in UV and very soft X-rays) and the expansion loss to $0.5 \times \times 10^{38}$ erg/sec (for equipartition conditions) leading to a total electron energy requirement of 2×10^{38} erg/sec. The loss of rotational energy from the pulsar is about 5×10^{38} erg/sec with an uncertainty of at least a factor of 3 due to uncertainties about the mass and detailed model of the neutron star. Comparing the energy requirement with the available energy we conclude that the efficiency for the acceleration of relativistic electrons cannot be much less than about 20 per cent. It follows from this that the ratio of the energy of protons and ions to that of electrons cannot be very large; the same qualitative conclusion is obtained from estimates of the total relativistic particle energy in the Nebula from the dynamics of the filamentary shell. This is in striking contrast with the cosmic-ray composition above 1 GeV and makes it improbable that objects like the Crab are the main sources of cosmic rays.

The extremely high efficiency for electron acceleration imposes severe restrictions on the acceleration mechanism. Two such mechanisms have been recently proposed. Gunn and Ostriker (1969) have shown that the intense electromagnetic radiation, which would be present in a vacuum around the pulsar, should accelerate electrons to about 10^{13} eV, sufficient to explain the X-rays from the Crab. The energy attained by a unit charge is proportional to $m^{1/3}$ and electrons would get about 8% of the energy of protons (or heavier particles). Alternatively in the model of Goldreich and Julian the electric fields around the pulsar would accelerate particles more or less electrostatically, probably resulting in a more equal distribution of the energy between electrons and positive particles. At the moment on the basis of these mechanisms we can perhaps understand the acceleration of the most energetic electrons. However the energy spectrum of the electrons – with most electrons having energies much smaller than the maximum – is not at all understood. But until a more self consistent model is forthcoming, it is difficult to say if this is a serious difficulty.

To summarize, it appears that the pulsar provides an adequate energy source for the Nebula and in a very qualitative way we may have the beginnings of a theory for the transformation of the rotational energy of the pulsar into the energy of relativistic particles and magnetic fields.

References

Goldreich, P. and Julian, W. H.: 1969, *Astrophys. J.* **157**, 869.
Gunn, J. E. and Ostriker, J. P.: 1969, *Astrophys. J.* **157**, 1395.

Discussion

J. P. Ostriker: Our discussion of the acceleration of particles in the nebula concerned particles near the star. We have now looked at the effect of waves on the filaments and other material they may encounter. The accelerations are much weaker, but they have a different dependence on mass, namely an inverse square.

F. C. Michel: Some progress has been made towards self-consistent calculations of the magnetic field structure. Weber and Davis have made such calculations showing how the solar magnetic field extends into interplanetary space. These calculations can be made relativistic to apply to the pulsar problem, and one readily concludes that, if the surface fields are as large as generally proposed, then the nebula fields would be of the order of 10^{-3} G.

L. Mestel: A self-consistent model of the magnetic field is likely to differ in at least one important aspect from the solar field. In the Goldreich-Julian model the Alfvén speed is comparable with or greater than c, so that the regime is 'relativistic' even when the velocities of co-rotation near the star are well below c. It then follows that the field is not pulled out by the outflowing gas, but stays quasi-dipolar out to distances approaching the light-cylinder. This is in contrast to the solar case, where those fieldlines that link up the Sun with the interplanetary medium acquire a nearly radial structure soon after emerging from the solar surface.

N. Visvanathan: Observations of linear polarisation in the pulses of the pulsar (in optical wavelength) and the polarisation observation around pulsar show the direction of rotation axis of the pulsar and the direction of polarisation in the nebula agree well. The implication is strong that the magnetic field, at least in its vicinity may have been produced by the pulsar itself in some stage of its history.

7.2 A MAGNETIC DIPOLE MODEL FOR THE CRAB EXPLOSION

J. P. OSTRIKER*

Princeton University Observatory, Princeton, N.J., U.S.A.

Dr. Ostriker suggested that the main features of the Crab Nebula, in particular the value of its magnetic field and the energy of the particles, could be derived very simply and directly from the known characteristics of the pulsar. The link was through the radiation from a spinning dipole. The following summary was prepared by the Editors.

The power radiated from a rotating dipole with a skew axis, as noted by Pacini in a discussion of neutron stars (and even earlier by Deutsch in connection with A stars) is proportional to $\omega^4 B^2 a^6$, where ω is angular velocity, B a field strength, and a the radius. This power is calculable directly from observations of ω and $\dot{\omega}$, given a value of the moment of inertia I from the neutron star theory. Since a and I do not vary during the evolution of the pulsar, we can assume that ω varies with time as

$$\frac{\omega^2}{\omega_0^2} = \left(1 + \frac{2t}{\tau_0}\right)^{-1}$$

where τ_0 is obtained from observations as

$$\tau_0 = \frac{\omega}{\dot{\omega}}.$$

Since the power radiated varies as ω^4, we can now follow the history of the power injected into the nebula through the dipole radiation.

The magnetic field B close to the pulsar is a dipole field. At larger distances it becomes a spiral, then it forms a spherical wave in which the field strength falls as R^{-1}. (This is possibly observable, providing a test of the theory.)

It is certain that energy is supplied continuously to the nebula. Woltjer has emphasised the requirement for injecting energy into the electrons which radiate X-rays and light. The electrons which radiate radio waves also require a continuous input, since their present energy of $10^{49 \pm 1}$ erg could not have been achieved after a large adiabatic expansion from a radius of say 10^{13} cm to 10^{18} cm. This would require an initial energy of 10^{54} erg.

Even the supernova event itself cannot be impulsive. When it becomes visible the thermal energy is of order 10^{49} erg and the radius 10^{15-16} cm. Again an adiabatic expansion cannot have taken place, since an energy of over 10^{54} erg would be required.

We therefore consider the whole process of expansion of the nebula as controlled by the energy output of the pulsar. The mass motions are consistent with this, if we equate kinetic energy to the loss of pulsar rotational energy.

* Alfred P. Sloan Fellow 1970–1971.

The model ignores only the initial phase of sudden contraction of a white dwarf core to a neutron star. The remaining outer parts of the star form the nebula, and the present characteristics are obtained from simple integrations involving the pressure in the cavity. The boundary conditions are unimportant, since, whatever processes of reflection or absorption occur, the cavity will be filled by a gas of relativistic particles with $\gamma = \frac{4}{3}$.

Numerical values are satisfactory if the initial mass of the star is about 10 to 12 $M_\odot$. The calculations give present values as follows: velocity 990^{+59}_{-30} km sec^{-1}, radius $2.6^{+0.3}_{-0.2} \times 10^{18}$ cm, magnetic field 7×10^{-5} G at the edge, energy in the cavity $1.6 \times \times 10^{49}$ erg.

Although the detailed processes are not understood, the general theory of magnetic dipole radiation provides the right values for relativistic particle energies and magnetic field. It is very attractive both for the Crab Nebula and for the extragalactic objects which are sources of strong synchrotron radiation.

A detailed solution will be found in the Astrophysical Journal dated March 1971.

Discussion

A. T. Moffet: Could you relate some of the features of your model to some of the observed features of the Crab Nebula?

J. P. Ostriker: The model is a shell, but if a bird's-nest is preferred there will be a scaling factor of order unity. The shell should be expanding at 990 km sec^{-1}, with radius 3.6×10^{18} cm. Within the shell the main field should fall off as r^{-1} to 7×10^{-5} G; its structure will have rotational symmetry, and it will define the polarization of the radiation. The particles giving optical radiation will fill the nebula, moving isotropically; the lowest energy particles giving radio will have a small zone of avoidance and will exist in the outer regions only.

F. D. Kahn: The magnetic field you describe in the nebula is associated with an electric field. Does this affect the nature of the synchrotron radiation produced by the relativistic electrons?

J. P. Ostriker: Yes, the radiation is not truly synchrotron radiation, but something we have called NIC (non-linear inverse Compton). It is synchrotron-like. J. E. Gunn and I have written a paper on the characteristics of NIC radiation which should appear soon.

F. D. Drake: Your model appears to predict a decay in the electromagnetic radiation from the nebula. Perhaps this is an observable quantity – can you tell us what it is.

J. P. Ostriker: Since the particles currently being accelerated in the Crab do not contribute to the observed radio radiation we would predict the usual result that changes are only due to the adiabatic losses.

7.3 PHYSICAL PROCESSES AND PARAMETERS IN THE MAGNETOSPHERE OF NP 0532

F. PACINI*

Center for Radiophysics and Space Research, Cornell University, Ithaca, N.Y. 14850, U.S.A.

Abstract. The Crab Nebula pulsar conforms to the model of a rotating magnetised neutron star in the rate of energy generation and the exponent of the rotation law.

It is suggested that the main pulse is due to electrons and the precursor to protons. Both must radiate in coherent bunches. Optical and X-ray radiation is by the synchrotron process.

The wisps observed in the Nebula may represent the release of an instability storing about 10^{43} erg and 10^{47-48} particles.

Finally, some considerations are made about the general relation between supernova remnants and rotating neutron stars.

1. Introduction

It is well known that the discovery of NP 0532 inside the Crab Nebula has played a decisive role in our understanding of the pulsar phenomenon.

This is the source with the shortest observed period (about 33 msec), compatible only with the idea that pulsars are related to rotating neutron stars (Gold, 1968).

The lengthening of its period corresponds to a loss of rotational energy close to 10^{38} erg sec^{-1}, about the same as the total luminosity of the entire Crab Nebula. As noted by several authors, this confirms the suggestion of an electrodynamic link between the rotation of neutron stars and the activity in supernova remnants (Pacini, 1967, 1968).

NP 0532 is the only pulsar known at present to be not only a radio source but also a strong emitter of infrared, optical, X-ray and, perhaps, gamma-ray radiation. The ratio between the high frequency emission (about $10^{35}-10^{36}$ erg sec^{-1}) and the radio output (about 10^{31} erg sec^{-1}) is such that one could almost call NP 0532 a radio-quiet pulsar. The pulses themselves account only for about one percent of the total release of rotational energy: they represent only a minor energy loss in the neutron star magnetosphere, probably by the same particles which are continuously injected into the Crab Nebula.

In the following, we consider some processes which are likely to be important in the magnetosphere of a rotating neutron star and a model for pulsar radiation. In Section 1 we review the electrodynamics of rotating neutron stars. In Section 2 we discuss the electromagnetic spectrum of NP 0532 and we connect the information obtained from this radiation with the more general pulsar electrodynamics. In Section 3 we present a tentative scheme for the origin of the wisps in the Crab Nebula. Finally, in Section 4, we discuss some aspects of the evolution of supernova remnants in relation to the evolution of rotating neutron stars.

* Now at Laboratorio di Astrofisica, Frascati (Italy).

Davies and Smith (eds.), The Crab Nebula, 394–406. All Rights Reserved.

2. Electrodynamics of Rotating Neutron Stars

Fast rotation and very strong magnetic fields are expected in a newly born neutron star. For illustrative purposes we consider a pre-supernova star with a radius $R \sim 10^{11}$ cm, angular rotation frequency $\Omega \sim 10^{-6}$ sec (rotation period $P \sim$ months), magnetic fields ranging between a few and, say, 10^4 G. If this star collapses to a typical neutron star radius $\sim 10^6$ cm, approximate conservation of angular momentum would lead to $\Omega \sim 10^4$ sec^{-1} and periods in the milliseconds range. Because of the very great electrical conductivity, the magnetic field also increases and leads to $B_0 \sim 10^{10}$–10^{14} G (Ginzburg, 1964, Woltjer, 1964). Note that the virial theorem applied to neutron stars sets the upper limits $\Omega \lesssim 10^4$ sec^{-1} and $B_0 \lesssim 10^{17}$ G*.

The interaction between the neutron star and the surrounding medium should therefore be dominated by fast rotation, strong magnetic fields and strong induced electric fields. The existence of the pulsar phenomenon suggests that the magnetic field is not symmetric with respect to the rotation axis (oblique rotator model).

Several features of the neutron star magnetosphere have been discussed and we shall consider only the most important points. More details are contained in the original papers and in some review articles (see, e.g., Cavaliere, 1969).

As first noted by Gold (1968), the neutron star can be surrounded by a corotating magnetosphere up to the speed of light distance $r_{cr} \sim c/\Omega$, provided that the magnetic energy exceeds the plasma energy up to this critical point. If the field is not sufficiently strong, corotation can be enforced only up to the 'speed of Alfvén distance' $r \sim v_A/\Omega$ (v_A is the Alfvén velocity).

Goldreich and Julian (1969) and Michel (1969) have shown that the neutron star should be surrounded by a dense plasma since the induced electric fields continuously extract charged particles from the surface of the star.

One can divide the near magnetosphere ($r < c/\Omega$) in two different parts. The particles attached to the field lines which do not cross the speed of light cylinder form the corotating magnetosphere. Near the star, the particles probably have an energy close to the escape energy (say, 10 up to about 100 MeV for protons). If however they move close to the critical distance, they should be relativistic.

The field lines which pass through the speed of light cylinder are open and the particles can stream out along them. As discussed by Goldreich and Keeley (1970), they can be accelerated by a component of the electric field parallel to the magnetic lines. The efficiency of this acceleration is uncertain because the plasma can partially neutralize the induced electric field. In the absence of space charges and radiation losses the Crab pulsar could inject particles with an energy up to about 10^{16} eV, but in real life the energies attained are likely to be much less than this upper limit. Since the mechanism is of electrostatic nature, protons and electrons would reach the same energy, unless radiation losses drastically modify this situation. We also recall that

* If $B_0 > 4 \times 10^{13}$ G one should however also take into account quantum effects (see, e.g., Chiu and Canuto, 1970a).

the protons and electrons move preferentially along different field lines (Goldreich and Julian, 1969).

The gyration around the magnetic lines cannot be excited simply by the accelerating forces since they act adiabatically on the gyration time scale. We speculate that a gyration will arise either because of collisions or because of transverse electric fields. Our picture for the motion of the particles includes a component along the field lines and a simultaneous gyration around them: we shall return to the consequences of this picture for the pulsar radiation in the next section.

In the case of an oblique rotator, retardation effects modify the electromagnetic field close to the critical distance: the neutron star radiates magnetic dipole radiation at the basic frequency Ω (Pacini, 1967, 1968; Ostriker and Gunn, 1969). The lines of force are swept back and the resulting tension corresponds to a torque slowing down the star. Under the assumption $B^2/8\pi > nmc^2\gamma$ (n is the plasma density) the effect of the outflowing plasma results in a similar (but quantitatively less important) torque. Apart from geometrical factors, the energy radiated is given by

$$L \sim \frac{B^2}{8\pi} c4\pi \left(\frac{c}{\Omega}\right)^2 \tag{1}$$

where $B = B(r_{cr})$. In a dipole field

$$L \sim \frac{1}{2c^3} B_0^2 a^6 \Omega^4 \tag{1'}$$

(B_0 is the surface field, a is the stellar radius).

An exact treatment of the problem (Deutsch, 1955) confirms this intuitive result but gives a numerical factor $\frac{2}{3}$ instead of $\frac{1}{2}$.

The loss of rotational energy corresponds to a slowing down law $\dot\Omega \propto \Omega^n$ where $n = 3$. The laws $L \propto \Omega^4$ and $\dot\Omega \propto \Omega^3$ imply a strictly dipolar geometry $B(r) \propto r^{-3}$, as one can immediately see from our simple derivation. In real life it might be necessary to take into account the plasma in the magnetosphere: the outflowing particles will tend to stretch the lines of force in a radial direction. For a purely radial field $B \propto r^{-2}$, one finds (with different coefficients) $L \propto \Omega^2$ and $\dot\Omega \propto \Omega$.

For the Crab pulsar, before the well known speed-up, the observed value was $n \sim 2.2$ (Richards et al., 1970), thus suggesting a dipole field slightly perturbed by the plasma. The rate of slowing down tells us (under the assumption of an electromagnetic torque) the strength of the magnetic field on the star. For most pulsars, the resulting value is around 10^{12} gauss, in agreement with the expectations.

In the far magnetosphere ($r > c/\Omega$) the magnetic field develops a toroidal component and is given by

$$B(r) = \frac{B_0 a^3}{c^2} \frac{\Omega^2}{r}. \tag{2}$$

For the Crab Nebula, $\Omega \sim 200$ sec^{-1}, $B_0 \sim 3 \times 10^{12}$ G, $r \sim 10^{18}$ cm: the resulting strength $B \sim 10^{-4}$ G matches the estimated value and suggests that also the nebular field is generated by the pulsar.

The particles escaping from the near magnetosphere can be accelerated further in the far wave region by the magnetic dipole radiation (Ostriker and Gunn, 1969). The mechanism is very efficient if the particles and the fields move in phase. For NP 0532, at $r \sim 10^8$ cm, $E = B = 10^6$ G and the particles can reach top energies around 10^{14} eV. A relatively dense plasma would however reduce the number of particles able to reach the maximum energy: the presence of such plasma appears unquestionable on several grounds.

First, its existence is suggested by the lowering of the slowing down index with respect to the pure dipole case. Second, an integration of the Crab Nebula injection spectrum (see, e.g., Shklovskii, 1969) corresponds to an outflow of about 10^{40} particles sec^{-1}. At the critical distance the density should therefore be $n \sim 10^{12}/\varepsilon$ cm^{-3} (ε is a geometrical factor introduced to take into account the possibility of an anisotropic injection). Finally, as we shall see in the next Section, a density between 10^{12} and 10^{14} cm^{-3} is indicated by the pulsar radiation. A description of the far wave acceleration based upon a quasi-vacuum approximation is therefore probably too optimistic.

Magnetic dipole radiation can accelerate particles but only a few of them probably reach the top energy $\sim 10^{14}$ eV. This would agree with the small proportion of very energetic electrons present in the Crab Nebula. As noted by Woltjer (1969), it is impossible to explain the spectrum of the Crab Nebula with a monochromatic injection of relatively few very high energy electrons which are then decelerated by radiation losses: the number of particles with energy $\lesssim 10^9$ eV exceeds in the nebula that of particles with energies $\gtrsim 10^{13}$ eV by a factor $\sim 10^6$. One should instead require an injection with widely different energies, roughly between 10^8 and 10^{14} eV. Both the near and the far field acceleration are likely to determine the injection spectrum.

Finally, we make a brief comment about the rapid decline of injection at energies less than about 100 MeV (see, e.g., Shklovskii, 1969). The cut-off occurs around the escape energy of protons from the neutron star and this might be more than a pure coincidence. In an electrostatic mechanism protons and electrons would acquire the same energy: if the electric field value is set by the protons and one can neglect deceleration effects one expects that also the electrons will have a minimum energy around 100 MeV.

3. The Electromagnetic Spectrum of NP 0532

The pulses could, in principle, provide some direct information about the neutron star magnetosphere. Their origin is however controversial and no general agreement exists on whether they are emitted close to the surface or whether they arise in a region distant about c/Ω from the star. The only obvious point is that they should come from a region where the magnetic energy density is not overcome by the plasma energy density: otherwise it would be difficult to imagine a well defined radiating sector of the magnetosphere.

Also, the very high brightness temperature of the radio emission implies a very coherent radiation mechanism. Different proposals have been investigated, such as coherent plasma effects (see, e.g., Ginzburg *et al.*, 1969), maser effects (Chiu and Canuto, 1970b), correlated motion of bunches of charges (Gold, 1968; Bertotti *et al.*, 1969; Komesaroff, 1970; Goldreich and Keeley, 1970; Pacini and Rees, 1970). As several authors have recognized, the incoherent high frequency emission from NP 0532 can instead be due to the usual synchrotron process: the small duty cycle $\delta \sim 10^{-2}$ of the pulsar demands an equally well collimated distribution of pitch angles for the motion of the particles around the field lines (Pacini and Rees, 1970).

It is often claimed that the pulsed nature of the source implies an emission within a few stellar radii, where the open field lines subtend a small angular sector. It is not clear to us whether this claim (at least when made a priori) is really justified. The distribution of the magnetic field close to the star can be much more irregular than a simple dipole. Also, it might be difficult to achieve a well collimated distribution of pitch angles close to the star since the expected very high plasma density could lead to collisions and smear out the collimation.

On the other hand the far wave solution of the stellar electromagnetic field clearly shows a regular, large scale structure (Deutsch, 1955). Because of the lower density, the collisions would not influence the distribution of pitch angles: a small pitch angle could perhaps result from adiabatic invariance. We shall therefore concentrate upon a model where the pulses are generated close to the critical distance and the radiation processes are related to the motion of those particles which move away from the star along the open field lines. At the end of this section, we shall also indicate some parameters which would apply if the pulses were generated within a few star's radii. Our treatment will follow the one given in earlier papers (Bertotti *et al.*, 1969; Pacini and Rees, 1970) with minor modifications and additions.

We have previously noted that the motion of particles in the open magnetosphere should probably involve two components: one component along the lines of force and a gyration around them. Since the field lines have an intrinsic curvature, the motion along them gives rise to a radiation process different from the synchrotron radiation resulting from the gyration. The two radiation processes are independent, since the Larmor radius is much smaller than the radius of curvature of the field lines.

We investigate first the generation of incoherent synchrotron radiation. At $r \sim c/\Omega \sim 10^8$ cm the strength of the magnetic field has dropped to about 10^6 G. The small duty cycle $\delta \sim 10^{-2}$ implies for the pitch angle $\Psi \lesssim 5 \times 10^{-2}$ rad; the corresponding upper limit for the perpendicular component of the magnetic field is $B_\perp \lesssim 5 \times 10^4$ G.

In a field of this order, the observed emission between the infrared and the X-ray frequencies demands a distribution of electron energies roughly between 10^8 and some 10^{10} eV. If the recent report (Vasseur *et al.*, 1970) of a gamma-ray emission will be confirmed, some electrons should have an energy $\gtrsim 10^{11}$ eV. The emitting volume is uncertain but the duty cycle again implies that only a fraction $\sim 10^{-2}$ of the speed of light circumference (about 10^9 cm) is involved in the radiation. If we assume an emitting volume $V \sim 10^{21}$ cm^3, the observed output requires an electron density close to 10^{12} cm^{-3}.

The usual lifetime of the electrons against radiation losses is small but irrelevant: while they radiate, the electrons are also accelerated by the electric fields. The energy gains can largely exceed the losses and the particles escape beyond the critical distance.

Two different explanations for the low frequency cut-off at $v < 8 \times 10^{14}$ Hz appear possible. O'Dell and Sartori, (1970) have noted that synchrotron radiation with small pitch angle Ψ has a turnover at a frequency $v \sim 10^6 \, B/\sin \Psi$. At lower frequencies the emission is due to the cyclotron process and declines steeply. With our previous parameters one would predict a cut-off at $v \sim 2 \times 10^{13}$ Hz but a smaller pitch angle would fit the observations.

On the other hand, Shklovskii (1970) has considered the possibility of synchrotron reabsorption: fields of the right order of magnitude ($B_\perp \sim 10^3$–10^4 G) are implied by the observations.

The possible extreme range of values $3 \times 10^3 \lesssim B_\perp \lesssim 5 \times 10^4$ G corresponds to a small uncertainty in the energy of the particles (roughly, $\gamma \propto B_\perp^{-1/2}$) but to a somewhat larger uncertainty in the plasma density (roughly, $10^{12} \lesssim n \lesssim 10^{14}$ cm^{-3}).

Despite this uncertainty, there is a very good agreement between the parameters inferred from the pulsar radiation and the requirements stemming from the link between the neutron star and the Crab Nebula. As pointed out in the previous section, a flux $\sim 10^{40}$ particles sec^{-1} across the speed of light distance gives there a density $n \sim 10^{12}/\varepsilon$ cm^{-3}. If the injection is isotropic $\varepsilon = 1$; if it only occurs along the field lines responsible for the pulsar radiation, $\varepsilon \sim 10^{-2}$.

The presence at $r \sim 10^8$ cm of particles with energies up to 10^{10}–10^{11} eV implies an acceleration mechanism in the near magnetosphere, such as the one suggested by Goldreich and Keeley (1970). Additional acceleration in the far field by magnetic dipole radiation can account for the presence in the Crab Nebula of some particles with energies up to about 10^{14} eV.

With the above parameters, the magnetic energy exceeds the plasma energy: the torque slowing down the star is therefore mostly electromagnetic. The outflowing, relatively dense, plasma can however partially stretch the field lines in a radial direction and account for a slowing down index less than 3.

Since the peak of synchrotron radiation is proportional to $B(r_{cr}) \propto P^{-3}$, slow pulsars should emit by this mechanism at infrared frequencies, rather than in the optical or X-ray band. The power radiated is $\propto n V \gamma^2 B^2$ (V is again the emitting volume): the dependency from the magnetic field implies a synchrotron luminosity $\propto P^{-6}$ but a stronger dependency should be expected since the acceleration of particles by slow pulsars is probably less efficient than in the Crab. Also, the synchrotron luminosity of NP 0532 should decrease by a quantity $\gtrsim 0.01$ magnitudes per year.

We consider next the origin of the coherent radio emission which we ascribe to the component of motion along the field lines. At a distance r from the star, the radius of curvature of these lines is $\varrho \sim (r\,c/\Omega)^{1/2}$ (see, e.g., Goldreich and Keeley, 1970). Close to the critical distance $\varrho \sim c/\Omega$ and therefore there is no difference between considering the radiation as due to the curvature of the field lines or to a general "corotation" at the basic frequency Ω.

Radio frequencies up to $v_{cr} \sim 10^8$–10^9 Hz will arise if the Lorentz factor $\gamma_\parallel$ corresponding to the parallel component of motion satisfies the relation

$$\frac{1}{2\pi}\frac{c}{\varrho}\gamma_\parallel^3 \gtrsim v_{cr}.\tag{3}$$

For NP 0532 this gives, roughly, $\gamma_\parallel \gtrsim 10^2$. A similar value $\gamma > \delta^{-1} \sim 10^2$ can be independently inferred from the requirement that the duty cycle is a consequence of the existence of an active sector of the magnetosphere and is not limited by the emission cone of the individual particles (otherwise the duty cycle would change with the frequency in a known way).

The brightness temperature of NP 0532 in the radio band is close to 10^{26} K but occasionally very strong pulses are observed with $T_b \sim 10^{30}$ K. This very high degree of coherence can arise because of strong correlations in the motion of particles and from the formation of charged bunches. The brightness temperature is limited by thermodynamics

$$kT_b \lesssim \text{(energy per bunch)}.\tag{4}$$

An upper limit to the size of the bunches is given by the wavelength of the emitted radiation, say ~ 1 up to about 10^2 cm. Note however that this limit refers only to the size along the visual line: in principle, the other sizes could be larger. The radio spectrum should largely reflect the degree of coherence at a given wavelength (i.e. the spectrum of correlations in the plasma), rather than the distribution of particle energies.

Since Relations (3) and (4) are only inequalities and we don't know the sizes of the bunches different from the one along the line of sight, the number of particles taking part in the radio emission cannot be reliably estimated. A priori, one can however expect that it will turn out to be smaller than the number of particles inferred from the optical and X-ray emission. First, only a small part of the particles responsible for the incoherent radiation is likely to have a sufficiently small pitch angle to give also $\gamma_\parallel \gtrsim 10^2$. Second, it is likely that there will be a partial neutralization inside the bunches: this would reduce the efficiency at which the bunches radiate coherently without affecting the incoherent emission.

With these uncertainties in mind, we examine for illustrative purposes two extreme cases.

We first consider quasi-spherical bunches with volume $L^3 \sim \lambda^3 \sim 10^2$ cm^3; also, we assume that the density and the energy of the plasma is close to the limits $n \sim 10^{12}$ cm^{-3} and $\gamma_\parallel \sim 10^2$. The number of particles per bunch is $\sim 10^{14}$ and the energy per bunch $\gtrsim 10^{10}$ erg. The upper limit to the brightness temperature is $\sim 10^{26}$ K. Note that the energy of the particles largely exceeds the electrostatic energy (about 10^6 erg): the formation of these bunches does not pose energetic problems.

The motion along the curved field lines causes a radiation per unit frequency (close to $\nu_{cr} \sim 10^8$ Hz)

$$p_\nu \sim \frac{e^2}{c} \frac{c}{\varrho} \gamma_\parallel n V \left(n L^3 \right). \tag{5}$$

The previous parameters can therefore easily account for the emitted power of NP 0532 (radio output $\sim 10^{31}$ ergs over a band $\sim 10^8 - 10^9$ Hz).

Alternatively, we can assume an 'effective density' $n = n_p - n_e \sim 10^7$ cm^{-3} which is of the same order as the density of space charge inferred at $r \sim c/\Omega$ by Goldreich and Julian (1969). This value does not contradict a total density $n_p \simeq n_e \simeq 10^{12} - 10^{14}$ cm^{-3} since it only refers to the net charge and, furthermore, only to the large scale distribution. We keep $\gamma_\parallel \sim 10^2$ (which is likely to be an underestimate). One finds then that the volume of the bunches should be around 10^{13} cm^3. Since the size cannot exceed the emitted wavelength, the bunches would be very elongated in a direction transverse to the line of sight.

Intermediate parameters are possible and the uncertainty in the number of particles involved in the radio emission is very large. Fortunately, in the case of NP 0532, the incoherent radiation does not suffer the same uncertainty and can be used to give a reliable estimate of the plasma parameters.

The optical pulses should be somewhat broader than the radio pulses (because of the additional contribution of the pitch angles): this agrees with the observations.

An interesting aspect of the radio emission is the occasional presence of very strong radio pulses: probably they arise when richer bunches are formed, which is energetically possible. Also, the main radio pulse is preceded by a precursor absent at optical and X-ray frequencies. We have already recalled in Section 2 that protons move along field lines different from those followed by electrons.[*] It seems therefore possible to account for the precursor in terms of radiation from the proton field lines. The protons could radiate at radio frequencies provided that they satisfy Equation (3) but they would be absent at optical and X-ray frequencies since proton synchrotron radiation would be negligible. It is interesting that the precursor shows much more linear polarization than the main pulse. This agrees with the picture since the radio emission is strongly polarized only when it falls at (or above) the critical frequency. If $\gamma_\parallel$ (protons) $\sim 10^2 \ll \gamma_\parallel$ (electrons) (as one should expect from an electrostatic mechanism) the protons would radiate close to the critical frequency and their emission would be very polarized. The electrons would instead be observed before the critical frequency, where the output can be magnified by coherence but the polarization would be small. Slow pulsars might be unable to reach $\gamma_\parallel$ (protons) $\sim 10^2$ if the efficiency of acceleration decreases with time: this agrees with the absence of a similar precursor in their emission.

---

[*] This statement should be interpreted in the sense that field lines at different latitudes have space charges of different sign and not in the sense that each field line carries one sign of particles.

Concerning the interpulse, probably it is emitted at the opposite side of the star: the time delay between the main pulse and the interpulse represents the difference in the light travel time.

Finally, we make some comments about the alternative scheme of an emission (by the same processes) close to the star. More details can be found in previous papers (Komesaroff, 1970; Pacini and Rees, 1970). The main point is that the low-frequency cut-off in the incoherent radiation can be neither cyclotron turn-over nor electron synchrotron reabsorbtion. The very high magnetic fields close to the star would imply a cut-off at frequencies much higher than observed. The only possibility of which we are aware (Pacini and Rees, 1970) is that the radiation could be due to protons instead of electrons. In this case the turn-over can be explained by synchrotron reabsorbtion in a field $B_\perp \sim 10^9$ G. The model requires a very high plasma energy density close to the star and the acceleration of relativistic particles should be essentially limited to the star's surroundings. Furthermore, the plasma energy exceeds the magnetic energy well before the critical distance. The torque slowing down the star would be due to the plasma and much of the electrodynamics discussed so far would be irrelevant.

In our opinion, the main problem with this alternative scheme (valid at least as long as no other explanation for the cut-off at $\nu \lesssim 8 \times 10^{14}$ Hz is available) is that it seems very unlikely that we can see only proton and no electron radiation. It seems to us that an emission close to the speed of light distance (either a bit inside or a few basic wavelengths c/Ω outside) is strongly supported by

(1) the agreement between the plasma density estimated from the radiation and the required output in the Crab Nebula of about 10^{40} electrons sec^{-1}.

(2) the explanation of the cut-off at $\nu < 8 \times 10^{14}$ requiring fields $B_\perp \sim 3 \times 10^3 - 5 \times 10^4$ G.

(3) the expected very small synchrotron power from slow pulsars.

(4) the interpretation of the precursor in terms of proton emission.

4. The Origin of the Wisps in the Crab Nebula

We present in this section some preliminary considerations about the origin of the wisps in the Crab Nebula. These bright features, discovered many years ago by Baade, have been recently investigated in great detail by Scargle (1969). They have a semi-periodical recurrence of activity, about once or twice per year, and involve the sudden release of about $10^{43 \pm 1}$ ergs from the center of activity in the Crab Nebula.

There is some evidence (Scargle and Harlan, 1970) that the wisps became active in the first weeks after the speed-up of NP 0532, thus suggesting a relation between the two phenomena. We accept, for the time being, this evidence which clearly awaits confirmation from future events.

The increase in rotational energy of the neutron star during the speed-up was around 10^{41} erg, clearly insufficient to explain the energetics of the wisps. Also, the change in the pulsar period $\Delta p/p \sim 2.5 \times 10^{-9}$ would produce only a very small change in the flux of energy from the neutron star. It is therefore worth investigating some other possibilities for the connection between the wisps and the speed-up.

The energetics of the wisps roughly matches instead the energy content of the near magnetosphere of NP 0532, since $B_0^2 a^3 \sim 10^{43}$ erg. One should therefore consider the possibility that the wisps arise, from time to time, because of a sudden instability releasing an amount of plasma energy comparable to the energy in the near magnetosphere.

As seen before, the particles moving along the open field lines gain much energy and escape continuously from above the magnetic poles; they give rise to the pulsar radiation and to a steady input of relativistic particles in the Crab Nebula. The situation is different in the corotating magnetosphere, around the rotation axis. Here the particles can accumulate and, close to the star, their individual energy is probably between 10 and, say, 100 MeV. Much of the magnetic energy is stored close to the star and can confine the particles up to the point when the magnetic and the plasma energies are of the same order. The total number of particles which can be confined is around 10^{47}–10^{48}, with a total mass $M \sim 10^{-9}$–$10^{-10} M_\odot$. When the limiting value has been reached, the system could explode and release the particles into the Crab Nebula.

One can evaluate the time interval between two successive events. The corotating magnetosphere is probably fed at about the same rate as the open magnetosphere, that is about 10^{40} particles sec^{-1} are injected into it. The time necessary to reach the critical value is therefore between several months and one year, in agreement with the frequency of the wisps.

The presence of a dense plasma can indirectly affect the continuous slowing down of NP 0532 by deforming the structure of the field in the near magnetosphere. This effect adds itself to the previously mentioned effect of the outflowing plasma. One expects that after the flare the magnetosphere will be less perturbed (until replenished again) and that the field be closer to a dipole. There is indeed some evidence that the slowing down index n became after the speed-up closer to the dipolar value $n=3$ (Drake, private communication).

A few days after the explosion, the plasma would have reached a distance $\sim 10^{15}$–10^{16} cm and one could observe a small increase in the pulsar dispersion measure. The rate of increase and its actual value are uncertain because they depend upon the expansion geometry and the plasma energy (a purely relativistic gas would clearly be less dispersive than a thermal gas); a tentative value is around $10^{16\pm1}$ electrons cm^{-2}. It is interesting that an increase of the dispersion measure was actually observed at Arecibo by Rankin and Roberts (private communication).

If the release of plasma is sufficiently well collimated with the rotation axis, and does not change appreciably the angular momentum of the star, the mass change in the corotating system 'star-magnetosphere' leads to a speed-up

$$\frac{\Delta\Omega}{\Omega} \sim \frac{\Delta M}{M} \sim 10^{-9}\text{–}10^{-10}. \tag{6}$$

It seems therefore possible to account for the observed increase in the pulsar frequency as a by-product of the same instability giving rise to the wisps.

Finally, the presence of a sinusoidal oscillation with period $P=77\pm7$ days was reported before the glitch (Richards *et al.*, 1970) but seems to have disappeared since (Duthie and Murdin, 1970). This would rule out an interpretation based upon planetary perturbations or mechanical precessions connected with the neutron star structure. It will be worth exploring an alternative explanation based upon the interaction between the skew stellar field and the axial field which would set up in the corotating magnetosphere. The disappearance of the plasma in the corotating magnotosphere for some time after the flare could account for the disappearance of the oscillations. One could expect that the oscillations will set in again in the coming months and that they represent a warning signal before a new glitch and wisp activity.

5. The Crab Nebula and the Other Supernova Remnants

We conclude with some general remarks about the relation between the activity and the morphology of supernova remnants and the evolution of the central neutron star.

It has become clear since the discovery of NP 0532 that the activity in the Crab Nebula follows the generation of high energy particles and of a relatively strong large scale magnetic field by the central neutron star. The Crab Nebula is however usually regarded as an 'unique' object: many other remnants show much less activity and are associated with a shell-structure, rather than a diffuse structure (see, e.g., Cas A).

This difference might stem from the absence in the other remnants of a neutron star and the present theory of supernova explosions is unable to predict the probability of finding a neutron star in a given SN remnant. Also, some attempt to find a bright pulsar in Cas A and other remnants have apparently failed.

On the other hand, both Cas A and the Tycho source SN 1572 are associated with an X-ray emission (see, e.g., Gorenstein, 1970). The spectral information is still insufficient to decide whether this radiation is non-thermal or whether it results from bremsstrahlung. If however the emission is due to the synchrotron mechanism, then there is the need for continuous injection of relativistic electrons, probably related to the presence of a rotating neutron star.

What kind of pulsars could one expect in these remnants? Also, why do they show a shell-like structure, rather than a diffuse structure like the Crab Nebula?

A possible answer to these questions has been given before (Cavaliere and Pacini, 1970; Pacini, 1970) and we recall here the arguments.

If the loss of rotational energy gives rise to the steady luminosity L of the remnant (like in the Crab Nebula), we have

$$L = \frac{2}{3c^3} B_0^2 a^6 \Omega^4 . \tag{7}$$

Also, a neutron star with magnetic field B_0 and age t has a period P

$$P^2(t) = P_0^2 + 2 \times 10^{-39} B_0^2 t \quad \text{(sec., G)} \tag{8}$$

where P_0 is the initial period and the numerical coefficients refer to a standard neutron star.

Under the assumption $P \gg P_0$, the previous equations determine the period and the magnetic field of the neutron star on the basis of the known age and total luminosity of the remnant. Both for Cas A and Tycho one finds 0.5 ± 0.3 sec and magnetic fields around 10^{14} G.

The relatively long period does not contradict the known ages (a few hundred years) since the very strong magnetic field causes a rapid slowing down. Note that the injection rate in Cas A changes by about 1% a year, in agreement with the observed decrease of the radio emission (usually ascribed to expansion losses).

Slow pulsars inside these remnants would be difficult to detect because of the weak radio emission (say, for analogy with other slow pulsars, 10^{28}–10^{30} erg sec^{-1}), but it seems extremely important to search for them and lower the present limits.* Because of the long period and the arguments given in Section 2, no pulsed optical or X-ray radiation should be expected.

The shell structure itself of these remnants could be related to the presence of a long period source. One can see from Equation (2) that a slowly rotating neutron star can generate only a weak extended field. Even if the central object produces relativistic particles, they will not radiate very strongly over most of the volume occupied by the remnant. Most of the emission would arise in the outer shell: here the magnetic field is not directly connected with the pulsar and can be stronger because of compression.

If the above point of view is correct, then one should conclude that the most important parameter determining the evolution of a SN remnant is the magnetic field strength on the surface of the central neutron star. Also, the apparently exceptional character of the Crab Nebula appears to be simply a consequence of the fact that, with a surface field around 10^{12} G, the central star can remain a long lasting producer of a relatively strong nebular field and high energy particles.

Finally, some neutron stars could be born with fields much less than 10^{12} G, say around 10^{10} G. In this case, as we have shown before (Pacini, 1970), the central neutron star would still have a period in the range ~ 10 msec after several million years, that is long after the SN remnant has disappeared. Since fast pulsars are probably associated with strong optical and X-ray emission, some of the existing X-ray sources could be old, but still very fast, pulsars.

We would not be surprised if it will turn out in the future that our present knowledge of the pulsar phenomenon is very incomplete, having missed slow pulsars in

---

* According to a private communication from Dr. A. Hewish (December 1969) the Cambridge search can put a limit for Cas A (pulsed flux) $\lesssim 1\%$ (total flux). Note that not even a pulsar like NP 0532 would have been detected against the strong nebular background.

some young shell-like SN remnants and old but fast rotating neutron stars hidden among some X-ray sources.

Acknowledgements

This work was supported partly by NSF under Grant GP-9621 and partly by the Office of Naval Research under contract N-00014-67-A-0077-0007. We are especially indebted to Drs. A. Cavaliere and M. Rees for useful discussions.

References

Bertotti, B., Cavaliere, A., and Pacini, F.: 1969, *Nature* 223, 1351.
Cavaliere, A.: 1970, 'Pulsars and SN Remnants', paper presented at the Meeting 'Pulsars and High Energy Activity in SN Remnants' (Rome, December 1969), in press.
Cavaliere, A. and Pacini, F.: 1970, *Astrophys. J.* (Letters) 159, L21.
Chiu, H. Y. and Canuto, V.: 1970a, preprint.
Chiu, H. Y. and Canuto, V.: 1970b, preprint.
Deutsch, A.: 1955, *Ann. Astrophys.* 18, 1.
Duthie, J. G. and Murdin, P.: 1970, preprint.
Ginzburg, V. L.: 1964, *Dokl. Akad. Nauk. USSR* 156, 43.
Ginzburg, V. L., Zheleznyakov, V. V., and Zaitsev, V. V.: 1969, *Astrophys. Space Sci.* 4, 464.
Gold, T.: 1968, *Nature* 218, 731.
Goldreich, P. and Julian, W.: 1969, *Astrophys. J.* 157, 869.
Goldreich, P. and Keeley, D. A.: 1970, preprint.
Gorenstein, P., Gursky, H., Kellogg, E. M., and Giacconi, R.: 1970, *Astrophys. J.* 160, 947.
Komesaroff, M. M.: 1970, *Nature* 225, 612 (1970).
Michel, F. C.: 1969, *Astrophys. J.* 158, 727.
O'Dell, S. and Sartori, L.: 1970, *Astrophys. J. Letters* 161, L43.
Ostriker, J. and Gunn, J.: 1969, *Astrophys. J.* 157, 1395.
Pacini, F.: 1967, *Nature* 216, 567.
Pacini, F.: 1968, *Nature* 219, 145.
Pacini, F.: 1970, 'Are There Different Classes of Pulsars?', paper presented at the Meeting 'Pulsars and High Energy Activity in SN Remnants' (Rome, December 1969), in press.
Pacini, F. and Rees, M.: 1970, *Nature* 226, 622.
Richards, D. W., Pettengill, G. M., Counselman III, C. C., and Rankin, J. M.: 1970, *Astrophys. J. Letters* 160, L1.
Scargle, J. D.: *Astrophys. J.* 156, 401.
Scargle, J. D. and Harlan, E. A.: *Astrophys. J. Letters* 159, L143.
Shklovskii, I. S.: 1969, *Supernovae*, Wiley and Sons, New York.
Shklovskii, I. S.: 1970, *Astrophys. J. Letters* 159, L77.
Vasseur, J., Paul, J., Parlier, B., Leroy, J. P., Forichon, M., Agrinier, B., Boella, G., Maraschi, L., Treves, A., Buccheri, R., and Scarsi, L.: 1970, *Nature* 226, 534.
Woltjer, L.: 1964, *Astrophys. J.* 140, 1309.
Woltjer, L.: 1970, 'Problems of the Crab Nebula', paper presented at the Meeting 'Pulsars and High Energy Activity in SN Remnants' (Rome, December 1969), in press.

7.4 THE NON-THERMAL CONTINUUM FROM THE CRAB NEBULA:

THE 'SYNCHRO-COMPTON' INTERPRETATION

M. J. REES

Institute of Theoretical Astronomy, Madingley Road, Cambridge, U.K.

Abstract. The continuum emission from the Crab Nebula may be radiation from relativistic electrons moving in the electromagnetic wave field radiated by the rotating magnetic dipole of the pulsar. This radiation, called Synchro-Compton radiation, would show ordering over the whole nebula, as is observed in measurements of polarisation. The properties of this radiation are described in an Appendix.

Studies of the 'oblique rotator' model, by Ostriker and Gunn (1969) and others, suggest that the rotational braking of pulsars may be primarily caused by emission of electromagnetic radiation at the rotation frequency. This paper will explore the possibility that such very low frequency (30 Hz) radiation emanating from NP 0532 plays the role conventionally ascribed to a large scale magnetic field, and that the continuum emission from the Crab Nebula is due to relativistic electrons moving in this wave. This suggestion was briefly discussed by Gunn (1970).

The rate of emission by an isotropic distribution of relativistic electrons moving in an electromagnetic field is determined by the electromagnetic energy density. It is therefore convenient to express the intensity L of the 30 Hz wave in terms of the magnetic field H_{eq} for which the wave energy density equals $H_{eq}^2/8\pi$. We find

$$H_{eq} = 1.8 \times 10^{-4} \left(\frac{L(30 \text{ Hz})}{5 \times 10^{38} \text{ erg sec}^{-1}} \right)^{1/2} \left(\frac{r}{10^{18} \text{ cm}} \right)^{-1} q \text{ G}. \tag{1}$$

This assumes that the wave propagates with velocity c, and ignores the fact that dipole emission would be twice as intense along the rotation axis as in the equatorial plane. The factor q (which is $\geqslant 1$) allows for possible reflection back into the nebula. For the Crab we expect the terms in brackets to be ~ 1. Thus, if $q \simeq 1$, the wave energy density is comparable with that of the weakest magnetic field ($\sim 10^{-4}$ G) permitted by energetic and dynamical considerations.

It is also useful to define an 'equivalent gyrofrequency' $\Omega/2\pi \simeq 3 \times 10^6 H_{eq}$ Hz, since the parameter $f = \Omega/\omega$, where ω is the wave frequency, determines the character of the relativistic particle orbits and of the radiation which these particles emit. Throughout the Crab Nebula we would expect $f \gtrsim 10$. In this situation a relativistic electron radiates at frequencies $\sim \gamma^2 \Omega$, as in the case of synchrotron radiation, and *not* $\sim \gamma^2 \omega$ as for inverse Compton emission. Further details of this radiation mechanism, – which we shall call 'synchro-Compton' emission – are given in the *Appendix*, but for the moment it is sufficient to know that the usual synchrotron formulae still, in general terms, apply, so that the standard inferences of the electron density and spectrum in the Nebula remain applicable.

Davies and Smith (eds.), The Crab Nebula, 407–413. All Rights Reserved.
Copyright © 1971 by the IAU.

Self-consistency demands that the plasma density within the Nebula should be low enough to allow the 30 Hz radiation to propagate. At first sight one might suspect that the formal plasma frequency $9 \times 10^3 \, n_e^{1/2}$ Hz would have to be below 30 Hz, which would lead to the exceedingly stringent condition that the electron density throughout the nebula be $\lesssim 10^{-5}$ cm^{-3}. However in the case of a 'strong' wave ($f > 1$) this condition can be relaxed somewhat (Ostriker and Gunn, 1969), so that a sufficient condition for the wave to propagate is

$$n_e \lesssim 10^{-5} f \quad \text{cm}^{-3}. \tag{2}$$

(The extra factor f occurs because, in order to reflect the wave, the electrons (moving at speeds $< c$) must be numerous enough to carry the induced current, and the latter is proportional to f. It is also easy to see that all particles exposed to a wave with $f > 1$ must be relativistic with $\gamma \gtrsim f$). The relativistic particles required to produce the observed continuum from the nebula all have $\gamma \gg f$. The propagation condition sets a limit on their density of

$$\int n(\gamma) \frac{\log \gamma}{\gamma} \, d\gamma \lesssim 10^{-5} \text{ cm}^{-3}. \tag{3}$$

(Zheleznyakov 1967), where $n(\gamma)$ is the differential electron spectrum. The main contribution is made by the particles of the lowest γ, but (3) is satisfied, with a factor ~ 10 to spare, by the particles with $\gamma \gtrsim 100$ whose density is directly inferred from observations above a few MHz. Since (2) is obviously not fulfilled by the general interstellar medium, the 30 Hz waves cannot penetrate beyond the boundary of the nebula – indeed the observable nebula would, in this picture, be delineated by the region which has been evacuated sufficiently for the wave to propagate. Also, the waves would not be able to penetrate the filaments in the nebula.

Before considering the synchro-Compton radiation mechanism in more detail, we shall briefly discuss how the general viewpoint suggested here affects current ideas on the acceleration and confinement of relativistic particles. Several other speakers at this symposium have argued that relativistic particles must be ejected continuously from the pulsar (and indeed Ostriker and Gunn have shown that the very strong wave in the vicinity of the speed of light cylinder constitutes an embarrassingly potent accelerator). For our considerations here, it is of course essential that *not all* the rotational energy of the pulsar should go directly into fast particles, but that a substantial fraction (~ 10 per cent at the very least) should escape as 30 Hz radiation. It is interesting to investigate the eventual fate of this wave energy. When the wave reaches the boundary of the nebula, only a fraction $\sim v_{\text{exp}}/c$ of its energy is used in pushing against the external medium, v_{exp} being the expansion velocity of the boundary. The bulk must be either reflected or absorbed. We shall show below that the high polarisation of the continuum from the nebula implies that the low frequency radiation must be ordered rather than random, and this precludes more than ~ 50 per cent reflection (so that, in (1), $q \lesssim 2$). This means that the energy must all be deposited in a thin 'skin' at the boundary. The densities are so low that there is no possibility of this energy being radiated thermally, so there seems no alternative to the view that it generates relativistic particles, probably mainly electrons. Thus the pulsar would be almost 100 per cent efficient in accelerating particles: whatever fraction escapes into the wave zone in the form of 30 Hz emission will produce particles at the boundary, or at the inner edges of filaments.

Even though the wave field simulates a stationary magnetic field as regards the radiation (except, as we shall see, in the important respect of circular polarisation) it is much less efficient for confining particles, since the orbits are basically straight lines. However even a very weak magnetic field, which is negligible as regards the emission mechanism, could confine the particles adequately if it were sufficiently tangled. Alternatively, the particles could be 'mirrored' at the boundary by the external interstellar field, and by the filaments if these contain a magnetic field.

The electromagnetic radiation emitted from an ideal spinning magnetic dipole would, in the equatorial plane, be completely linearly polarised, the electric vector lying perpendicular to the plane. At higher latitudes, the wave would be elliptically polarised, and along the rotation axis the polarisation would be purely circular. As discussed in the *Appendix*, the polarization and propagation direction of the 30 Hz wave would determine the polarisation properties of the synchro-Compton radiation that is actually observed. We cannot, however, compare the model with the polarisation data on the Crab without assuming the orientation of the dipole. In certain pulsar models the existence of an interpulse indicates that the observer is located close to the equatorial plane. Guided by this, let us suppose that the rotation axis of NP 0532 is precisely in the plane of the sky. Then, provided that (3) is satisfied by a large enough margin that the effects of the medium are negligible, the synchro-Compton radiation from the equatorial plane should be linearly polarised parallel to the rotation axis. The direction of polarisation will be similar at other latitudes (even along a line of sight intercepting the rotation axis there will be a linearly polarised contribution). Both the optical and the radio observations show that the direction of polarization is in fact fairly constant over the inner part of the Crab Nebula. It is therefore tempting to take this as supporting evidence for the synchro-Compton model. The observed polarization angle would then imply that the rotation axis is aligned in a NW–SE direction. It will be interesting to compare this with the orientation predicted by various pulsar models. Regions close to the rotation axis would contribute *circularly polarized* radiation (see the Appendix) and we would expect a circular polarization of a few per cent at radio, optical and X-ray wavelengths. However the polarization will have opposite senses in the two halves of the nebula, and so the net circular polarisation from the whole nebula may be very low.

Near the edges of the nebula, we may plausibly expect the evacuation to be less effective, and the effects of the medium more important. In this situation, the 30 Hz wave would suffer refraction, and, if the density increases outward, there will be a tendency for it to be deflected tangentially. Since, as is shown in the appendix, the synchro-Compton emission is polarised at right angles to the propagation direction of the 30 Hz wave when the refractive index is significantly less than 1, one may thus be able to account for the observed polarisation vectors normal to the boundary (especially around the bays). The reduced group velocity when $\mu < 1$ increases the wave energy density, and this tends to enhance the synchro-Compton emission from regions where (2) and (3) are only marginally satisfied.

The gross features of the linear polarisation thus support the synchro-Compton interpretation of the radiation. When one recalls the difficulties of accounting for a large scale ordered magnetic field in the Crab Nebula*, the attractiveness of a theory which removes the need for such a field altogether becomes even greater. Perhaps the most crucial test of the general scheme would be the detection of circular polarisation (of both senses) from regions in the nebula. This would be inexplicable in a standard synchrotron picture. On the other hand, the fact that circular polarisation is less easily smeared out than linear polarisation means that the *absence* of circular polarisation at, say, the one per cent level would pose a severe problem for the attractive and widely-held view that much of the pulsar's energy is radiated at the rotation frequency (or low harmonics thereof). Further high-resolution observations of the continuum will obviously help to test the model further. Also, the details of the process whereby the 30 Hz wave is absorbed at the edge of the nebula deserve greater theoretical study, with a view to determining the likely energy spectrum of the resulting relativistic electrons.

Acknowledgements

Valuable discussions with Drs J. P. Ostriker, J. E. Gunn, F. C. Michel and V. L. Trimble are gratefully acknowledged.

Appendix

We summarize here some useful results pertaining to the motion of charged particles in low frequency electromagnetic waves and the properties of the resulting radiation. Although detailed derivations will not be given, and some results merely quoted, anyone familiar with the usual theory of synchrotron and inverse Compton emission should be able to confirm them without difficulty.

We consider particles with various energies (characterised by the Lorentz factor γ) in an electromagnetic wave of frequency ω propagating in the direction $\mathbf{k}$. As in the text, we define Ω as the gyrofrequency in a magnetic field with the same energy density as the wave. The nature of the orbits, and of the emitted radiation, depends on the parameter $f = \Omega/\omega$. First we discuss the behaviour of a test particle in a wave propagating through a vacuum, and then (in (II)) the modifications arising from the presence of a 'cold' plasma which causes the refractive index at frequency ω to depart appreciably from unity.

---

* If, as was suggested first by Piddington (1957), the field were amplified as a result of being tightly wound, the scale of the field reversals would be so low that (a) the associated current densities would be too high to be carried by the available particles; and (b) the emission mechanism would not be standard synchrotron radiation, because of the rapidly-reversing field.

I. TEST PARTICLE

(a) When $f \ll 1$ we have ordinary Compton (or inverse Compton) scattering. A particle released from rest into the wave oscillates *non*-relativistically ($v/c \lesssim f$) in a plane perpendicular to the propagation direction $\mathbf{k}$. In a linearly polarised wave the particle oscillates along a line in the direction of the $\mathbf{E}$ vector; in a circularly polarised wave it executes circular motion. Particles scatter the wave in accordance with the standard Thomson formula. Relativistic particles – or indeed any particle for which $v/c \gg f$ – move basically in straight lines, but the wave induces transverse oscillations with wavelength $2\pi c/\omega \, (1 - \mathbf{v} \cdot \mathbf{k}/c)^{-1}$ around the mean path (and in a frame sharing the mean velocity, particles would execute *non*-relativistic transverse oscillations). The scattered radiation due to relativistic particles is beamed in the direction of v, and its spectrum peaks at a frequency $\sim \gamma^2 \omega (1 - \mathbf{v} \cdot \mathbf{k}/c)$. The spectrum of the scattered radiation cuts off sharply above this frequency, but there is a low-frequency tail $\propto v^1$ contributed by photons which, in the moving frame, are scattered almost into the backward direction. This is the standard case of 'inverse' Compton scattering, in which both linear and circular polarisation are largely preserved (Bonometto *et al.*, 1970).

(b) When $f \gtrsim 1$ the wave is strong enough to impart relativistic speeds to a charge released from rest. The $\mathbf{v} \times \mathbf{B}$ term then cannot be neglected, so the motion is not (as in (a) above) restricted to a plane perpendicular to $\mathbf{k}$. A particle would acquire a typical γ of up to $\sim f^2$ (Jory and Trivelpiece, 1958). If the low frequency wave were plane polarised, particles would move in the plane defined by $\mathbf{k}$ and $\mathbf{E}$. The radiation from these particles would typically be at frequencies $\sim \gamma^2 \Omega$, analogously to synchrotron radiation. The low frequency tail of the spectrum would be roughly proportional to $v^{1/3}$ (again as in the synchrotron case) though the exact spectrum would depend on the particular orbit. Even in a strong electromagnetic wave with $f \gg 1$, particles with $\gamma \gg f$ move in wavy lines, the angular excursions from the direction of the mean v being $\sim f/\gamma$. The peak frequency of the radiation emitted by such a particle, however, will be $\sim \gamma^2 \Omega$, and *not* $\sim \gamma^2 \omega$ as in the standard inverse Compton case. The reason for the close analogy with synchrotron emission despite the very different character of the orbits is that the radiation is always beamed in a cone of angle $\sim \gamma^{-1}$, and the particle turns through this angle in a distance small compared with $2\pi c/\omega$, consequently 'seeing' a quasi-static field over the relevant period. Another way to understand this result is to transform to a frame moving with the particle's mean velocity. The Lorentz factor of this frame is $\sim \gamma f^{-1}$, and with respect to it the particle moves *relativistically* with Lorentz factor $\sim f$ (in contrast to the situation for $f \ll 1$, when the mean velocity has Lorentz factor γ and the oscillatory component of the motion is non-relativistic). Below the frequency $\sim \gamma^2 \Omega$ where the spectrum peaks, the slope is $\frac{1}{3}$ over a frequency range $\sim f^3$, but below $\sim f^{-3} \gamma^2 \Omega$ the slope steepens to ~ 1 as for inverse Compton emission.

The polarization of synchro-Compton radiation can be visualised qualitatively if one bears in mind that, when $\gamma \gg 1$, the only relevant electrons are those moving almost directly towards us, and that the $\mathbf{E}$ and $\mathbf{v} \times \mathbf{B}$ contributions to the transverse acceleration are of comparable importance, As regards *linear* polarisation the results closely resemble those of Bonometto *et al.* (1970) for the standard inverse Compton case. If the low frequency wave is linearly polarised, the synchro-Compton emission will be linearly polarised in the direction of the projected E-vector. (In the frame sharing its mean motion, each particle traces out a figure-of-eight orbit lying in a plane). When the low frequency wave is *circularly* polarised, the non-uniform part of the motion of a particle with $\gamma \gg f$ is relativistic circular motion, with Lorentz factor $\sim f$, in a plane perpendicular to v. This motion would give rise to synchrotron-type radiation concentrated in a 'fan' at angles $\pi/2 \pm (\sim f^{-1})$ to $\mathbf{v}$. This radiation would be circularly polarised in opposite senses on the two sides of the plane of the orbit. In the transformation to the rest frame, the factor $(1 - v/c \cos \theta)$ in the Doppler formula favours the emission from the *forward* hemisphere by a factor $\sim (1 + f^{-1})/(1 - f^{-1})$, which leads to a net circular polarisation $\sim f^{-1}$. Therefore synchro-Compton emission by electrons of *all* energies can possess circular polarisation of order f^{-1}. This contrasts with the γ^{-1} dependence found by Legg and Westfold (1968) for synchrotron radiation, which leads to an undetectably small degree of circular polarization in most astronomical contexts.

II. INFLUENCE OF MEDIUM

The above remarks have all referred to a test particle. However if the density of 'cold' particles (by which is meant merely those less energetic than the one under consideration) is such that (3) is satisfied by less than, say, an order of magnitude, then the refractive index would be significantly less than unity. This situation allows an important new effect: namely that the synchro-Compton radiation can be linearly polarised even when the low frequency wave is itself *un*polarised. To understand this effect, consider the situation when $\mathbf{k}$ is at right angles to the line of sight. The only electrons which concern us are, of course, those whose velocities $\mathbf{v}$ are directed almost towards us. The transverse acceleration of these electrons is caused by the components of the E and B field *perpendicular* to the $\mathbf{k}$–$\mathbf{v}$ plane; the E field yields synchro-Compton emission polarised perpendicular to the plane, and the B-field gives radiation polarised *in* the plane. For an unpolarised low frequency wave propagating in a vacuum, these contributions are equal, so the synchro-Compton emission is unpolarised. However the E/B ratio varies as μ^{-1}, when μ is the refractive index. Therefore, when $\mu < 1$, as it is for a plasma, the synchro-Compton emission will be polarised in a direction *perpendicular* to $\mathbf{k}$. The possible degree of polarization is $\sim (E^2 - B^2)/(E^2 + B^2) \simeq (1 - \mu^2)/(1 + \mu^2)$. Thus if (3) is only satisfied by a factor of 10, one might expect ~ 10 per cent linear polarisation perpendicular to the propagation direction of the low-frequency wave.

It is also possible to derive this result by transforming to a frame sharing the mean motion of the particle, as we did in I – one finds that when $\mu \neq 1$ the wave appears polarised in this frame even when it is unpolarised in the rest frame.

References

Bonometto, S., Cazzola, P., and Saggion, A.: 1970, *Astron. Astrophys.* **7**, 292.
Gunn, J. E.: 1970, *Publ. Astron. Soc. Pacific* **82**, 538.
Jory, H. R. and Trivelpiece, A. W.: 1968, *J. Appl. Phys.* **39**, 3053.
Legg, M. P. C. and Westfold, K. C.: 1968, *Astrophys. J.* **154**, 499.
Ostriker, J. P. and Gunn, J. E.: 1969, *Astrophys. J.* **157**, 1395.
Piddington, J.: 1957, *Australian J. Phys.* **10**, 530.
Zheleznyakov, V. V.: 1967, *Soviet Astron.* **11**, 33.

Discussion

P. Stewart: It is not obvious that permittivity has any meaning in intense radiation fields.

M. Rees: I agree that the problem is complicated when the intensity of the wave is high, and it was mainly for that reason that I avoided being too quantitative. However, all that is necessary in order for the 'synchrotron' radiation to be polarised perpendicular to the propagation direction of the 30 Hz radiation is that E/B should exceed its vacuum ratio, and I believe that this will be true even for a *strong* low frequency wave.

J. E. Felten: I did not understand your diagram of the 'bay' – what happens to the polarisation of the radiation scattered from a ray which is travelling outward in a direction normal to the boundary of the bay?

M. Rees: There will indeed be some 30 Hz radiation that is incident precisely normally on the boundary. However, there will be a tendency for most of it to be curved away towards the tangential direction as it moves into a region of increasing density (and increasing refractive index), and a consequent tendency for the observed radiation to be polarised perpendicular to the boundary.

J. A. Roberts: Do you envision the optical line emitting filaments as being outside the volume swept to electron densities $< 10^{-5} \, \gamma \, \mathrm{cm}^{-3}$.

M. Rees: The 30 Hz radiation would be absorbed on the inner edge of filaments, and would not propagate through them. If the absorbed energy goes into relativistic electrons, this may be relevant to the enhanced radio emission from the region of some filaments which was reported by Mr. Wilson on Wednesday.

Note added in proof. In response to the suggestion made in this paper, Landstreet and Angel (*Nature* **230**, 103 (1971)) searched optically for circular polarization from the Crab Nebula. Their reported upper limit of $\sim 0.05\%$ (which refers to two regions along the major axis of the nebula) is hard to reconcile with the predictions of the synchro-Compton model. It may imply that the 30 Hz waves from NP 0531 do not penetrate beyond the wisps in the nebula.

7.5 MASER THEORY OF PULSAR RADIATION

HONG-YEE CHIU

Institute for Space Studies, Goddard Space Flight Center, NASA, New York, N.Y., U.S.A.

Abstract. In this paper we present an account of a theory of pulsar radio emission. The emission mechanism is via a maser amplification process. This theory avoids the difficulty of coherent plasma emission, that the bandwidth of radiation must be less than $\frac{1}{2} \lambda$. The high brightness radio temperature and the insensitivity of pulsar radio flux to pulsar periods can be easily accounted for.

1. Introduction

In 1968 Hewish and Miss Bell discovered the existence of pulsating celestial radio sources (Hewish *et al.*, 1968). These radio sources emit bursts of radio emission in the 100–1000 MHz frequency region with a $v^{-\alpha}$ type spectrum ($\alpha \sim 2$) at regular intervals (periods) ranging from a fraction of a second to a few seconds. Subsequent work by other workers resulted in the discovery of more pulsars (Maran and Cameron, 1969).

The period lies between $\frac{1}{30}$ sec to 3.7 sec. In cases where accurate determination of the period can be made, it is found that the period P lengthens at a rate $dP/dt = 10^{-12}$ to 10^{-16}, and in no case does the period shorten regularly. The lengthening of the period shows that the time keeping mechanism is rotation and not oscillation (which is the cause for variability of some stars), for in the case of rotation the period lengthens as a result of dissipation of rotational energy and in the case of oscillation, a linearized theory will give rise to a constant period independent of the amplitude and when nonlinear effects are considered, the period decreases as the amplitude is damped, such as the case of a pendulum.

If the basic time keeping mechanism is rotation then the value of the period gives an upper limit to the size of the pulsars. The centrifugal force per unit mass at the equator of a rotating star is $\omega^2 R$ and the gravitational force is GM/R^2 when R is the radius, ω the rotational angular velocity, M the mass of the star. In order that the rotation be stable, it is necessary that $\omega^2 R < (GM/R^2)$, which gives $R < 100$ km for the case of the Pulsar NP 0532 which has the shortest period of $\frac{1}{30}$ sec.

Individual pulses are detectable in the Arecibo 1000 ft radio telescope. It is found that the signal strength and structure changes from pulse to pulse, and in the case of NP 0532 microtime structure with a time constant as short as 100 μsec has been detected. If the dimension of the emitting region is R, then the time structure of the signal emitted cannot have a time constant shorter than the light transit time t_c which is R/c. For NP 0532, $t_c \sim 100$ μsec and this means that $R < 30$ km (Drake and Craft, 1968).

No known astronomical objects other than asteroids are known to have a physical dimension less than 1000 km (the white dwarf has a minimum theoretical dimension of 10^8 cm $= 10^3$ km) except the neutron star, which was predicted by Landau in 1932 soon after the discovery of the neutron. At high density the Fermi energy of the electrons will be high enough so that the following reaction will proceed favorably

$$e^- + p \rightarrow n + v \tag{1}$$

where p and n may be free or bound to a nucleus. The inverse reaction is forbidden because the available electron states are occupied. As (1) proceeds the elements become richer and richer in neutrons. At a density of 10^{12} g/cm and above, few nuclei can exist and all but a few protons are converted into neutrons. A small fraction of matter remains in the proton and electron state so that the inverse of (1) is prevented. Stable neutron stars have densities between 10^{14} to 10^{16} g/cm^3; the upper limit is dependent on the detailed structure of nuclear interaction, which is poorly known at high densities. The radius of a neutron star is of the order of 10 km.

The neutron stars are results of stellar collapse which also gives rise to the supernova phenomenon. A pulsar (with the shortest period of $\frac{1}{30}$ sec) is found at the center of the Crab Nebula, a well known remnant of a supernova which flared up in 1054 A.D. Many theories predict that the star remnants of supernovae are neutron stars, hence, in all probability, pulsars are neutron stars. This pulsar NP 0532 is the youngest one on record and is the only one which has been found to emit not only in the radio frequency but also in the optical and X-ray spectrum ($10^{13} \rightarrow 10^{18}$ Hz). The division of energy flux in the radio, optical, and X-ray regions are roughly: 10^{31} erg sec^{-1}, 10^{34} erg sec^{-1}, and 10^{37} erg sec^{-1} respectively.

The neutron stars are themselves objects of immense interest. First, during the formation of a neutron star, a normal star suffers a change of radius from 10^{11} to 10^6 m, representing a compression ratio of 10^5. The angular momentum conservation law $I\omega = MR^2\omega =$ constant, requires that the period of rotation be increased as R^{-2}. The rotation period of a normal star such as the sun is 26 days, and after collapse a period of rotation as short as 1 msec may be achieved. (The centrifugal force at the equator gives a limiting rotation period of about 1 msec.) The rotational energy of a neutron star U_R, is roughly:

$$U_R \approx \tfrac{1}{2} I\omega^2 \approx \tfrac{1}{2} MR^2\omega^2 \approx 7 \times 10^{46} \, (M/\odot) \, (R/10 \text{ km})^2 \, P^{-2} \text{ ergs} \tag{2}$$

where

$$\odot = \text{solar mass} = 2 \times 10^{33} \text{ g}.$$

Many stars possess magnetic fields of the order of 10^3 G or greater. During compression, the magnetic flux lines of a plasma are conserved and increase as R^{-2}. If the loss of flux lines is small during formation, the magnetic field of a neutron star may be as high as 10^{13} G.

Finally, another piece of observational evidence supports the rotational neutron star hypothesis of pulsars. The nuclei in the surface layer of a neutron star can form crystalline-like lattices (resulting in a smaller Coulomb energy) and the melting temperature of these crystalline structures is of the order of 10^9 K or more for densities between $10^7 \rightarrow 10^{13}$ g/cm^3. The temperature of neutron stars is usually quite low

($\ll 10^9$ K), hence, the outer layer of neutron stars is made of crystalline solid. As the rotation of the star slows down, the equilibrium configuration is changed and strains are present in the crystalline layer. When the strain exceeds the stress limit of the crystalline layer, sudden resettlement (starquake) will take place. This will cause a sudden change in the moment of inertia of the star, resulting in a change of period as well as a change in the phase of the pulse called a glitch. Glitches have been observed in two cases: the case of Vela pulsar and the case of the Crab pulsar. The glitch phenomenon, though easily explained in the rotating neutron star model, is virtually impossible to explain in oscillating models.

2. The Magnetic Field of the Neutron Star

A field as strong as 10^{12} G is hard to contemplate, but there is good evidence that such fields exist. As Pacini (1968) and Ostriker and Gunn (1969) suggested, a rotating neutron star with a magnetic field can emit electromagnetic radiation of frequency the same as that of the rotation. The energy of emission comes from the energy of rotation U_R. The rate of radiation dU/dt in the case of a rotating dipole is:

$$\frac{dU}{dt} = -\frac{2\bar{M}\sin^2\theta\omega^4}{3c^2} \tag{3}$$

where $\bar{M}$ is the magnetic dipole moment, which can be approximated by

$$\bar{M} \approx \frac{\phi^2 R^2}{4\pi^2} = \frac{B_p^2 R^6}{4} \tag{4}$$

where ϕ is the flux line through the pole ($\phi \sim \beta_p R^2/\pi^2$ where β_p is the average field at the pole), θ is the angle of inclination of the dipole with respect to the rotational axis.

Equating the rate of loss of rotation of energy to dU/dt, it is found that

$$\frac{d\Omega}{dt} = -\frac{\phi^2\sin^2\theta R^2\omega^3}{6\pi^2 Ic^2} = 2\pi\frac{dP}{dt} \tag{5}$$

$$\Omega = \Omega_0 (1 + 2t/\tau_0)^{-1/2}.$$

For the Crab Nebula,

$$\omega \sim 100 \text{ sec}^{-1}, \qquad \theta \approx \pi/4, \qquad R \approx 10 \text{ km}, \qquad B_p \approx 10^{12} \text{ G} \tag{6}$$

This gives

$$\frac{d^2P}{dt^2} \approx 1.3 \times 10^{-24} \text{ sec}^{-1}$$

which agrees with observations.

The frequency of the emitted dipole radiation is the same as the pulsar rotation frequency. Such a radiation field is similar to the field in accelerators, and can accelerate particles to high energy. This explains the puzzling fact that the Crab nebula appears to be still active in producing high energy particles (which are needed to account for radio, optical, and X-ray emission from the nebula). Without the classical dipole radiation it will be very hard if not impossible to explain the rather large rate of loss of rotational energy of the pulsar. Analysis of other pulsars whose values of dP/dt are known all yield a field of the order of 10^{12} G. It thus seems well established that neutron stars may have fields as strong as 10^{12} G.

Goldreich and Julian (1969) pointed out that in the rest frame of the rotating magnetic neutron star there is a component of the electric field parallel to the magnetic H such that the invariant $\mathbf{E} \cdot \mathbf{H} \neq 0$. The field strength at the surface is

$$E \sim H \frac{v}{c} \sim 10^{10} \, (H/10^{13} \text{ G}) \, (v_r/10 \text{ km/sec}) \quad \text{volt/cm} \tag{7}$$

where v_r is the rotational velocity at the magnetic pole. This field will accelerate electrons and ions to stream out of the surface. As will be seen below, this electric field is also important to account for the radio (and optical and X-ray) emission from pulsars.

The properties of an electron gas in an intense magnetic field have been extensively studied (Chiu, 1970). First, the motion of electrons in a strong field is quantized in the direction perpendicular to the field and the usual expression of the kinetic energy of an electron is replaced by the equation

$$\varepsilon = mc^2 \left[1 + (p_z/mc)^2 + 2n \, (H/H_q) \right]^{1/2} - mc^2 \tag{8}$$

where the field is taken to be in the z direction, n is the quantum number characterizing the size of the classical orbit, $H_q = m^2 c^3/e\hbar = 4.4 \times 10^{13}$ G. Therefore, the momentum in the direction perpendicular to the field, $p_\perp$, is quantized by the relation:

$$p_\perp^2/m^2 c^2 \to 2n \, (H/H_q). \tag{9}$$

In the nonrelativistic limit we have

$$\varepsilon = p_z^2/2m + n \, (H/H_q) \, mc^2. \tag{10}$$

The value of $(H/H_q) mc^2$ is 11.6 keV if $H = 10^{12}$ G. If the electron kinetic energy is less than $(H/H_q) mc^2$, the only allowable state for the electron is the state $n=0$, and then

$$\varepsilon = p_z^2/2m \tag{11}$$

which is the expression for the energy of a one-dimensional particle. The density of state for a one-dimensional electron is

$$\varrho_1 \, (\varepsilon) = \frac{2 \, dp}{h \, d\varepsilon} = \frac{2m}{h} \, p_z^{-1}. \tag{12}$$

For an electron in the absence of field

$$\varrho_3 \, (\varepsilon) = \frac{2}{h^3} \frac{d^3 p}{d\varepsilon} = \frac{2m}{h^3} \, p \int d\Omega \tag{13}$$

where $d\Omega$ is the solid angle element in the direction of the emerging particle. Therefore for a one-dimensional particle $\varrho_1(\varepsilon) \to \infty$ as $p \to 0$, and for a three-dimensional particle $\varrho_3(\varepsilon) \to 0$ as $p \to 0$.

This one-dimensional behavior has been observed in low temperature experiments and this gives rise to the so-called de Hass–Van Alphen effect. As a result of Equation (9) there are only two radiation processes for 'free' electrons in a magnetic field: (a) the spontaneous transition of an electron from a state n to $n' \neq n$; (b) the Coulomb de-excitation of an electron from a state n to n' (no restriction on n' but the energy must be conserved). (a) is the quantized version of the classical synchrotron radiation, the emission gives rise to lines and the minimum energy of emission is $H/H_q mc^2$ (nonrelativistic case) or $(H/H_q) mc^2 \cdot mc^2/\varepsilon$ (relativistic case). The life time of electrons is generally very short; it is

$$\tau_s = \left[\frac{2}{3} \frac{e^2}{\hbar c} \frac{mc^2}{\hbar} \frac{\varepsilon}{mc^2} \left(\frac{H}{H_q} \right)^2 \right]^{-1}. \tag{14}$$

For $H \sim 10^{13}$ G, $\varepsilon \sim 1$ MeV, $\tau_s \ll 10^{-15}$ sec.

The bremsstrahlung radiation has been calculated by Goldman and Oster (1964), but their result is very difficult to apply. We have recently calculated the nonrelativistic bremsstrahlung radiation rate in a magnetic field (Canuto and Chiu, 1970). In this calculation the electron wave function in a magnetic field in conjunction with a Green's function appropriate for the field is used. This calculation is therefore expected to be valid for nonrelativistic electrons with energies up to, say, one or two hundred keV, in fields up to a fraction of $H_q = 4.414 \times 10^{13}$ G, for frequencies not in the vicinity of the plasma frequency of the Larmor frequency of the electrons or ions.

The expression for the transition probability for the transition at forward angle is (Canuto and Chiu, 1970)

$$W = W_0 C_1(\lambda)(p\omega)^{-1} \quad (\text{cm}^3 \text{ sec}^{-1}) \tag{15}$$

where $(\varepsilon = E/mc^2)$

$$W_0 = Z^2 \alpha^3 \pi^2 N_i \lambdabar_c^3 \hbar c^2 (H/H_q)^{-2} = 2.09 \times 10^{-37} Z^2 N_i (H/H_q)^{-2}$$
$$(\text{erg cm}^2 \text{ sec}^{-1})$$

N_i is the ion density and Z the average atomic number, and

$$C_1(\lambda) = e^\lambda (1 + \lambda) E(\lambda) - 1, \quad \lambda = (\lambdabar_c^2/2\omega_H) \left[\left(\frac{\omega}{v} \right)^2 + d^{-2} \right] \tag{16}$$

$$E(\lambda) = \int_\lambda^\infty x^{-1} \exp(-x) \, dx \xrightarrow{\lambda \to 0} \ln(\gamma\lambda)^{-1} \quad \gamma = 1.781\,02\ldots \tag{17}$$

where ω_H is the Larmor frequency, v the electron velocity, α is the Debye length. This term d^2 arises from electron screening. (We assume the validity of classical electron screening theory.) In the regime of interest $\lambda \ll 1$ we can write

$$W = W_0 (p\omega)^{-1} \ln(\gamma\lambda)^{-1}. \tag{18}$$

The effect of a dense plasma is to alter the relation between the photon frequency ω and the wave vector $\mathbf{k}$ of the photon.

The dielectric constant is the same for a classical as well as a quantum electron gas, provided $\omega \ll \omega_H$ where ω_H is the Larmor frequency. The dielectric constant in a magnetic field is a tensor quantity. In the direction along the magnetic field the radiation can be split circularly into two circular polarization components, the ordinary component (denoted by a subscript o) and the extraordinary component (denoted by a subscript x):

$$n_x = 1 - \frac{\omega_p^2}{\omega^2} \frac{\omega}{\omega - \omega_H} - \frac{\omega_I^2}{\omega^2} \frac{\omega}{\omega + \Omega_H} \quad \text{(right hand polarization)}$$

$$n_o = 1 - \frac{\omega_p^2}{\omega^2} \frac{\omega}{\omega + \omega_H} - \frac{\omega_I^2}{\omega^2} \frac{\omega}{\omega - \Omega_H} \quad \text{(left hand polarization)}. \tag{19}$$

ω_p and ω_I are the plasma frequencies of the electron gas and ion respectively, ω_H and Ω_H the Larmor frequency for electrons and ions. For $H = 10^{12}$ G, $\omega_H = 10^{19}$ and $\Omega_H = 10^{17}$, $\omega_p = 10^{16}\sqrt{\varrho}$, $\omega_I \approx 10^{14}\sqrt{\varrho}$, where ϱ is the density. Hence, for $\varrho \lesssim 10^6$ g/cm^3, n_x and n_o are close to unity for $\omega \sim 10^{10}$.

3. Radiation Mechanisms for Pulsars (Chiu and Canuto, 1970)

From the upper limit of the radiating region of pulsars ($< 3 \times 10^6$ cm) and the energy radio flux of emission (10^{31} ergs/sec) the flux of radiation I_v at the emitting region is about 10^{18} ergs/cm^2 sec. The frequency of emission is at $\sim 10^3$ MHz. Assuming a bandwidth $\Delta f = f$, we find that the temperature T_B of a black body with an equivalent emissivity in the same frequency regime is given by:

$$I_\omega \sim 2kT_B \left(\frac{\omega}{2\pi}\right)^2 c^{-2} \tag{20}$$

or:

$$T_B = \frac{I_\omega c^2}{2k(\omega/2\pi)^2}.$$

Therefore, for pulsars the brightness temperature is about 10^{24} K. If this radiation is emitted by random processes then T_B is the minimum temperature of the emitter. The unduly large value of T_B requires that the process of emission cannot be an incoherent process.

Several laboratory devices can give rise to large brightness temperature in a relatively cool medium. These are lasers or Klystrons. In a Klystron an energetic electron beam bunched in space mingles with an electromagnetic wave with such a phase relationship that the electrons are decelerated by the electric field of the electromagnetic wave, and thereby the electrons give their kinetic energy to the electromagnetic wave. In this process the electrons must have such a spatial distribution that they are always decelerated, otherwise the electromagnetic wave will give energy to the electrons. For this reason, only electromagnetic waves of wavelengths within a bandwidth $\frac{1}{2}\lambda$ are amplified.

This emission mechanism is also referred to as coherent plasma emission and has been considered as a possible emission mechanism for pulsars. There are a number of reasons why this mechanism may not work. First, the coherence of a plasma extends only over a spatial region of $\frac{1}{2}\lambda$; this also implies that the bandwidth of emission is of the order of $\frac{1}{2}\lambda$. Observation shows that the bandwidth extends from 50–1000 MHz, much beyond this bandwidth. Second, the intensity of radiation depends critically on the plasma density. Since the density of plasma surrounding the star depends on the structure of the magnetosphere which depends critically on the period P, it is expected that the intensity of radiation should also depend on the period of rotation. However, observation shows that there is no such correlation. Third, observation shows that the optical and X-ray emission come from the same emission region. Therefore, the same plasma must be also responsible for optical and X-ray emission.

It is not possible to explain the (so far) unique case of optical and X-ray emission of NP 0532 except by imposing very artificial conditions. The other mechanism that we shall discuss here avoids all these difficulties. In the case of a laser, emission takes place between two atomic states whose energies are $\varepsilon_1, \varepsilon_2$ ($\varepsilon_1 < \varepsilon_2$). Coherence is achieved via a stimulated emission process, and the condition for coherent stimulated emission is such that the system has a negative temperature. Let the population of these two states be N_2 and N_1. Then an excitation temperature T_e may be defined by the equation

$$\frac{N_2}{N_1} = e^{-(\varepsilon_2 - \varepsilon_1)/kT_e}. \tag{21}$$

The condition for $T_e < 0$ is that $N_2 > N_1$. This case is referred to as population inversion.

The mass absorption coefficient $K(\omega)$ with stimulated emission is given by

$$K(\omega) = N_1 \frac{\sigma c}{\varrho} (1 - N_2/N_1) \tag{22}$$

where σ is the absorption cross-section, ϱ is the density. $K(\omega)$ therefore becomes negative when $N_2 > N_1$. The radiative transfer equation

$$\frac{dI}{ds} = - K(\omega)\varrho I \tag{23}$$

will have an exponentially growing solution

$$I_\omega = I_0 \exp \int [- K(\omega)]\varrho \, ds. \tag{24}$$

Radiation thus can be amplified. However, a laser operated between two atomic states again only gives rise to line emission. In order that a broad band emission be produced, it is necessary that the negative absorption coefficient extends over a finite frequency range. To see how this condition can be fulfilled, we can rewrite Equation (21) as follows:

$$\frac{N_2 - N_1}{N_1} = e^{-(\varepsilon_2 - \varepsilon_1)/kT_e} - 1.$$ (25)

Let us consider a continuum and let $N_2 \rightarrow N_1$. Then we can write $N_2 - N_1 = dN$, $\varepsilon_2 - \varepsilon_1 = d\varepsilon$, and expand the exponential. We find

$$\frac{dN}{N} = -\frac{d\varepsilon}{kT_e}$$ (26)

or:

$$\frac{dN}{d\varepsilon} = -\frac{N}{kT_e}.$$ (27)

The requirement for a negative temperature is then $dN/dE > 0$. (Note: N refers to the occupation number per state and not the number of particles per unit energy range.) However, since the condition $dN/dE > 0$ cannot be maintained throughout the entire energy range (otherwise the total number of particles will necessarily be infinite) somewhere dN/dE will become less than zero. The condition for a negative absorption coefficient is more complicated than that for discrete states. The expression for the absorption coefficient $K(\omega)$ for an electron gas in the lowest Landau level is

$$K(\omega) = \frac{2m\hbar\omega}{\varrho} \int_0^\infty gh^{-1} \, dp \left[f(p) \frac{\partial}{\partial p} \frac{\mathcal{W}(p)}{p + p'} + f(-p) \frac{\partial}{\partial p} \frac{\mathcal{W}(p)}{p - p'} \right]$$ (28)

where $\mathcal{W}(p)$ is the transition probability for the absorption of a photon of energy ω by an electron of momentum p, p' is the electron momentum after absorption of a photon, $f(p)$ is the electron distribution function, g is a statistical weight factor of the electron (g is a function of the magnetic field) ϱ is the density. $\mathcal{W}(p)$ is related to $W(p)$ [Equation (15)] via the detailed balance theorem which for a one-dimensional particle is simply

$$g\mathcal{W}(p) \, dp = gW(p) \, dp', \qquad dp'/dp = p/p'.$$ (29)

Amplification of radiation (maser effect) is possible if $K(\omega) < 0$. The condition for this is not straightforward, as is clear from the form of Equation (28). However, if the electron distribution function is asymmetrical, e.g., $f(p) \gg f(-p)$ (this corresponds to a population inversion in terms of a coherent streaming motion in the $+p$ direction with respect to the medium) then the condition for amplification of radiation is just:

$$\frac{\partial}{\partial p} \frac{\mathcal{W}(p)}{p + p'} < 0$$ (30)

or $\mathscr{W}(p)$ must not rise faster than the first power of p. As is clear from the form of W for the $0\rightarrow0$ transition, Equation (15), this condition is completely fulfilled. Hence, it is possible to have a negative emission coefficient.

4. A Simple Model (Chiu and Canuto, 1970)

We choose $f(p)$ a delta function at p_m $[f(p)=A\delta(p-p_m)]$ where A is a normalization constant. ($A=g^{-1}N_e$ where N_e is the electron density.) We then have

$$\varrho K(\omega) = -2m\hbar c^{-1}N_e W_0 p_m^{-3}\ln(\gamma\lambda)^{-1}. \tag{31}$$

The absorption coefficient is therefore negative and amplification of radiation takes place. The amplification is frequency independent, and covers the entire range of radio spectrum.

The radiation transfer equation is:

$$\frac{dI(\omega, s)}{ds} = -I(\omega, s)\varrho K(\omega) \tag{32}$$

whereas the spontaneous emission term is neglected. Equation (32) then gives

$$I(\omega, s) = I(\omega, 0)\exp\left\{\int_0^s p_m^{-3}\delta\ln[\gamma\lambda]\right\} \tag{33}$$

$$\delta = 2m\hbar c^{-1}N_e W_0$$

and s is the path length over which amplification takes place.

Before we proceed further, we will discuss a very important effect in the theory of light and microwave amplification by stimulated emission. This is the effect of saturation of amplification. As it turns out, the amplification process is so efficient that the saturation phenomena takes place in all cases. The saturation phenomenon gives rise to non-linear effects and in effect places the *average* brightness temperature within a range compatible with observations.

If a small signal is fed into an amplifier (such as an audio amplifier), the output power is proportional to the input signal. However, when the output power reaches the inherent power limit of the amplifier, further increase in the input signal level will not increase the output power. When this happens, the amplifier no longer amplifies the input signal linearly, and this amplifier is said to have reached saturation.

The same saturation effect also exists in amplification by stimulated emission.* Let us consider a two-level system. The reasoning can be easily extended to continuum energy level systems. Let there be n_2 particles in the upper energy state 2 and let n_1 be the number of particles in the lower state 1. In the absence of stimulated radiation, the lifetime of the state 2 is τ_2. Let $N=n_2-n_1$ be the difference between particles in the upper and in the lower states, and population inversion is achieved when $N>0$.

---

* This is very extensively discussed in literature; see, for example, Troup (1963).

Let N_0 be the steady state value of N. (N_0 depends on the pumping rate and other properties of the system.) The rate equation for N in the absence of stimulated radiation is:

$$\frac{dN}{dt} = \frac{N_0 - N}{\tau_2}. \tag{34}$$

Now let us introduce stimulated radiation and let the transition probability via stimulated emission be P_{21}. The rate of change of N due to stimulated emission is simply $-2NP_{21}$. (The factor 2 arises from the fact that each transition from the state 2 to the state 1 changes the value of N by 2). The rate Equation (34) now becomes

$$\frac{dN}{dt} = \frac{N_0 - N}{\tau_2} - 2NP_{21} \tag{35}$$

A steady state obtains when N_s satisfies

$$\frac{N_0 - N_s}{\tau_2} - 2N_sP_{21} = 0 \tag{36}$$

or when

$$\frac{N_s}{N_0} = [1 + 2P_{21}\tau_2]^{-1}. \tag{37}$$

P_{21} is proportional to the intensity of stimulated radiation, $I(\omega, s)$. When $I(\omega, s)$ becomes large such that $2P_{21}\tau_2 \sim 1$, the value of N_s begins to decrease appreciably from the original value N_0 and this laser amplifier no longer amplifies exponentially; the 'gain' decreases with increased output. Further increase in $I(\omega, s)$ will only cause a further decrease in the gain and, finally, as N_s approaches zero, the absorption coefficient being proportional to N, also becomes zero and the medium becomes transparent to radiation.

We can obtain the saturated intensity from the condition for the onset of saturation:

$$2P_{21}\tau_2 \approx 1. \tag{38}$$

P_{21} is the transition probability for emission per photon state and is easily seen from the Boltzmann equation for the photon distribution function to be just

$$P_{21} = F(\omega)\varrho K(\omega) \tag{39}$$

where $F(\omega)$ is the number of photons per state.

τ_2^{-1} is the transition probability for spontaneous emission, or collision, whichever is greater. The collision rate is about 10^{10} per sec. The transition probability for spontaneous emission is much greater. The expression for τ_2^{-1} for spontaneous emission is

$$\tau_2^{-1} = N_eW(p) = 2.09 \times 10^{-37} Z^2N_eN_i(p\omega)^{-1}(H/H_q)^{-2}C_1(\lambda). \tag{40}$$

Numerically, if we set $N_e = 10^{24}$ cm^{-3}, $N_i = N_e/26$ (ion composition at the surface),

$p=p_m=0.01\ mc$ (see Section 8), $H/H_q=0.02\ (H\sim10^{12}\ \mathrm{G})\ \omega\approx10^9$ Hz, we find

$$\tau_2^{-1}\approx10^{27}.\qquad(41)$$

The value of $\varrho\,K(\omega)$ which will give rise to saturation is about 40. Hence, from Equation (39) we obtain the lower limit for the saturated value of $F(\omega)$:

$$F(\omega)\sim2\times10^{25}.\qquad(42)$$

The brightness temperature T_B is: (cf. Equation (20))

$$F(\omega)=\frac{kT_B}{\hbar\omega}\cong10^2T_B\quad(\omega\sim10^9).\qquad(43)$$

Comparing Equation (42) with Equation (43), we find that at saturation the lower limit of the brightness temperature is about 10^{23} K. However, this value of T_B is an underestimate, for Equation (39) gives the condition for the onset of saturation, and when saturation fully takes place, due to non-linear effects, the value of T_B may be several orders of magnitude greater than its value at the onset of saturation. The intensity of pulsar radiation is therefore independent on the period of pulsars, being a function of other parameters (such as the magnetic field, etc.).

5. Polarization of Radiation

In previous papers (Chiu and Canuto, 1969c; Chiu, Canuto and Fassio-Canuto, 1969; Chiu and Canuto, 1969d; Chiu, 1969) we have suggested that the propagation of radiation is strongly affected by the magnetic field resulting in a beamed emission. In this section we will discuss the dielectric properties of a plasma in a magnetic field.

As the frequency of radiation is small compared with the Larmor frequency of the ions and the electrons, as well as the plasma frequency of the electrons and the ions, we can use the dielectric constant computed for a cold ionic neutral plasma in a magnetic field. It turns out that quantum mechanical calculations (Kelly, 1964) and classical calculations coincide in this limit:

The dielectric tensor is (assuming that the magnetic field is in the z-direction) (Stix, 1969):

$$\varepsilon_{\alpha\beta}=\begin{pmatrix}S & -iD & 0\\ iD & S & 0\\ 0 & 0 & P\end{pmatrix}\qquad(44)$$

where

$$R=1-\frac{\omega_p^2}{\omega^2}\frac{\omega}{\omega-\omega_H}-\frac{\omega_I^2}{\omega^2}\frac{\omega}{\omega+\Omega_H}\qquad(45)$$

$$L=1-\frac{\omega_p^2}{\omega^2}\frac{\omega}{\omega+\omega_H}-\frac{\omega_I^2}{\omega^2}\frac{\omega}{\omega-\Omega_H}$$

$$P=1-\left(\frac{\omega_p}{\omega}\right)^2-\left(\frac{\omega_I}{\omega}\right)^2,\qquad 2D=R-L,\qquad 2S=R+L\qquad(46)$$

$$\omega_p^2 = 4\pi e^2 N_e/m \qquad \omega_I^2 = 4\pi e^2 Z N_z/m$$
$$\omega_H = eH/mc \qquad \Omega_H = Z\omega_H m/M_i \qquad (47)$$

ω_p and ω_I are plasma frequencies of the electron gas and the ion gas, respectively, and ω_H and Ω_H the corresponding Larmor frequencies.

Generally speaking, the photon propagation in a doubly refractive medium such as Equation (44) can be analyzed into two modes, the ordinary and the extraordinary mode of propagation. The propagation of these two modes can be easily studied for the direction $\theta = 0$ and $\theta = \frac{1}{2}\pi$, and can be studied numerically at other angles. Here we will describe the cases $\theta = 0$ and $\theta = \frac{1}{2}\pi$, respectively.

A. ALONG THE MAGNETIC FIELD $\theta = 0$

In this direction these two modes are right-handed (R) and left-handed (L) circularly polarized, and the refractive indices for these two modes are: $(o = \text{ordinary}, x = \text{extra-ordinary})$

$$n_x^2 = R \qquad (48)$$
$$n_o^2 = L. \qquad (49)$$

The emissivity of the x- and the o-modes in the $0 \to 0$ bremsstrahlung transition is:

$$\text{Ordinary mode: } W^{(o)} = L^{1/2} W_0 (p\omega)^{-1} C_1 (\lambda) E_1^{-2} \qquad (50)$$
$$\text{Extraordinary mode: } W^{(x)} = R^{1/2} W_0 (p\omega)^{-1} C_1 (\lambda) E_2^{-2} \qquad (51)$$

where

$$E_1^{-1} = (H/H_q) [\hbar\omega/mc^2 - H/H_q - \tfrac{1}{2}L (\hbar\omega/mc^2) \\ - L^{1/2} (\hbar\omega/mc^2) (p/mc)]^{-1} \qquad (52)$$

$$E_2^{-1} = (H/H_q) [\hbar\omega/mc^2 + H/H_q + \tfrac{1}{2}R (\hbar\omega/mc^2) \\ - R^{1/2} (\hbar\omega/mc^2) (p/mc)]^{-1}. \qquad (53)$$

Now we will substitute into these equations quantities pertinent to a neutron star: $H = 10^{12}$ G, $N_e = 10^{24}$. We find $\omega_p \simeq 10^{16}$, $\omega_I \simeq 10^{14}$, $\omega_H \simeq 10^{19}$, $\Omega_H \simeq 10^{16}$. If we choose $\omega = 10^9$ Hz, we find that $E_1^{-1} = E_2^{-1} = 1$. For R and L we can expand equations (45) and (46) into power series in ω/ω_H and ω/Ω_H. For a charge neutral plasma, for which Equation (47) is valid, we find that the first nonvanishing term in the expansion gives:

$$R = 1 + \frac{\omega_p^2}{\omega_H^2} + \frac{\omega_I^2}{\Omega_H^2} \simeq 1 \qquad (54)$$

$$L = 1 - \frac{\omega_p^2}{\omega_H^2} - \frac{\omega_I^2}{\Omega_H^2} \simeq 1 \qquad (55)$$

and hence

$$W^{(o)} (p) = W^{(x)} (p) = W_0 (p\omega)^{-1} C_1 (\lambda) \qquad (56)$$

Therefore these two modes have the same emissivity as that in vacuum.

We therefore conclude that the radiation emerging from the dense plasma of the star along the magnetic field has both right- and left-hand polarization and has the same intensity. It is not known whether these two modes can combine into a linearly polarized beam or these two modes will propagate separately to yield an unpolarized beam.

B. PROPAGATION PERPENDICULAR TO THE FIELD, $\theta = \frac{1}{2}\pi$.

In the perpendicular direction there is only one mode of propagation; this is the extraordinary mode. The refractive index for the ordinary mode is simply $1 < (\omega_I/\omega)^2 + (\omega_p/\omega)^2$, and propagation of this mode is forbidden. For the extraordinary mode the refractive index is:

$$n_x^2 = RL/S \approx 1 \tag{57}$$

and propagation is the same as in vacuum. This mode has one hundred percent linear polarization whose direction is perpendicular to both the wave vector **k** and the magnetic field H. There is another mode of emission whose electric field is along the magnetic field. Our computation shows that this mode is important in vacuum. However, since this mode cannot propagate in a dense plasma, it is not considered here.

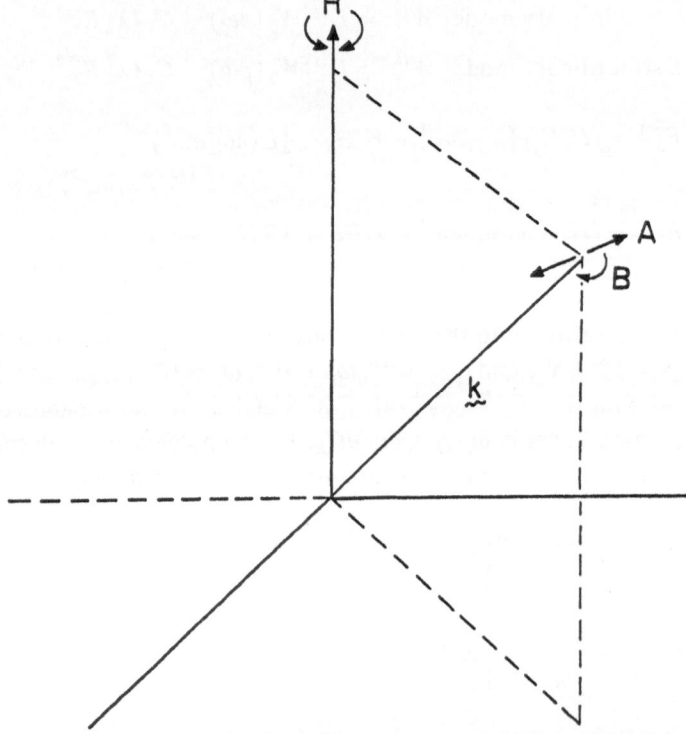

Fig. 1.   Polarization of the bremsstrahlung radiation in a strongly magnetized medium. **k** = propagation vector, H = magnetic field. A is the linearly polarized component perpendicular to the field; B is the circularly polarized component of the extraordinary component.

C. ALONG AN ARBITRARY DIRECTION

As it turns out the refractive index for the extraordinary mode is always close to unity for the field strength and plasma frequency we have quoted. The emergent radiation has an elliptical polarization which can be decomposed into a circular polarization plus a linear polarization. The polarization of the radiation from bremsstrahlung process that can emerge from the medium can either be circularly polarized, or linearly polarized, depending on the emissivity associated with these two polarization states. However, from simple arguments it seems that linear polarization is favored over circular polarization.

Figure 1 shows the directions of polarization of the emerging radiation. It is clear that if we look into the direction of the magnetic field, the radiation will appear unpolarized, assuming that these two modes (o and x) propagate and become amplified independently of each other. However, if we look at the medium at an oblique angle, the radiation will appear partially linearly polarized and the degree and inclination of polarization with respect to a fixed plane of rotation will depend on the relative orientation of the emitting surface with respect to the observer. This is illustrated in Figure 2. Thus, in general, we expect that pulsar radiation will be

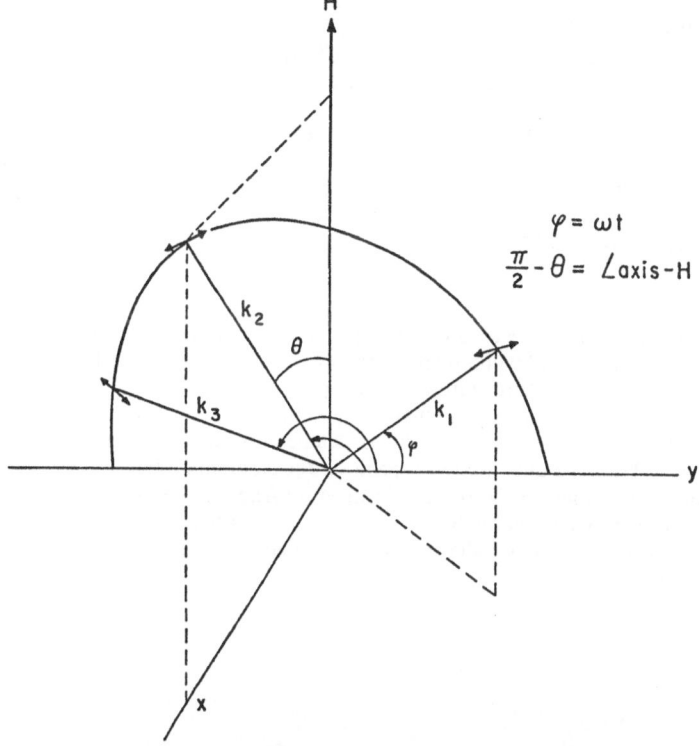

Fig. 2. The relative orientation of polarization in the observer's plane for a pulsar whose rotational axis is perpendicular to the line of sight and the magnetic field is inclined at an angle $\pi/2 - \theta$ with respect to the rotational axis.

linearly polarized and the polarization strength and orientation will change through the cycle. Observation shows that in some cases pulsar radiation is found to be polarized and the orientation and the degree of polarization change with the phase of the pulse (Radhakrishnan and Cooke, 1969). We are now in the process of extending our computation of the emissivity of radiation at an arbitrary angle so that a comparison of our theory with observation can be made.

6. Conclusion

In this paper we have presented a model for pulsars. This model is based on the behavior of electrons in an intense magnetic field. While the model has not developed into a full quantitative treatment, it successfully arrived at a mechanism for emissions of intense radio emission. This mechanism avoids the usual difficulty of coherent plasma emission, that is the emission be limited to a bandwidth of $\frac{1}{2} \lambda$. A full account of this model is being published (Chiu and Canuto, 1971).

References

(For a more complete bibliography see Chiu (1970))

Canuto, V. and Chiu, H. Y.: 1970, *Phys. Rev.* **2A**, 518.
Chiu, H. Y.: 1970, *Publ. Astron. Soc. Pacific* **82**, 487.
Chiu, H. Y. and Canuto, V.: 1971, *Astrophys. J.*, **163**, 577.
Drake, F. D. and Craft, H. D.: 1968, *Nature* **220**, 231.
Goldman, P. and Oster, L.: 1964, *Phys. Rev.* **136A**, 602.
Goldreich, P. and Julian, W. H.: 1969, *Astrophys. J.* **157**, 869.
Hewish, A., Bell, S. J., Pilkington, J. D. H., Scott, P. F., and Collins, R. A.: 1968, *Nature* **217**, 709.
Maran, S. P. and Cameron, A. G. W.: 1969, *Earth Extraterrest. Sci.* **1**, 1.
Ostriker, J. P. and Gunn, J.: 1960, *Astrophys. J.* **157**, 1395.
Pacini, F.: 1968, *Nature* **219**, 145.

Discussion

C. Heiles: At 430 MHz, T_B for NP 0532 is $\sim 10^{28}$ K in the average and $\gtrsim 10^{31}$ K for strong pulses. At 100 MHz these numbers would be up by more than two orders of magnitude.

H. Y. Chiu: The brightness temperature I gave represents the typical value for a rotation at 1 sec period. When saturation occurs the brightness temperature depends on other factors, and can go as high as 10^{31} K. I would expect that the brightness temperature should increase at lower frequencies.

G. Chanmugam: Does the quantisation depend on the assumption that the magnetic energy goes as the scalar product of μ and H? If so is this justified for fields $\sim 10^{12}$ G?

H. Y. Chiu: The theory we developed is based on the solution of the Dirac equation of a free electron in a magnetic field, and is valid even when the field is greater than 10^{13} G.

RADIATION MECHANISMS OF THE PULSAR

8.1 THE RADIATION MECHANISM IN PULSARS

F. G. SMITH

University of Manchester, Nuffield Radio Astronomy Laboratories, Jodrell Bank, U.K.

Abstract. The properties of the radio pulses from pulsars are described. The formation of pulses seems to be a geometric phenomenon, for which the beaming due to relativistic motion of the source is the only candidate at present. Values of the volume emissivity and surface flux density indicate that all the particles near the source must have relativistic energies.

It will be interesting, I think, to recall that in 1955 when we had a conference here there was a paper by Oort and Walraven which was on the polarization and radio emission mechanism of the Crab Nebula. At that conference we did perhaps begin to understand what was going on in the nebula, but also at that Conference there were papers by Shklovsky on the radiation from extragalactic objects, from the Galaxy and other objects, and, speaking for myself, I really did not understand these papers which nevertheless contained the key to the radiation mechanism from the discrete sources – the synchrotron radiation. I was hoping that Ginzburg would be here and would give a talk, which even if I did not understand it would give me the key to the radiation from the pulsars, but we have to manage without. I think that he might have helped because I think that the kind of plasma wave mechanism which he and Zheleznyakov and others have been discussing is perhaps the way into the problem. However, instead of discussing the theory I am going to substitute for Ginzburg's talk by giving the best summary I can manage of what I think are the important observations.

The most important thing about the pulsars is that they are periodic and that there is attached firmly to the rotating object a location containing the source. This location is defined by the integrated pulse profile. In Figure 1 are examples of the various well known profile shapes you can get by adding many pulses from a single pulsar. Now the shape of the profile I am going to call 'the window', and I will say that each pulsar has a fixed window through which it radiates, like a lighthouse has a rotating turret with a hole cut in which a lens is placed. The angular width of this window is not obviously dependent on period; it remains in the range of about 1–4% of the period, so that 1–4% of 2π gives the angular width. The pulse shape and width are independent of frequency to a first order. We know, of course, that for 1133 separation of the peaks does depend on frequency and we know that details in the shape of others, such as 2045, or one of those with multiple pulses, certainly do change somewhat with frequency but the phenomenon is basically broadband and well represented in the shapes of Figure 1. Next I want to emphasise, and indeed the demonstration yesterday of live pulses from 0329 emphasised it to you, that this is not the typical shape of an individual pulse. For some pulsars the individual pulses appear as discrete individuals with a typical shape and they also have a typical polarization pattern inside them. It looks very much as if the more complicated pulses which you sometimes see are made up of a series of different individual pulses.

Davies and Smith (eds.), The Crab Nebula, 431–440. All Rights Reserved.
Copyright © 1971 by the IAU.

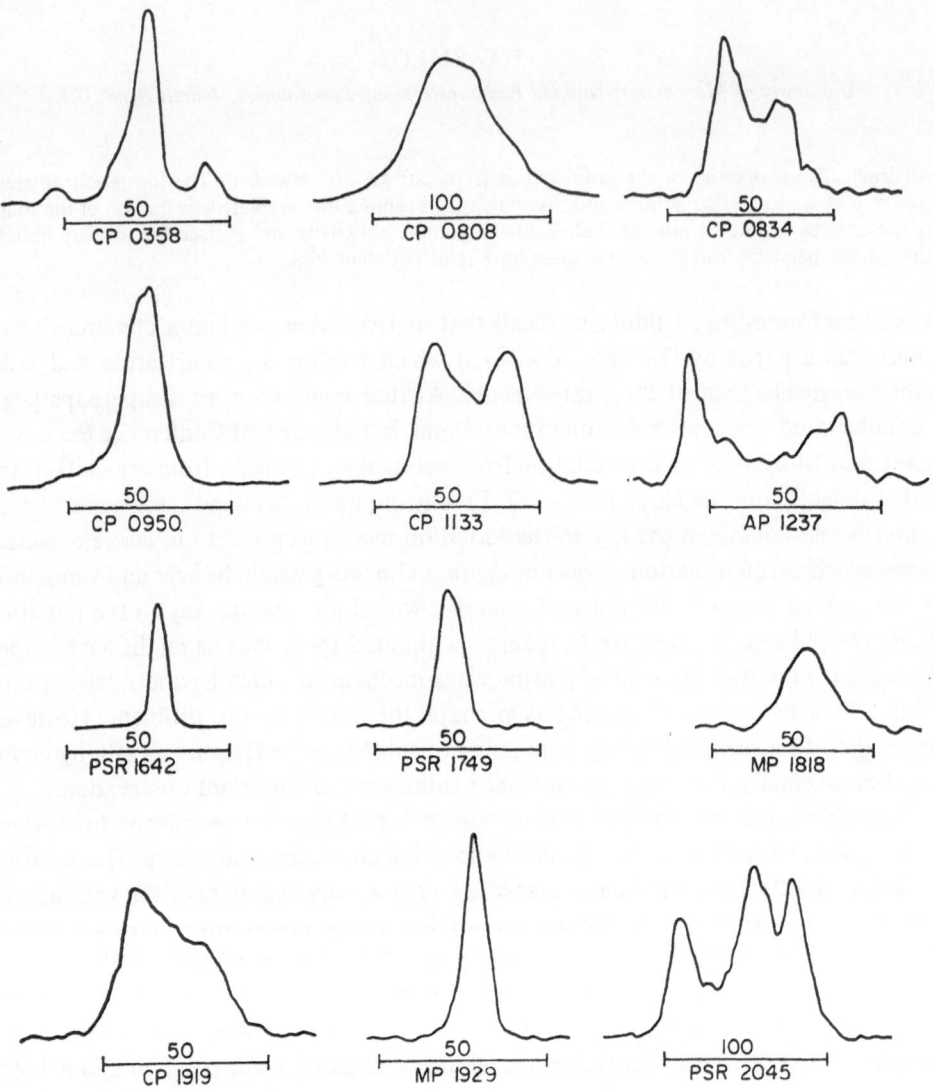

Fig. 1.   Integrated pulse profiles at 408 MHz.

Wherever a single pulse appears in the window, it tends to have the same width. If you have a double profile I believe it is made up of two general locations inside which the pulses are liable to crop up. I believe that we must explain two things, the window and the individuals.

And if you are talking about a radiation beamwidth it should be the beamwidth derived from the width of the individuals, while the locations on the rotating object are determined by the window. It is as though you have a lighthouse with a turret in which the lens is rather poorly mounted so that it can shake about. If you want to describe the lens you think of what happens when a single beam goes past you.

If you want to describe the turret you add lots of appearances of the beam which tells you within what range the lens must be seated. So there are two angular widths to talk about.

The lens analogy is not bad because you can imagine the lens moving about rather slowly in its turret, corresponding to what we call pulse drift. A recognisable pulse appears on separate rotations and drifts across the window. They drift at definite rates in a given pulsar and a pulsar may have more than one rate. Not all pulsars show this clearly, but it could be that every pulsar has a rate and you do not easily see it because the pulses tend to die during one rotation, or move across so fast that they do not reappear. The source has a definite location within the window and it usually moves in a forward direction.

Now, of course, I have spoken about pulsars in general and I ought to interpolate the question: is the Crab pulsar the same? It is very difficult to say because we are unable to observe the individual pulses in such detail on many of these pulsars. For the Crab we are able, fortunately, to look at a few individuals because it has this peculiarity that it produces occasionally very strong individual pulses, and we believe that these individual pulses are very similar to those seen in other pulsars. We do not know really how to interpret the relationship between optical and radio pulses in the Crab Nebula. At first sight the envelopes, the integrated profiles, are very similar. It is surprising that you have got two sharp peaks of, say, a millisecond wide separated by 13 msec, and this happens equally at both ends of the spectrum. There are differences. I put it to you that those differences may be no greater than the differences which you find across the radio spectrum in any of the other pulsars. So let us say it is a very rough approximation that the integrated pulse shapes are much the same over the whole of the spectrum in the Crab. Obviously that needs a bit more argument and the sort of way to tackle it is to see whether the details of the individuals are the same as the envelope in the optical range as well as the radio range. That is where I found it very interesting to hear the results from Hegyi, who showed in fact that the optical pulses do follow all the same pattern and it is the sum of many similar identical pulses which is producing that profile. That may be a difference between the optical and radio behaviours.

Whether there is circular polarization in the individual pulses from the Crab is not yet proven, but there are after all other pulsars which do not produce much circular so that I do not think we can say whether that proves whether the Crab is the same or different from other pulsars. The individual strong pulses from the Crab, like those of other pulsars, behave in much the same way over a wide range of radio frequencies. If you get a strong pulse arriving early in the window of one frequency, then you get a strong pulse arriving early on all radio frequencies. Again this is a sweeping statement and we have not a lot of evidence. The evidence which we heard about differences between 111 and 74 MHz could be interpreted as saying that at least half the pulses do behave that way; for the other half we have to be careful about scintillation and so forth, and I feel they are perhaps really broadband also. So let us say that individual pulses are producing the same sort of intensity over a very wide frequency range.

How wide? It may be that the spectrum is not quite as wide as we think of when we can actually see pulses over several octaves. It could be that the typical width of a pulse, at half power points, is only an octave or so. There is some evidence that some of the pulsars are producing radiation which decreases towards the lower frequencies. For example, I think 0329 which you saw on the oscilloscope does peak somewhere about 400 MHz, which is fortunate because otherwise we would not have been able to demonstrate it.

The state of polarization is extremely important. It is elliptical in general, and you saw that it changes through an individual pulse. When it changes through an individual it changes only slightly by reversing hand once, or by a swing of position angle less than one radian. Examples can be seen in Figure 2. There is usually as much circular as linear in a typical pulse. When you add many pulses, the reversal of hand, which is variable, means that you lose the circular in most pulsars, and this has perhaps led many people to neglect this most important circular polarization in their theories. I would emphasise again that individual pulses are typically elliptically polarized.

The first thought about explaining this is that it might be synchrotron radiation from a beam of electrons being bent round a magnetic field and radiating in the forward direction. If you make the whole system rotate, the observer cuts across a fan beam. You see circular, linear, then circular. The degree of circular is very high, especially at the edge. This explanation looks right at first but what is wrong with it is the next observational fact, that the polarization is independent of frequency for both the integrated pulses and the position angles you get out of them. It also refers to individual pulses as Graham showed yesterday. If you photograph a single pulse, and manage to catch it at two frequencies which requires a little bit of slight of hand, you find that it has got the same pattern of polarization in it. So it is a very broadband phenomenon – you are not going to get away with it with a highly tuned maser or any other narrow band mechanism.

The next thing that we should say is that if it is synchrotron radiation, the width of the pulse should depend on frequency; and it does not. It should depend on frequency as $v^{-1/3}$, in the simple illustration I gave you, or if you take a spectrum of electrons in the usual way you can make it into $v^{-1/2}$, but you cannot make it stay still except by using a long tail of the synchrotron radiation way above the critical frequency when you must ask whatever has happened to all the radiation at the lower frequencies. So I think that this is impossible. You could therefore say that the width of the beam is not made by the $1/\gamma$ natural width of a beam of electrons but is made by a range of velocity vectors of the electrons spread over a range of angles which determines your beamwidth. This sort of theory is, indeed, a very attractive one because you can imagine your bunch of field lines and electrons going out, and Komesaroff has shown you can get nice beamshapes and nice polarization swings as in Radhakrishnan's theory. Others have taken up this idea as well. You can get a very nice arrangement of beamwidths but if you do that you lose your circular polarization because you are convolving the odd angular function with the wider angular beam. So you can produce the correct Stokes I, but not the correct Stokes V. If you have got circular

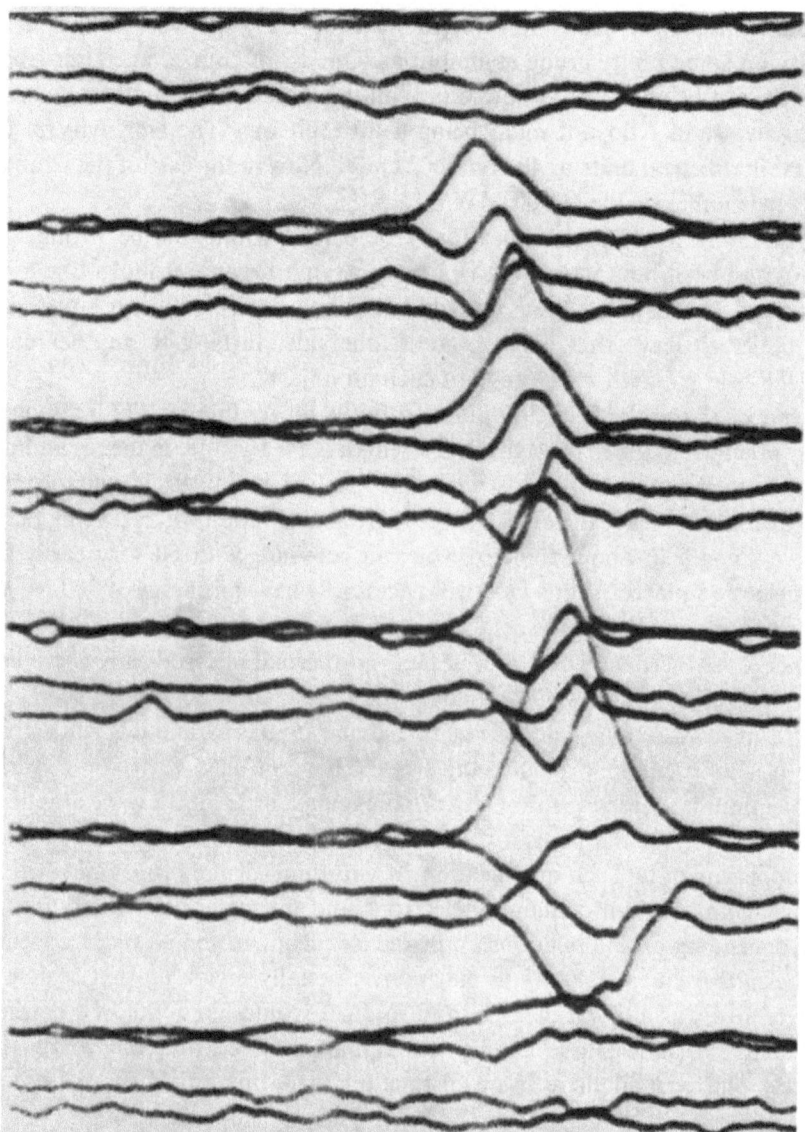

Fig. 2. A sequence of individual pulses from P 0329, recorded with a polarimeter at 408 MHz. The four traces record the Stokes parameters I, V, Q and U for each pulse.

you cannot have a beam which does not change its frequency on a simple synchrotron theory alone – that is out of the question.

I turn now to brightness temperature and intensity. You have already heard that the brightness temperatures exceeds 10^{28} K. The total intensity and volume emissivity are even more remarkable. Let us estimate the size of the source. It is fair to say that this looks like an object which is within the velocity of light circle of the pulsar, so I have just guessed that it might be 0.1 R across where the radius R is equal to $c\omega^{-1}$.

I have done this for the Crab and I have done it for 0329, and worked out the volume emissivity on some pretty crude assumptions. One is, of course, uncertain about the effect of the fan beam. I would not go into all this because the numbers rather speak for themselves and I do not mind being a little bit out. The emissivity of 0329 is 400 W/m^3 in practical units or 40 erg/sec^{-1} cm^{-3}. Now in the case of the Crab pulsar, the answer is unbelievable at 100 MW/m^3.

Now there are many possibilities for error here. I could well be wrong – I wish Ginzburg had been here instead. It is a bit worrying because it looks like a nuclear power station. On the other hand, it might tell us how to make nuclear power stations. Now I prefer to leave that department to one side, turning to another question: what is the field strength at the edge of such an object?

You can work out the field strength E from the power flux $E^2/377$ if you use these nice easy practical units. In 0329 the field strength is 5×10^6 V/m in the radiation which we observe to leave this object. And in the Crab it is 7×10^8 V/m in the radiation which we observe to leave the object. Since there are radiation waves leaving the source with a wavelength of about a metre, you can convince yourself very easily that an ambient electron placed in this field will necessarily have an energy of 700 MeV in an oscillatory form, and therefore spends most of its time in a highly relativistic way even if it wants to be thermal. There are, in fact, no thermal electrons anywhere near the radiator. Therefore if you wish to analyse the propagation conditions there, it is no use in writing ω/ω_P or ω/ω_L, unless you put some gammas in, and it is worse than that because it is an oscillatory gamma and I really have no idea what you ought to do. Now I mention that because although I am plugging a little bit for wave amplification, if you look at any of the papers which have so far appeared on it the propagation factors appear with no such qualification. If you think of the propagation of a radio wave through a medium containing electrons and magnetic field, there are various regimes depending on the relationship of the frequency to the plasma frequency and to the Larmor frequency, and you may conventionally divide up that regime into a number of areas, as a CMA diagram and so on. I would like to know what happens to that diagram under these conditions. It may well be that the whole thing is completely blurred and there is no distinction to be made any more between the different modes of propagation.

Now just let me say a little bit about the idea of making a large wave. You may say in fact that if you have got to make a great big wave you could say 'some waves are born great, some achieve greatness and others have greatness thrust upon them'. 'Born great', by that I would mean you have got synchrotron radiation with bunches in it, and I do not think you ought to work that way because of this polarization and beamwidth problem. 'Some achieve greatness', well I think that is something like a maser; you have got to be careful with the maser because you have to get an adequate bandwidth. Now we heard that such things might be possible from Dr. Chiu; I would only worry a little bit that his radiation mechanism is near the surface and he has therefore got to produce a beamwidth with some configuration of field lines and I just do not know how he is going to do that and still produce the right circular and elliptical polarization. But it may be there is a broadband maser. 'Greatness thrust upon them', I think that would be a reasonable description of the enormous plasma waves that must be present, and the enormous plasma waves coupling with electro-magnetic waves may indeed thrust greatness upon them. So we ought to look carefully first of all whether the energy from the pulsar can be converted easily into plasma waves, and I understand from the work of Tsytovich, Kaplan, and others that there will be indeed a rapid conversion of the energy of any relativistic particles into plasma waves, so doubtless they can exist. There is also the possibility that those waves will become rapidly isotropic and I do not know what happens to their magnification, and their coupling process, when they are isotropic. When they are not isotropic and there is a strong magnetic field, one can refer to both papers by Tsytovich and Kaplan. Now let me just say that there is however one particular worry about amplification. Amplification tends to pick out one mode. If modes are clearly dis-tinguishable, and of that I am not so sure, then you are liable to get one mode (one hand of circular if you like) amplified by, say, e^{30} and the other one amplified by e^{28}, say. In which case you get a very strong circularly polarized wave coming out because there is an e^2 between them. You do not need much difference between the amplifi-cation factors on two modes before you get full polarization, which of course is an advantage because we have got full polarization; the only trouble is that it tends to switch on one mode only and observation shows a smoothly changing elliptical polarization across the pulses. It is not all one mode as in the OH lines, for example. We have a beam of OH line coming straight towards us; it is circularly polarised; we are not allowed to look round the side of the beam but we have a strong suspicion it is going to be circular all over that beam. This is not the case here, so we should give attention to the mode selection process.

The coupling between plasma waves and radiation is greatest near the gyro-frequency, and the coupling can even be strong in directions which are perpendicular to the magnetic field. This means that there can be an amplification over a large solid angle and not just a small beam. It seems to me that in all these circumstances to try and construct a beam with these attributes by a diffraction-type process is hopeless because we have got such a wide range of wavelengths to deal with. Furthermore, in the Crab we have evidently got to do something rather similar in the optical range and the radio range to get our beam. And now I am going to describe a way of beaming the radiation which has nothing to do with any directive processes of amplification or diffraction, and which allows the radiation process to be nearly isotropic. I suggest to you that this process is nothing to do with electromagnetism, it is pure geometry. And the geometry that I am going to suggest to you is a pretty crude thing. It is already published (Smith, 1970); it does not tell you anything about the radiation mechanism but it does not need to. The radiator, a mass of plasma waves – porridge or kasha, or whatever the Russians would call it – is radiating out in all directions, and it is being driven round by the rotation of the pulsar on a radius 0.9 of the distance of the light circle; so it is moving at 0.9c. When it comes towards you you get a γ^2 amplification of the radiation and you get a pulse which looks as though it is a

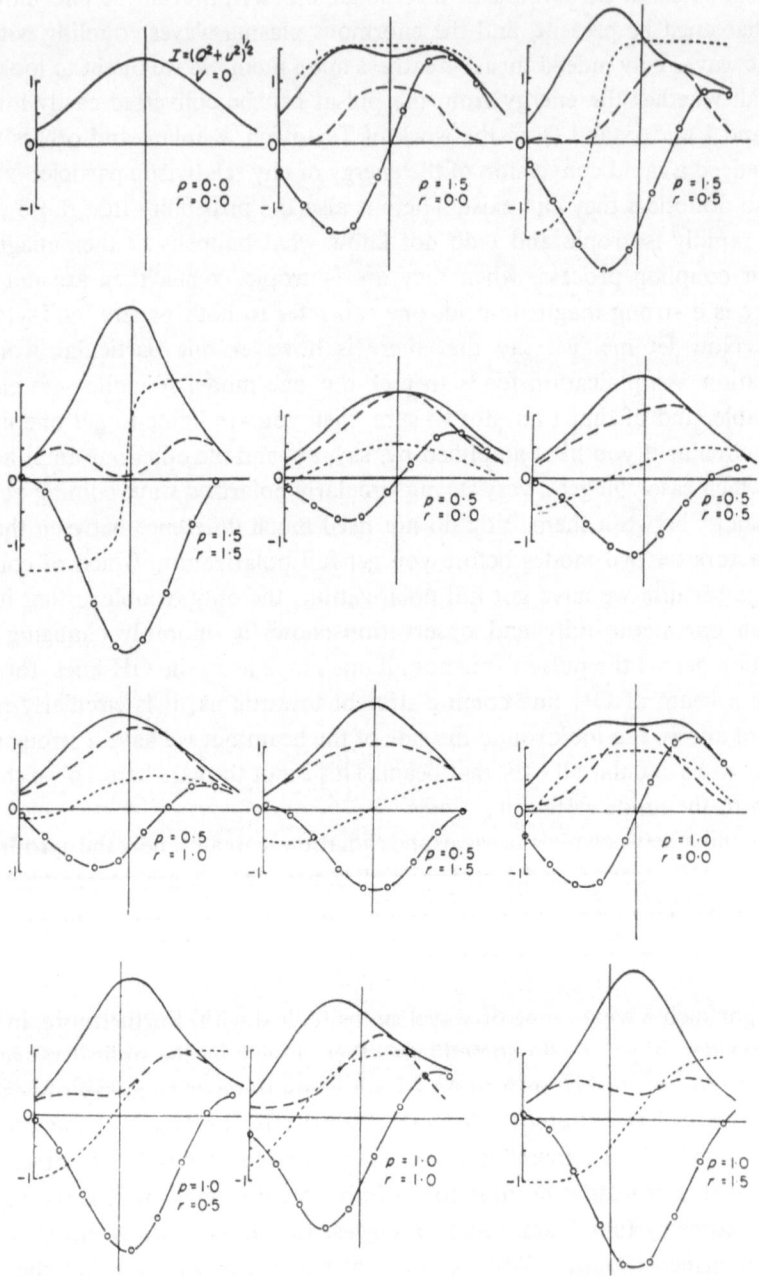

Fig. 3.   Computed pulse shapes, using a cyclotron polar diagram and relativistic pulse formation.
The parameters I, V, $(Q^2 + U^2)^{1/2}$ and $\tan^{-1}Q/U$ are shown for various orientations of the polar
diagram relative to the pulsar rotation axis (from Smith, 1970).

directive radiation sweeping across you. The width depends on the gamma, i.e. on the v/c ratio, but the shape is more or less independent of that. If you are not actually in the equatorial plane, you get a pulse of about the same width but not as high. You roughly get a beamwidth in the N–S direction which is about $1/\gamma^2$. If you add polarization to this you have a model of the radiator. My best model of the radiator is to take a circulating electron, or whatever it is, going round a magnetic field and just look at nearly isotropic radiation with the same polar diagram as cyclotron radiation. In fact I took the cyclotron polar diagram with a maximum of radiation at the poles (circular), and linear polarization at the equator. The whole thing is whizzing round the pulsar, but I do not know the attitude of it. I have got two parameters to find the direction of the pole in relation to the pole of the pulsar and I have explored all those parameters to produce theoretical pulse shapes, including Stokes parameters. These are published in my paper and some are shown in Figure 3. If you look long enough at an oscilloscope screen showing all the Stokes parameters you will find all the theoretical pulse shapes. I do not know that it really proves anything but it does, in fact, allow you to explain pulse shapes and polarizations with a polar diagram that is nearly isotropic, and this may be a help to theorists who wish to make a model of this most fantastic radiator.

Reference

Smith, F. G.: 1970, *Monthly Notices Roy. Astron. Soc.* **149**, 1.

Discussion

M. M. Komesaroff: Can your model explain the fact that, at least in the case of some pulsars, e.g. PSR 0833 — 45, the linear polarization is the same at identical phases of consecutive pulses, whereas the circular polarization, if present, must vary in a random way from pulse to pulse.

F. G. Smith: If the polarization is 100% then there can be virtually no circular polarization. This must be the case for PSR 0833 — 45. On my model it is possible to get linear polarization and no circular round one particular great circle of the emission polar diagram. Let the radiation to be emitted by a circular motion of an electron with an axis coinciding with the rotation axis of the pulsar, then the radiation from the equator will be linearly polarized with the same plane as seen from anywhere on the equator. Only a great circle which is inclined to the equator will contain any circular polarization. If the circular polarization is variable it means that its angle is variable, but that there is an average great circle which defines the linear polarization.

A. T. Moffett: I would like to point out that the emissivities and brightness temperatures encountered in pulsars are not staggeringly high by terrestrial standards. A large Klystron amplifier has peak emissivity of about 10^{12} W m^{-2}, and its brightness temperature might be about 10^{30} K if you look at its output waveguide.

F. G. Smith: This may not prove a thing but I suspect that you can obtain the same high energy densities also in micro circuits.

F. D. Drake: In considering the radiation mechanism, one must take into account the broadening of pulse width at lower radio frequencies, as observed in all pulsars at Arecibo. This effect, especially the observed increase in separation of peaks in a double peaked pulse shape, must be intrinsic to the pulsars in general, and so is not just an occasional peculiarity in behaviour.

More importantly, I think there is a compelling case that the particles are bunched. This of course automatically provides an amplification mechanism. As emphasized by Pacini, there are two basic frequencies present. The rotation frequency, 30 Hz, requires a quite unreasonable γ to provide the observed frequencies. The gyro-frequency, some 10^{12} Hz, is too high to give any radio emission, but likely provides the optical and shorter wavelength radiation. To get the radio frequency radiation in your model, which is very close to Gold's, one must have the particles in bunches, whose scale is of the order of a metre, since these bunches sequentially radiate towards the observer as the velocity vectors sequentially point in that direction. The observed radiation field contains radio frequencies – this is a compelling argument for bunching.

Goldreich has recently shown that bunching will occur in a stream of radiating particles. The leading member of a proto-bunch, created by statistics, experiences a greater radiation reaction driving it to the rear. This process continues to produce a major bunching.

The presence of circular polarization may put no demands upon the emission mechanism. It could result from magnetoionic effects working on radiation initially linearly polarized.

F. G. Smith: I would first like to comment on the suggestion made by Gold that the radiation is generated by the peripheral motion of the electrons. In my model the radiation is not generated by the peripheral motion but it exists within the rotating frame of reference. The peripheral motion has the sole function of concentrating the radiation into a beam. I think there is a real difference between these two approaches. In my model the peripheral motion has a γ of only $2\frac{1}{2}$ or 3, whereas in Gold's model it must be some hundreds or more. The second point which you made suggesting that the gyro-frequency has nothing to do with the radio frequency is not obviously true. If the particles have very high energies then it may be that these are identical. It would, of course, be necessary to explain why the radiation is primarily at the gyro-frequency, but this may of course be due to the bunching.

You rightly emphasize the observed changes in pulse width which are most noticeable at low frequencies. I would, however, emphasize the opposite phenomenon that over a very wide range of frequencies above about 100 MHz, most pulsars show very little change in width.

Your final point concerning circular polarization does offer a possibility, but again I think that to invoke magneto-ionic theory will be very difficult since the polarization seems to be so independent of frequency.

J. V. Jelley: Your blob cannot surely be all that near to the velocity of light circle, because when viewed tangentially to the orbit, the pulse lasts say $1/10$ of the period. If this is the case, then for an orbit near the velocity of light circle, the true dimensions of your blob, in the rotating frame, will be a long arc in azimuth.

F. G. Smith: There are two different times to be considered. Firstly, the pulse window represents an arc which can contain a source of radiation. Secondly, a single isotropic source of radiation within this arc can give a pulse whose duration is determined by γ. The duration is in fact of the order $1/\gamma^3$ times the period.

8.2 PULSAR MODELS AND RADIATION PROPERTIES

V. RADHAKRISHNAN

Division of Radiophysics, CSIRO, P.O. Box 76, Epping, N.S.W., Australia

Dr. Radhakrishnan outlined a geometrical theory of a rotating pattern of field lines, which would account for the pattern of polarisation observed in the integrated radio pulses from pulsars.

The following summary has been prepared by the Editors.

Any theory for the radiation mechanism in pulsars requires a description of the properties of the region in terms of brightness temperatures, polarisation, spectrum, and directivity. There is a close relation between the observed polarisation properties and the directivity, since the sweep of position angle within the pulses is independent of frequency and must be a purely geometrical effect related to rotation. Since the position angle can change rapidly through a large range, the radiation must come from a pole, presumably a magnetic pole. The source of the radiation must be well within the velocity of light cylinder, since the field lines must be close to a dipole configuration.

Other so-called intrinsic properties of the radiation not accounted for by the above model are rapid pulse-to-pulse variations, variable circular polarization, and marching subpulses. All of these properties are more pronounced at lower radio frequencies. This suggests strongly an origin in modulation or propagation phenomena. It is hypothesized that all of the above variable characteristics are imposed on the radiation in the circumstellar region by a circulating diffracting screen permeated by a moderately strong magnetic field.

Discussion

F. G. Smith: You suggest that the sweep of polarization angle implies that the source is seen near the pole of a magnetic field. This is not necessarily so. The sweep of angle is an intrinsic part of my model which allows the source to be seen from any angle, including from the equator. Secondly, I again emphasize that if you are to obtain circular polarization by the propagation effect then you must take care that this effect is not frequency selective. Further, you must make the polarization change during a single pulse just sufficiently for one complete change of Stokes parameters to occur, no more and no less.

J. M. Cohen: If the Goldreich-Julian procedure is repeated for an oblique rotator, it turns out that the radiation zone continues right up to the surface of the star (in some sense). It is necessary to use the full radiation solution in order to match the boundary conditions of the neutron star surface. A near-field approximation does not work.

J. P. Ostriker: If the radiation is attributed to bunching of electrons, then there is no frequency dependence. However there might be some screening which is frequency dependent.

F. D. Drake: The picture of pulsar emission as radiation from coherent bunches of particles at radio frequencies, and ordinary synchrotron radiation from the same particles individually producing the optical and higher frequencies, can explain pulsar variability. For the radio emission we depend upon the construction of a relatively small number of bunches of particles which vary in size. There can be large statistical variations in these leading to large radio variations. However since the particles radiate at optical frequencies independently, there would be no significant variation.

V. Radhakrishnan: I can suggest an observational test of this idea. It implies that at higher frequencies the mean polarization will approach that observed optically. In particular, circular polarization will decrease and the marching sub-pulses should disappear. The detailed structure of the pulse will, in general, become smoothed out.

F. C. Michel: By this model one would expect the pulse intensity to be flat at maximum when the line of sight travels very close to the pole. However in the case of the Crab pulsar, the maximum appears to be sharp at all resolutions.

V. Radhakrishnan: The model is primarily intended to explain the pattern of polarization, but this does not exclude the possibility of obtaining almost any pulse shape as the line of sight sweeps across the emitting region.

8.3 LOW MODE COHERENT SYNCHROTRON RADIATION AND PULSAR MODELS

BERNARD J. EASTLUND

Division of Research, U.S. Atomic Energy Commission, Washington, D.C., U.S.A.

Abstract. The radio pulses from pulsars are considered to be formed from the beamed radiation from relativistic electrons. This radiation is in low order harmonics of the gyro frequency, giving pulse shapes and polarisations which agree with observation.

A possible connection between the shapes of radio pulses and the emission of optical pulses, via the electron energy, leads to a suggested means of searching for visible pulsars with improved sensitivity.

The theory for pulsar radiation described in the present paper is based on a model in which an oblique rotator emits coherent synchrotron radiation. In this model, illustrated in Figure 1, the knife-like angular pattern of coherent synchrotron radiation from charged particles trapped in a magnetosphere is converted into a time dependent pulse at the location of an observer by rotation of the star.

Work with this model to date (Maran and Cameron, 1968; Eastlund, 1968, 1970a, b) has concentrated on a comparison of individual and average RF pulses with a detailed computation of pulse shapes based on the well known single particle synchrotron emission formula given by Equation (1) (Bekefi, 1966).

$$P_m = \frac{e^2 m^2 \omega_L^2}{8\pi\varepsilon_0 c}\left[\left(\frac{\cos\theta - \beta_{\|}}{\sin\theta}\right)^2 J_m^2(m\beta_\perp \sin\theta) + \beta_\perp^2 J_m'^2(m\beta_\perp \sin\theta)\right] \qquad (1)$$

where P_m = the intensity of synchrotron mode, m.

$$\omega_L = \frac{m\omega_0}{(1 - \beta_{\|}\cos\theta)}$$

$\beta_\perp = v_\perp/c$

$\beta_{\|} = v_{\|}/c$

m_0 = mass of the charged particle

$\gamma = (1 - \beta^2)^{-1/2}$

θ = the angle between the observer and the magnetic field direction.

The transformation between angular dependence and time dependence is made using the following expression:

$$P_m(\theta) = P_m\left(\frac{2\pi}{\tau}t\right) \qquad (2)$$

where τ is the pulse repetition rate.

For relativistic electrons, Nodvick and Saxon (1954) have shown that the single

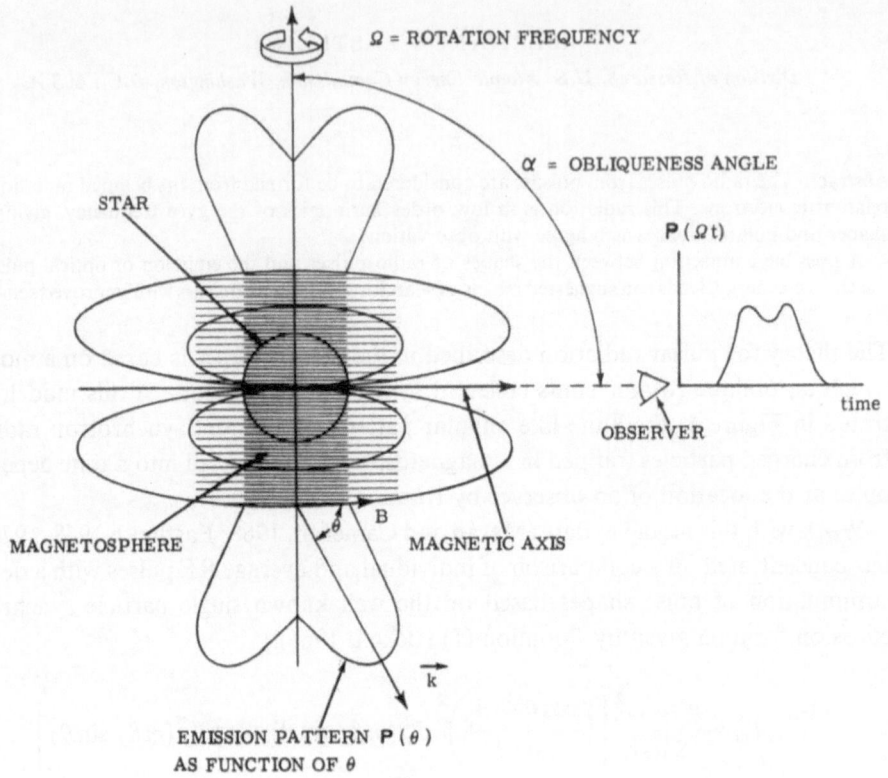

EMISSION PATTERN $P(\theta)$
AS FUNCTION OF θ

Fig. 1.

particle spectrum of Equation (1) is related to the intensity of coherent radiation as follows:

$$P = N^2 \sum_{m=1}^{\infty} P_m f_m \qquad (3)$$

where $f_m = e^{-m^2\psi^2/4}$
            $N =$ the number of electrons phase bunched over an angle ψ on a cyclotron orbit.

This equation is presented to allow some discussion of the expected coherent synchrotron spectrum and its relation to optical spectra. It is assumed that details of the magnetosphere and an exact calculation of the coherent effect will be much more complicated than the prediction of Equation (3).

To conserve time and permit a more detailed presentation of new results the previous work based on the above equations will not be described in detail in this paper.

According to Equation (1), the pulse shape is a function of the synchrotron mode number, m; the components of the particle velocity perpendicular, $\beta_\perp$ and parallel, $\beta_\parallel$ to the magnetic field. The value of the magnetic field does not effect the pulse shape. Figure 2 shows the predicted pulse intensity and polarization (Stokes parameters I, $L=(Q^2+U^2)^{1/2}$ and V are calculated assuming an obliqueness angle $\alpha=90°$ with an observer located in the star's equatorial plane) for fixed particle energy (8 MeV) and synchrotron mode $m=110$ with variable $\beta_\parallel$. (Mode 110 corresponds roughly to a receiver frequency of 25 MHz based on Figure 7 of Eastlund, 1970a.) Note the asymmetry in the pulse shapes as $\beta_\parallel$ increases. There is evidence of such asymmetric behavior in data from many pulsars (Clark and Smith, 1969; Drake and Craft, 1968). Detailed study of such behavior may indicate the mirroring of particles trapped in the pulsar magnetosphere (Eastlund, 1970b).

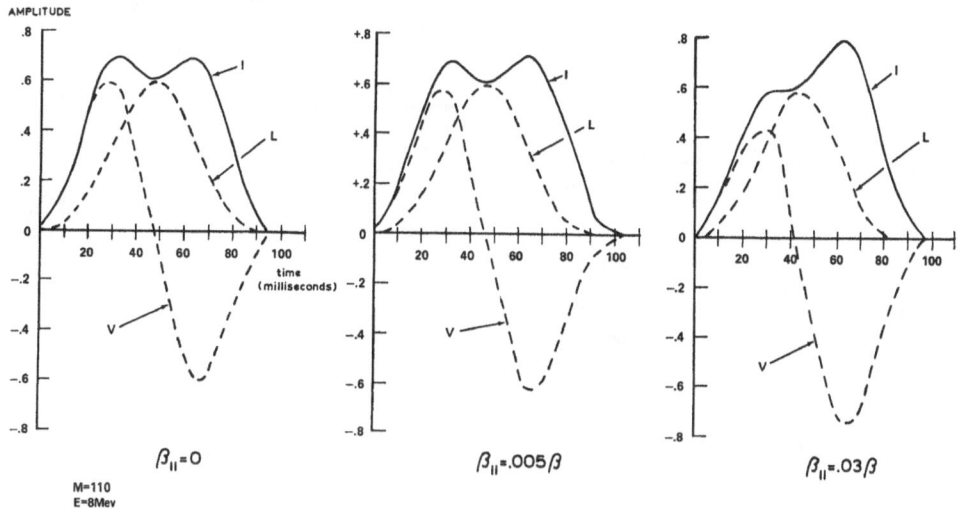

Fig. 2.

If charged particles are trapped at some radius, R, in the magnetosphere of a neutron star they may be heated by one or more mechanisms and thus exhibit a change of energy with time. Equation (1) is used to compute the pulse shapes that would be observed at a *fixed receiver frequency* if the energy of the charges (γ) were changing with time in a region of constant magnetic field, B. If γ increases with time, the fundamental synchrotron mode will vary as $eB/\gamma m_0$. Thus, if the radio receiver is turned to a fixed frequency, then if γ changes to a value γ_2 from γ_1 a synchrotron mode m_2 is detected that is related to the initial value m_1 by the expression $m_2=m_1\gamma_2/\gamma_1$.

Thus, as γ is changed in the computation, the pulse shape is computed for the new mode that corresponds to a constant receiver frequency. (Assumed equal to some initial value, m.) The results of such a computation for the case where $\beta_\parallel=0$ predicts a rather distinctive behavior for the pulse shapes. The peak to peak pulse separation for a double peaked pulsar is predicted to increase from zero to a saturation value with increasing energy. This result is illustrated in Figure 3 for a value of $\tau = 1.871$ sec (CP 1133). The solid line represents the predicted peak to peak separation for a fixed receiver frequency as a function of energy. Note that as the energy of the emitting charges increases; (a) the overall pulse width, $\Delta\tau_2$ at the half power point decreases; (b) the ratio of the peak to the minimum of the double pulse *increases* and (c) the amplitude A_1 increases. Each synchrotron mode, (i.e. receiver-frequency) will generate a different curve – as illustrated by the dashed lines. The lower limit of the energy was arbitrarily set at 2 MeV to conserve computation time.

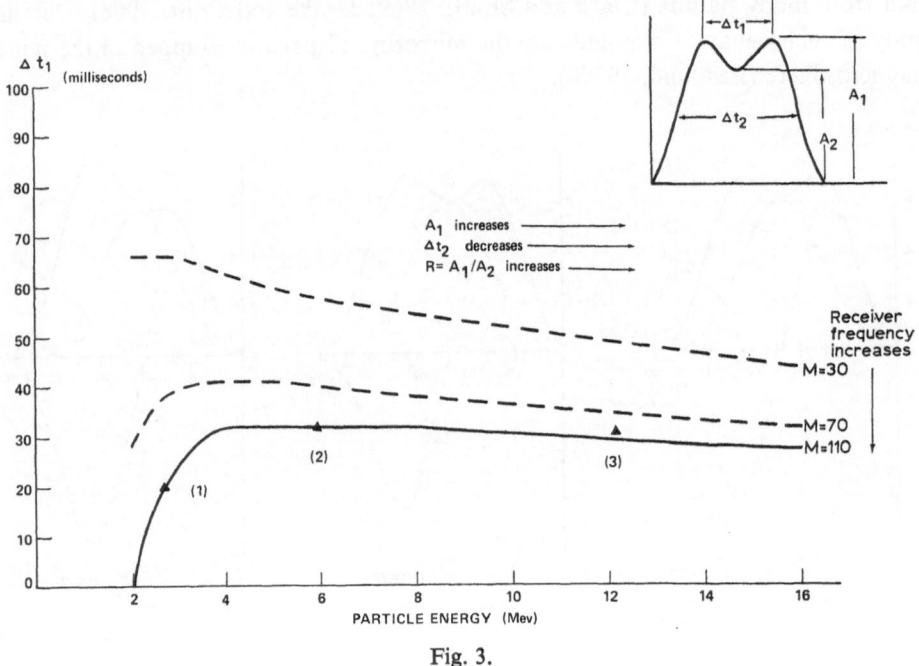

Fig. 3.

Such behavior seems to have been observed in sequential pulse data from CP 1133 (Craft, 1970). The peak to peak separation of three sequential pulses (1), (2) and (3) are plotted as triangles on Figure 3. (Note that due to the saturation of Δt_1 with energy, that the location of pulse (2) and (3) relative to the energy co-ordinate is somewhat arbitrary.) The data is compared with theoretical pulse shapes in Figure 4.

The upper row of curves are theoretically predicted shapes for three different particle energies. The lower curves are the three sequential pulses from Craft (1970) taken at a receiver frequency of 196.5 MHz. (A better comparison will be possible when Equation (1) can be completed for values of $m > 950$.) Based on this data it appears that the rate of change of energy is close to 4.0 MeV/sec. (The pulses occurred 1.187 sec apart.) There is thus a possibility that this data represents a *direct observation* of *increasing particle* energy. Much additional comparison with data remains before the details of such energy gain are understood. For example, the 4th pulse in the

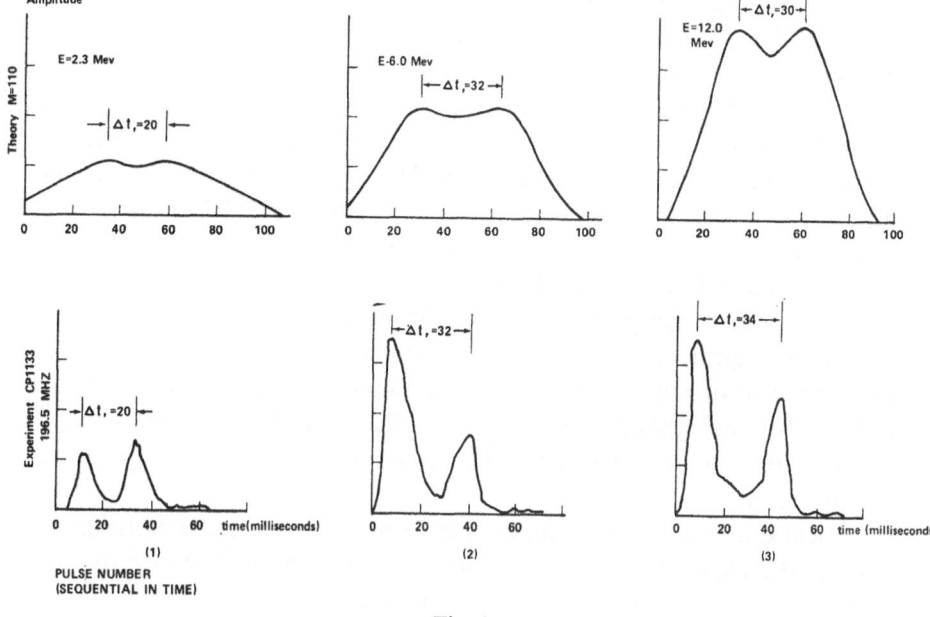

Fig. 4.

sequence was not observed. This could indicate that the energy emitted in coherent radiation by the bunched charges either exceeded the rate of energy gain or destroyed the coherence responsible for the emission. In general, the data from CP 1133 indicates a varying value for $d\gamma/dt$.

If such a heating effect is indeed occurring in some region of the magnetosphere, then this may have some consequences as far as the optical behavior of pulsars is concerned. The critical frequency of synchrotron emission (the frequency near the maxima of the spectrum) is given by the expression (Bekefi, 1966):

$$\omega_c = \frac{3}{2}\frac{eB}{m_0}\gamma^2.$$

(4)

Thus, there will be appreciable optical emission only for large values of γ. This suggests a technique for finding optical signals in pulsars other than NP 0532. The individual pulse shapes of sequential pulses must be recorded and the energy of the charges emitting the radiation determined from the pulse shape. (The *shape* gives a more accurate value for γ, than the amplitude because the amplitude is a function of γ, N^2 and other factors affecting transmission of the signal.) If a set of simultaneous measurements are recorded, then only those recordings taken during pulses with large γ should be added to obtain the optical signal. Such a measurement would require synchronization of an optical and radio telescope but could enhance the signal to noise ratio for optical searches by recording signals *only* during periods of optical activity. Note that the large values of γ do not seem to occur with any predictable or regular pulse spacing.

A crude power balance calculation can be performed on the basis of a theory by Grawe (1969) for RF heating in a magnetic mirror. The observed value of dE/dt of 4×10^6 eV/sec can be accounted for by a heating flux of 700 to 7000 W/m^2 of RF power incident on the region of the magnetosphere in which the emitting charges are trapped. (The theories of Chiu and Occhionero (1969) or Kaplan and Tystovich (1970) could provide mechanisms for such a flux of RF power.) If the emitting region is at a radius of 2×10^8 cm, then a steady state power of 10^{16}–10^{17} W is indicated. This is within an order of magnitude of the observed 10^{18} W based on a beaming assumption and the data of Ekers and Moffet (1969).

In conclusion, the single particle formula, through detailed comparison with pulsar data seems to indicate effects due to mirroring and heating of charged particles. A method for detecting optical signals in other pulsars is suggested. Exact calculations of single particle pulse shapes as a function of γ, $\beta_\perp$ and $\beta_\parallel$ are under way at the present time for synchrotron modes from the RF to the optical region and should be available soon.

The present calculations, based on the exact solution of the polar emission pattern for single particle synchrotron radiation represents merely the first step on the road to a complete theory. It is hoped that such computations will be found useful by others working on more complete analysis of pulsar phenomena.

References

Bekefi, G.: 1966, *Radiation Processes in Plasma*, John Wiley and Sons, New York.
Chiu, H. Y. and Occhionero, F.: 1969, *Nature* **223**, 1113.
Clark, R. R. and Smith, F. G.: 1969, *Nature* **221**, 724.
Craft, H. D., Jr.: 1970, 'Radio Observation of the Pulse Profiles and Dispersion Measures of 12 Pulsars'. Ph.D. Thesis, Cornell University (Fig. No. 5.28).
Drake, F. D. and Craft, H. D.: 1968, *Nature* **220**, 231.
Eastlund, B. J.: 1968, *Nature* **220**, 1293.
Eastlund, B. J.: 1970a, *Nature* **225**, 430.
Eastlund, B. J.: 1970b, to be published in the proceedings of the Conference on Pulsars and High Energy Activity in Supernovae Remnants, Rome, 18–20 December 1969.
Ekers, R. D. and Moffet, A. T.: 1969, *Nature* **220**, 756.
Grawe, H.: 1969, *Plasma Physics* **11**, 151.
Kaplan, S. A. and Tystovich, V. N.: 1970, Lebedev Physical Institute, Preprint No. 16, Moscow.
Maran, S. P. and Cameron, A. G. W.: 1968, *Physics Today*, August.
Nodvick, J. S. and Saxon, D. S.: 1954, *Phys. Rev.* **96**, 180.

8.4 RADIATION FROM PULSARS

I. LERCHE*

Enrico Fermi Institute and Department of Physics, University of Chicago, Chicago, Ill., U.S.A.

Abstract. We suggest that an oscillating interface, carrying a steady current between a rotating, radiating, magnetic dipole and an external plasma is the source of pulsar radiation produced over a broad band of high frequencies (10^8–10^{18} Hz). The efficiency of the radiation mechanism is so high that the oscillation energy goes mainly into radiation in each cycle of the oscillation.

Recently Pacini (1968) and Gunn and Ostriker (1969) have shown that a rotating neutron star, with a magnetic dipole axis oriented perpendicular to the spin axis, provides a reasonable and plausible explanation of the basic 'clock' mechanism for pulsars. Pacini has pointed out that the radiation pressure due to the oscillating field pushes any material outward, thereby creating a cavity dominated by the radiation, and surrounded by plasma on the outside. The radiation sweeps all material out of the cavity so that the radiation in the cavity is describable to a first approximation by a vacuum field. Gunn and Ostriker demonstrated that a charged particle placed in such a vacuum field is rapidly accelerated to relativistic energies as it is swept outward by the radiation pressure.

This paper is concerned with the interesting physical consequences for radiation output from the interface between the external plasma and the radiation field assuming the particles at the interface are relativistic, so that they can respond to the basic pulsing low frequency (~ 30 Hz) electromagnetic radiation at a velocity close to c.

The boundary of the cavity terminates the radiation field, with external plasma pressure balancing the radiation pressure. At the interface the stress of the vacuum dipole radiation field is transferred to the particles in the plasma by the Lorentz forces of a surface current. The dipole radiation pattern rotates with the neutron star at an angular frequency Ω. Thus the radiation pressure on the interface oscillates about its mean value with a frequency Ω, and so the position of the interface oscillates about its mean value with a frequency Ω. Thus the steady current in the interface, which transfers the stress of the dipole field to the plasma outside the cavity, oscillates perpendicularly to its direction along the interface with an angular frequency Ω. *Such an oscillating current sheet radiates energy.*

In order to illustrate the manner in which high frequency radiation is produced in such a system, consider the particular situation of an infinite, plane, current sheet in vacuum carrying a steady current of density j_0 per unit length which points in the x-direction. This particular example contains most of the essential physics of the conversion mechanism. The problem can also be considered from the point of view of a finite thickness, partially coherent, interface and/or an independent particle picture. Since all three points of view yield essentially the same results (apart from

---

* Alfred P. Sloan Foundation Fellow.

numerical factors of order unity) we have concentrated here on the point of view which most succinctly illustrates the essential physics of the basic problem.

Take the current sheet to move sinusoidally in the y direction at an angular frequency Ω and with amplitude y_0.

Then Maxwell's equations are

$$c^2 \nabla^2 \mathbf{A} - \frac{\partial^2 \mathbf{A}}{\partial t^2} = 4\pi c j_0 \hat{x} \, \delta [y - y_0 \sin \Omega t], \tag{1}$$

which gives the electric field, $\mathbf{E} = -c^{-1} \, \partial \mathbf{A}/\partial t$, as

$$\mathbf{E} = \hat{x} \pi j_0 c^{-1} \sum_{n=1}^{\infty} J_n [n \Omega y_0/c] \exp [in\Omega (y - ct)/c], \quad y > y_0. \tag{2}$$

[For details concerning this and other points in the paper we refer the reader to the original papers (Lerche, 1970a, b, c).] Then the time averaged Poynting vector is

$$\mathbf{P} = \hat{y} (\pi j_0^2/8c) \sum_{n=1}^{\infty} J_n [n \Omega y_0/c]^2, \quad y > y_0; \tag{3}$$

$$\equiv \hat{y} (\pi j_0^2/16c) [(1 - \Omega^2 y_0^2/c^2)^{-1/2} - 1].$$

Now the relativistic particles move nearly at c. So simple arguments based on the forces involved indicate that y_0 could be as large as c multiplied by half a period of the oscillation, $\pi c/\Omega$. But if y_0 were as large as $\pi c/\Omega$, then the radiation output would be infinite. So energy considerations require that y_0 be less than c/Ω. With radiation damping reducing y_0 to something less than c/Ω we write $\Omega y_0 = c(1 - \frac{1}{2}\varepsilon^2)$. [We are unable to solve the dynamical equations of motion for a relativistically oscillating interface, including radiation reaction. We would, of course, be extremely interested to see calculations related to this problem.]

For $\varepsilon \ll 1$ the total power radiated is

$$P = \frac{\pi j_0^2}{16c\varepsilon} \quad \text{erg cm}^{-2} \text{ sec}^{-1}. \tag{4}$$

We expect that ε adjusts itself so that the power emitted, P, equals the power available, $P_0 \equiv cB(r_c)^2/16\pi$, where r_c is the distance of the interface from the neutron star. A very crude estimate of r_c can be obtained as follows. The balance of particle pressure with radiation pressure at r_c gives $nmc^2 \mathscr{E} = B_*^2 R_*^6 \Omega^4/(8\pi c^4 r_c^2)$, where B_* is the surface field of the neutron star, R_* its radius, $mc^2 \mathscr{E}$ the mean relativistic particle energy and n the number density of particles at r_c. If we also take the displacement current to be $O(nec)$ at r_c then we find $n(r_c) \approx \Omega^2 \mathscr{E} m e^{-2} \simeq 10^2$ cm^{-3}, and $r_c R_*^{-1} \simeq \simeq eB_* R_*^3 \Omega/(8mc^2 \mathscr{E}) \approx 10^5$. We have assumed, with Gunn and Ostriker, that $B_* \simeq 10^{12}$ G, $\Omega \simeq 10^2$ sec^{-1}, $R_* \simeq 10^6$ cm, $\mathscr{E} \simeq 10^6$. Also the wave zone of the dipole field starts at about $r \simeq 10^2 R_*$ for the numbers given. Thus the plasma-radiation interface is far out in the wave zone of the dipole field. And the interface is probably $O(c/\Omega)$ thick, which is a cyclotron radius at r_c.

From the balance of Lorentz force with dipole stress $j_0 B(r_c) = cB(r_0)^2/8\pi$ and $P \lesssim P_0$, we have $\varepsilon \gtrsim \frac{1}{64}$ for an infinitely thin current sheet. The differential spectrum of the radiation is, for $\omega \gg \Omega$,

$$\frac{dP}{d\omega} = \frac{\pi j_0^2}{8c\Omega} J_{\omega/\Omega} \left[\frac{\omega}{\Omega} \left(\frac{\Omega y_0}{c} \right) \right]^2, \tag{5}$$

or, using the asymptotic expansion of the Bessel function for $\omega \gg \Omega$ and $\Omega y_0/c \approx 1$, we have

$$\frac{dP}{d\omega} \simeq \frac{\Gamma\left(\frac{1}{3}\right)^2 j_0^2}{16 \cdot 2^{1/3} \cdot 3^{1/6} \pi c \Omega^{1/3} \omega^{2/3}} \quad \text{erg cm}^{-2} \ \text{sec}^{-1} (d\omega)^{-1}, \tag{6}$$

for $\omega \lesssim 3\Omega\varepsilon^{-3} \equiv \omega_{\max}$; and where $\omega = n\Omega$, $n \gg 1$.

Thus the differential power spectrum varies as $\omega^{-2/3}$. The larger ε^{-1} becomes the higher is the upper limit to the emission frequency. There is no radiation for $\omega > \omega_{\max}$. But for an infinitely thin current sheet, encompassing the total current j_0, the emission mechanism is too efficient. With $\varepsilon = \frac{1}{64}$ ($\omega_{\max} \simeq 7000 \ \Omega$) it would be difficult to obtain X-ray emission since 10 keV radiation ($\omega_{\max} \gtrsim 10^{19} \ \text{sec}^{-1}$) requires $\varepsilon^{-1} \sim 10^6$.

However it should be remembered that the current sheet is $O(c/\Omega)$ thick, with a distribution of current across the interface which gives partial incoherence. Since we do not have a detailed description of the interface structure and dynamics, including radiation reaction, the problem of partial, or complete, incoherence across a finite thickness interface remains unsolved. About all we can do is to introduce a phenomological constant, μ, into the formula for emitted power and write $P = \mu \pi j_0^2/16c\varepsilon$, and $dP/d\omega = \mu \times [dP/d\omega \ \text{from Equation (5)}]$. Here μ, which is small compared to unity, represents the incoherent nature of the finite thickness interface for emission at frequencies well in excess of Ω. And now $\mu = 64\varepsilon$ for balance of power emitted with power available. The effect of the incoherent nature (for $\omega \gg \Omega$) of the finite thickness interface is to increase the spectral range of the emission (by allowing ε^{-1} to be of the order of 10^6 for 10 keV pulsed X-ray emission, for example) but to decrease the power at a given frequency by μ. Thus increasing the spectral range by a factor 10^{14} (ε^{-1} going from 64 to 10^6) decreases the power at a given frequency by a factor 2×10^4.

For example, a finite thickness interface with the current per unit area, **j**, oscillating as

$$\mathbf{j}(y, t) = \hat{x} \int_{-\infty}^{\infty} J(Y_0) \, \delta\{y - Y_0 - y_0 \sin[\Omega t + \varphi(Y_0)]\} \, dY_0,$$

with $y_0\Omega = c(1 - \frac{1}{2}\varepsilon^2)$ and $\varphi(Y_0)$ a *random* phase, also gives $dP \propto \omega^{-2/3} \, d\omega$. The power output is proportional to $\int_{-\infty}^{\infty} J(Y_0)^2 \, dY_0$, with $\int_{-\infty}^{\infty} J(Y_0) \, dY_0 = j_0$.

This is to be contrasted with the emission from an infinitely thin current sheet, carrying the current per unit length j_0, for which the power output is proportional to j_0^2. The $\omega^{-2/3}$ dependence of the differential power output is a property of the $\sin(\Omega t)$

motion. The *level* of power at a fixed frequency is a property of the structure of the current distribution in the interface. Only if the motion differs from $\sin(\Omega t)$ does the shape of the power spectrum change. For example if each point in the interface moves with $y = Y_0 + y_0(t)$ with $d^2 y_0/dt^2 \propto \delta$-function during each period and $y_0(t) = y_0(t + 2\pi/\Omega)$ then $dP \propto \omega^{-2} d\omega$, assuming random phase between neighboring points.

It has not escaped our attention that a finite thickness interface can explain several other features of pulsar emission. Elsewhere (Lerche, 1970b, c) we show that such an interface can be used to explain pulse structure, finite pulse width (transit time delay of radiation through the interface), polarization 'swing' (rotation of Lorentz current through the interface) through a pulse, progressive 'march' of individual features through many pulses, and highly directional emission. Since this paper is concerned with the basic physics of the emission process rather than detailed models of the interface structure, we do not believe it is appropriate to enter into a discussion of these points here. Instead we refer the interested reader to the original papers (Lerche, 1970a, b, c.)

Extensive observations of pulsars at many frequencies have been made. [In order to keep the number of references under control we shall refer generically to the 'Progress Report on Pulsars' by Maran and Cameron (1969) and the 'Review of Theories of Pulsars' by Chiu (1970). The interested reader is referred to both for the chronology of pulsar observations and theories. The Maran-Cameron report was up to date as of March, 1969; the Chiu report was up to date as of May, 1969.] Very preliminary results of infra-red observations of NP 0532 have also been reported (Neugebauer *et al.*, 1969). They are not included here in view of their preliminary nature.

In order to apply our results to actual pulsars we choose to discuss pulsar NP 0532 as an illustrative case.

In the case of NP 0532 the basic oscillation frequency is $\Omega = 2\pi$ 30 sec^{-1} corresponding to a pulsing period of $\frac{1}{30}$ sec. The time averaged emission spectrum has been seen with a bandwidth extending from the radio (408 Mc/sec $-$ 1720 Mc/sec) through optical (3400 Å–8300 Å) through X-ray (1.5 keV–10 keV). The intensity levels seen in these bands are

$$P(\text{X-ray}):P(\text{optical}):P(\text{radio}) \simeq 7000:100:1,$$

and the differential power spectrum, $dP/d\omega$, is proportional to $\omega^{-\alpha}$ with $\alpha = 0.9 \pm 0.2$ in the radio region, $\alpha = 0.5 \pm 0.2$ in the optical region, $\alpha = 1 \pm 0.2$ in the X-ray region (Maran and Cameron, 1969; Oke, 1969, Gorenstein *et al.*, 1968, Bradt *et al.*, 1969).

The overall shape of the theoretical spectrum (Equation 6) from radio to X-ray agrees with the observations since the theory predicts $\alpha = 0.7$. Upon integration Equation (6) the theory also predicts that, in the bandwidths given above, the power levels should be in the ratios

$$P(\text{X-ray}):P(\text{optical}):P(\text{radio}) = 1600:80:1.$$

which, within a factor 4 or so, are in accord with the experimentally determined values.

Also, if we *assume* $\omega_{max} = 10$ keV/$\hbar$, then $\varepsilon^{-1} \simeq 10^6$. The upper cut-off frequency is very sensitive to the choice of ε and, conversely, ε is not a very sensitive function of the upper cut-off frequency. No optical emission has yet been seen from any pulsar other than NP 0532. These null observations then suggest that ω_{max} for such pulsars is less than about 10^{15} sec^{-1}.

The motion and radiation described above encompass the essential physics of the conversion mechanism.

We wish to stress that, within the framework of the Pacini and Gunn-Ostriker rotating magnetic neutron star picture, consideration of the radiation-plasma interface is of prime importance. For the underlying neutron star only provides the basic 'clock'; the high frequency radiation observed originates in, and at, the interface through the motion and dynamics of the surface current produced by the Lorentz forces transferring the dipole stress to the external plasma.

It is this fact we wish to emphasize rather than the particular model of an infinitely thin current sheet oscillating sinusoidally in vacuum, and the detailed illustration of this situation as applied to NP 0532. The agreement of this simple model with the observations is clearly encouraging, but is obviously less than a full resolution of the interface dynamics, motion and radiation.

However we point out that the essential physics of the interface motion is captured in the simple illustrative situation described above, and while more complex models of the motion, structure and dynamics of the interface may alter the 'fine grain' details of the radiation output, they will not change the basic physics involved.

In short, we believe that the physical mechanism outlined in this paper, and discussed in detail elsewhere (Lerche, 1970a, b, c), operates for pulsars, but that our numerical values for power output, etc., are probably good to a factor 10 or so only. The spectral index of 0.7 is probably good to a factor of order unity – the precise value depending on a detailed model for the interface structure and motion.

Acknowledgements

I am grateful to Professor S. Chandrasekhar for his encouragement and help in matters concerning the realization of this paper. I also wish to thank Professor E. N. Parker for many stimulating discussions concerning the basic physical mechanism. This work was supported by the United States Air Force, Office of Aerospace Research under contract F-19628-69-C-0041.

References

Bradt, H., Rappaport, S., Mayer, W., Nather, R. E., Warner, B., McFarlane, M., and Kristian, J.: 1969, *Nature* **222**, 723.
Chiu, H.-Y.: 1970, The Crab Nebula Symposium, Flagstaff, *Publ. Astron. Soc. Pacific* **82**, 487.
Gorenstein, P., Kellogg, E. M., and Gursky, H.: 1969, *Astrophys. J.* **156**, 315.
Gunn, J. E. and Ostriker, J. P.: 1969, *Nature* **221**, 454.
Lerche, I., 1970a, *Astrophys. J.* **159**, 229.

Lerche, I.: 1970b, *Astrophys. J.* **160**, 1003.

Lerche, I.: 1970c, *Astrophys. J.* **162**, 153.

Maran, S. P. and Cameron, A. G. W.: 1969, 'Progress Report on Pulsars', first part of a report on the Fourth Texas Symposium on Relativistic Astrophysics, Dallas, Texas, December 1968.

Neugebauer, G., Becklin, E. E., Kristian, J., Leighton, R. B., Snellen, G., and Westphal, J. A.: 1969, *Astrophys. J.* **156**, L115.

Oke, J. B.: 1969, *Astrophys. J.* **156**, L49.

Pacini, F.: 1968, *Nature* **219**, 145.

Discussion

R. N. Manchester: Is the transition at which the radiation occurs the same as the one discussed by Rees earlier – that is, the boundary of the nebula.

I. Lerche: No, my interface is not at the distance of Rees'. If I remember correctly Rees' interface is at the outside edge of the nebula, whereas mine is at 10^{11} cm from the pulsar.

M. J. Rees: I do not at all understand what is supposed to provide the huge external pressures which prevents your current layer from being pushed out at least to the edge of the Nebula.

I. Lerche: Either the excess plasma pressure, if it is non-uniform, or an imbedded magnetic field in the plasma will hold the interface at its mean position.

F. D. Drake: In order to produce the detailed pulse shapes which are stable over years, how does your model maintain a fixed complex small scale current structure in the emitting region?

I. Lerche: I don't know! All I know is that the total current per unit length must be of order $cB/4\pi$ always. Why it should be distributed with a fixed small scale structure throughout the interface I do not know.

8.5 RADIATION BEAMING IN PULSARS

V. CANUTO

Institute for Space Studies,
Goddard Space Flight Center, NASA, New York, N.Y., U.S.A.

It is usually considered that the beaming of the radiation coming out of a pulsar has to be strictly connected with the mechanism producing the radiation itself. We want to show that even when the emitting mechanism gives rise to an isotropically distributed radiation, the presence of a strong magnetic field will automatically beam the radiation preferentially along the magnetic field line rather than in any other direction. We have computed the Compton scattering and from that the opacity K_H (K_0 is the opacity for zero field). In Figure 1 the ratio K_H/K_0 is given vs. θ, the angle between the propagation vector and the magnetic field axis. H_q is a critical magnetic field numerically equal to 4.41×10^{13} G; N_e is the electron density. For the ordinary wave the opacity is reduced at $\theta=0$, while it is unaffected at $\theta=\pi/2$ where $K_H \to K_0$.

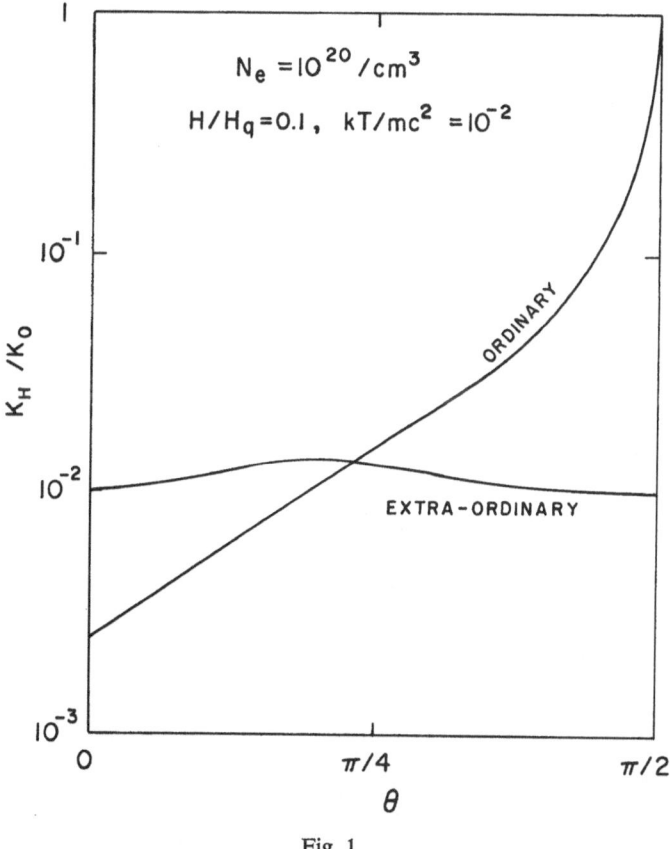

Fig. 1.

Davies and Smith (eds.), The Crab Nebula, 455–456. All Rights Reserved.
Copyright © 1971 by the IAU.

Even at $\theta = \pi/4$ the ratio K_H/K_0 is still $\simeq 10^{-2}$, and a good beaming is still present. The values of the parameters are proper for a neutron star surface. It is to be noticed that the ratio K_H/K_0 is of the order of $(\omega/\omega_H)^2$ or $[(kT/mc^2)/(H/H_q)]^2$. One therefore can conclude that the presence of a magnetic field itself assures the beaming of radiation along the field lines.

8.6 LOW-FREQUENCY CUTOFFS IN SYNCHROTRON SPECTRA AND THE OPTICAL SPECTRUM OF NP 0532*

STEPHEN L. O'DELL and L. SARTORI

*Department of Physics and Center for Space Research,
Massachusetts Institute of Technology, Cambridge, Mass., U.S.A.*

Abstract. We point out the existence of a natural low-frequency cutoff in the spectrum of a synchrotron source. The turnover frequency is $eB/mc \langle \sin \psi \rangle$, where ψ is the pitch angle for the electron motion. The spectrum below the turnover goes as ν^n, where n is the index in the electron energy distribution. For the Crab pulsar, a model with $B \sim 10^6$–10^7 G and $\langle \sin \psi \rangle \sim 10^{-3}$–$10^{-2}$ provides a plausible explanation for the turnover observed in the infrared.

It is commonly believed that without self-absorption, the low-frequency spectrum of a non-thermal source can be no steeper than $\nu^{1/3}$. This assertion is based on the fact that the spectrum of a single particle behaves as $\nu^{1/3}$ at low frequencies. However, that behaviour does not extend to zero frequency, but cuts off abruptly at the fundamental frequency

$$\nu_0 = \nu_B / \gamma \xi^2 \tag{1}$$

where ν_B is the cyclotron frequency and ξ is the sine of the pitch angle. (Actually the spectrum is discrete at the low-frequency end but with a distribution in energy this will never be noticed.)

If one takes account of this cut-off, and folds in the usual power law distribution in electron energy, $dN/d\gamma \sim \gamma^{-n}$, one finds that the low-frequency spectrum in fact has the form

$$I_\nu \sim \nu^n \sim \nu^{2\alpha+1} \tag{2}$$

where α is the index that characterizes the high-frequency spectrum. The turnover frequency is approximately

$$\nu_t \sim \nu_B / \bar{\xi} \tag{3}$$

where $\bar{\xi}$ is an average pitch angle for the emitting particles. For quasi-isotropic distributions in pitch angle, ν_t is just the cyclotron frequency, but if small pitch angles are predominant, the turnover is shifted to higher frequency.

Notice further that below the turnover, radiation of frequency ν comes only from electrons whose fundamental frequency is lower than ν; these must have $\gamma > \gamma_{\min}$, where

$$\gamma_{\min}(\nu, \xi) = \nu_B / \nu \xi^2. \tag{4}$$

That is to say, the low-frequency end of the spectrum comes only from electrons of

* Research supported in part by NASA and also in part by the National Science Foundation.

high energy. Low-energy electrons (if they are present) contribute only in the neighborhood of v_t. (For small pitch angles, 'low' energy means $\gamma \lesssim \xi^{-1}$; any such electron is nonrelativistic in the frame in which its motion is circular.)

This effect, which we call cyclotron turnover, must occur in every synchrotron source. (If the electron energy distribution should be truncated at the low end, this merely flattens the spectrum near the turnover.) However, cyclotron turnover must compete with synchrotron self-absorption in cutting off the spectrum. Clearly, the effect which sets in at higher frequency will be the one observed. If the angular size of the source and the typical pitch angles are known, one can say unequivocally which effect dominates. The condition is

$$C(\alpha) \, v_m^4 S (v_m)^{-2} \, \theta^4 \xi^{-2} \begin{array}{ll} < 1 & \text{self-absorption} \\ > 1 & \text{cyclotron turnover} \end{array} \tag{5}$$

where v_m is the observed turnover frequency, $S(v_m)$ is the observed flux at the turnover frequency, and $C(\alpha)$ is a slowly varying function of the spectral index. Briefly, what this says is that we calculate $B_\perp$ assuming self-absorption, and use that field to find v_t using equation (3). If the v_t so determined is higher than v_m, the assumption of self-absorption is inconsistent and only cyclotron turnover is a possible explanation. (It must be verified in any case that the field implied by the favored mechanism is reasonable for the physical object involved.)

Remark: the standard formulae for synchrotron self-absorption have been used. It is not obvious that these are valid in the case of very small pitch angles; we are looking into this point.

We are considering the applicability of cyclotron turnover to a number of objects. I shall discuss here only the Crab pulsar, for which $v_m \approx 8 \times 10^{14}$ Hz (Oke, 1969; Neugebauer *et al.*, 1969.) The precise location of the turnover is quite uncertain because of large and uncertain reddening corrections. The angular size is not known, but one can set a reasonable upper limit by assuming that the source region is roughly within the light cylinder, and subtends at the center of the neutron star an angle no greater than $2\pi\delta$, where δ is the observed duty cycle. This makes $\theta \lesssim 10^{-14}$ rad $\approx \approx 2 \times 10^{-9}$ sec of arc.

For emission close to the star, criterion (5) favors selfabsorption. However, the field so determined is much too small and such a model is not tenable. For emission near the light cylinder, cyclotron turnover is favored for all $\xi \lesssim 0.1$. This value is about the most the pitch angles could be; otherwise the duty cycle of the pulsar could not be as small as observed. Several arguments suggest that the relevant values of ξ should be considerably smaller than 0.1. Among them are:

> acceleration preferably along field lines
> adiabatic invariance
> radiative losses deplete the supply of high-ξ electrons.

It therefore seems to us that cyclotron turnover is the more likely explanation for the turnover in the pulsar spectrum, although selfabsorption is not strictly excluded.

For the pulsar, Equation (3) gives

$$B/\bar{\xi} \approx 3 \times 10^8 \text{ G} \tag{6}$$

so with ξ between 10^{-2} and 10^{-3} one gets fields between 3×10^5 and 3×10^6 G, which are reasonable values. We can obtain another relation between B and ξ and thereby fix both of them uniquely if we are willing to make the following additional speculation. The X-ray spectrum of the pulsar appears to bend down at $\nu \sim 10^{18-19}$ Hz. The data are not certain, but the optical and soft X-ray spectra are flatter than those of the nebula, and cannot continue very far with the same index before they would cross the nebular spectrum. Assume that this break is due to a lifetime effect in the pulsar, i.e., that $\tau_{\text{rad}}(\nu_e) \sim d/c$, where ν_e is the break frequency and d measures the extent of the emission region. One then obtains the following relation:

$$B\bar{\xi} \approx 10^{15} d^{-2/3} \nu_e^{-1/3}. \tag{7}$$

Putting $\nu_e = 10^{18}$ Hz and $d =$ radius of light cylinder $= 2 \times 10^8$ cm gives

$$B = 10^6 \text{ G}$$
$$\bar{\xi} = 3 \times 10^{-3}. \tag{8}$$

These numbers are quite encouraging. Notice they come purely from analysis of the spectrum, with no independent assumptions about the field. I should also remark that these results depend very weakly on the parameters involved; B in fact varies as $\nu_t^{1/2} \nu_e^{-1/6} d^{-1/3}$. One cannot carry out a similar analysis if one assumes the turnover is due to self-absorption, since both ν_t and ν_e then involve the same combination $B\bar{\xi}$. One can only check the two relations for consistency, but because of the fourth-power dependence of B on θ such a check is not very sensitive. The values of B and $\bar{\xi}$ given by Equation (8) are in fact almost the same as those deduced by Shklovsky (1970) who interpreted the turnover as self-absorption. Shklovsky obtained the value of B by assuming the canonical value at the surface, and a dipole dependence on r. However, according to Equation (5), selfabsorption is inconsistent for those values of the parameters. The electrons radiating in the vicinity of the maximum are non-relativistic in the frame that moves with the gyrocenter, and therefore the standard synchrotron theory is inapplicable.

References

Neugebauer, G., Becklin, E. E., Kristian, J., Leighton, R. B., Snellen, G., and Westphal, J. A.: 1969 *Astrophys. J. Letters* **156**, L115.
Oke, J. B.: 1969, *Astrophys. J. Letters* **156**, L49.
Shklovsky, I. S.: 1970, *Astrophys. J. Letters* **159**, L77.

8.7 PULSAR (AND X-STAR) EMISSION FROM THE MAGNETOSPHERE OF A COLLAPSED STAR

BRUNO COPPI

Massachusetts Institute of Technology, Cambridge, Mass., U.S.A.

and

ATTILIO FERRARI

Laboratorio di Cosmo-geofisica del CNR and Istituto di Fisica dell'Università, Torino, Italy

Abstract. We propose nonthermal plasma mechanisms to account for the most evident physical characteristics of pulsar emission, including the production of X- and γ-rays and the acceleration of high-energy particles. Special reference to the Crab Nebula pulsar (NP 0532) is made and an application of the same model to other non-thermal X-ray sources is suggested.

1. Quasi-Equilibrium of the Plasma Magnetosphere

Following the arguments proposed by several authors (Gold, 1968; Large *et al.*, 1968), we assume that the condensed stars associated with pulsars are strongly magnetized and rapidly rotating as a consequence of a gravitational collapse and of angular momentum and magnetic flux conservation. It becomes then natural to assume that the star is surrounded (inside the 'speed of light' cylinder $R_c = c/\omega_0$, ω_0 being the angular velocity of rotation) by a relatively dense plasma with a small charge separation since, if this were not the case, unrealistically large electric fields (of the order of $u_0 B$, u_0 being the velocity required for co-rotation and B the magnetic induction) would develop along the magnetic field lines (Goldreich and Julian, 1969).

As a consequence of the high value of the magnetic field B ($B_{surf} \gtrsim 10^{10}$ G), the Langmuir frequency ω_{pe} is much smaller than the electron gyrofrequency Ω_e. Taking $B_0 = 10^{10}$ G and $n_0 = 10^{18}$ cm^{-3} as reference values, we have in fact $\Omega_e/\omega_{pe} = 3.1 \times 10^3$ $(B/B_0)(n_0/n_e)^{1/2}$.

In these conditions, considering, for instance, mildly relativistic electrons with energy perpendicular to the magnetic field, they will tend to lose it by synchrotron radiation in a typical time $\tau \approx 2.58 \times 10^{-12} (B_0/B)^2$ sec. As a consequence of radiation losses, these electrons have a tendency to acquire an anisotropic distribution function in velocity space, with more particles having velocity parallel to the magnetic field than perpendicular. In this case the plasma is subject to microinstabilities which tend to restore the distribution function isotropy.

Thus, by taking into account the effects of collective phenomena (such as micro-instabilities) and interparticle collisions, we can assume that the plasma surrounding a strongly magnetized collapsed star is in macroscopic quasiequilibrium. A stationary state of this type, in the presence of fluctuations excited by the electron distribution function anisotropy in a strong magnetic field, may be described by the so-called quasilinear approximation (Shapiro and Shevchenko, 1968). The possibility of an

Davies and Smith (eds.), The Crab Nebula, 460–470. All Rights Reserved.

equilibrium (Chandrasekhar, 1961) is particularly important for cases in which the magnetic configuration is not symmetric around the star's axis of rotation, and which are suitable for pulsar models (e.g. Gold, 1968; Eastlund, 1968). The lack of rotational symmetry is, in fact, indicated by the observation of pulses and the presence of an 'equilibrium' by the remarkably stable pulse pattern.

Assuming, for simplicity, that the plasma under consideration is composed of electrons and ions of charge Z, the equilibrium of the plasma region inside the 'speed of light' cylinder can be approximately represented by a set of proper moment equations. In particular the electron-momentum balance equation can be written as

$$
\mathbf{E}' = - \mathbf{u}'_e \times \frac{\mathbf{B}}{c} - \frac{m_e}{e} \left[\tilde{\mathfrak{f}}_e \cdot \left(\mathbf{u}'_e - \mathbf{u}'_i \frac{Zn_i}{n_e} \right) \right.
$$
$$
\left. + \frac{1}{m_e n_e} (\nabla \cdot \mathfrak{P}_e - \mathbf{S}_e^{(p)}) + \mathbf{u}'_e \cdot \nabla \mathbf{u}'_e + 2\boldsymbol{\omega}_0 \times \mathbf{u}'_e - \mathbf{g} \right]. \tag{1}
$$

Here primed quantities refer to the frame co-rotating with the star; g indicates gravity, n density, $\mathbf{u}$ average velocity and subscripts e and i refer to electrons and ions. $\mathbf{S}_e^{(p)}$ represents the momentum source that must be taken into account if, for instance, particle pair creation by high energy photons occurs. The tensor $\tilde{\mathfrak{f}}_e$ is an 'effective' collision frequency, representing the momentum exchange between electrons and ions, and $\mathfrak{P}_e$ is the relativistic electron-pressure tensor. Both of these tensors incorporate the effects of particle collisions, collective modes and kinetic processes involving detailed consideration of velocity space, that is knowledge of the electron distribution function $f_e(\mathbf{r}, \mathbf{v})$. A similar equation can be derived for the ions and by properly summing the two, one obtains the total momentum conservation equation and, consequently, the small current density $\mathbf{J}'_\perp$ required to maintain the equilibrium transverse to the magnetic field

$$
\mathbf{J}'_\perp = (\mathbf{B}/B^2) \times \left\{ e\mathbf{E}' (n_e - Zn_i) + \sum_{j=i,\, e} m_j n_j \right.
$$
$$
\left. \times (\mathbf{u}'_j \cdot \nabla \mathbf{u}'_j + 2\boldsymbol{\omega}_0 \times \mathbf{u}'_j - \mathbf{g}) + \nabla \cdot \mathfrak{P}_j - \mathbf{S}_j^{(p)} \right\}. \tag{2}
$$

The equilibrium parallel to $\mathbf{B}$ can be described by the $\mathbf{B}/B$ component of Equation (1), which as usually done in plasma physics, can be written as

$$
E'_\parallel = \eta_{an} J'_\parallel \tag{3}
$$

$J'_\parallel$ being the longitudinal current density and η_{an} the anomalous, or nonclassical resistivity that takes into account all relevant collisional and turbulent plasma processes (Kadomtsev and Pogutze, 1968; Coppi and Mazzucato, 1969). We recall, in fact, that in many significant cases η_{an} has been found experimentally (Artsimovich et al., 1968) to be orders of magnitude higher than the classical resistivity based on the momentum exchange in electron-proton collisions (see Figure 1).

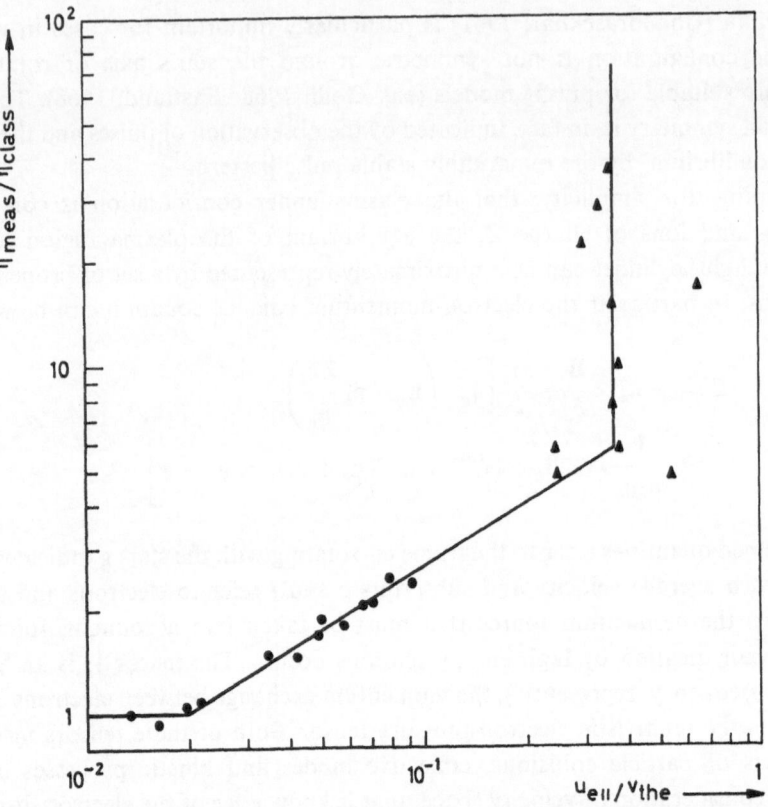

Fig. 1.   Ratio of the measured plasma resistivity (η_{meas}) to the theoretical one (η_{class}) evaluated from the effects of interparticle collisions only. The experimental points were obtained from data on plasmas confined in strong magnetic field with $u_{e\parallel}$ being the electron flow velocity along the magnetic field and v_{the} the electron thermal velocity. See Coppi and Mazzucato (1969).

The relationship between $J'_\parallel$ and $\mathbf{J}'_\perp$ is given by the charge conservation equation

$$\nabla \cdot \mathbf{J}' = 0. \tag{4}$$

Here we have to assume that $\mathbf{J}' \simeq \alpha \mathbf{B}$, to lowest order in order not to have unrealistic forces $\mathbf{J}' \times \mathbf{B}$. Poisson's equation reads

$$\nabla \cdot \mathbf{E}' = 4\pi e \left(Z n_i - n_e\right) + \nabla \cdot \left(\mathbf{u}_0 \times \frac{\mathbf{B}}{c}\right), \tag{5}$$

where $\mathbf{u}_0 = \boldsymbol{\omega}_0 \times \mathbf{r}$. Notice that the electric field seen in the inertial frame is

$$\mathbf{E} = \mathbf{E}' - \mathbf{u}_0 \times \mathbf{B}/c. \tag{6}$$

Therefore, the plasma could be in exact co-rotation (Chandrasekhar, 1961; Goldreich and Julian, 1969) with the star magnetic field only if $E'_\parallel = 0$, a circumstance that is clearly not possible since, as we have seen, collective, collisional and inertial effects are necessary to the existence of the plasma quasi-equilibrium.

Noting that the field:

$$|\mathbf{u}_0 \times \mathbf{B}/c| \simeq 1.2 \times 10^{10} \left[\omega_0 Br/(\omega^0 B_0 r_0)\right] \text{ V/cm} \tag{7}$$

where $r_0 = 2 \times 10^6$ cm and $\omega^0 = 2 \times 10^2$ rad/sec are reference values applicable to the Crab Nebula pulsar, we argue that a very small lag between particles and magnetic field exists corresponding to $\mathbf{E}'$, and in particular $E'_\parallel$ being non zero although much less than $|\mathbf{u}_0 \times \mathbf{B}/c|$.

Very small values of $E'/E_\perp$ which for $n_e \approx n_0$ correspond to a relative charge separation $1 - Zn_i/n_e \approx 10^{-7}$ as indicated by Equation (5), are consistent with a prevalence of the electromagnetic force terms in Equation (1) and in the total momentum conservation equation. Therefore, we can concentrate on looking at the plasma aspect of the problem and consider, in more detail, the equilibrium along a typical open line of force emanating from a pole or a spot of the star.

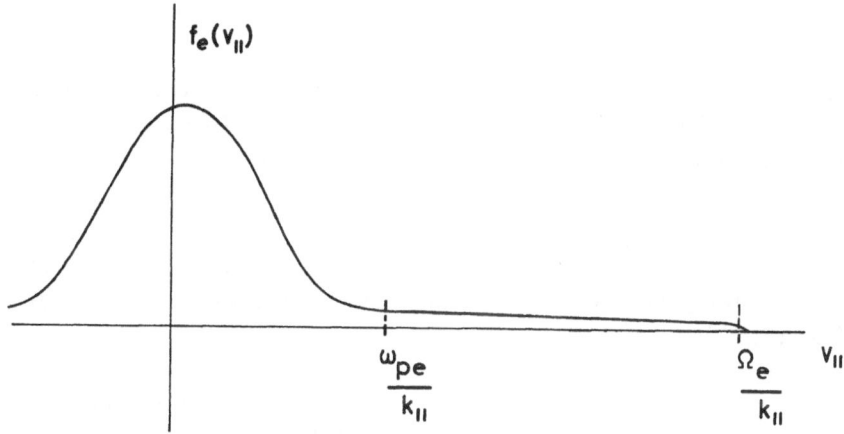

Fig. 2.   Example of a typical 'runaway' distribution, with regions of particle-wave resonance indicated.

Some points have to be emphasized in relation with the existence of this equilibrium.

(1) An important limit for $E'_\parallel$ is the runaway critical field $E_{run} = v_{ei} m_e v_{the}/e$, where v_{ei} is the average classical electron-ion collision frequency and v_{the} the electron thermal velocity. The ratio $E'_\parallel/E_{run}$ is, in fact, significant in relation to the tail of superthermal electrons that characterizes the electron distribution function $f_e(v_\parallel)$, $v_\parallel$ being the electron velocity along $\mathbf{B}$. In other words, for not-too-low values of $E'_\parallel/E_{run}$ the typical profile of $f_e(v_\parallel)$ that results as a consequence of electron collisions and collective modes is of the type represented in Figure 2. We recall for this that in magnetically confined plasmas (Stodiek, 1969) with temperatures $1 \lesssim T_e \lesssim 10$ eV, electrons of about 200 keV with density $\approx 10^{-2} n_e$ have been observed for $10^{-1} < E_\parallel/E_{run} < 1$.

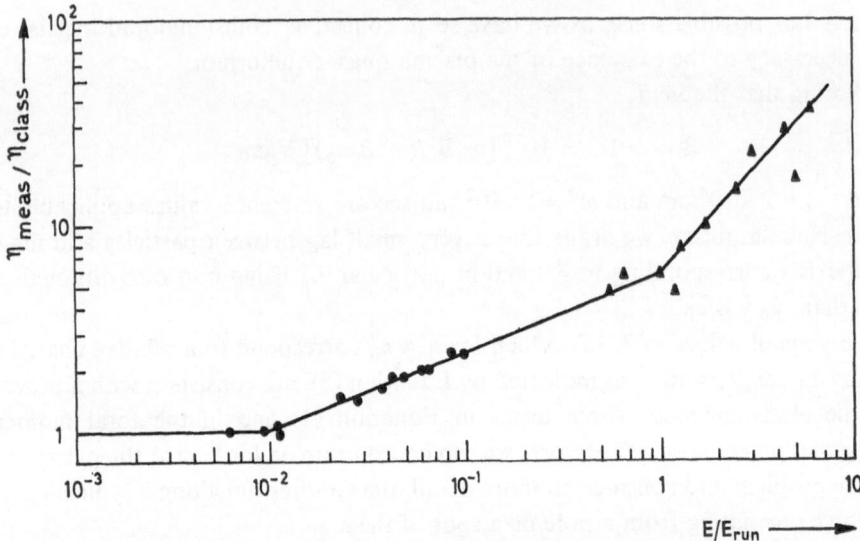

Fig. 3   The same ratio $\eta_{\text{meas}}/\eta_{\text{class}}$ as in Figure 1 versus $E_\parallel/E_{\text{run}}$, $E_\parallel$ being the electric field parallel
to the magnetic field and E_{run} the critical 'runaway' field.

Moreover, for plasmas with $\omega_{pe} < \Omega_e$, the anomalous resistivity η_{an} has been found
to become much larger than $\eta_{cl} = v_{ei} m_e/(n_e e^2)$ for $E_\parallel \gtrsim E_{\text{run}}$ (see Figure 3). A numerical evaluation of the runaway field is

$$E_{\text{run}} \simeq 3.1 \times 10^4 \, \alpha n_e T_0/(n_0 T_e) \text{ V/cm}, \tag{8}$$

where $T_0 = 10^6$ K has been taken as typical electron temperature in the vicinity of
the star following current neutron star models, and $\alpha = 1 - 0.13 \log_{10}(n_e/n_0) +$
$0.26 \log_{10}(T_e/T_0)$.

(2) The longitudinal energy gained by the electrons under the influence of $E'_\parallel$ is
scattered in the transverse (to **B**) direction by electron-proton collisions or by collective
effects such as the following (e.g. Coppi *et al.*, 1969). A plasma electrostatic (longitudinal) wave with frequency $\omega = \omega_{pe}(k_\parallel/k)$, **k** being the wave vector, is involved in
particle-wave resonances producing the energy and momentum exchange

$$\hbar\omega + \Delta\varepsilon_j = 0, \qquad \hbar\mathbf{k} + \Delta\mathbf{p}_j = 0 \tag{9}$$

where $\Delta\varepsilon_j$ is the energy variation and $\Delta\mathbf{p}_j$ the momentum variation for particles of
species j, while $\hbar\omega$ and $\hbar k$ refer to the wave. Now, $\Delta\varepsilon_j = \Delta\varepsilon_{\parallel j} + \Delta\varepsilon_{\perp j}$, with $\Delta\varepsilon_{\parallel j} =$
$= m_j v_\parallel \Delta v_\parallel$, $\Delta\varepsilon_{\perp j} = n^0 \hbar\Omega_j$, n^0 being an integer. In addition, $\Delta p_{\parallel j} = m_j \Delta v_\parallel$, while $\Delta p_{\perp j}$
is taken up by the magnetic field. Therefore, the wave-electron resonance condition is:

$$\omega + n^0\Omega_e - k_\parallel v_\parallel = 0. \tag{10}$$

Now, since $\omega \ll \Omega_e$, Equation (10) shows that the energy exchanged with the wave is
negligible for $n^0 \neq 0$, and the main effect is a transfer of longitudinal electron energy
into transverse energy. Notice that a plasma wave of the considered frequency can

exist, without being strongly damped, only for $\omega \gg k_\parallel v_{the}$. Therefore, the electrons participating in the resonance (10) have to be highly energetic and belong to the distribution tail. In particular, the relevant theory requires that $(\partial f_e/\partial v_\perp)_{v_\parallel} = \Omega_e/k_\parallel$ be sufficiently large for the process represented by Equation (10) not to be overcome by the ordinary Landau damping corresponding to $n^0 = 0$ and involving particles with $v_\parallel = \omega/k_\parallel \ll \Omega_e/k_\parallel$ (Figure 2). Considering the high values that Ω_e/ω_{pe} can have for the case under consideration, it is clear that the resonating particles have to be relativistic and in this case Equation (10) becomes roughly $\gamma k_\parallel c \approx \Omega_{e0}$, where Ω_{e0} is the rest mass gyrofrequency.

(3) The observed high anomalous resistivity in laboratory plasma experiments with $\omega_{pe} < \Omega_e$ is attributed to wave-particle interactions of the type discussed above or of other types associated, for instance, with streaming instabilities. Therefore, a large $E_\parallel$ can be compatible with acceptable values of $J_\parallel$.

(4) The current $J_\parallel$ can also excite low-frequency modes with $\omega < \omega_{pe}$ and $\omega < \Omega_i$, Ω_i being the ion gyrofrequency. For this we may consider, among other possibilities, modes of the ion-sound wave type, which depend on $T_e \neq 0$, have a frequency $\omega \approx k_\parallel \sqrt{(T_e/m_p)}$, are electrostatic and are associated with finite electron thermal conductivity (Coppi and Mazzucato, 1969) or with electron Landau damping along the same direction.

Finally we remark that the driving electric field can be related to the strong slippage of the plasma with respect to the rotating magnetic field in the far out regions (Goldreich and Julian, 1969), and is induced by the star rotation. So this is the ultimate energy source for all the emission processes.

2. Plasma Radiation Mechanisms

From the discussion given above it follows that two main classes of emission processes can be considered. One is related to the existence of a sizeable tail of superthermal electrons and therefore to not-too-low values of $E_\parallel/E_{run}$. Another is associated with the main body of the electron distribution and can be effective at very low values of $E_\parallel/E_{run}$ [for instance $E_\parallel/E_{run} \approx (m_e/m_p)^{1/2}$].

We recall for this that the appearance of 'runaway' electrons in a collisional non-turbulent plasma can be represented by a sharp function of $E_\parallel/E_{run}$, typically of the form (Gurevich, 1961)

$$n_s \approx n_e \, e^{-E_{run}/E_\parallel}. \tag{11}$$

Then, considering the first class of processes, we notice that:

(a) X-ray emission by mildly relativistic electrons (with $\gamma = \varepsilon/mc^2 \simeq 1$) is related to the basic role of the frequency Ω_e in plasma collective modes. This emission can, in fact, be the result of direct cyclotron radiation by those electrons that have acquired transverse energy by a scattering process.

(b) γ-ray emission, possibly due to a scattering process similar to that for X-ray emission can result from synchrotron radiation by highly relativistic electrons.

(c) Optical emission is expected to follow along with X-ray emission. For instance if longitudinal high frequency plasma modes are excited, a frequency decay process toward lower frequencies and coupling to electromagnetic modes take place (Kaplan and Tsytovich, 1969).

(d) A nonlinear frequency decay process can also contribute to the infrared and upper band of the radio spectrum. In addition, a direct coupling of the longitudinal waves (having $\omega \approx k_\parallel \omega_{pe}/k$ and owing their excitation to the anisotropy of the tail of superthermal electrons) with transverse waves can be relevant to this range or even to the optical, if $n_e \approx n_0$.

(e) A small fraction of the total electron population accelerated by $E_\parallel$ will be able to escape all collisional and collective interactions and to attain very high energies in the direction of the magnetic field, so that a process for direct particle acceleration is also to be considered.

(f) In reference to Equation (11) we may argue that a collapsed star, because of its initial conditions and of aging which implies slowing-down and probably decreasing T_e, can reach a point where the ratio $E_\parallel/E_{run}$ is no longer sufficient to maintain an adequate population of fast particles to make the particle wave resonance with $n^0 \neq 0$ prevail over ordinary Landau damping. This corresponds to the resonance $\omega - k_\parallel v_\parallel = 0$ $(n^0 = 0)$, which generally involves a considerably larger number of particles (Fig. 2).

Referring to the second class of emission processes, we consider the low-frequency longitudinal modes that are excited by $J_\parallel$. These modes, after propagating outward along the magnetic field lines, can couple (in regions of lower density) to longitudinal plasma waves with frequency $\sim \omega_{pe}$, and then with transverse waves (Ginzburg and Zaitsev, 1969) (in the radio band) that are expected to be strongly polarized. So this emission mechanism is characterized by propagation direction clearly related to that of the magnetic field, and does not exclude the possibility that the emitted wave be linearly or circularly polarized.

3. Application of Plasma Model to Pulsars

A plasma model of the type given above can be applied more specifically to the interpretation of the Crab Nebula pulsar emission and to other pulsars and X-ray sources. Then, on the basis of the physical processes discussed in the previous section, we will divide the pulsar spectrum in high-energy and low-energy regions in agreement with observational evidence from NP 0532. Neugebauer *et al.* (1969) have in fact measured a break-down of the spectrum in the infrared region, so that two distinct peaks appear in the radio region and in the X (or possibly γ) region.

A. HIGH-ENERGY EMISSION

The Crab Nebula pulsar, because of its large angular velocity and of its recent formation can be assumed to have an electron parallel distribution function with a sizeable tail of superthermal electrons. We notice that the corresponding lack of isorotation of the plasma with the star's magnetic field leads to a reasonably small variation of the rotational velocity over the scale length of the fields.

We also recall from the previous sections that runaway electrons with high parallel energy can undergo pitch-angle scattering or by plasma collective modes and then acquire perpendicular energy. This will be lost by cyclotron radiation at the harmonics of the electron cyclotron frequency $\Omega_e = 1.76 \times 10^{17} \, \gamma^{-1} \, (B/B_0)$ (γ being the relativistic factor, $\gamma = \varepsilon/m_0 c^2$), which for mildly relativistic electrons is in the X-range.

As pointed out earlier, we consider the optical emission as strictly correlated with the higher energy emission either by means of a decay process of higher frequency modes or by direct coupling of Langmuir waves with electromagnetic plasma modes. The efficiency of the decay process does not have to be high, since the observations for the Crab pulsar give $W_{opt}/W_x \approx 10^{-3}$. As for the direct coupling, we notice that the plasma waves giving electron pitch-angle scattering have frequency $\omega = (k_{\parallel}/k) \, \omega_{pe}$; so, if the density is sufficiently high $(n_e \gtrsim n_0)$, optical emission is produced and this is evidently correlated with X-rays.

γ-ray emission, which has been tentatively identified recently (Vasseur et al., 1970; Kinzer et al., 1971; Hillier et al., 1971) obviously involves highly relativistic particles. So one possibility is that these particles emit by synchrotron radiation and this be maintained by pitch-angle scattering processes. Another possibility comes from considering (see previous sections) the relatively small population of runaway electrons which can escape all the effects of interparticle collisions and collective phenomena. So, γ-rays can be thought of as being originated by the single particle acceleration process along open field lines emanating from poles or spots not aligned with the rotation axis.

B. LOW-ENERGY EMISSION

We refer, for the radio emission, to low-frequency modes which can be excited by the current $J_{\parallel}$ as indicated in the previous section.

These longitudinal waves do not require the presence of runaway electrons. They can propagate towards outer regions of decreasing density where they can couple with Langmuir waves and electromagnetic plasma waves propagating transversely to the magnetic field. In these conditions it is not difficult to reconcile propagation along the magnetic field of waves with frequency $\omega < \omega_{pe}$ with the observed linear polarization for the Crab pulsar and several others. In particular, we expect that the lower the frequency the further is the distance to which the electrostatic low-frequency modes have to propagate before coupling to electromagnetic modes. Since the beam of open lines of force widens with increasing distance from the star pole or spot, the process we have proposed may contribute to the observed progressive pulse wideing as the (radio) frequency decreases (Drake, 1969).

On the subject of linear polarization we recall that the optical pulses from NP 0532 (Crab), as the radio pulses from PSR 0833 (Vela) exhibit a wide rotation of the angle polarization within the pulse (Wampler et al., 1969). We may explain this observation assuming that in these cases the coupling of electrostatic modes with electromagnetic waves occurs close to the star where the magnetic field changes direction over a relatively short distance. Thus the radiation that is received at each instant within the pulse may come from a sequence of different points along a given magnetic tube of flux, where the magnetic field direction undergoes a finite change (such as for instance, 60°). The electromagnetic modes which propagate perpendicularly to this direction have their electric field strongly correlated to it and reflect its variation.

We also recall that the single radio pulses from several pulsars have been observed to have a large circularly polarized component (Smith, 1970) with equal probability of left hand and right hand polarization. In these cases part of the received wave energy should be carried along the received star's magnetic field lines by plasma electromagnetic waves which, as is well known, are circularly polarized in either direction.

Thus far, NP 0532 is the only observed pulsar emitting over a wide range of frequencies. Attempts to detect optical or X-emission from other pulsars have failed. On the other hand, the period of PSR 0833 (Vela) is less than three times longer than that of NP 0532 and yet PSR 0833 exhibits only radio emission. So, although there is no reason to believe that the two pulsars were born with the same physical characteristics, the possibility of a sharp aging process, with a short X and optical phase followed by a relatively long radio-emission phase, has to be considered. For this we recall that X-ray and optical emission was related to the existence of a sizeable tail of superthermal electrons, and that the appearance of this tail is represented by a sharp function of $E_\parallel / E_{\text{run}}$, such as that given by Equation (11).

Thus, to explain the transition from the case of NP 0532 to that of PSR 0833, we assume that for the latter star, because of its initial conditions and aging (which implies slowing down of rotation and probably decrease of T_e), the ratio $E_\parallel / E_{\text{run}}$ is no longer sufficient to maintain an adequate population of fast particles and ensure the excitation of high-frequency modes and X-ray and optical emission. As for the emitted spectrum, for instance in the X-ray region, this can be obtained from the evaluation of the quasi-equilibrium distribution function and of the rate of pitch-angle scattering by the quasi-linear theory (Shapiro and Shevchenko, 1968) of the anomalous Doppler effect represented by the resonance (10).

Again for the other regions of emission, optical and radio, it is possible, by evaluating the matrix elements of the various forms of couplings, to arrive at an estimate of the spectrum. This requires a detailed analysis that is in progress, but given the variety of modes which can be excited, there does not seem to be particularly strong difficulties to arrive at a fit of the available experimental data.

4. High-Energy Particles

A mechanism for production of highly relativistic particles can be proposed by identifying them with the 'runaway' particles which escape the effects of collisions and of collective phenomena and undergo almost free acceleration along the magnetic field.

It is important at this point to recall that an electromagnetic wave of frequency ω_0 is emitted outside the light speed cylinder from the rotation of a non-axisymmetric magnetic configuration (Deutsch, 1955; Pacini, 1968). This wave tends to sweep all charged particles from the light speed cylinder and our model provides a continuous injection of particles that can be transported, without relevant energy loss, to distances of the order of the Crab Nebula extended X-ray source ($\simeq 10^{18}$ cm). This process involves a very efficient acceleration of the particles, which are forced to remain almost in phase with the electromagnetic wave, as has been checked by numerical computations (Ferrari and Trussoni, 1971).

The resulting high-energy particles can emit synchrotron radiation over a wide area in the nebular magnetic field. The extended X-ray, optical and radio source can, in fact, be the result of such a process.

We also recall that the electromagnetic wave radiation and the related particle acceleration most likely represents the main energy loss of the rotating neutron star. For the Crab pulsar, assuming $B_{surf} \simeq 10^{12}$ G, this energy loss can amount to $\approx 10^{38}$ erg/sec and, with reasonable values of the parameters of the star, the period lengthening of that pulsar can be explained (Pacini, 1968; Gunn and Ostriker, 1969).

5. X-Stars and Cosmic-ray Sources

Finally, we point out the possibility of interpreting other X-ray sources in terms of rotating collapsed bodies. Non-thermal emission mechanisms of the type outlined previously could then be applied to the interpretation of point X sources with a non-thermal spectrum. These would differ from NP 0532 mainly by the absence of the regular pulse pattern and of a nebula in the vicinity of the star which make the source point-like to the observer. The absence of observed pulsed emission can be attributed to a magnetic configuration that is symmetric around the rotation axis, or never points its beams of enhanced emission towards the Earth, while a high steady background exists.

Particles that are able to escape the effects of collisions and plasma collective effects, undergoing almost free acceleration by $E_\parallel$ close to the star, and not affected by a nebula will contribute to the high-energy tail of the cosmic-ray spectrum. The combined high values of rotation and magnetic field of a collapsed star can therefore provide a direct mechanism for high-energy particle acceleration, with the consequences that a sizeable cosmic-ray anisotropy may be present at very high energies.

Acknowledgements

It is a pleasure to thank A. Treves for his criticism and a number of valuable suggestions. One of us (A.F.) wishes to thank for hospitality the Massachusetts Institute of Technology, where this work was started.

References

Artsimovich, L. A., Babrovoskii, G. A., Mirnov, S. V., Rasumova, K. A., and Strelkow, V. S.: 1967, *Soviet Atomic Energy* **22**, 325.

Chandrasekhar, S.: 1961, *Hydrodynamic and Hydromagnetic Stability*, Oxford University Press.

Coppi, B. and Mazzucato, E.: 1971, *Phys. Fluids* **14**, 134.

Coppi, B., Rosenbluth, M. R., and Sudan, R. N.: 1969, *Ann. Phys.* **555**, 207.

Deutsch, A.: 1955, *Ann. Astrophys.* **18**, 1.

Drake, F. D.: 1969, paper presented at the 131st AAS Meeting, New York.

Eastlund, B. J.: 1968, *Nature* **220**, 1293.

Erber, T.: 1966, *Rev. Mod. Phys.* **38**, 626.

Ferrari, A. and Trussoni, E.: 1971, *Nuovo Cimento Letters*, Serie 2, **1**, 137.

Ginzburg, V. L. and Zaitsev: 1969, *Nature* **222**, 230.

Gold, T.: 1968, *Nature* **218**, 713.

Goldreich, P. and Julian, W. H.: 1969, *Astrophys. J.* **157**, 869.

Gurevich, A. V.: 1961, *Soviet Phys. JETP* **12**, 904.

Hillier, R. R., Jackson, W. R., Murray, A., Redfern, R. M., and Sale, R. G.: 1971, *Astrophys. J.* **162**, L177.

Kadomtsev, B. B. and Pogutze, O. P.: 1968, *Soviet Phys. JETP* **26**, 1146.

Kaplan, S. A. and Tsytovich, V. N.: 1969, *Soviet Phys. Usp.* **12**, 42.

Kinzer, R. L., Noggle, R. C., Seeman, N., and Share, G. H.: 1971, *Nature* **229**, 187.

Large, M. I., Vaughan, A. E., and Mills, B. Y.: 1968, *Nature* **220**, 340.

Neugebauer, G., Becklin, E. E., Kristian, J., Leighton, R. B., Snellen, G., and Westphal, J. A.: 1969, *Astrophys. J. Letters* **156**, L115.

Ostriker, J. P. and Gunn, J. E.: 1969, *Astrophys. J.* **157**, 1395.

Pacini, F.: 1968, *Nature* **219**, 145.

Shapiro, V. D. and Shevchenko, V. I.: 1968, *Sov. Phys. Usp.* **27**, 635.

Smith, F. G.: 1970, this symposium, Paper 8.1, p. 431.

Stodiek, W.: 1969, private communication.

Vasseur, J., Paul, J., Parlier, B., Leray, J. T., Forichon, N., Agrinier, B., Boella, G., Maraschi, L., Treves, A., Buccheri, R., and Scarsi, L.: 1970, *Nature* **226**, 536.